DEVELOPMENT OF THE NERVOUS SYSTEM

SECOND EDITION

DEVELOPMENT OF THE NERVOUS SYSTEM

SECOND EDITION

DAN H. SANES
THOMAS A. REH
WILLIAM A. HARRIS

Amsterdam   Boston   Heidelberg   London   New York   Oxford
Paris   San Diego   San Francisco   Singapore   Sydney   Tokyo

Academic Press is an imprint of Elsevier

Elsevier Academic Press
30 Corporate Drive, Suite 400, Burlington, MA 01803, USA
525 B Street, Suite 1900, San Diego, California 92101-4495, USA
84 Theobald's Road, London WC1X 8RR, UK

This book is printed on acid-free paper. ♾

Permissions may be sought directly from Elsevier's Science & Technology Rights
Department in Oxford, UK: phone: (+44) 1865 843830, fax: (+44) 1865 853333,
E-mail: permissions@elsevier.co.uk. You may also complete your request on-line
via the Elsevier homepage (http://elsevier.com), by selecting
"Customer Support" and then "Obtaining Permissions."

Library of Congress Cataloging-in-Publication Data
Application submitted

British Library Cataloguing in Publication Data
A catalogue record for this book is available from the British Library

ISBN 13: 978-0-12-618621-5
ISBN 10: 0-12-618621-9
ISBN 13: 978-0-12-369447-8 (CD-ROM)
ISBN 10: 0-12-369447-7 (CD-ROM)

For all information on all Elsevier Academic Press publications
visit our Web site at www.books.elsevier.com

Printed in the United States of America
07 08 09 10 9 8 7 6 5 4 3

To our families

Contents

Preface to the First Edition

The human brain is said to be the most complex object in our known universe, and the billions of cells and trillions of connections are truly wonders of enormous proportions. The study of the way that the cellular elements of the nervous system work to produce sensations, behaviours, and higher order mental processes has become a most productive area of science. However, neuroscientists have come to realize that they are studying a moving target: growth and change are integral to brain function and form the very basis by which we can learn anything about it. As the behavioral embryologist George Coghill pointed out, "Man is, indeed, a mechanism, but he is a mechanism which, within his limitations of life, sensitivity and growth, is creating and operating himself." To understand the brain, then, we need to understand how this mechanism arises and the ways in which it can change throughout a lifetime.

The construction of the brain is an integrated series of developmental steps, beginning with the decision of a few early embryonic cells to become neural progenitors. As connections form between nerve cells and their electrical properties emerge, the brain begins to process information and mediate behaviors. Some of the underlying circuitry is built into the nervous system during embryogenesis. However, interactions with the world continuously update and adapt the brain's functional architecture. The mechanisms by which these changes occur appear to be a continuation of the processes that sculpt the brain during development. Since the text covers each of these developmental steps, it is relatively broad in scope.

An understanding of the development of the nervous system has importance for biologists in a larger context. Studies of development have led to insights into the evolutionary relationships among organisms. The dogma of phylogeny and ontogeny of the last century has been superseded by a deeper understanding of the ways in which evolutionary change can be effected through changes in development. The brain is no exception to these rules. We should expect that insight into the evolution of that which makes us most human will be gained from an appreciation of how developmental processes are modified over time.

The goal of this text is to provide a contemporary overview of neural development for undergraduate students or those who have some background in the filed of biology. This intent is not compatible with a comprehensive review of the literature. A recent MEDLINE search of publications in the field of neural development [(neural or neuron or nervous) and (development or embryology or maturation)] yielded 56,840 papers published between 1966 and 1999. We admit, up front, to having read only a fraction of these papers or of the thousands that were published before 1966. As a practical matter, we made use of authoritative books, contemporary review articles, hallway conversations, and e-mail consultations to select the experiments that are covered in our text. Even so, we expect that important contributions have been missed inadvertently. Therefore, advanced students will find themselves quickly turning to specialized texts and reviews. Another compromise that comes from writing an undergraduate biology book well after the onset of the revolution in molecular biology is that all subjects now have a rather broad cast of molecular characters. In addition, the most instructive experiments on a particular class of molecules have often been performed on nonneural tissue. Even if we chose to cover only the genes and proteins whose roles have been best characterized in the nervous system, most chapters would run the risk of sounding like a (long) list of acronyms. Therefore, we charted a compromise between the need to update students and our strong inclination to hold their attention. The book does not contain exhaustive lists of molecular families, and the most current articles must serve as an appendix to our text.

Among the many scientists who helped us through discussions, unpublished findings, or editorial comment are (in alphabetical order) Chiye Aoki, Michael Bate, Olivia Bermingham-McDonogh, John Bixby, Sarah Bottjer, Martin Chalfie, Hollis Cline, Martha Constantine-Paton, Ralph Greenspan, Voker Hartenstein, Mary Beth Hatten, Christine Holt, Darcy Kelley, Chris Kintner, Sue McConnell, Ilona Miko, Ronald Oppenheim, Thomas Parks, David Raible, Henk Roelink, Edwin Rubel, John Rubenstein, David Ryugo, Nancy Sculerati, Carla Shatz, and Tim Tully.

Preface to the Second Edition

The human brain—perhaps the most complex object in our universe—is composed of billions of cells and trillions of connections. It is truly a wonder of enormous proportions. Although we are far from a thorough understanding of our brains, study of the way that the cellular constituents of the nervous system, the neurons and glia, work to produce sensations, behaviors, and higher order mental processes has been a most productive area of science. However, more and more, neuroscientists are realizing that we are studying a moving target-growth and that changes are integral to brain function, forming the very basis for learning, perception, and performance. To comprehend brain function, then, we must understand how the circuits arise and the ways in which they are modified during maturation. Santiago Ramón y Cajal, one of the founders of modern neuroscience, was able to make his remarkable progress in studies of the cellular makeup of the nervous system in large part because of his work with the embryonic brain, choosing to study "the young wood, in the nursery stage . . . rather than the . . . impenetrable . . . full grown forest."

The construction of the brain is an integrated series of developmental steps, starting with the decision of a few early embryonic cells to become neural progenitors and nearing completion with the emergence of behavior, which is the scope of this book. Interactions with the world continuously update and adapt synaptic connections within the brain, and the mechanisms by which these changes occur are fundamentally a continuation of the same processes that sculpted the emerging brain during embryogenesis.

Studies of development have also led to insights about the evolutionary relationships among organisms. The dogma of phylogeny and ontogeny of the last century has been superseded by our current deeper understanding of the ways in which evolutionary change can be effected through changes in development. The brain is no exception to these rules, and we can expect that much insight into the evolution of that which makes us most human will be gained from an appreciation of how developmental processes are modified over time.

The goal of this text is to provide a contemporary overview of neural development both for undergraduate students and for those who have some background in the field of biology. This intent is not compatible with a comprehensive review of the literature. In the first edition, we noted that there were about 54,000 papers published in this field between 1966 and 1999. Another 25,000 have appeared during the past 4 years (using the search string "neural or neuron or nervous" and "development or embryology or maturation" and 2000:2004). We charted a compromise between the need to update students and our strong inclination to hold their attention. The book does not contain exhaustive lists of molecular families, and the most current review articles must serve as an appendix to our text. Since the text does not encompass many exciting areas of research, students will find themselves quickly turning to specialized texts and reviews.

Among those who helped us through discussion and editorial comment are: Chiye Aoki, Michael Bate, Carla Shatz, Ford Ebner, Edward Gruberg, Christine Holt, Lynne Kiorpes, Vibhakar Kotak, Tony Movshon, Ron Oppenheim, Sarah Pallas, Sheryl Scott, Tim Tully, and Lance Zirpel.

Finally, we acknowledge our editor, Johannes Menzel, with particular gratitude, for his help, advice, and perseverance.

Dan H. Sanes
Thomas A. Reh
William A. Harris
July 2005

1

Neural Induction

DEVELOPMENT AND EVOLUTION OF NEURONS

Almost as early as multicellular animals evolved, neurons have been part of their tissues. Metazoan nervous systems range in complexity from the simple nerve net of the jellyfish to the billions of specifically interconnected neuron assemblies of the human brain. Nevertheless, the neurons and nervous systems of all multicellular animals share many common features. Voltage-gated ion channels are responsible for action potentials in the neurons of hydras, as they are in people. Synaptic transmission between neurons in nerve nets is basically the same as that in the cerebral cortex in humans (Figure 1.1). This book describes the mechanisms responsible for the generation of these nervous systems, highlighting examples from a variety of organisms. Despite the great diversity in the nervous systems of various organisms, underlying principles of neural development have been maintained throughout evolution.

It is appropriate to begin a book concerned with the development of the nervous system with an evolutionary perspective. The subjects of embryology and evolution have long shared an interrelated intellectual history. One of the major currents of late-nineteenth-century biology was that a description of the stages of development would provide the key to the path of evolution of life. The phrase "ontogeny recapitulates phylogeny" was important at the start of experimental embryology (Gould, 1970). Although the careful study of embryos showed that they did not resemble the adult forms of their ancestors, it is clear that new forms are built upon the structures of biological predecessors. One aim of this book is to show how an understanding of the development of the nervous

system will give us insight into its evolution. It is also wise to remember, as Dobzhansky pointed out, that "nothing in biology makes sense except in the light of evolution."

EARLY EMBRYOLOGY OF METAZOANS

The development of multicellular organisms varies substantially across phyla; nevertheless, there are some common features. The cells of all metazoans are organized as layers. These layers give rise to the various organs and tissues, including the nervous system. These layers are generated from the egg cell through a series of cell divisions and their subsequent rearrangements (Figure 1.2). The egg cells of animals are typically polarized, with an "animal pole" and a "vegetal pole." This polarity is often visible in the egg cell, since the vegetal pole contains the yolk, the stored nutrient material necessary for sustaining the embryo as it develops. Once fertilized by the sperm, the egg cell undergoes a series of rapid cell divisions, known as cleavages. There are many variations of cleavage patterns in embryos, but the end result is that a large collection of cells, the blastula, is generated over a relatively short period of time. In many organisms the cells of the blastula are arranged as a hollow ball, with an inner cavity known as a blastocoel. Those cells at the vegetal pole will ultimately develop as the gut, whereas those at the animal pole will give rise to the epidermis and the nervous system. Cells in between the animal and vegetal poles will generate mesodermal derivatives, including muscles and internal skeletal elements. The rearrangement of this collection of cells into the primary (or germ) layers is called

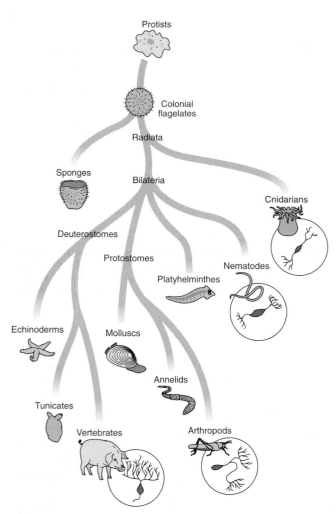

FIGURE 1.1 Neurons throughout the evolution of multicellular organisms have had many features in common. All animals other than colonial flagellates and sponges have recognizable neurons that are electrically excitable and have long processes. The Cnidarians have nerve networks with electrical synapses, but synaptic transmission between neurons is also very ancient.

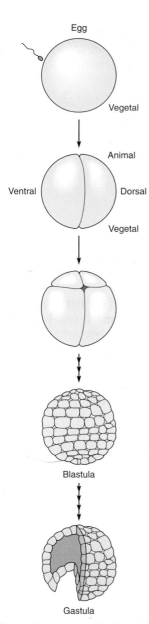

FIGURE 1.2 The early processes of animal development follow a conserved pattern; after fertilization, a series of cleavage divisions divide the egg into a multicellular blastula. The animal and vegetal poles represent an initial asymmetry in the oocyte, and the second axis, dorsal-ventral in this example, is established after fertilization. The process of gastrulation brings some of the cells from the surface of the embryo to the inside and generates the three-layered structure common to most multicellular animals.

gastrulation. Gastrulation can occur via a variety of mechanisms, but all result in an inner, or endodermal, layer of cells, an outer layer of cells, the ectoderm, and a layer of cells between the two other layers, known as the mesoderm (Gilbert and Raunio, 1997). The middle layer can be derived from either the ectoderm (ectomesoderm) or the inner layer (endomesoderm). During the process of gastrulation, the cells of the mesoderm and endoderm move into the inside of the embryo, often at a single region, known as the blastopore. Once the endoderm and mesoderm are inside the ball, they usually obliterate the blastocoel and form a new cavity, the archenteron, or primitive gut. Animals can be divided in two on the basis of whether the mouth forms near the point of this blastopore (in protostomes) or at a distant site (in deuterostomes). Once

these three primary germ layers are established, the development of the nervous system begins. A more detailed description of the development of the other organ systems is beyond the scope of this text. Nevertheless, one should keep in mind that the development of the nervous system does not take place in a vacuum, but is an integral and highly integrated part of the development of the animal as a whole.

The next three sections will deal with the embryology of several examples of metazoan development: Cnidarians (hydra); nematode worms (*Caenorhabditis elegans*); insects (*Drosophila melanogaster*), and several vertebrates (frogs, fish, birds, and mammals). The development of these animals is described because they have been particularly well studied for historical and practical reasons. However, one should take these examples as representative, not as definitive. The necessity of studying many diverse species has become critical to the understanding of the development of any one species.

DERIVATION OF NEURAL TISSUE

The development of the nervous system begins with the segregation of neural and glial cells from other types of tissues. The many differences in gene expression between neurons and muscle tissue, for example, arise through the progressive narrowing of the potential fates available to a blast cell during development. The divergence of neural and glial cells from other tissues can occur in many different ways and at many different points in the development of an organism. However, the cellular and molecular mechanisms that are responsible for the divergence of the neural and glial lineages from other tissues are remarkably conserved.

Hydra

The first generalization that can be made concerning neural segregation is that the nervous system is derived from the ectodermal germ layer in all triploblastic (three-germ-layered) organisms. Most of the organisms we will discuss in this book are triploblastic; that is, they have three distinct primary layers. However, neurons are present in more primitive diploblastic (two-layered) organisms such as jellyfish and hydras. The jellyfish and hydras are among the organisms that belong to the cnidarian phylum. These animals are among the most primitive multicellular animals, with no defined organs and only a tissue level of organization for the different cell types. The freshwater hydra is one of the more well-studied examples of the phylum (Figure 1.3A). Hydras have an outer layer, the epidermis, and an inner epithelium, the gastrodermis (Figure 1.3B), and between these layers is an extracellular matrix similar in composition to the basement membranes of other animals. The gastric cavity has a single opening that serves as both a mouth and an anus, and the hydra uses the surrounding ring

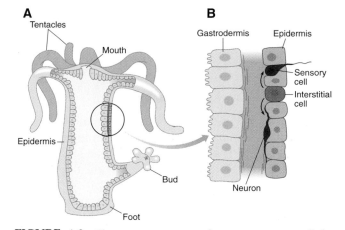

FIGURE 1.3 The nervous system shares a common cellular lineage with the ectoderm. In very simple animals, like the hydra, the neurons are derived from a precursor in the epidermis, known as the interstitial cell, which can generate both neurons and other sensory cells.

of tentacles to capture food. The nervous system of hydras and jellyfish is composed of bipolar neurons organized as a network. The neurons coordinate the activity of the animal via voltage-gated channels, action potentials, and chemical and electrical synaptic transmission. Thus, the basic features of the nervous system have been around for at least 600 million years, and appear to have been present in animals ancestral to all metazoans except sponges.

Given the many similarities neurons have had since they arose in evolution, it is worthwhile to consider how they develop in these most primitive animals. Cnidarians can reproduce either asexually or sexually, and most biologists are familiar with the asexual budding of hydras (Figure 1.3A). A bud forms as an evagination from a region of the body wall known as the bud zone. The bud elongates over the next two days and then separates from the parent organism. The neurons, like all the cells of these animals, arise from multipotent progenitor cells in the epidermal layer, known as the interstitial cells. The interstitial cells are a heterogeneous collection of true stem cells and progenitor cells differentiating along various cell-specific pathways. Are these interstitial cells similar to the precursors of the neurons in other phyla? Unfortunately, although the development of the nervous system has been described in some detail, little is known of the cellular and molecular mechanisms that give rise to neurons in these animals. Therefore, it is difficult to make direct comparisons between the Cnidarians and other metazoans. However, genes related to the proneural genes (see below) of *Drosophila* and vertebrates have been discovered in hydra (Grens et al.,

1995), although much more study of the Cnidarians is necessary to determine the degree to which the mechanisms for neurogenesis are common to all multicellular animals.

C. elegans

The development of *C. elegans*, a nematode worm, highlights the shared lineage of the epidermal and neural cell fates. These animals have been studied primarily because of their simple structure (containing only about a thousand cells), their rapid generation time (allowing for rapid screening of new genetic mutants), and their transparency (enabling lineage relationships of the cells to be established). These nematodes have a rigid cuticle that is made of collagenous proteins secreted by the underlying cells of the hypodermis. The hypodermis is analogous to the epidermis of other animals, except that it is composed of a syncytium of nuclei rather than of individual cells. They have a simple nervous system, composed of only 302 neurons and 56 glial cells. These neurons are organized into nerve cords instead of the nerve net of the jellyfish. The nerve cords are primarily in the dorsal and ventral sides of the animals, but there are some neurons that run along the lateral sides of the animal as well. The nematodes move by a series of longitudinal muscles, and they have a simple digestive system.

Nematodes have long been a subject for developmental biologists' attention. Theodore Boveri studied nematode embryology and first described the highly reproducible pattern of cell divisions in these animals in the late 1800s. Boveri's most famous student, Hans Spemann, whose work on amphibian neural induction will be described below, worked on nematodes for his Ph.D. research. The modern interest in nematodes, however, was motivated by Sydney Brenner, a molecular biologist who was searching for an animal that would allow the techniques of molecular genetics to be applied to the development of metazoans (Brenner, 1974).

Because of the stereotypy in the pattern of cell divisions, the lineage relationships of all the cells of *C. elegans* have been determined (Sulston et al., 1983). The first cleavage produces a large somatic cell, the AB blastomere, which gives rise to most of the hypodermis and the nervous system and the smaller germline P cell, which in addition to the gonads will also generate the gut and most of the muscles of the animal (Figure 1.4). Subsequent cleavages produce the germ cell precursor, P4, and the precursor's cells for the rest of the animal: the MS, E, C, and D blastomeres (Figure 1.4), and these cells all migrate into the interior of the

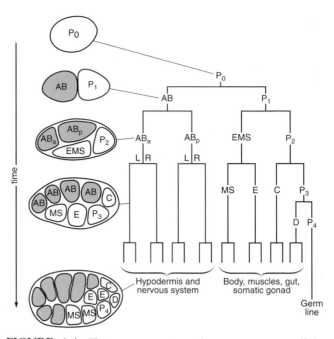

FIGURE 1.4 The nervous system shares a common cellular lineage with the ectoderm. The cell divisions that generate the *C. elegans* nematode worm are highly reproducible from animal to animal. The first division produces the AB blastomere and the P1 blastomere. The germ line is segregated into the P4 blastomere within a few divisions after fertilization. The subsequent divisions of the AB blastomere go on to give rise to most of the neurons of the animal, as well as to the cells that produce the hypodermis—the epidermis of the animal.

embryo, while the AB-derived cells spread out over the outside of the embryo completing gastrulation (Figure 1.5). The next phase of development is characterized by many cell divisions and is known as the *proliferation phase*. Then an indentation forms at the ventral side of the animal marking the beginning of the morphogenesis stage, and as this indentation progresses, the worm begins to take shape (Figure 1.5). At this point, the worm has only 556 cells and will add the remaining cells (to the total of 959) over the four larval molts. The entire development of the animal takes about two days.

The neurons of *C. elegans* arise primarily from the AB blastomere, in lineages shared with the ectodermally derived hypodermis. An example of one of these lineages is shown in Figure 1.5. The Abarpa blastomere can be readily identified in the 100-min embryo through its position and lineal history. This cell then goes on to give rise to 20 additional cells, including 9 neurons of the ring ganglion. The progeny of the Abarpa blastomere, like most of the progeny of the AB lineage, lie primarily on the surface of the embryo prior to 200 min of development. At this time, the cells on the ventral and lateral sides of the embryo move

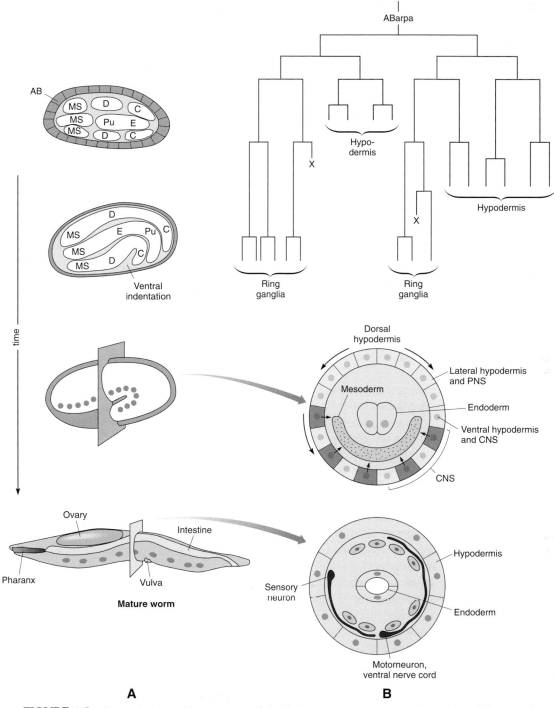

A   **B**

FIGURE 1.5 The next phase of development of the *C. elegans* worm also highlights the shared lineages of hypodermis and neurons. A. During gastrulation, the MS, E, C, and D blastomeres all migrate into the interior of the embryo, while the progeny of the AB blastomeres spread out over the external surface. Once the embryo starts to take form, sections through the embryo show the relationships of the cells. The neurons are primarily derived from the ventrolateral surface, through the divisions of the AB progeny cells. As these cells are generated, they migrate into the interior and form the nerve rings. B. A typical lineage is also shown. The Abarpa blastomeres undergo five rounds of division, to generate 9 neurons and 10 hypodermal cells. Neural lineages are shown in red.

inside and become the nervous system, whereas the AB progeny that remain on the surface spread out to form the hypodermis, a syncytial covering of the animal. Most of the neurons arise in this way; of the 222 neurons in the newly hatched *C. elegans*, 214 arise from the AB lineage, whereas 6 are derived from the MS blastomere and 2 from the C blastomere.

Drosophila

The development of the fruit fly, *Drosophila*, is characteristic of the "long germ band" arthropods. Unlike the embryos of the nematode and the leech, where cleavage of the cells occurs at the same time as nuclear divisions, the initial rounds of nuclear division in the *Drosophila* embryo are not accompanied by corresponding cell divisions. Instead, the nuclei remain in a syncytium up until just prior to gastrulation, three hours after fertilization. Prior to this time, the dividing nuclei lie in the interior of the egg, but they then move out toward the surface and a process known as cellularization occurs, and the nuclei are surrounded by plasma membranes. At this point the embryo is known as a cellular blastoderm.

The major part of the nervous system of *Drosophila* arises from cells in the ventrolateral part of the cellular blastoderm (Figure 1.6, top). Soon after cellularization, the ventral furrow, which marks the beginning of gastrulation, begins to form (Figure 1.6, middle). At the ventral furrow, cells of the future mesoderm fold into the interior of the embryo. The process of invagination occurs over several hours, and the invaginating cells will continue to divide and eventually will give rise to the mesodermal tissues of the animal. As the mesodermal cells invaginate into the embryo, the neurogenic region moves from the ventrolateral position to the most ventral region of the animal (Figure 1.6). The closing of the ventral furrow creates the ventral midline, a future site of neurogenesis. On either side of the ventral midline is the neurogenic ectoderm, tissue that will give rise to the ventral nerve cord, otherwise known as the central nervous system (Figure 1.6). However, it is worth noting that a separate neurogenic region, known as the procephalic neurogenic region, gives rise to the cerebral ganglia or brain.

Drosophila neurogenesis then begins in the neurogenic region; they enlarge and begin to move from the layer into the inside of the embryo (Figure 1.6). At the beginning of neurogenesis, the neurogenic region is a single cell layer; the first morphological sign of neurogenesis is that a number of cells within the epithelium begin to increase in size. These larger cells then undergo a shape change and squeeze out of the epithelium. This process is called *delamination* and is shown

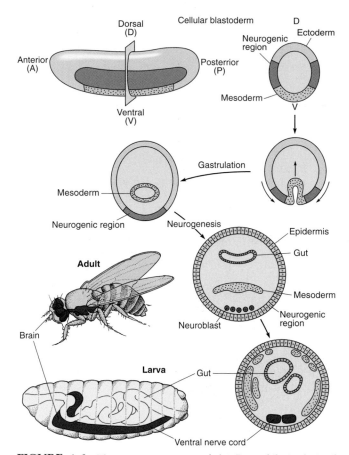

FIGURE 1.6 The nervous system of the *Drosophila* is derived from the ventrolateral region of the ectoderm. The embryo is first (top) shown at the blastoderm stage, just prior to gastrulation. The region fated to give rise to the nervous system lies on the ventral-lateral surface of the embryo (red). The involution of the mesoderm at the ventral surface brings the neurogenic region closer to the midline. Scattered cells within this region of the ectoderm then enlarge, migrate into the interior of the embryo, and divide several more times to make neurons and glia. These neurons and glia then condense into the ganglia of the mature ventral nerve cord (or CNS) in the larva and the adult.

in more detail in Figure 1.7. The cells that delaminate are called *neuroblasts* and are the progenitors that will generate the nervous system. In the next phase of neurogenesis, each neuroblast divides to generate many progeny, known as *ganglion mother cells* (GMCs). Each GMC then generates a pair of neurons or glia. In this way, the entire central nervous system of the larval *Drosophila* is generated. However, the *Drosophila* nervous system is not finished in the larva, but rather additional neurogenesis occurs during metamorphosis. Sensory organs, like the eyes, are generated from imaginal discs, small cellular discs in the larva that undergo a tremendous amount of proliferation during metamorphosis to generate most of what we recognize as an adult fly.

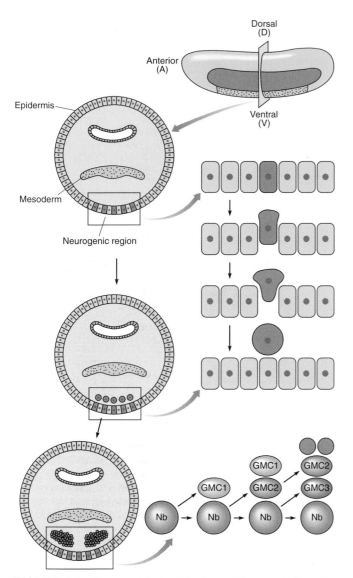

FIGURE 1.7 The neuroblasts of the *Drosophila* separate from the ectoderm by a process known as delamination. The neuroblasts enlarge relative to the surrounding cells and squeeze out of the epithelium. The process occurs in several waves; after the first set of neuroblasts has delaminated from the ectoderm, a second set of cells in the ectoderm begins to enlarge and also delaminates. The delaminating neuroblasts then go on to generate several neurons through a stereotypic pattern of asymmetric cell divisions. The first cell division of the neuroblast produces a daughter cell known as the ganglion mother cell, or GMC. The first ganglion mother cell divides to form neurons, while the neuroblast is dividing again to make another GMC. In the figure, the same neuroblast is shown through its successive stages as Nb, while the GMCs are numbered successively as they arise.

Vertebrates

Vertebrate embryos undergo a fundamentally similar process of early development, though at first appearance they seem to be quite different. In this section we will review the development of several

different vertebrates: amphibians, fish, birds, and mammals. In all of these animals, multiple cleavage divisions generate a large number of cells from the fertilized oocyte. However, while gastrulation in all of these animals is basically conserved, the details of the cellular movements during this phase can look quite different.

Amphibian eggs are like those of many animals in that the egg has a distinct polarity with a nutrient-rich yolk concentrated at the "vegetal" hemisphere and a relatively yolk-free "animal" hemisphere. After fertilization, a series of rapid cell divisions, known as cleavages, divides the fertilized egg into blastomeres. The cleavage divisions proceed less rapidly through the vegetal hemisphere, and by the time the embryo reaches 128 cells, the cells in the animal half are much smaller than those of the vegetal half (Figure 1.8). The embryo is called a *blastula* at this stage. The process by which the relatively simple blastula is transformed into the more complex, three-layered organization shared by most animals, is called *gastrulation*. During this phase of development, cells on the surface of the embryo move actively into the center of the blastula. The point of initiation of gastrulation is identified on the embryo as a small invagination of the otherwise smooth surface of the blastula, and this is called the *blastopore* (Figure 1.8). In amphibians the first cells to invaginate occur at the dorsal side of the blastopore (Figure 1.8), opposite to the point of sperm entry. As described below, these cells have a special significance to the development of the nervous system. The mechanism of involution is complex, and it appears that a small group of "bottle" cells initiate the process by changing shape and creating a discontinuity in the surface.

The involuting cells lead a large number of cells that were originally on the surface of the embryo to the interior (Figure 1.8). The part of the blastula that will ultimately reside in the interior of the embryo is called the *involuting marginal zone* (IMZ). Most of these cells will give rise to mesodermally derived tissues, like muscle and bone. The first cells to involute crawl the farthest and produce the mesoderm of the anterior part of the animal (i.e., the head). The later involuting IMZ cells produce the mesoderm of more posterior regions, including the tail of the tadpole. At this point in development, the neural plate of the vertebrate embryo still largely resembles the rest of the surface ectoderm. However, shortly after its formation, the neural plate begins to fold onto itself to form a tubelike structure, the neural tube (Figure 1.9). Much more will be said about the neural tube and its derivatives and shape changes in the next two chapters. For now, suffice it to say that this tube of cells gives rise to nearly all the neurons and glia

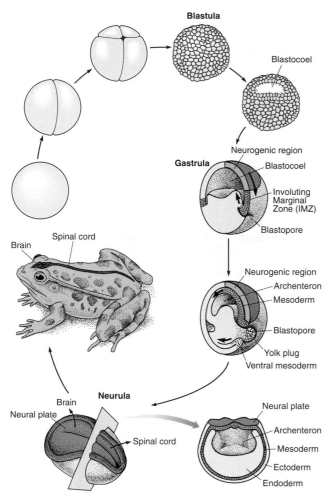

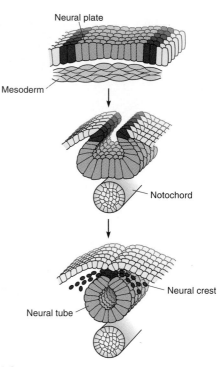

FIGURE 1.8 The development of the central nervous system, brain, and spinal cord in a frog embryo is shown from the egg cell to the adult. After a series of cleavage divisions produce a blastula, a group of cells known as the involuting marginal zone, or IMZ, grow into the interior of the embryo at a point known as the blastopore. This process of gastrulation is shown in two cross sections. The involuting cells go on to form mesodermal tissues (blue) and induce the cells of the overlying ectoderm to develop into neural tissue, labeled as the neurogenic region (red). After the process of neural induction, the neurogenic region is known as the neural plate and is now restricted to giving rise to neural tissue. A cross section of the embryo at the neural plate stage shows the relationships between the tissues at this stage of development. The neural plate goes on to generate the neurons and glia in the adult brain and spinal cord.

FIGURE 1.9 The neural plate (light red) rolls up into a tube separating from the rest of the ectoderm. The involuting cells condense to form a rod-shaped structure—the notochord—just underneath the neural plate. At the same time, the neural plate begins to roll up and fuse at the dorsal margins. A group of cells known as the neural crest (bright red) arises at the point of fusion of the neural tube.

of vertebrates. Another source of neurons and glia is the neural crest, a group of cells that arises at the junction between the tube and the ectoderm (Figure 1.9). The neural crest is the source of most of the neurons and glia of the peripheral nervous system, whose cell bodies lie outside the brain and spinal cord. This tissue is unique to vertebrates, and has the capacity to generate many diverse cell types; we will have more to say about neural crest in later chapters.

The complex tissue rearrangements that occur during gastrulation in the amphibian occur in other vertebrates in fundamentally the same way. However, the details of these movements can be quite different. Much of the difference in cell movements lies in differences in the amount of yolk in the egg. Fish and bird embryos have a substantial amount of yolk; since the cleavage divisions proceed more slowly through the yolk, these animals have many more cleavage divisions in the animal pole than in the vegetal pole. In zebrafish embryos, the blastomeres are situated at the top of the egg, and as development proceeds these cells divide and spread downward over the surface of the yolk cells in a process known as epiboly. At 50% epiboly, when the spread reaches the equator, there is a transient pause as the process of gastrulation begins at the future dorsal margin of the embryo, which is at this point called the *shield* (Figure 1.10). The shield begins to thicken as gastrulation commences. Prospective mesodermal cells delaminate, move inside the ectodermal layer, and begin migrating back toward the animal pole. The rest of the ectoderm then continues its migration to the vegetal pole until the yolk is completely enveloped at 100% epiboly. As the dorsal meso-

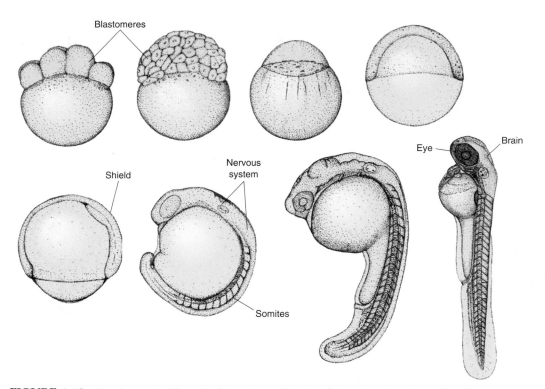

FIGURE 1.10 Development of the zebrafish embryo. The zebrafish embryo develops primarily on top of a large ball of yolk. The cleavages are confined to the dorsal side. After multiple cleavage divisions, the cells migrate over the yolk in a process called epiboly. Once the cells have enclosed the yolk, the development of the nervous system proceeds much like that described for the frog. (After Kimmel et al., 1995)

dermal cells migrate toward the animal pole on the shield side, the ectoderm above becomes committed to a neural fate and a definitive neural plate begins to form. One of the best things about the zebrafish embryos, as far as developmental biologists are concerned, is their extraordinary optical transparency.

Avian embryos are an extreme example of a "yolky" egg. We all know how much yolk is in a chicken egg. As a result, when the single large egg cell begins to divide, the cell cleavage divisions do not penetrate into this yolk but are restricted to the relatively yolk-free cytoplasm at the animal pole. These cleavages lead to a disc of cells, called the *blastodisc*, which is essentially floating on the yolk. The invagination of mesoderm occurs in this disc through a blastopore-like structure known as the "primitive streak." During this invagination, future mesoderm cells migrate into the interior of the embryo (Figure 1.11).

What about mammalian embryos, which have essentially no yolk and derive all their nourishment from the placenta? The cleavage divisions of mammalian embryos are complete (Figure 1.12), and the resulting cells are equal in their potential; there is no obvious animal or vegetal pole. However, after a sufficient number of divisions, when the blastula

forms, there are cells on the inside of the ball, called the *inner cell mass*, that produce the embryo, whereas the cells on the outside of the ball make the placenta and associated extra-embryonic membranes. Even though they lack yolk, mammalian embryos undergo a process of gastrulation that is similar to the avian embryo in that the developing mesodermal cells migrate through the primitive streak to reach the interior (Figure 1.12). The primitive streak runs along the anterior-posterior axis of the embryo, and the ectoderm laying above the ingressing mesodermal cells becomes the neural plate and subsequently the neural tube, much like that described above for the other vertebrate embryos.

INTERACTIONS WITH NEIGHBORING TISSUES IN MAKING NEURAL TISSUE

The three basic layers of the embryo—the endoderm, mesoderm, and ectoderm—arise through the complex movements of gastrulation. These movements also create new tissue relations. For example, after gastrulation in the frog, presumptive mesoderm

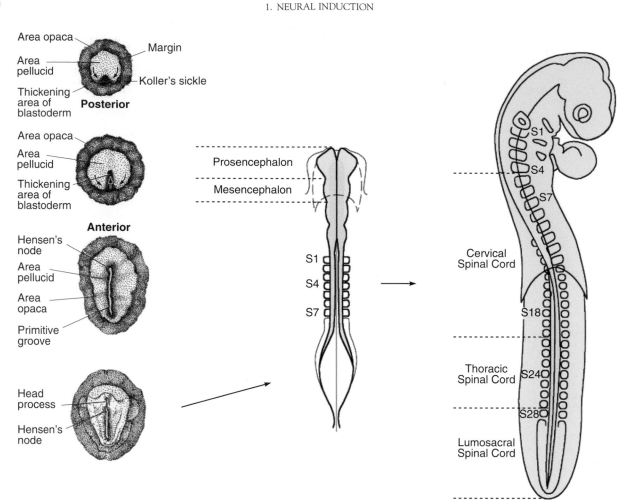

FIGURE 1.11 Development of the chick embryo. The blastoderm (area opaca) sits on top of the large yolk and is the result of a large number of cleavage divisions. At the start of gastrulation, cells move posteriorly (arrows) and migrate under the area opaca. The embryo begins to elongate in the anterior-posterior axis, and the region where the cells migrate underneath the area opaca is now called the primitive groove and then primitive streak. Details of the cellular movements are shown in the cutaway view. The cells migrate into the blastocoel to form the mesoderm. At the anterior end of the primitive streak an enlargement of the streak is called *Hensen's node*.

now underlies the dorsal ectoderm. A large number of experimental studies in the early part of the twentieth century revealed that these new tissue arrangements were of critical importance to the development of a normal animal. By culturing small pieces of embryos in isolation, it was possible to determine the time at which each part of the embryo acquired its character or fate (Figure 1.13). When the dorsal ectoderm was cultured in isolation prior to gastrulation, the cells differentiated into epidermis, while when roughly the same piece of tissue was isolated from gastrulating embryos, the piece of ectoderm now differentiated into neural tissue, including recognizable parts of the brain, spinal cord, and even eyes. These results led Hans Spemann, a leading embryologist of the time, to speculate that the ectoderm became fated to generate

neural tissue as a result of the tissue rearrangements that occur at gastrulation (Hamburger et al., 1969). One possible source of this "induction" of the neural tissue was the involuting mesoderm, known at the time as the archenteron roof. As noted above, the involuting tissue is led into the interior of the embryo by the dorsal lip of the blastopore. To test the idea that the involuting mesoderm induces the overlying ectoderm to become neural tissue, Spemann and Hilde Mangold carried out the following experiment (Figure 1.14). The dorsal lip of the blastopore was dissected from one embryo and transplanted to the interior of another embryo, and the latter embryo was allowed to develop into a tadpole. Spemann and Mangold found that an entire second body axis, including a brain, spinal cord, and eyes developed from the ventral side of the

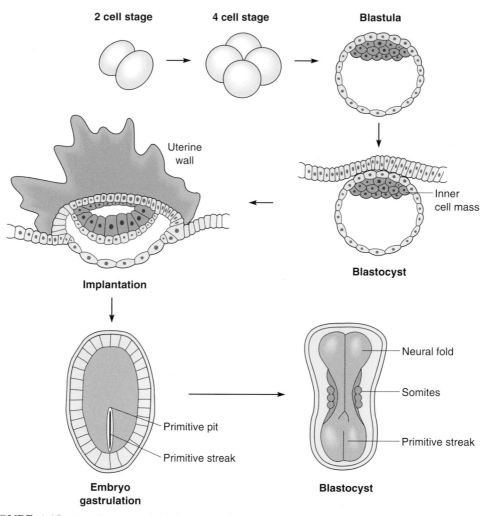

FIGURE 1.12 Development of the human embryo. The initial cleavage divisions are symmetric and produce apparently identical blastomeres. There is not much yolk in the mammalian embryo, since most nutrients are derived from the placenta. After multiple cleavage divisions, the embryo is called a blastocyst and develops a distinct inner cell mass and an outer layer of cells. The inner cell mass will develop into the embryo, while the outer cells will contribute to the placenta. After implantation, the embryo begins to elongate, and a section through the amniotic cavity shows that the human embryo develops a primitive streak much like that present in the chick embryo. The primitive streak is a line of cells migrating into the blastocoel that will form the mesoderm, and the neural tube will form from the ectoderm overlying the involuting mesoderm. The tube rolls up and forms the brain and spinal cord in a process much like that described for the other vertebrates.

embryo, where neural tissue does not normally arise. To determine whether the new neural tissue that developed in these twinned embryos came from the dorsal lip tissue, they transplanted the dorsal blastopore lip from the embryo of a normal pigmented frog into the embryo of a nonpigmented strain of frogs. They found that the second body axis that resulted was made of mostly nonpigmented cells, indicating that it came largely from the host blastula, not the transplanted dorsal lip. Thus, the grafted blastopore cells have the capacity to induce neural tissues from a region of the ectoderm that would normally not give

rise to a nervous system. In addition to the neural tissue in these embryos, they found that mesodermally derived structures also contributed to the twinned embryo. They concluded that the dorsal lip acts not only as a neural inducer but also as an "organizer" of the entire body axis. As a result of these experiments, this region of the embryo is known as the Spemann organizer.

In the years following these initial studies of Spemann and Mangold, several embryologists tried to further characterize the induction process, as well as to identify the inducing principle or factor. One of the

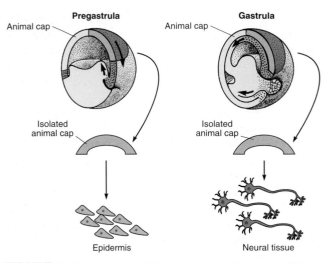

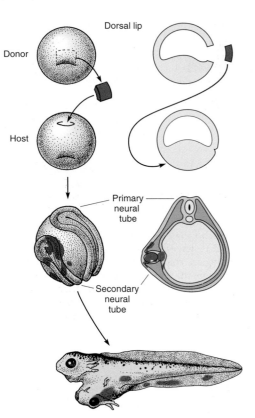

FIGURE 1.13 Isolation of fragments of embryos at different stages of development demonstrates when tissue becomes committed to the neural lineage. If the animal cap is isolated from the rest of the embryo (left), the cells develop as epidermis, or skin. If the same region of the embryo is isolated a few hours later, during gastrulation (right), it will develop into neural tissue (shown in the figure as red neurons). Experiments like these led to the idea that the neural lineage arises during gastrulation.

FIGURE 1.14 Spemann and Mangold transplanted the dorsal lip of the blastopore from a pigmented embryo (shown as red) to a non-pigmented host embryo. A second axis, including the neural tube, was induced by the transplanted tissue. The transplanted dorsal blastopore lip cells gave rise to some of the tissue in the secondary axis, but some of the host cells also contributed to the new body axis. They concluded that the dorsal lip cells could "organize" the host cells to form a new body axis, and they named this special region of the embryo the organizer.

first realizations that came from these additional studies was that "the organizer" has subdivisions, each capable of inducing specific types of differentiation. Holtfreter subdivided the organizer region into pieces, and, using the same transplantation strategy, he found that when more lateral aspects of the dorsal lip were used, tails were induced, whereas when more medial regions of the organizer region were transplanted, heads were induced. In an attempt to more precisely define the heterogeneity of the region, Holtfreter also cultured small bits of the dorsal lip and found that that these develop into more or less well-defined structures, such as single eyes or ears! As Holtfreter succinctly summarizes: "even at the gastrula stage the head organizer is not actually an equipotential entity, but is subdivided into specialized inductors although distinct boundaries between them do not seem to exist."

Neural induction does not appear to act solely through a vertical signal passed from the involuting mesoderm; there is also evidence that a neural inducing signal can be passed through the plane of the ectoderm. As noted above, when blastulas are placed in hypotonic solutions just prior to gastrulation, the IMZ cells fail to involute, and instead evaginate to produce an exogastrula. Under these conditions, the process of signaling between the mesoderm and the ectoderm should be blocked, since involuting mesoderm is no

longer underneath the ectoderm. Surprisingly, however, some neural induction does appear to take place. Although this was difficult to determine in Holtfreter's time, the use of antibodies and probes for neural-specific proteins and gene expression clearly shows that organized neural tissue forms from such exogastrulated ectoderm (Holtfreter, 1939; Ruiz i Altaba, 1992). This so-called planar induction can also be demonstrated in a unique tissue combination invented by Ray Keller that bears his name, the Keller sandwich. In this preparation, the presumptive neural ectoderm and the dorsal lip are dissected from two embryos and sandwiched together. In these, the mesoderm does not move inside of the sandwich, but rather extends away from the neurectoderm like an exogastrula (Keller et al., 1992; Figure 1.15). Only a thin bridge of tissue connects the mesoderm with the neural ectoderm, but nevertheless, extensive and patterned neural development occurs in these cultures (Figure 1.15).

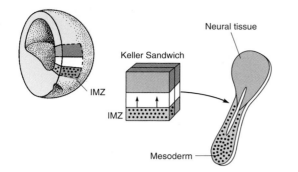

FIGURE 1.15 Planar neural induction can be contrasted from vertical neural induction by Keller sandwiches. The organizer region, including the IMZ cells, along with some of the surrounding ectodermal tissue, can be cultured in isolation and will not involute when two IMZ regions are placed back-to-back. The tissue undergoes morphological changes similar to those that occur during gastrulation, except the tissue extends rather than involutes. Nevertheless, neural tissue (red) is induced in the attached ectoderm, indicating that the signals for neural induction can be passed through the small region that connects the mesodermal cells and the ectodermal cells.

THE MOLECULAR NATURE OF THE NEURAL INDUCER

Early efforts to define the chemical nature of the neural inducer were unsuccessful. In the initial attempts at characterization, Bautzman, Holtfreter, Spemann, and Mangold showed that the organizer tissue retained its inductive activity even after the cells had been killed by heat, cold, or alcohol. Holtfreter subsequently reported that the neuralizing activity survived freezing, boiling, and acid treatment; however, the activity was lost at temperatures of 150°C. Several embryologists then set out to isolate the active principle(s) in the dorsal lip of the blastopore using the following three approaches: (1) extracting the active factor from the dorsal blastopore cells; (2) trying out candidate molecules to look for similar inductive activities; and (3) testing other tissues for inductive activities.

The initial attempts at direct isolation of the inducing activity from the blastopore lip cells resulted in failure, largely because of the small amounts of tissue that could be obtained and the limited types of chemical analyses available at the time. From the initial report in 1932 to the late 1950s, over one hundred studies tried to characterize the neural inducing activity. The search for the neural inducer was one of the major preoccupations of developmental biologists in this period. Whereas one group reported that the active principle was lipid extractable, another would report that the residues were more active than the

extracts. To obtain more tissue to work with, several investigators screened a variety of adult tissues for similar inducing activities. Although some found a certain degree of specificity, liver and kidney being the most potent neural inducers, others found that "fragments from practically every organ or tissue from various amphibians, reptiles, birds, and mammals, including man, were inductive" (Holtfreter, 1955). Perhaps most disconcerting to the investigators at the time were the results from the candidate molecule approach. Some of the factors found to have neuralizing activity made some sense: polycyclic hydrocarbon steroids, for example. However, other putative inducers, such as methylene blue and thiocyanate, most likely had their effects through some toxicity or contamination.

In the early 1980s, a number of investigators began to apply molecular biological techniques to study embryonic inductions. The first of these studies attempted to test for factors that would trigger the process of mesoderm induction in the frog. As described above, the frog embryo is divided into an animal half and a vegetal half; the animal half will ultimately give rise to neural tissue and ectodermal tissue, while the vegetal half will give rise primarily to endoderm. The mesoderm, which will ultimately go on to make muscle and bone and blood, arises in between these two tissues, from the cells around the embryo's equator (see Figure 1.16). It has been known for many years, based on the work of Peter Nieuwkoop, that the formation of the mesodermal cells in the equatorial region requires some type of interaction between the animal and vegetal halves of the embryo. If this animal half or "cap" is isolated from the vegetal half of the embryo, no mesodermal cells develop. However, when Nieuwkoop (1973, 1985) recombined the animal cap with the vegetal half, mesodermal derivatives developed in the resulting embryos (Figure 1.16). He postulated that a signal from the vegetal half of the embryo induced the formation of mesoderm at the junction with the animal half of the embryo. The identification of the molecular basis for this induction has been the subject of intense investigation, and the reader is referred to more general textbooks on developmental biology for the current model of this process.

At the same time these studies of mesodermal inducing factors were taking place, a number of investigators realized that the animal cap assay might also be a very good way to identify neural inducers. Not only do isolated animal caps fail to generate mesodermal cells, but they also fail to develop into neuronal tissue. Several factors added to animal caps caused the cells to develop into neural cells as well as mesoder-

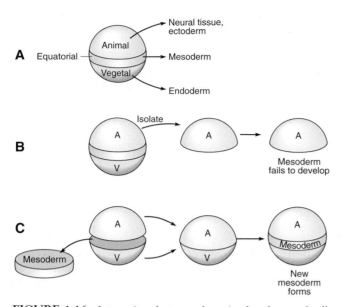

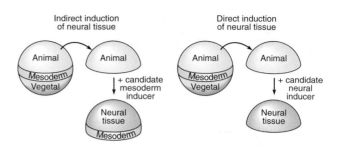

FIGURE 1.16 Interactions between the animal and vegetal cells of the amphibian embryo are necessary for induction of the mesoderm. A. The regions of the amphibian embryo that give rise to these different tissue types are shown. The animal pole gives rise to epidermal cells and neural tissue, the vegetal pole gives rise to endodermal derivatives, like the gut, while the mesoderm (blue) arises from the equatorial zone. B. If the animal cap and vegetal hemispheres are isolated from one another, mesoderm does not develop. C. If the equatorial zone is removed from an embryo and the isolated animal and vegetal caps are recombined, a mesoderm forms at a new equatorial zone.

FIGURE 1.17 Indirect neural induction versus direct neural induction. The organizer transplant experiments show that the involuting mesoderm has the capacity to induce neural tissue in the cells of the animal cap ectoderm. When assaying for the factor released from mesoderm that is responsible for this activity, it was important to distinguish between the direct and indirect induction of neural tissue when animal caps were treated with a candidate factor. In the first example (left), mesoderm (blue) is induced by the factor, and then neural tissue (red) is induced by the mesoderm. Thus, both mesoderm and neural genes are turned on in the animal caps. However, in the case of a direct neural inducer (right), neural genes are turned on (red), but mesoderm-specific genes are not expressed.

mal tissue. However, since the organizer at the dorsal lip of the blastopore is made from mesoderm, it was not clear whether the neural tissue that developed in animal caps was directly induced by the exogenous factor, or, alternatively, whether the factor first induced the organizer and subsequently induced neural tissue. (Figure 1.17). Therefore, to refine the assay to look for direct neural induction, studies concentrated on identifying factors that would increase the expression of neural genes without the concomitant induction of mesoderm-specific gene expression.

The animal cap assay was used for the isolation of the first candidate neural inducer. Richard Harland and his colleagues (Lamb et al., 1993; Smith et al., 1993) used a clever expression cloning system to identify a neural inducing factor (Figure 1.18). The cloning was done by taking advantage of the fact that UV-irradiated frog embryos fail to develop a dorsal axis, including the nervous system, and instead develop only ventral structures. Nevertheless, the transplantation of a dorsal blastopore lip from a different embryo can restore a normal body axis to the UV-treated embryo, indicating that the UV embryo can still respond to the neural inducing factor(s). Furthermore, injection of mRNA from a hyperdorsalized embryo can

also restore a normal body axis. Harland's group took advantage of this fact and used pools of cDNA isolated from the organizer region to rescue the UV-treated embryos. By dividing the pools into smaller and smaller collections, they isolated a cDNA that coded for a unique secreted protein, which they named Noggin. When Noggin was purified and supplied to animal caps, it was capable of specifically inducing neural genes, without inducing mesodermal genes. Noggin mRNA is expressed in gastrulating embryos, by the cells of the dorsal lip of the blastopore, precisely where the organizer activity is known to reside. Injection of Noggin mRNA into UV-treated embryos at the four-cell stage can restore body axis and even hyperdorsalize the embryos to give bigger brains than normal.

At the same time that the Noggin studies were being done, other labs were using additional approaches to identify other neural-inducing molecules. DeRobertis was interested in identifying genes that were expressed in the dorsal blastopore lip organizer region. They isolated a molecule they named Chordin. Like Noggin, this is a secreted protein that is expressed in the organizer during the period when neural induction occurs (Figure 1.19). Overexpression of Chordin in the ventral part of the embryo causes a secondary axis, similar to goosecoid. Thus, chordin appears to be similar to noggin as a putative neural inducer.

A third candidate neural inducer was identified by Melton and his colleagues, as a previously identified reproductive hormone known as *follistatin* (Hemmati-Brivanlou et al., 1994). In the reproductive system, fol-

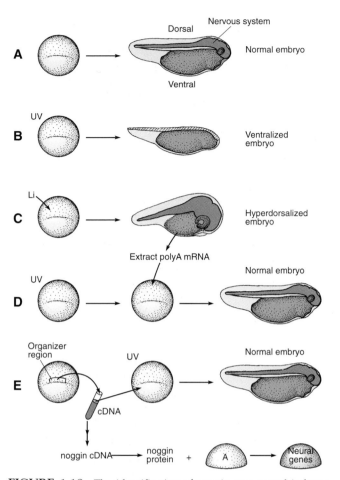

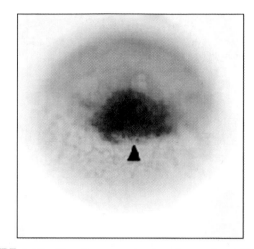

FIGURE 1.19 Chordin expression in a frog embryo. The arrow points to the organizer region at the dorsal lip of the blastopore, and the blue label shows the cells of the pre-gastrula that express the gene chordin. (From Sasai et al., 1994)

FIGURE 1.18 The identification of noggin as a neural inducer used an expression cloning strategy in *Xenopus* embryos. A. Normal development of a *Xenopus* embryo. B. UV light treatment of the early embryo inhibits the development of dorsal structures by disrupting the cytoskeleton rearrangements that pattern the dorsal inducing moelcules prior to gastrulation (ventralized). C. Lithium treatment of the early embryo has the opposite effect; the embryo develops more than normal dorsal tissue (i.e. hyperdorsalized). D. If messenger RNA is extracted from the hyperdorsalized embryos and injected into a UV-treated embryo, the messages encoded in the mRNA can "rescue" the UV-treated embryo and it develops relatively normally. E. Similarly, cDNA from the organizer region of a normal embryo can rescue a UV-treated embryo. The noggin gene was isolated as a cDNA from the organizer region that could rescue the UV-treated embryo when injected into the embryo, and subsequently, recombinant protein was made from this cDNA and shown to induce neural tissue from isolated animal caps, without any induction of mesodermal genes. Neural tissue is shown in red in all panels.

listatin works as a regulatory factor by binding to and inhibiting activin, a member of the TGT-β family of proteins that controls FSH secretion from the pituitary gland. During a screen for mesoderm-inducing factors, Melton found that activin could act as a mesoderm inducer. To study the mechanism of activin action on mesoderm induction, he constructed a truncated

activin receptor that when misexpressed in embryos would interfere with normal endogenous activin signaling (Hemmati-Brivanlou and Melton, 1994) (Figure 1.20). To the surprise of these investigators, interfering with activin signaling not only disrupted normal mesoderm development, but it also induced the cells of the animal cap to develop as neurons without any additional neural-inducing molecule. They proposed that activin—or something like it—normally inhibits neural tissue from differentiating in the ectoderm. They also suggested that perhaps neural induction occurred by inhibiting this neural inhibitor; in other words, that the Spemann organizer secretes factors that antagonize a neural inhibitor. These results led to the idea that neural tissue is in some way the default state of the ectoderm and that it must be actively inhibited by activin-like proteins of the TGF-β family. The idea that the ectoderm is actively inhibited from becoming neural tissue has some additional support. In 1989, two groups (Godsave and Slack, 1989; Grunz and Tacke, 1989) reported that dissociation of the animal cap cells prior to neural induction resulted in most of the cells differentiating as neurons (Figure 1.21). Taking these lines of evidence together, it became clear that molecules that could inhibit activin signaling would make good neural inducers. Since follistatin was already known to inhibit TGF-β signaling from the studies of these factors in the reproductive system, Melton and colleagues tested whether follistatin could act as a neural inducer. They found that indeed follistatin could cause a secondary axis when misexpressed, and recombinant follistatin could induce neural tissue from animal caps. Follistatin is also

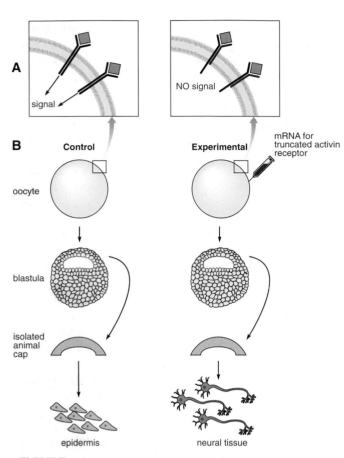

FIGURE 1.20 Expression of a truncated activin receptor blocks normal signaling through the receptor and induces neural tissue. A. The normal activin receptor transmits a signal to the cell when activin binds the receptor and it forms a dimer. The truncated activin receptor still binds the activin (or related TGF-β), but now the normal receptor forms dimers with the truncated receptor. Lacking the intracellular domain to signal, the truncated receptor blocks normal signal transduction through this receptor. B. Oocytes injected with the truncated activin receptor develop to the blastula stage, and when the animal caps were dissected from these embryos, they developed into neural tissue without the addition of a neural inducer. This result indicated that inhibiting this signaling pathway might be how neural inducers function.

expressed in the organizer region of the embryo at the time of neural induction, like chordin and noggin.

CONSERVATION OF NEURAL INDUCTION

Even more fascinating than the identification of three candidate neural-inducing factors in a relatively short period of time is that these three factors may all act by a related mechanism and that this mechanism appears to be at least partially conserved between ver-

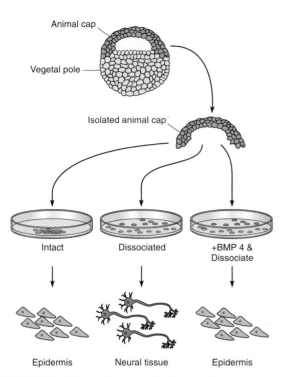

FIGURE 1.21 Dissociation of animal cap cells prior to gastrulation causes most of them to differentiate into neurons in culture. Animal caps can be cultured intact (left) or dissociated into single cells by removing the Ca^{+2} ions from the medium (middle and right). If the intact caps are put into culture, they develop as epidermis (left). If the dissociated animal cap cells are cultured, they develop into neurons (red; middle). This result supports the hypothesis that the neural fate is actively suppressed by cellular associations in the ectoderm. If the cells are dissociated and then BMP is added to the culture dish, the cells do not become neurons, but instead act as if they are not dissociated and develop as epidermis (right).

tebrates and invertebrates (Figure 1.22). Analysis of chordin's sequence revealed an interesting homology with a *Drosophila* gene called *short gastrulation* or *sog*. *Sog* is expressed in the ventral side of the fly embryo, and mutations in this gene in *Drosophila* result in defective dorsal-ventral patterning of the embryo. In null mutants of *sog*, the epidermis expands and the neurogenic region is reduced. And, like chordin, microinjection of *sog* into the nonneurogenic region of the embryo causes the formation of ectopic neural tissue. Thus, *sog* seems to be the functional homolog of chordin. At this point the advantages of fly genetics were important. From analysis of other *Drosophila* mutants, it was possible to show that *sog* interacts with a gene called *decapentaplegic*, or *dpp*, a TGF-like protein related to the vertebrate genes known as bone-morphogenic proteins, BMPs. *Dpp* and *sog* directly antagonize one another in *Drosophila*. Mutations in *dpp* have the opposite phenotype as *sog* mutations; in *dpp*

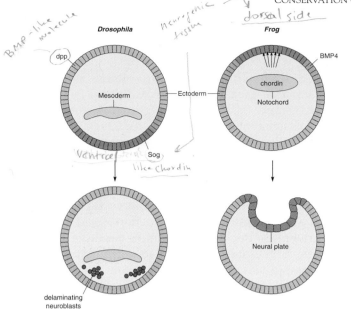

[handwritten annotations: "BMP-like molecule", "neurogenic tissue", "dorsal side", "Ventrolateral", "like chordin"]

FIGURE 1.22 Vertebrates and invertebrates use similar molecules to pattern the dorsal-ventral axis. The *Drosophila* embryo in cross section resembles an inverted *Xenopus* embryo. As described in Figure 1.6, the neurogenic region is in the ventral-lateral *Drosophila* embryo, whereas in the vertebrate embryo, the neural plate arises from the dorsal side. In the *Drosophila*, a BMP-like molecule, dpp, inhibits neural differentiation in the ectoderm, and in the vertebrate embryo, the related molecules, BMP2 and BMP4, suppress neural development. In *Drosophila*, sog (short gastrula) promotes neural development by inhibiting the dpp signaling in the ectoderm in this region, while in the *Xenopus*, a related molecule, chordin (chd), is one of the neural inducers released from the involuting mesodermal cells and in an analogous way inhibits BMP signaling, allowing neural development in these ectodermal cells.

to one another was first proposed by Geoffry Saint-Hillaire from comparative anatomical studies, and this appears to be confirmed by these recent molecular studies (Figure 1.22). Second, the antagonistic mechanism between sog and dpp in the fly also led to the hypothesis that the various neural inducers might work through a common mechanism, the antagonism of BMP4 signaling. The following three key experiments all indicate that this is indeed the case. First, BMP4 will inhibit neural differentiation of animal caps treated with chordin, noggin. or follistatin. Second, BMP4 will also inhibit neural differentiation of dissociated animal cap cells. Third, antisense BMP4 RNA causes neural differentiation of animal caps without addition of any of the neural inducers. The dominant-negative activin receptor induction of neural tissue can also be understood in this context, since the activin receptor is related to the BMP4 receptor, and additional experiments have shown that the expression of the truncated receptor also blocks endogenous BMP4 signaling (Wilson and Hemmati-Brivanlou, 1995).

Do all three of these neural inducers act equivalently to inhibit BMP4 signaling? Biochemical studies have demonstrated that chordin blocks BMP4-receptor interactions by directly binding to the BMP4 with high affinity. Noggin also appears to bind BMP4 with an even greater affinity, while follistatin can bind the related molecules BMP7 and activin. Therefore, it is likely that at least these three neural inducers act by blocking the endogenous epidermalizing BMP4, thereby allowing neural differentiation of the neurogenic ectoderm (Piccolo et al., 1996) (Figure 1.23).

The studies described in *Xenopus* embryos have provided evidence that these factors are capable of inducing neural tissue, but it is more difficult technically to determine whether these factors are required for neural induction in *Xenopus*. To study the requirement for BMP inhibition in neural induction, several labs have examined animals that have mutations in one or more of the putative neural inducer genes. Zebrafish with mutations in the chordin gene have reductions in both neural tissue and in other dorsal tissues (Schulte-Merkerr et al., 1997). In mice, targeted deletions have been made in the genes for follistatin, noggin, and chordin. Although deletion of any one of these genes has only minor effects on neural induction, elimination of both noggin and chordin has major effects on neural development. Figure 1.24 shows the nearly headless phenotype of these animals. The cerebral hemispheres of the brain are almost completely absent. Nevertheless, some neural tissue forms in these animals. Thus, while antagonism of the BMP signal via secreted BMP

mutants the neurogenic region expands at the expense of the epidermis, and ectopic expression of *dpp* causes a reduction in neural tissue. These *Drosophila* studies motivated studies of the distribution of the BMPs at early stages of *Xenopus* development, and a similar pattern has emerged. BMP4 is expressed throughout most of the gastrula, but at reduced levels in the organizer and neurogenic animal cap. As expected, recombinant BMP4 can suppress neural induction by chordin, the vertebrate homolog of Sog.

The studies of sog/chordin and dpp/BMP4 lead to two conclusions. First, it appears that the dorsal-ventral axis of the developing embryo uses similar mechanisms in both the fly and the vertebrate. However, as discussed in the previous section, the neural tissue in the vertebrate is derived from the dorsal side of the animal, while the neurogenic region of the fly is on the ventral side (DeRobertis and Sasai, 1996; Holley et al., 1995). The idea that the vertebrate and arthropod body plans were inverted with respect

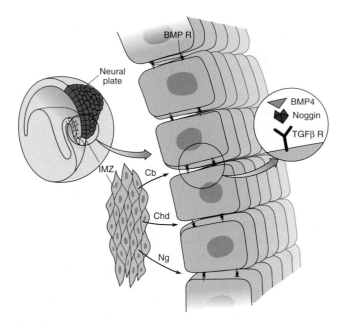

FIGURE 1.23 The current model of neural induction in amphibian embryos. As the involuting mesodermal cells of the IMZ release several molecules that interfere with the BMP signals between ectodermal cells. Ceberus, chordin, noggin, and follistatin all interfere with the activation of the BMP receptor by the BMPs in the ectoderm and thereby block the anti-neuralizing effects of BMP4. In other words, they "induce" this region of the embryo to develop as neural tissue, ultimately generating the brain, spinal cord, and most of the peripheral nervous system.

antagonists is clearly required for the development of much of the nervous system, other factors are likely involved.

Experiments with chick embryos and ascidians indicate that at least one of these additional factors is likely a member of the fibroblast growth factor (FGF) family of signaling molecules. In the chick embryo, Streit et al. (2000) found that neural induction actually occurs prior to gastrulation. Moreover, blocking FGF signaling with an FGF receptor inhibitor, called SU5402, prevented this early phase of neural induction. Evidence from the ascidian embryo further supports the role of FGF in neural induction. Ascidians are not vertebrates, but before becoming a sessile adult, they have a "tadpole" intermediate form that resembles a simple vertebrate-like larva, with a notochord and dorsal neural tube. In this animal, BMP antagonists like chordin and noggin do not appear to be involved in the induction of the neural tube. Instead FGF is the critical factor, as for the chick. Does FGF act to antagonize the BMP repression of the neural fate in ectoderm, Noggin, and Chordin? All of the ways in which these two different signaling pathways interact with one another are not yet clear. Nevertheless, there is evidence that the downstream pathway components activated by FGF inhibit BMP inhibition by phosphorylating Smad proteins (see Box). In addition, a

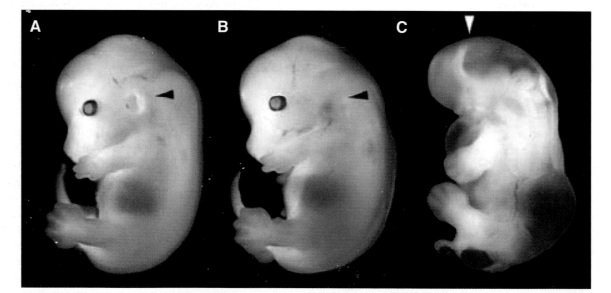

FIGURE 1.24 Loss of noggin and chordin in developing mice causes severe defects in head development. Left, wild-type mouse embryo, middle loss of noggin only. Right, loss of both noggin and chordin. From ref with permission. Note that only mild defects are present in mice deficient in only noggin, but the head is nearly absent when both genes are knocked out. (From Bachiller et al., 2000)

BOX

BMP AND WNT: AN INTRODUCTION TO SIGNALING IN DEVELOPMENT

Once an organism develops more than one cell, the cells need ways to communicate with one another. In multicellular organisms several key molecular signals have evolved. These signals are frequently proteins that are released from one cell and bind to receptors on adjacent cells. This binding of the factor and receptor causes a change in the cell that is mediated through a series of intracellular transducing molecules known as the signal transduction cascade. Changes in the molecules of the signal transduction cascade ultimately cause changes in gene expression. Although at first there may seem to be a bewildering number of signals and receptors described in this book, there are really under 10 different basic types of signaling systems used for nearly all of the cell–cell interactions in development. We assume that the reader has a basic understanding of these signaling systems, some of the most critical signaling systems will be highlighted in various chapters.

BMPs are members of a very large family of proteins, known as the TGF-beta family of factors, since the first protein of this group discovered was Transforming Growth Factor-beta. These proteins range in size from 10 kD to 30 kD and have a characteristic structure, known as the cystein-knot. They bind to a receptor that is a heterodimer, composed of two type I receptor subunits and two type II receptor subunits. The type II receptor subunits are *kinases*; that is, they can add a phosphate group (phosphorylation) onto specific serine or threonine amino acids on the adjacent type I receptor subunit proteins. The phosphorylation of the serine or threonines on the type I receptors causes the further phosphorylation of another group of proteins, known as *R-smads*. The phosphorylated R-Smads then form complexes with closely related co-Smads. This complex moves to the cell's nucleus and binds to specific sequences in the cell's DNA and activates nearby genes. The specific DNA sequences are known as BMP *response elements*, and they occur in the promoter regions of genes that are expressed when this pathway is activated by the BMP.

Another signaling protein critical in the regulation of development is the wnt pathway. The name wnt comes from the rather circuitous route that led to the discovery of this class of proteins. In the 1980s, it was discovered that a gene necessary for wing development in flies (the *Wingless* gene) and a gene that was activated in certain forms of cancer caused by viral integration *(Int-1)* were homologs. The contraction of wingless and Int, led to the name wnt. It has turned out that there are many of these wnt proteins in vertebrates; in humans there are 16 members of this protein family.

Wnts are secreted molecules, but they are typically associated with the cell membrane and diffuse only limited distances from the cell that secretes them. The wnt proteins bind to a receptor called frizzled, an integral membrane protein, with seven transmembrane domains. Along with the frizzled receptor, there is a second component to the receptor complex, the LRP protein. When wnt is bound to its receptor complex, an intracellular protein, β-catenin, associates with several other proteins, including Axin, *GSK3*β, and APC. This complex is continually degraded, and so exists only transiently in the

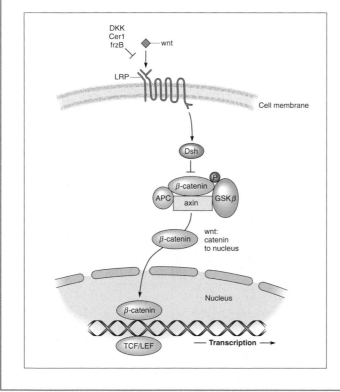

BOX (cont'd)

BMP AND WNT: AN INTRODUCTION TO SIGNALING IN DEVELOPMENT

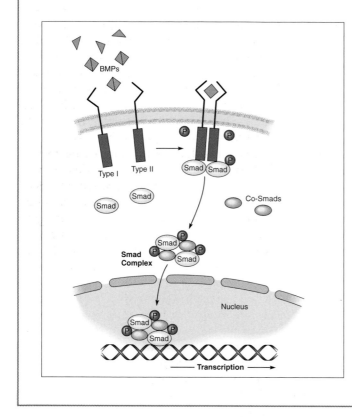

cell. However, when *wnt* binds to frizzled, a protein called Disheveled blocks the degradation of the complex and causes the β-catenin to accumulate. As the β-catenin accumulates, some of it moves to the nucleus, where it forms a complex with a different protein, TCF. The β-catenin/TCF complex can bind to DNA at specific sequences and activate target genes.

There are many similarities between the wnt and BMP signaling pathways. Both rely on cell-surface receptors to cause a change in cytoplasmic components of the cell that eventually reach the nucleus to cause a change in gene transcription. In addition, there are several natural inhibitors of these pathways; for the BMP pathway, follistatin, noggin, and chordin can interfere with the activation of the pathway by blocking BMP from binding to the receptor, and for the wnt pathway, *cerberus*, FrzB, and Dkk prevent activation, most likely by blocking the wnt from accessing the receptor. Throughout this book we will see that different signal-receptor systems are involved in nearly all developmental events, and while the types of proteins and details of the transduction cascades may vary, the fundamental features of all these pathways are similar.

Smad-interacting protein called Sip1 is activated in the neural plate by FGFs (Akai and Storey, 2004).

The model of neural induction that has emerged from many lines of investigation is quite gratifying and at the same time somewhat surprising. Despite all of the experiments that were done to study the embryology of neural induction by a great many investigators, it has only been recently that we have appreciated that the development of neural tissue in the ectoderm of a gastrulating embryo is actively inhibited by BMPs. Neural "induction" is actually the reversal of this inhibition (Figure 1.23). What has also emerged from these studies is that the process of neural induction is coupled to the process of axis specification, and the "neural inducers" have more general effects on defining the dorsal axis of the embryo. Moreover, despite the apparent redundant requirement for the BMP antagonists in CNS induction, in all of the single and compound knockout animals that have been analyzed to date, the posterior nervous system develops relatively

normally. Therefore, it is possible that antagonism of BMP in more posterior regions of the embryo may require other types of inhibition, such as antagonism of the downstream signaling cascade or Smad pathway.

INTERACTIONS AMONG THE ECTODERMAL CELLS IN CONTROLLING NEUROBLAST SEGREGATION

The generation of neurons from the neurogenic region of both vertebrates and invertebrates typically involves an intermediate step: a neural precursor cell is first produced, and this cell goes on to produce many neurons. The neural precursor is capable of mitotic divisions, whereas the neuron itself is usually a terminally postmitotic cell. In the previous section, we saw that in both *Drosophila* and *Xenopus* the antagonism of

BMP/dpp was critical in defining the neurogenic region of the embryo. This section describes some of the genes important in the next stage of nervous system development, the formation of neuroblasts in these neurogenic regions. Once again, the mechanisms are conserved in vertebrates and invertebrates, and so these mechanisms are very ancient ones. The production of neural precursor cells will be described first in *Drosophila*, where the mechanisms are best understood. Though the same mechanisms also appear to regulate this process in vertebrates, less is known about this class.

In *Drosophila,* as described in the previous section, the neural precursors or neuroblasts form by a process that starts with their delamination: certain cells within the neurogenic region enlarge and begin to move to the inside of the embryo. Next, each neuroblast divides to generate many progeny, known as *ganglion mother cells* (GMCs). Each GMC then generates a pair of neurons or glia (Figure 1.8). The neuroblasts form from the neurogenic ectoderm in a highly stereotyped array, and each neuroblast can be assigned a unique identity based on its position in the array, the expression of a particular pattern of genes, and the particular set of neurons and glia that it generates (Doe, 1992). The first neuroblasts to form are arranged in four rows along the anterior-posterior axis and in three columns along the medio-lateral axis (Figure 1.25). The types of neurons and glia generated by a particular neuroblast depend on its position in the array, and so each neuroblast is said to have a unique identity. The next waves of neuroblasts to form are also organized in rows and columns, adjacent to the preceding waves of neuroblasts. The genes involved in controlling the identity of several of the neuroblasts have been defined and will be discussed in Chapter 4; however, at this point the mechanisms that control the segregation of the neuroblasts from the ectoderm will be described.

Among the molecules that are intimately involved in the segregation of the neuroblasts from the other epidermal cells are the members of the *achaete scute* gene complex. The *achaete scute* genes were identified for their effects on the development of the bristles, or chaete, on the fly, each of which contains a sensory

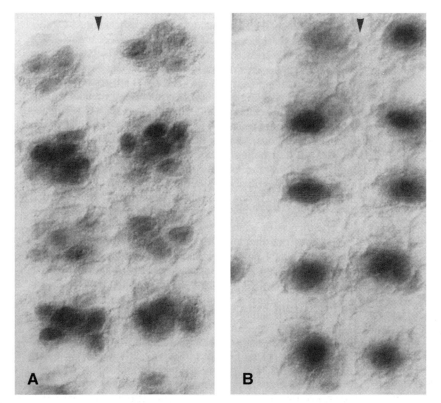

FIGURE 1.25 Neuroblast segregation in the *Drosophila* neurogenic region proceeds in a highly patterned array. A. In this embryo stained with an antibody against *achaete-scute (as-c)* protein, clusters of proneural cells in the ectoderm express the gene prior to delamination. B. A single neuroblast develops from each cluster and continues to express the gene. The other proneural cells downregulate the *as-c* gene. (From Doe, 1992)

neuron. These genes are organized in a complex of four (*achaete; scute; asense;* and *lethal of scute*) at a single locus in *Drosophila* (Alonso and Cabrera, 1988). Deletion of this locus results in the absence of most of the neuroblasts in the fly, in both the central and peripheral nervous systems (Cabrera et al., 1987), while animals with extra copies of these genes have ectopic neurons and sense organs (Brand and Campos-Ortega, 1988). Since these genes are required for the formation of neurons from the epidermal cells, they have been called the "proneural genes" (Figure 1.26) An additional *achaete scute*-related gene, *atonal*, is a proneural gene for the internal chordotonal sensory organs and the eye (Jarman et al., 1993).

How do the proneural genes function in neuroblast segregation? The proneural genes code for transcription factors of a particular class, known as the **b**asic-**h**elix-**l**oop-**h**elix, or *bHLH*, family. These proteins bind to specific short stretches of DNA, known as E-boxes, in the promoters of target genes, and activate their transcription. Proneural *bHLH* transcription factors are members of a broader class of tissue-specific transcription factors (class B). Other tissues, like muscle, also have class B transcription factors, but instead of activating neural genes, the muscle-specific bHLH proteins activate muscle-specific genes. The first of these that was discovered was called *MyoD*, for myogenic determination factor. These tissue-specific class B transcription factors bind DNA as dimers; their dimer partners are similar, but more ubiquitously expressed, *bHLH* genes, known as class A. In *Drosophila* the class A gene is called *daughterless* (Da), named for its role in the sex determination process (Caudy et al., 1988). The dimerization occurs through one of the helices (Figure 1.27), while the basic region is an extension of the other helix and interacts with the major groove of the DNA. The *achaete scute* transcription factors are thought to bind to E-boxes in the promoter regions of neuroblast-specific genes and activate their transcription, maintaining that cell as a neuroblast.

The process of neuroblast formation requires that a precise number of cells from the neurogenic region delaminate. Just prior to the delamination of the neuroblasts, *achaete* is expressed in a group of four to six epidermal cells (Skeath and Carrol, 1992; Figure 1.25). A combination of upstream regulatory genes act on the promoter regions of the *achaete scute* genes to induce their expression in these regularly spaced clusters of cells along the neurogenic region (Skeath et al., 1992). The four to six cells that express *achaete scute* are known as the proneural cluster, and they all have

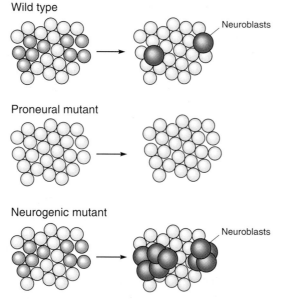

FIGURE 1.26 Neurogenic genes and proneural genes were first identified in the *Drosophila* due to their effects on neural development. In the wild-type embryo (top), only one neuroblast (red) delaminates from a given proneural cluster in the ectoderm. However, in flies mutant for proneural genes (middle), like *achaete scute*, no neuroblasts form. By contrast, in flies mutant for neurogenic genes (bottom), like *Notch* and *Delta*, many neuroblasts delaminate at the positions where only a single neuroblast develops in the wild-type animal. Thus, too many neurons delaminate—hence the name "neurogenic."

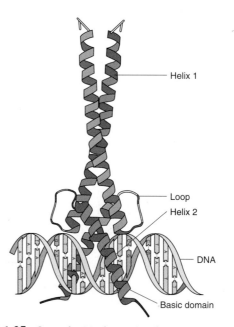

FIGURE 1.27 Several critical proteins that are necessary for the development of specific cell and tissue types are members of the *bHLH* transcription factor family of molecules. *bHLH* denotes the "**b**asic-**h**elix-**l**oop-**h**elix" structure of these molecules. The *bHLH* transcription factors dimerize via their first helix and interact with DNA via their second helix and their basic region.

the potential to form neuroblasts. Under normal circumstances, only a single cell from each cluster will delaminate as a neuroblast. The cell that delaminates to form the neuroblast continues to express the *achaete scute* proneural genes, while all of the other cells in the proneural cluster downregulate their expression of *achaete scute* (Figure 1.25).

In experiments designed to determine whether interactions among the cells were necessary for singling out one of the cells of the proneural cluster for delamination as the neuroblast, Taghert et al. (1984) used a laser microbeam to destroy the developing neuroblast at various times during its delamination (Figure 1.28). They found that ablation of the delaminating neuroblast with a laser microbeam causes one of the other cells in the cluster to take its place and delaminate. These results led to the idea that the expression of *achaete scute* genes, and hence neuroblast potential, is regulated by a system of lateral inhibition. The cell that begins to delaminate maintains its *achaete scute* expression and suppresses proneural gene activity function in the other cells of the cluster. The other cells then remain as epidermal cells, while the *achaete scute*-expressing cell delaminates as the neuroblast.

The mechanisms by which the cell that ultimately develops as the neuroblast is singled out from the original cluster have been the subject of intensive investigation. Studies of this process have uncovered a unique signaling pathway, which may underlie lateral inhibitory processes in many regions of the embryo. Molecules that act prominently in this process are the Notch receptor and one of its ligands, Delta. Notch is a large transmembrane protein characterized by an extracellular portion with a large number of repetitive domains, known as EGF-repeats because of their simi-

larity to the cysteine-bonded tertiary structure of the mitogen epidermal growth factor, EGF. However, despite this apparent structural similarity to an extended peptide mitogen, Notch has no apparent mitogenic activity, but rather acts to bind two ligands with somewhat similar structures, Delta and Serrate (Fehon et al., 1990). These proteins are expressed not only in the nervous system, but also in many other areas of the embryo where lateral inhibitory interactions define tissue boundaries. In fact, Notch and Delta, and the additional ligand, Serrate, were named for their effects on wing development, where lateral inhibition is also mediated by these molecules and is necessary for the proper development of wing morphology.

The Notch/Delta pathway is critical for singling out the neuroblast from the proneural cluster, and the fate of the cells in the neurogenic region depends on their level of Notch activity. Low Notch receptor activity in one of these cells causes it to become a neuroblast, while high activity results in the cell adopting an epidermal fate. In Notch null mutants, nearly all of the cells in the neurogenic region become neuroblasts; as a result, the Notch null embryos have defects in the epidermis (Figure 1.26). A similar phenotype occurs in Delta null mutants. Because of the phenotypes the mutant animals show, the Notch and Delta genes have been termed *neurogenic*. Of course, activation of this system actually has the opposite effect and suppresses neuroblast formation.

One model for how the Notch/Delta signaling pathway mediates the lateral inhibitory interactions among the cells of the proneural cluster is shown in Figure 1.29. Central to this model is the idea that *achaete scute* transcription factors the drive expression of *Delta*. Initially, all of the *achaete scute*-expressing cells also express an equal amount of Notch and Delta (Hartly et al., 1987). If, by a stochastic process, the central cell in the proneural cluster expresses more *achaete scute* than the others, this cell will then concomitantly express a higher level of Delta than the other cells of the cluster. When the Delta in the central cell activates Notch on the neighboring cells, it suppresses their *achaete scute* expression and further downregulates their Delta expression, preventing them from differentiating as neuroblasts. In this way, only a single neuroblast develops from the proneural cluster at a particular location in the fly.

How does Notch activation lead to the suppression of *achaete scute* in a neighboring cell? In the past few years a considerable amount of effort has gone into working out the signal transduction cascade for the Notch/Delta signaling system, and so a reasonable answer can now be given to this question. The identification of the downstream signaling components of

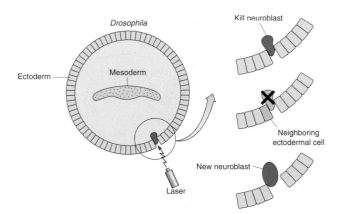

FIGURE 1.28 Ablation of the delaminating neuroblast with a laser microbeam directed to the ventral neurogenic region of the fly embryo causes a neighboring ectodermal cell to take its place. This experiment shows that the neuroblast inhibits neighboring cells from adopting the same fate via the mechanism of lateral inhibition.

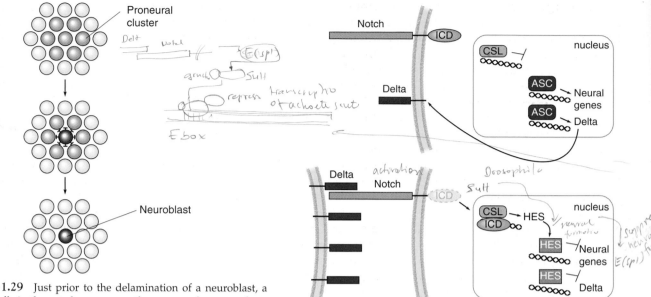

FIGURE 1.29 Just prior to the delamination of a neuroblast, a group of cells in the ectoderm express the proneural genes, *achaete* and *scute* (light red); this group is the proneural cluster. Shortly after, one of the cells near the center expresses a higher level of the proneural genes and through a process of lateral inhibition begins to block proneural gene expression in the cells around it. Finally, only this one cell is left expressing the proneural genes, and it delaminates to form the neuroblast (red).

FIGURE 1.30 The lateral inhibitory mechanism involves the neurogenic proteins, Notch and Delta. A cell expresses achaete scute genes (ASC), and Notch is inactive (top). The ASC activates its own transcription to maintain its expression and also acts as a transcription factor to activate the expression of downstream neural-specific genes and the Notch ligand Delta. This cell would then become a neuroblast. However, if a neighboring cell expresses more Delta and thus activates the Notch pathway in the cell (bottom), the activated Notch leads to a proteolytic release of the intracellular part of the Notch molecule. The piece of Notch that has been chopped off is known as the ICD (for intracellular domain). The Notch ICD interacts with another molecule, Suppressor of Hairless (SuH, also known as CSL), and together they diffuse to the nucleus and act as transcription factors to turn on the expression of another gene, called *Hairy* or *Enhancer of Split* (or more simply HES). The HES proteins are repressors of ASC gene transcription, and so they block further neural differentiation and reduce the levels of Delta expression. Thus, this cell is suppressed from the neural fate because of Notch activation by a neighboring cell that expressed more Delta.

the Notch pathway took advantage of other *Drosophila* mutants, as well as biochemical analysis of the homologous pathway in vertebrates. As described above, Notch is a transmembrane protein, with many EGF-repeats in the extracellular domain and a distinct intracellular domain, which contains a repeating motif known as cdc10/ankyrin repeats. The cdc10/ankyrin repeats in the intracellular domain of Notch are critical for the signal transduction via this system. Mutations in Notch that delete the cdc10/ankyrin repeats in the intracellular domain affect Notch signaling. Moreover, another neurogenic protein in *Drosophila*, Suppressor of hairless (SuH), has been shown to specifically bind to the cdc10/ankyrin repeats of Notch (Fortini and Artavanis-Tsakonis, 1994). A current hypothesis for how signal transduction works in this system is as follows (Figure 1.30). When Notch is not bound to Delta, SuH is tethered to the cytoplasm, bound to the cdc10/ankyrin repeats of Notch. The activation of Notch by binding to Delta causes SuH to be released from Notch. SuH can then be translocated to the nucleus where it acts as a transcriptional activator at genes with the specific DNA sequence GTGGAA in their promoters.

The genes activated by SuH should suppress neuroblast formation, since Notch activation in the epi-

dermal cells prevents them from delamination. This expectation has been confirmed by the demonstration that SuH directly regulates the transcription of another class of *bHLH* proteins, the *enhancer of split complex* (E(spl)) of genes. The *E(spl)* genes code for proteins that are similar to the proneural *bHLH* proteins of the *achaete scute* class, but instead of acting as transcriptional activators, they are strong repressors of transcription. There are seven *E(spl)* genes in *Drosophila*, and their expression patterns overlap considerably, so they are thought to be at least partly redundant. The proteins coded by these genes form heterodimers with the achaete scute proteins through their dimerization domains, but they can also form homodimers. The E(spl) proteins may directly interfere with *achaete scute*-mediated transcription, and in addition these proteins bind to nearby sites on the promoter and recruit additional repressor proteins, such as groucho. As might be expected by their function as transcriptional repressors, the E(spl) proteins are expressed in the cells surrounding the delaminating neuro-blast, and they act to prevent cells from adopting the neural fate.

E(spl) proteins can be added to the mechanism of Notch/Delta lateral inhibition in the following way (Figure 1.30; Bailey and Posokony, 1995). As described above, Notch activation by Delta causes SuH translocation to the nucleus. Specific sequences in the promoters of E(spl) genes bind the SuH protein, resulting in the expression of E(spl) in these cells. The E(spl) proteins, along with groucho, bind to the E-boxes in the *achaete-scute* gene promoter and repress transcription. Therefore, the amount of *achaete-scute* protein in the cells with active Notch declines, and the cell loses its potential to delaminate as a neuroblast. In addition, since Delta expression is activated by *achaete-scute* proteins, the reduction in the expression of achaete-scute protein in the cells with activated Notch also results in a reduction of Delta expression in the same cells. Thus, the Notch/Delta pathway provides a positive feedback pathway for neuroblast formation and a molecular mechanism for lateral inhibitory interactions among cells.

Virtually all of the molecules described in the segregation of neuroblasts in *Drosophila* have vertebrate homologs that are required for similar roles in the vertebrate. However, it is important to note that the Notch/Delta signaling pathway is one of the fundamental mechanisms by which neighboring cells become different in metazoans. Throughout development, there are numerous examples of cells with a common lineage differentiating into different cell types. It is likely that Notch/Delta signaling functions in most, if not all, of these cases of symmetry-breaking.

NOTCH, DELTA, AND ACHAETE SCUTE GENES IN VERTEBRATES

Vertebrate homologs of the *achaete scute* genes have been identified in several species. The first of these to be discovered was named *Mash1*, for **a**chaete **s**cute **h**omolog-1 (Johnson et al., 1990). In fact, vertebrates have many more members of this family than do *Drosophila*. These genes are expressed in the developing nervous system in distinct subsets of neural progenitor cells. These genes have the same *bHLH*

structure as the *Drosophila achaete scute* genes and can act as transcriptional activators as heterodimers with vertebrate daughterless homologs, E12 and E47. These correlative data have all supported a role for *achaete scute*-like genes in the vertebrate that may be analogous to that in the *Drosophila*. In addition, similar genes have been identified in *C. elegans* and even Cnidarians, and so this seems to be a very ancient system for the segregation of neuroblasts from the epidermis.

As in *Drosophila*, the proneural *bHLH* genes are both necessary and sufficient for nervous system formation in vertebrates. There are over 20 *bHLH* transcription factors expressed in the developing CNS of vertebrate embryos. Some of these proteins are expressed throughout the developing brain and spinal cord, like Neurogenin and NeuroD1. However, others in this family are expressed specifically in particular regions of the nervous system. *Ath5,* a homolog of the *Drosophila* atonal gene, is expressed specifically in the developing retina, for example. To determine whether genes of this class have proneural activity in vertebrates, the genes can be experimentally overexpressed in developing *Xenopus* embryos. Overexpression of *NeuroD1* in this assay causes cells in the ectoderm on the flank of the embryo to develop as neurons instead of epidermis (Lee et al., 1995) (Figure 1.31). Deletion of any single *bHLH* gene from the mouse using knockout technology has often not produced a clear answer as to their requirement during development; however, it is thought that this is due to a considerable redundancy and overlap in their expression. In this case, *Ath5* provides a good example. *Ath5* is normally expressed only in the retina, the sensory region of the eye, and it is only expressed during the time when the precursor cells of the eye are developing into a specific class of neurons, the retinal ganglion cells. Deletion of *Ath5* from mice by homologous recombination causes the specific loss of nearly all the retinal ganglion cells (Brown et al., 1998). Thus, we can see that in this case the proneural gene is required not for the initial stages of neural tube formation, but rather for neurons to form from the progenitor cells. Although this is somewhat different from the analogous process in *Drosophila*, the conservation of homologous genes in the earliest stages of neural development suggests that some aspects of the developmental decision to form neurons via proneural *bHLH* genes are conserved between vertebrates and invertebrates.

What about the system of lateral inhibition, Notch and Delta? Here the vertebrate nervous system appears to be somewhat different from the fly. Instead of a single Notch receptor, vertebrates have at least four Notch genes, three of which are expressed in parts

WT   + neuroD

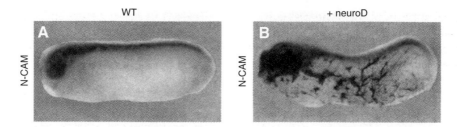

FIGURE 1.31 Overexpression of the proneural *bHLH* gene *NeuroD* in frog embryos leads to the formation of a large number of ectopic neurons (right) all over the side of the animal. These neurons are never present in the normal animal (left). Neural tissue expresses the neural cell adhesion molecule (NCAM), and both embryos have been strained for this protein. (From Lee et al., 1995)

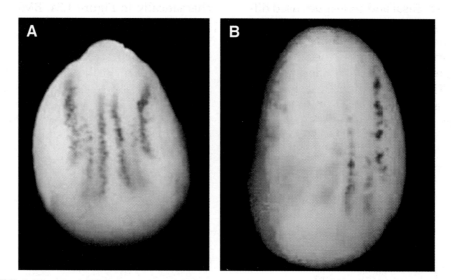

FIGURE 1.32 Notch ICD expression in frog embryos blocks neural differentiation. In this experiment, the intracellular domain of the Notch receptor was overexpressed in the frog embryo. A. A normal neural plate stage frog embryo labeled with a neuron marker, Neuron-Specific-Tubulin. There are three purple stripes of neurons on either side of the midline, and these are the primary neurons of the spinal cord. They are already beginning their differentiation at neural plate stages, before the neural tube has rolled up. B. A neural plate stage embryo that has been injected at the two-cell stage in the left blastomere with the Notch intracellular domain. All of the neurons on the left side of the embryo have failed to develop, although the right side develops the three normal stripes of primary neurons. (From Wettstein et al., 1997; Chitnis et al., 1992)

of the developing nervous system. In addition, vertebrates also have several ligands for Notch receptors, including two Delta ligands and two Serrate-like ligands, also known as *Jagged* in mammals (Kintner and Weinmaster, 2003). The Notch signaling system is active during many of the cell fate decisions made in the developing embryo, not just those in the nervous system. Therefore, experimental manipulations of this pathway often have widespread consequences during development. Nevertheless, the findings have been consistent with a model of Notch/Delta function that is similar to that discovered in *Drosophila*. For example, overexpression of activated Notch1 in *Xenopus* embryos prevents neural differentiation; the progenitor cells remain undifferentiated and do not form neurons (Figure 1.32). The various Notch receptors have been eliminated in mice and cause significant disruption in development. Examination of the nervous system in Notch1 −/− mouse embryos shows that neurons prematurely differentiate in the neural tube. This result further supports the model that Notch/Delta interactions inhibit neuron formation from the progenitor cells, by inhibiting the proneural genes as described above for *Drosophila*. As for the *bHLH* genes, the Notch/Delta signals are used in vertebrates in a process analogous to that of *Drosophila*.

LINKING INDUCTION TO PRONEURAL ACTIVITY

Finally, we would like to link up the processes of neural induction, covered earlier in the chapter, with the proneural *bHLH*/Notch pathway genes discussed in the second half of the chapter. To look for links between these processes, Sasai and colleagues used differential screening for genes upregulated in the animal caps after chordin treatment. One family of genes found in this type of screen, members of the *Sox* family, were found to be expressed very early in neural plate, soon after induction. The *Sox* proteins are transcription factors that contain a high mobility group (HMG) domain and bind to DNA directly. *SoxD* is a gene

expressed prior to neural induction in the ectoderm, but it becomes restricted to the neural ectoderm at midgastrulation. Overexpression of *SoxD* causes neural differentiation in animal caps as well as in nonneural ectoderm in the embryo, similar to that observed from *NeuroD* (Mizuseki et al., 1998) (Figure 1.33). Another *Sox* gene, *Sox2*, has a more restricted pattern of expression, and is expressed in the presumptive neural ectoderm from the late blastula onwards. Sasai and colleagues have found that overexpression of a dominant negative form of *Sox2* blocks the effects of neural-inducing factors, and inhibits nervous system formation. The relative position of *Sox2* or *SoxD* in the overall mechanism of neural induction is shown schematically in Figure 1.34. BMPs activate the Smad pathway, which inhibits *SoxD/2* gene expression. Noggin, chordin, or follistatin block this pathway and allow *SoxD/2* expression, which leads to proneural gene expression in the cells. FGF receptor activation may be an additional direct activator of *SoxD/2* expression. Although the details of the relationships among these genes will likely change as developmental biolo-

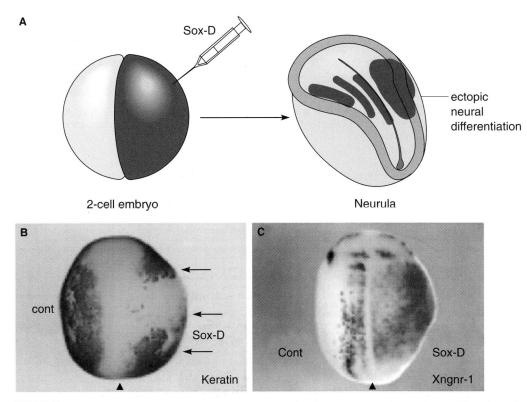

FIGURE 1.33 *SoxD* overexpression causes ectopic neural differentiation. A. *SoxD* is injected into the right blastomere of a two-cell embryo (red), and the embryo is allowed to develop to the neural plate stage. B. Epidermis fails to develop in a large region of the right side as indicated by the lack of keratin (arrows). C. In place of epidermis on the right side of the embryo, a large region of ectopic neurons develops (labeled with the proneural gene *Neurogenin* (*Xngnr-1*). (From Mizuseki et al., 1998)

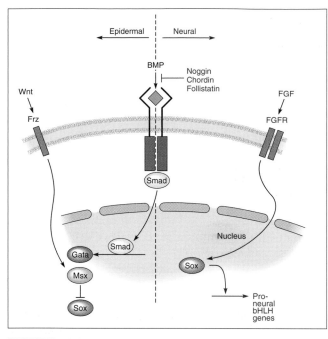

FIGURE 1.34 The current model of neural induction. The default condition of the ectoderm is to make epidermis, through the activation of the BMP pathway. BMP in the ectoderm stimulates the receptor to activate a set of intracellular proteins (Smad; see Box) that regulate transcription of genes such as *Gata* and *Msx*. The resulting *Msx* and *Gata* proteins inhibit *Sox* transcription, and the tissue becomes epidermis. If the BMP signal is blocked by one of the many inhibitors in the organizer, the Smad pathway is inactivated, and *Sox* are made in the cells. The *Sox* can then directly activate the proneural gene transcription to cause the cells to develop as nervous system. FGF may provide an alternate pathway that directly activates *Sox* expression, or might also work to inhibit the Smad signal (see Box).

gists continue to refine the model, the overall scheme from BMP antagonism to proneural gene expression is beginning to take shape.

SUMMARY

Classical embryology led to the conclusion that the nervous system arises during development through an inductive conversion of ectodermal cells to neural tissue via the Spemann organizer. In the past 15 years, a concerted effort has identified the molecular basis of this neural tissue induction as an inhibition of a tonic BMP-mediated repression. In both *Drosophila* and *Xenopus*, the antagonism of the TGF-beta-like factors BMP/dpp are critical in defining the neurogenic region of the embryo, and a number of "neural inducers," BMP antagonists, are found in the organizer. The antagonism of BMP is both necessary and sufficient for nervous system induction; however, in *Drosophila*, the delamination of neuroblasts from the neurogenic region requires a second signaling system, the Notch/Delta system, to regulate the expression and function of a proneural *bHLH* class of transcription factors. The Notch/Delta/proneural system is conserved in vertebrates, thus, it is likely that the overall system of early neural "induction" and commitment to neural tissue is largely conserved across all animals.

2

Polarity and Segmentation

REGIONAL IDENTITY OF THE NERVOUS SYSTEM

Like the rest of the body in most metazoans, the nervous system is regionally specialized. The head looks different from the tail, and the brain looks different from the spinal cord. There are a number of basic body plans for animals with neurons, and in this section, we will consider how the regional specialization of the nervous system arises during the development of some of these animals. At least some of the mechanisms that pattern the nervous system of animals are the same as those that pattern the rest of the animal's body. Similarly, many different types of tissues play key roles in regulating the development of the nervous system.

In the vertebrate embryo, most of the neural tube will give rise to the spinal cord, while the rostral end enlarges to form the three primary brain vesicles: the prosencephalon (or forebrain), the mesencephalon (or midbrain), and the rhombencephalon (or hindbrain) (Figure 2.1). The prosencephalon will give rise to the large paired cerebral hemispheres, the mesencephalon will give rise to the midbrain, and the rhombencephalon will give rise to the more caudal regions of the brainstem. The three primary brain vesicles become further subdivided into five vesicles. The prosencephalon gives rise to both the telencephalon and the diencephalon. In addition to generating the thalamus and hypothalamus in the mature brain, important features of the diencephalon are the paired evaginations of the optic vesicles. These develop into the retina and the pigmented epithelial layers of the eyes. The mesen-

cephalon remains as a single vesicle and does not expand to the same extent as the other regions of the brain. The rhombencephalon divides into the metencephalon and the myelencephalon. These two vesicles will form the cerebellum and the medulla, respectively.

The most caudal brain region is the rhombencephalon, the region that will develop into the hindbrain. At a particular time in the development of this part of the brain, the rhombencephalon becomes divided into segments, known as rhombomeres (see below). The rhombomeres are regularly spaced repeating units of hindbrain cells and are separated by distinct boundaries. Since this is one of the clearest areas of segmentation in the vertebrate brain, study of the genes that control segmentation in rhombomeres has received a lot of attention and will be discussed in detail in the next section as a model of how the anterior-posterior patterning of the nervous system takes place in vertebrates.

The insect nervous system is made up of a series of connected ganglia known as the ventral nerve cord. In many insects, the ganglia fuse at the midline. The segmental ganglia of the ventral nerve cord are not all identical, but rather vary from anterior posterior in the number and types of neurons they contain. The insect brain is composed of three regions, known as the protocerebrum, the deutocerebrum, and the tritocerebrum (Figure 2.2). The compound eyes connect through the optic lobes to the rest of the brain. Thus, as in the vertebrate, there are quite distinct regional differences along the anterior-posterior axis of the insect nervous system, and so there must be mechanisms that make one part of the nervous system different from another part.

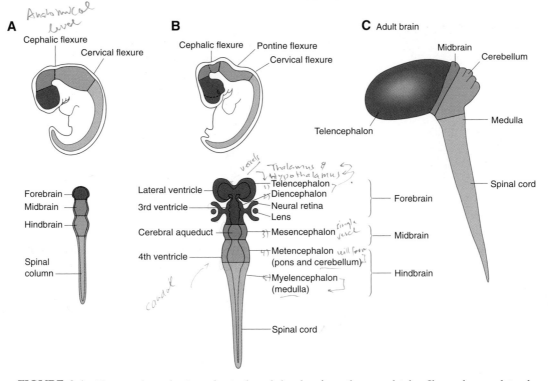

FIGURE 2.1 The vertebrate brain and spinal cord develop from the neural tube. Shown here as lateral views (upper) and dorsal views (lower) of human embryos at successively older stages of embryonic development (A,B,C). The primary three divisions of the brain (A) occur as three brain vesicles or swellings of the neural tube, known as the forebrain (prosencephalon), midbrain (mesencephalon), and hindbrain (rhombencephalon). The next stage of brain development (B) results in further subdivisions, with the forebrain vesicle becoming subdivided into the paired telencephalic vesicles and the diencephalon, and the rhombencephalon becoming subdivided into the metencephalon and the myelencephalon. These basic brain divisions can be related to the overall anatomical organization of the mature brain (C).

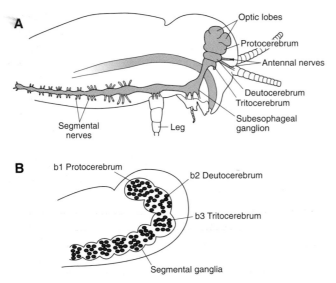

FIGURE 2.2 The brain of the *Drosophila* develops from extensive neuroblast delamination in the head. A. Three basic divisions of the brain are known as the protocerebrum, the deutocerebrum, and the tritocerebrum. B. These divisions are similar to the segmental ganglia in that they are derived independently from delaminating neuroblasts in their respective head segments. However, they later fuse together and along with the optic lobes form a complex network.

THE ANTERIOR-POSTERIOR AXIS AND *HOX* GENES

In both vertebrates and invertebrates, the mechanisms that control the regional development of the nervous system are dependent on the mechanisms that initially set up the anterior-posterior axis of the embryo. Much more is known about these mechanisms in the *Drosophila* embryo, and so this will be described first; however, it appears that many of the same genes are involved in the specification of the anterior-posterior axis in the vertebrate.

The anterior-posterior axis of the fly is primarily set up by the distribution of two molecules: a transcription factor known as *bicoid*, localized in the anterior pole of the embryo, and a gene that codes for an RNA-binding protein called *Nanos*, localized primarily in the posterior pole of the embryo (Driever and Nusslein-Volhard, 1988). The mRNAs for these genes are localized in their distribution in the egg prior to fertilization by the nurse cells in the mother. Shortly after fertilization, these mRNAs are translated, resulting in oppos-

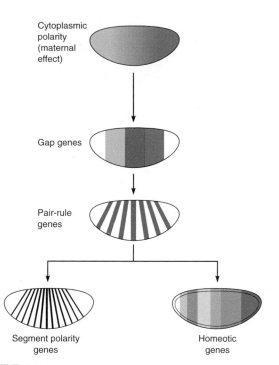

Cytoplasmic polarity (maternal effect)

Gap genes

Pair-rule genes

Segment polarity genes

Homeotic genes

FIGURE 2.3 The unique positional identity of the segments in *Drosophila* is derived by a program of molecular steps, each of which progressively subdivides the embryo into smaller and smaller domains of expression. The oocyte has two opposing gradients of mRNA for the maternal effect genes; bicoid and hunchback are localized to the anterior half, while caudal and nanos messages are localized to the posterior regions. The maternal effect gene products regulate the expression of the gap genes, the next set of key transcriptional regulators, which are more spatially restricted in their expression. Orthodenticle (otd), for example, is a gap gene that is only expressed at the very high concentrations of bicoid present in the prospective head of the embryo. Specific combinations of the gap gene products in turn activate the transcription of the pair-rule genes, each of which is only expressed in a region of the embryo about two segments wide. The periodic pattern of the pair-rule gene expression is directly controlled by the gap genes, and along with a second set of periodically expressed genes, the segment polarity genes determine the specific expression pattern of the *homeotic* genes. In this way, each segment develops a unique identity.

ing protein gradients of the two gene products (Figure 2.3). The levels of these two proteins determine whether a second set of genes, the gap genes, are expressed in a particular region of the embryo. The gap genes, in turn, control the striped pattern of a third set of genes, the pair rule genes. Finally, the pattern of expression of the pair rule genes controls the segment-specific expression of a fourth set of genes, the segment polarity genes. This developmental hierarchy progressively divides the embryo into smaller and smaller domains with unique identities (Small and Levine, 1991; Driever and Nusslein-Volhard, 1988). This chain of transcriptional activations produces the reproducible pattern of segmentation of the animal (Figure 2.3).

At this point in the development of the fly, the anterior-posterior axis is clearly defined, and the embryo is parceled up into domains of gene expression that correspond to the different segments of the animal. The next step requires a set of genes that will uniquely specify each segment as different from one another. The genes that control the relative identity of the different parts of *Drosophila* were discovered by Edward Lewis (1978). He found mutants of the fly that had two pairs of wings instead of the usual single pair. In normal flies, wings form only on the second thoracic segment; however, in flies with a mutation in the *ultrabithorax* gene, another pair of wings forms on the third thoracic segment. These mutations transformed the third segment into another second segment. Mutations in another one of these homeotic genes—*antennapedia*—causes the transformation of a leg into an another antenna. Elimination of all of the *hox* genes in the beetle, *Tribolium*, results in an animal in which all parts of the animal look identical (Stuart et al., 1993) (Figure 2.4). Analysis of many different types of mutations in this complex have led to the conclusion that, in insects, the homeotic genes are necessary for a given part of the animal to become morphologically different from another part.

The *Homeobox* genes in *Drosophila* are arranged in a linear array on the chromosome in the order of their expression along the anterior-posterior axis of the animal (Figure 2.5). A total of eight genes are organized on the chromosome as two complexes, the Antennapaedia (ANT-C) and Bithorax (BX-C) clusters (Duboule and Morata, 1994; Gehring, 1993). The *Homeobox* genes code for proteins of the homeodomain class of transcription factors and were the original members of this very large set of related molecules. All of the *Homeobox* proteins have a sequence of approximately 60 highly conserved amino acids. Like other types of transcription factors, the *Homeobox* proteins bind to a consensus sequence of DNA in the promoters of many other genes (Gehring, 1993; Biggin and McGinnis, 1997).

How do these genes control segmental identity in *Drosophila*? A good example is the mechanism by which the *BT-X* genes control abdominal segment identity. Insects have three pairs of legs, one on each of the thoracic segments, but none on the abdominal segments. The products of the *BT-X* gene complex are responsible for suppressing the formation of legs on the abdominal segments by the repression of a key regulatory gene necessary for leg formation, the distal-less gene. Although this kind of simple regulatory interaction occurs for some aspects of segmental identity, the *Homeobox* gene products bind to a rather short core DNA sequence of just four bases, and there are likely to be many genes that contain the sequence

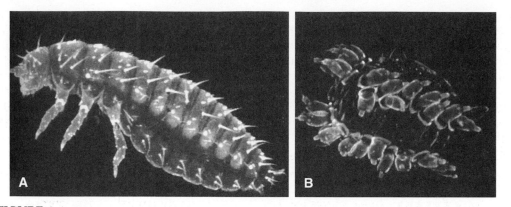

FIGURE 2.4 Elimination of the *Hox* gene cluster in the *Tribolium* beetle results in all segments developing an identical morphology. A shows the normal appearance of the beetle, and B shows an animal without a *Hox* gene cluster. The normal number of segments develop, but all of the segments acquire the morphology of the antennal segment, showing the importance of the *Hox* genes in the development of positional identity in animals. (Reproduced from Stuart et al., 1993, with permission)

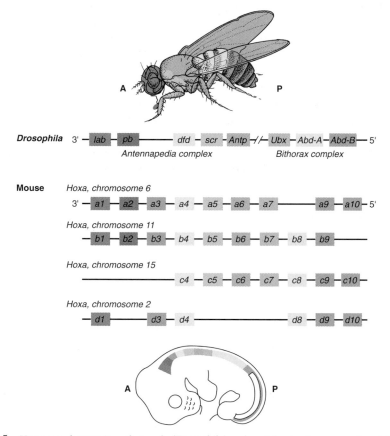

FIGURE 2.5 *Hox* gene clusters in arthropods (*Drosophila*) and vertebrates (mouse embryo) have a similar spatial organization and similar order along the chromosomes. In *Drosophila*, the *Hox* gene cluster is aligned on the chromosome such that the anterior most expressed gene is 3′ and the posterior most gene is 5′. In the mouse, there are four separate *Hox* gene clusters on four different chromosomes, but the overall order is similar to that in arthropods: the anterior to posterior order of gene expression is ordered in a 3′ to 5′ order on the chromosomes.

in their promoters, and thus are potentially regulated by *Homeobox* genes. In fact, any change in the morphology of a particular segment is likely to require the coordinated activation and suppression of numerous genes. For example, it has been estimated that between 85 and 170 genes are likely regulated by the *Ubx* gene (Gerhart and Kirschner, 1997). In addition, the *Homeobox* genes interact with other transcription factors to enhance their DNA-binding specificity.

Another striking feature of the *Homeobox* genes is their remarkable degree of conservation throughout the phyla. Organized *Homeobox* clusters similar to those found in *Drosophila* have been identified in nearly all the major classes of animals, including Cnidarians, nematodes, arthropods, annelids, and chordates. Figure 2.5 shows the relationship between the *Drosophila Homeobox* genes and those of the mouse. There have been two duplications of the ancestral *Hox* clusters to produce the A, B, C, and D clusters in the mammal. In addition, there have also been many duplications of individual members of the cluster on each chromosome, to produce up to 13 members. In mammalian embryos, the *Hox* genes are expressed in specific domains. As in *Drosophila*, the *Hox* gene position on the chromosome is correlated with its expression along the anterior-posterior axis. By aligning the mammalian *Hox* genes with their *Drosophila* counterparts, it is possible to infer the organization of the *Hox* clusters in the common ancestor between the phyla (Figure 2.5). In mice and other vertebrates, *Hox* genes in the same relative positions on each of the four chromosomes, and similar to one another in sequence, form paralogous groups. For example, *Hoxa4*, *Hoxb4*, *Hoxc4*, and *Hoxd4* make up the number 4 paralogous group.

HOX GENE FUNCTION IN THE NERVOUS SYSTEM

The function of the *Hox* genes in controlling the regional identity of the vertebrate nervous system has been most clearly investigated in the hindbrain. The vertebrate hindbrain provides the innervation for the muscles of the head through a set of cranial nerves. Like the spinal nerves that innervate the rest of the body, some of the cranial nerves contain axons from motor neurons located in the hindbrain, as well as sensory axons from neurons in the dorsal root ganglia. However, we will primarily be concerned with the motor neurons for the time being. The cranial nerves of an embryo are shown in Figure 2.6. As noted above, during embryonic

development, the hindbrain undergoes a pattern of "segment formation" that bears some resemblance to that which occurs in the fly embryo. In the developing hindbrain, the segments are called rhombomeres (Figure 2.6). The rhombomeres give rise to a segmentally repeated pattern of differentiation of neurons, some of which interconnect with one another within the hindbrain (the reticular neurons) and some of which project axons into the cranial nerves (Lumsden and Keynes, 1989). Each rhombomere gives rise to a unique set of motoneurons that control different muscles in the head. For example, progenitor cells in rhombomeres 2 and 3 make the trigeminal motor neurons that innervate the jaw, while progenitor cells in rhombomeres 4 and 5 produce the motor neurons that control the muscles of facial expression (cranial nerve VII) and the neurons that control eye muscles (abducens nerve, VI), respectively. Rhombomeres 6 and 7 make the neurons of the glossopharyngeal nerve, which controls swallowing. Without differences in these segments, animals would not have differential control of smiling, chewing, swallowing or looking down. Clearly, rhombomere identity is important for our quality of life.

How do these segments become different from one another? The pattern of expression of the paralogous groups of *Hox* genes coincides with the rhombomere boundaries (Figure 2.6), and in fact the expression of these genes precedes the formation of obvious morphological rhombomeric boundaries. Members of paralogous groups 1–4 are expressed in the rhombomeres in a nested, partly overlapping pattern. Group 4 genes are expressed up to the anterior boundary of the seventh rhombomere, group 3 genes are expressed up to and including rhombomere 5, while group 2 genes are expressed in rhombomeres 2–5. These patterns are comparable in all vertebrates. As discussed below, loss of a single *Hox* gene in mice usually does not produce the sort of dramatic phenotypes seen in *Drosophila*. This is probably because of overlapping patterns of *Hox* gene expression from the members of the four paralogous groups. When two or more members of a paralogous group are deleted, say *Hoxa4* and *Hoxb4*, then the severity of the deficits increases. The deficits that are observed are consistent with the *Hox* genes acting much as they do in arthropods. That is, they control the relative identity of a region of the body.

As noted above, studies of *Hox* genes in neural development have concentrated on the hindbrain. Several studies have either deleted specific *Hox* genes or misexpressed them in other regions of the CNS and examined the effect on hindbrain development. Only a few examples will be given. Elimination of the *Hoxa1* gene from mice results in animals with defects in the

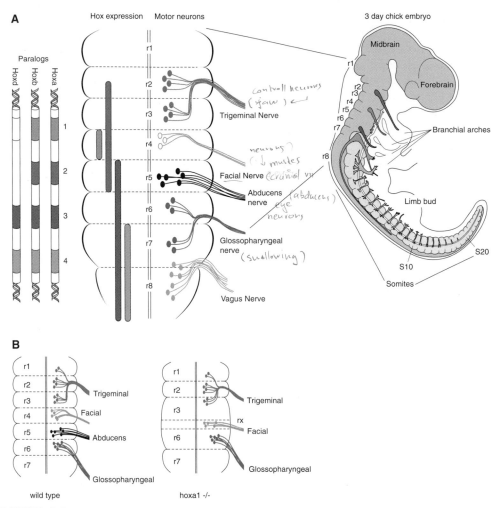

FIGURE 2.6 Rhombomeres are repeated morphological subdivisions of the hindbrain. A. The rhombomeres are numbered from the anterior-most unit, *r1*, just posterior to the midbrain (mesencephalon), to the posterior most unit, *r7*, at the junction of the hindbrain with the spinal cord. The members of the *Hox* gene cluster are expressed in a 3' to 5' order in the rhombomeres. The segmentation in this region of the embryo is also observed in the cranial nerves, and the motoneurons send their axons through defined points at alternating rhombomeres. B. Rhombomere identity is determined by the *Hox* code. *Hox* gene knockouts in mice affect the development of specific rhombomeres. Wild-type animals have a stereotypic pattern of motoneurons in the hindbrain. The trigeminal (V) cranial nerve motoneurons are generated from *r2* and *r3*, while the facial nerve motoneurons are produced in *r4* and the abducens motorneurons are produced by *r5*. Deletion of the *Hoxa1* gene in mice causes the complete loss of rhombomere 5 and a reduction of rhombomere 4 (rx). The abducens motoneurons are lost in the knockout animals, and the number of the facial motoneurons is reduced.

development of rhombomeres and the neurons they produce (Lufkin et al., 1993; Gavalas et al., 2003). Specifically, the rhombomere 4 domain is dramatically reduced and does not form a clear boundary with rhombomere 3. Rhombomere 5 is completely lost, or fused with rhombomere 4, into a new region called "rx." The abducens motoneurons fail to develop in these animals, and the facial motor neurons are also defective. However, some of the neurons derived from this region of the hindbrain now begin to resemble the trigeminal motor neurons (Figure 2.6). Thus, when *Hoxa1* is lost from the hindbrain, rhombomere 4 and 5

are partly transformed to a rhombomere 2/3 identity. Thus, at least at some level, the *Hox* genes of mice appear to confer regional anterior-posterior identity on a region of the nervous system in a manner similar to the homeotic genes of *Drosophila*.

Earlier in this section, we showed a picture of an arthropod that had no *Hox* genes; all segments were essentially identical. Is this true of vertebrates? What would the hindbrain look like without *Hox* genes? Studies in both *Drosophila* and vertebrates have found that the specificity of the *Hox* genes for promoters on their downstream targets is significantly enhanced through their interactions with the *Pbx* and *Meis* homeodomain proteins. Moens and her colleagues (Waskiewisz et al., 2002) have taken advantage of this interaction to ask what the hindbrain would look like without any functional *Hox* code. By eliminating the *Pbx* genes from the hindbrain of the zebrafish with a combination of genetic mutation and antisense oligonucleotide gene inactivation, they have found that the "ground state" or default condition of the hindbrain is rhombomere number 1. Embryos lacking both *Pbx* genes that are normally expressed in the hindbrain during rhombomere formation lose rhombomeres 2–6, and instead these segments are transformed into one long rhombomere 1 (Figure 2.7).

The remarkable conservation of *Hox* gene functioning in defining segmental identity in both *Drosophila* embryos and vertebrate hindbrain prompts the question whether similar mechanisms upstream of *Hox* gene expression are also conserved. As discussed above, a developmental cascade of genes—the gap genes, the pair-rule genes, and the segment polarity genes—parcel up the domains of the fly embryo into smaller and smaller regions, each of which has a unique *Homeobox* expression pattern. Does a similar mechanism act in the vertebrate brain to control the expression of the *Hox* genes? Although the final answers are not yet known, there are several key observations that indicate vertebrates may use somewhat different mechanisms to define the pattern of *Hox* expression.

One of the first signaling molecules to be implicated as a regulator of *Hox* expression was a derivative of vitamin A, retinoic acid (RA). This molecule is a powerful teratogen; that is, it causes birth defects. Retinoic acid is a common treatment for acne, and since its introduction in 1982, approximately one thousand malformed children have been born. The most significant defects involve craniofacial and brain abnormalities. The way in which RA works is as follows: RA crosses the cell membrane to bind a cytoplasmic receptor (Figure 2.8). The complex of RA and the retinoic acid receptor (RAR) moves into the nucleus, where it can regulate gene expression through interaction with a specific sequence in the promoters of target genes (the retinoic acid response element, or RARE). In the normal embryo, there is a gradient of

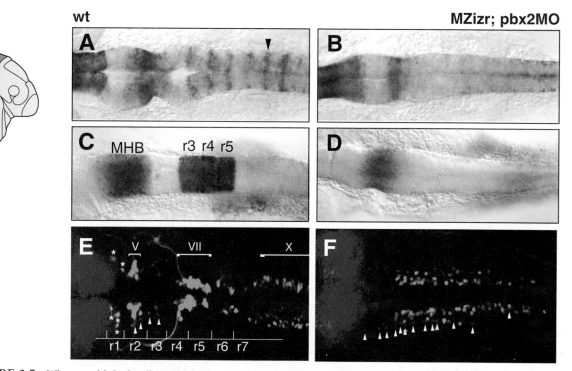

FIGURE 2.7 What would the hindbrain look like without *Hox* genes? By eliminating the *pbx* genes from the hindbrain of the zebrafish with a combination of genetic mutation and antisense oligonucleotide gene inactivation, Moens and colleagues found that the "ground state" or default condition of the hindbrain is rhombomere number 1. To the right is a drawing of the fish for orientation, with the hindbrain highlighted in red. A, C, and E show the wild-type embryo, and panels B, D, F show the mutant embryo hindbrain. In embryos lacking both *pbx* genes all segments are transformed into one long rhombomere 1, and both the specific gene expression seen in rhombomeres 3, 4, and 5 (D) and the diversity of neurons that form in the hindbrain (E) are lost in the mutant. (Modified from Waskiewicz et al., 2002)

RA signalling system

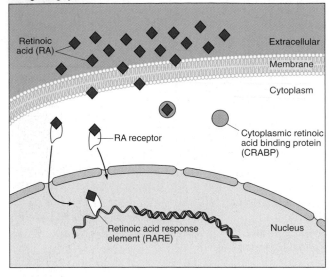

RA in posterior mesoderm at neurula stage

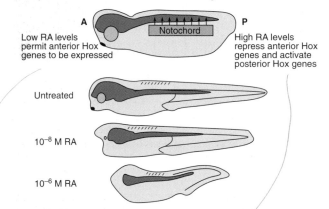

FIGURE 2.8 Retinoic acid signaling is important for the anterior-posterior pattern of *Hox* gene expression. RA crosses the cell membrane to bind a cytoplasmic receptor. The retinoic acid receptor (RAR) translocates into the nucleus where it can regulate gene expression through interaction with retinoic acid response element (RARE). RA levels are about 10 times higher in the posterior region of *Xenopus* embryos, and RA-treated embryos typically show defects in the anterior parts of the nervous system. When embryos are exposed to increasing concentrations of RA, they fail to develop head structures and the expression of anterior genes is inhibited.

RA concentration, with RA levels about 10 times higher in the posterior region of *Xenopus* embryos. When *Xenopus* embryos are treated with RA, they typically show defects in the anterior parts of the nervous system. When embryos are exposed to increasing concentrations of RA, they fail to develop head structures (Figure 2.8), and the expression of anterior *Hox* genes is inhibited (Durston et al., 1989).

Do the teratogenic effects of RA have anything to do with the control of regional identity in the CNS? In fact, it has been known for some time that retinoic acid can induce the expression of *Hox* genes when added to embryonic stem cells. With low concentrations of RA added to the ES cells, only those *Hox* genes normally expressed in the anterior embryo are expressed, while at progressively higher concentrations of RA, more posteriorly expressed *Hox* genes are expressed in the cell line (Simeone et al., 1991). Targeted deletion of the RARs produces defects similar to those observed from pharmacological manipulation of this pathway (Chambon et al., 1995). Finally, both the *Hoxa1* and the *Hoxb1* promoters have RAREs, and these elements are both necessary and sufficient for the rhombomere-specific pattern of expression of these genes. These facts all point to the importance of RA signaling in hindbrain development, but where does the gradient of RA come from in normal embryos? Early models of gradient formation invoked a highly expressing source of the signal and a declining gradient from the source, possibly "sharpened" by an active degradation mechanism. Evidence from several labs now indicates that the source of the RA is the mesoderm that lies immediately adjacent to the neural tube. The so-called paraxial mesoderm contains enzymes that synthesize the RA, and this then diffuses into the hindbrain neural tube to activate the pattern of *Hox* gene expression. The fact that the nonneural tissue outside the developing nervous system can have such a critical impact in its formation reminds us that the nervous system does not develop in a vacuum, but rather many important aspects of its development rely on interactions with adjacent nonneural tissues.

Overall, the similarity of body segmentation in *Drosophila* and hindbrain rhombomere development in vertebrates has led to a rapid understanding of both processes. However, the development of other regions of the vertebrate nervous system does not rely so heavily on the same mechanisms. Instead, other types of transcription factors control the development of the more anterior regions of the brain. In the next sections we will review how divisions in these other brain regions arise.

SIGNALING MOLECULES THAT PATTERN THE ANTERIOR-POSTERIOR AXIS IN VERTEBRATES: HEADS OR TAILS

The overall organization of the anterior-posterior axis of the nervous system in vertebrates is coupled with earlier events in axis determination and neural induction. As noted in Chapter 1, evidence from

Spemann and others demonstrated that there may be separate "head" and "tail" organizers. This fact suggests that the very early inductive signals for neural development also influence the A-P axis. In a now classic experiment, Nieuwkoop (see Chapter 1) transplanted small pieces of ectodermal tissue from one embryo into a host at various positions along the anterior-posterior axis. In all cases, the transplanted cells developed anterior neural structures. However, when the cells were transplanted in the caudal neural plate, posterior structures, such as spinal cord, also developed. Therefore, he concluded that the initial signal provided by the organizer is to cause ectodermal cells to develop anterior characteristics, known as the "activator," while a second signal is required to transform a portion of this neural tissue into hindbrain and spinal cord, known as the "transformer."

Several more recent lines of evidence are consistent with the activator-transformer hypothesis. For example, the neural inducers that have been identified (e.g., noggin, chordin, follistatin) produce primarily anterior brain structures when added to animal caps (see Chapter 1). Also, as described in Chapter 1, targeted deletion of putative neural inducers, such as the noggin/chordin double knockout mouse, results in headless mice. At the present time, three molecular pathways have been implicated as contributing to the "transformer" activity. As described above, retinoic acid treatment can posteriorize embryos and is almost certainly responsible for the patterning of the hindbrain *Hox* gene expression. Other groups have found that there is an endogenous AP gradient of *wnt/beta-catenin* activity in the embryo, with the highest levels in the posterior of the embryo (Kiecker and Niehrs, 2001).

Several lines of evidence support the hypothesis that development of head structures and brain neural tissue requires the inhibition of not only the BMP signaling pathway, but also the inhibition of the *wnt* pathway (Glinka et al. (1997); Figure 2.9). When dominant-negative *wnt8* was injected into *Xenopus* embryos along with the truncated BMP receptor, complete ectopic axes, including head structures, were formed. In addition, several inhibitors of the *wnt* pathway are expressed in the organizer region. One of the first factors specifically implicated in head induction was called cerberus, after the three-headed dog that guards the gates of Hades in Greek mythology. Injection of cerberus into *Xenopus* embryos causes ectopic head formation without the formation of trunk neural tissue (Baumeester et al., 1996). A second *wnt* inhibitor, known as *frzB*, is a member of a family of proteins that are similar to the receptors for the *wnt* proteins, known as frizzleds. The *frzB* proteins work by binding to the *wnt* proteins and preventing them from binding to their

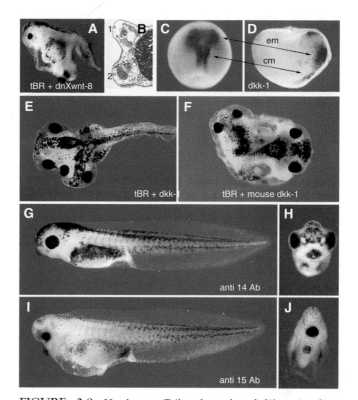

FIGURE 2.9 Heads vs. Tails: the role of Wnt signaling. Antagonism of *Wnt* signaling is important for head induction in frog embryos. A,B. Injection of four-cell embryo with both the truncated BMP receptor (tBR) and a dominant-negative form of *wnt8* (*dnXwnt8*) causes frog tadpoles to develop a second head. B shows a section through such a tadpole revealing both the primary and secondary brains. C,D. Expression of *dkk-1* in late gastrulae (stage 12) *Xenopus* embryos. *In situ* hybridization of embryo whole-mount (C) and section (D). Embryos are shown with animal side up, blastopore down. Arrows point to corresponding domains in C and D. The endomesoderm (em) is stained in a wing-shaped pattern, and most posterior expression is in two longitudinal stripes adjacent to the chordamesoderm (cm). E,F. Injection of either *Xenopus* or mouse *Dkk-1* into the blastomeres of a four-cell-stage frog embryo causes an extra head to develop as long as the truncated BMP (tBR) receptor is co-injected. G-J. *Dkk-1* is required for head formation. Stage 9 embryos were injected with antibody (Ab) into the blastocoel and allowed to develop for three days. G,H. Embryos injected with a control (anti-14) antibody show no abnormalities. An anterior view is shown on the right. I,J. Embryos injected with anti-*dkk1* (anti-15) antibody show microcephaly and cyclopia. An anterior view is shown on the right. Note that trunk and tail are unaffected. (Modified from Glinka et al., 1997; Glinka et al., 1998)

signaling receptor. Injection of extra *frzB* into *Xenopus* embryos also causes them to form heads larger than normal. A third *wnt* inhibitor was isolated by a functional screen similar to that described for noggin (see above); in this case, Niehrs and colleagues injected the truncated BMP receptor (tBMPR) along with pools from a cDNA library and looked for genes that would cause complete secondary axes, including heads, only when co-injected with the tBMPR. They identified a gene that was particularly effective in inducing head structures, *dickkopf,* for the German word meaning big-head or stubborn (Glinka et al., 1998). These three *wnt* inhibitors are reminiscent of the BMP inhibitors described above, in that they are expressed in the organizer region during the time when the inductive interactions are taking place, and they all have head-inducing activity, particularly when combined with a BMP inhibitor (Figure 2.9).

The evidence that there are indeed several putative *wnt* inhibitors in the organizer is good support for the model that a co-inhibition of *wnt* and BMP signals leads to induction of the anterior neural structures, that is, the brain. In fact, the cerberus protein can inhibit both the *wnt* and BMP pathways. Additional support for the model has recently been obtained from studies of mice in which the mouse homolog to *dickkopf, dkk1,* has been deleted via homologous recombination. The mice lacking *dkk1* alone are similar to the compound noggin/chordin knockout mice described above: they lack head and brain structures anterior to the hindbrain (Mukhopadhyay et al., 2001; Figure 2.10). Synergy between the BMP antagonist, noggin, and the *wnt* antagonist, *dkk1,* can be seen by producing mice with a single allele of each of these genes. Although the loss of a single allele of either of these genes has no discernible effect on mice, the loss of a single allele of both of these genes causes severe head and brain defects, similar to those animals that have lost both alleles of the *dkk1* gene. Similarly, knocking out the *wnt* inhibitor *dkk1* leads to a headless embryo, and in the zebrafish mutant *masterblind*—where there

is a loss of axin, a component of *wnt* inhibitory signaling pathway—no forebrain develops. Taken together, the studies in mice show that the wnt and BMP antagonists work together to bring about the correct induction and pattern of the nervous system.

The third class of molecules that has been proposed as a "transformer" is FGF. FGFs are able to act as neural inducers and in addition are able to induce posterior gene expression in animal caps that have undergone "neural induction" using a BMP antagonist. Although the specific FGF necessary for the endogenous transforming activity is not known, several members of this family are expressed in early development. Although the relative contributions of FGF, *wnt,* and RA signaling pathways for A-P axis specification in the brain are not clear, work by Kudoh et al. (2003) indicates that these factors may all converge on a common pathway. Both FGF and *wnt* signals suppress expression of *cyp26,* an enzyme involved in retinoic acid metabolism. Without this enzyme, the levels of RA rise in the anterior of the embryo, which could lead to posteriorization.

How do these signals—FGF, RA, and *wnt*—direct the development of the different brain regions? As noted above, the *Hox* genes are critical in the development of rhombomere identity; however, two other homeodomain transcription factors—*Otx2* and *Gbx2*—are necessary for a more fundamental division of the brain, the division between the hindbrain and the forebrain (Joyner et al., 2000). At late gastrula/early neural plate stages in the frog, one can already see these genes expressed in domains adjacent to one another: *Gbx2*-expressing cells extend from the poste-

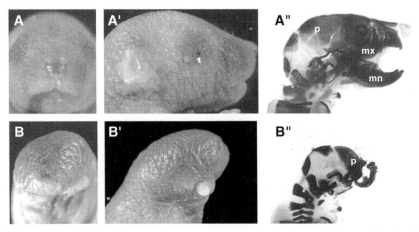

FIGURE 2.10 *Dkk1* and noggin cooperate in head induction. Mice in which one allele for the genes for both *Dkk1* and *Nog* have been deleted have severe head defects. Frontal (*A,B*) and lateral (*A',B'*) views of wild-type (*A,A'*) and mutant (*B,B'*), newborn animals. Lateral view of skeletal preparations from wild-type (*A"*) and severe mutant (*B"*) newborn heads reveal loss of maxillar (mx), mandibular (mn), and other bones anterior to the parietal bone (p). (Modified from del Barco Barrantes et al., 2003)

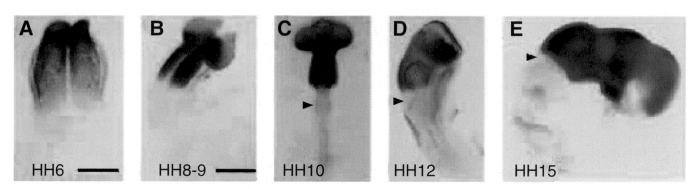

FIGURE 2.11 Expression of *otx2* reflects the basic division between the rostral brain and the hindbrain and spinal cord. *Otx2* expression at various stages of embryonic development in the chick brain. *Otx2* is expressed in the anterior neural plate (A) and remains expressed in most of the brain throughout development (B–E). The arrowhead points to the midbrain–hindbrain boundary. (From Millet et al., 1996)

rior end of the brain to the midbrain/hindbrain border, while *Otx2* has the complementary pattern of expression, from the midbrain/hindbrain border to the anterior-most part of the brain (Figure 2.11). Direct evidence that shows these genes are critical for this fundamental division of the CNS into anterior and posterior compartments come from mouse gene targeting experiments. Deletion of the *Otx2* gene in mice results in animals without a brain anterior to rhombomere 3 (Figure 2.12; Matsuo et al., 1995; Acampora et al., 1995). In mice without the *Gbx2* gene, the converse result is observed: the mice lack the hindbrain region (Millet et al., 1999; Wassarman et al., 1997). These genes are initially induced in this region by another type of transcription factor, known as *Xiro*. One current model is that *Xiro* activates both *Otx2* and *Gbx2*, which then cross-repress one another to create a sharp border between them (Glavic et al., 2002). This type of cross repression of transcription factors is a widely used mechanism for the generation of distinct boundaries between expression domains in the embryo. As we shall see in the next section, the midbrain/hindbrain boundary becomes an important organizing center in its own right.

ORGANIZING CENTERS IN THE DEVELOPING BRAIN

The division between the metencephalon and the mesencephalon appears to be a fundamental one. This boundary is a major neuroanatomical division of the mature brain as well; the metencephalon gives rise to the cerebellum, and the mesencephalon gives rise to the midbrain (superior and inferior colliculi) (Figure 2.13). But in addition to the important neural struc-

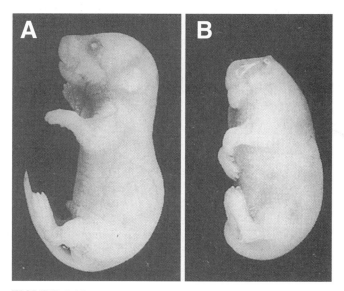

FIGURE 2.12 *Otx2* is required for the formation of the mouse head. A dramatic illustration of the importance of the *otx2* gene in the development of the mouse forebrain and rostral head. If the gene is deleted using homologous recombination, embryos without either allele of the gene fail to develop brain regions rostral to rhombomere 3, a condition known as anencephaly. Since many of the bones and muscles of the head are derived from neural crest, which also fails to form in these animals, the animals lack most of the head in addition to the loss of the brain. (From Matsuo et al., 1995)

tures derived from this region, the midbrain/hindbrain border (or mesencephalon/metencephalon border) has a special developmental function. Like the Spemann "organizer" of the gastrulating embryo, the midbrain/hindbrain border expresses signaling molecules that have an important organizing influence on the development of the adjacent regions of the neuroepithelium.

The idea that specific regions of the neural tube act as organizing centers for patterning adjacent regions was

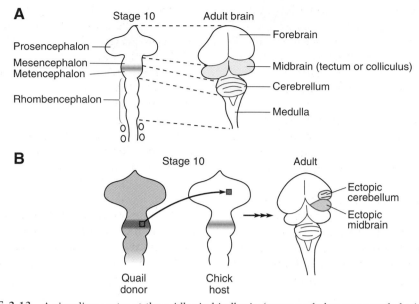

FIGURE 2.13 A signaling center at the midbrain-hindbrain (mesencephalon-metencephalon) boundary organizes this region of the brain. A. During normal development, the region of the midbrain-hindbrain junction expresses the homeodomain transcription factor *engrailed* (red), and this region of the neural tube contains the progenitors of the midbrain (tectum) and the cerebellum. B. To determine whether these parts of the neural tube were restricted in their potential at this time in development, Alvarado-Mallart et al. transplanted a small piece of the quail metencephalon (red) to the forebrain of a similarly staged chick embryo. Cerebellum still developed from the metencephalon transplants, but in addition, the transplanted tissue had induced a new mesencephalon to develop from the adjacent forebrain neural tube cells.

first put on a firm molecular basis through studies of the midbrain/hindbrain border. In a series of experiments designed to test the state of commitment of this part of the neural tube, Alvarado-Mallart and colleagues transplanted small pieces of the neuroepithelium from the midbrain/hindbrain border of chick embryos to similarly staged quail embryos (Alvarado-Mallart, 1993). Grafting between these two species allows the investigator to follow the fate of the transplanted cells. Although the chick and quail cells behave similarly and integrate well together in the tissues, molecular and histological markers can be used to tell them apart after histological processing. When the presumptive metencephalon region was transplanted from a quail to the metencephalon of a chick embryo, the transplanted cells developed as cerebellum. When cells from the mesencephalon were transplanted to a corresponding region of the chick embryo, the cells developed into midbrain structures, like the optic tectum (or superior colliculus). However, when cells from the metencephalon were transplanted to the forebrain, not only did cerebellum still develop from the metencephalon transplants (Figure 2.13) but, surprisingly, the transplanted tissue "induced" a new mesencephalon to develop in the forebrain. In other words, the small piece

of hindbrain neural tube was able to re-pattern the more anterior regions of the neural tube to adopt more posterior identities. This experiment is reminiscent of the organizer transplant of Spemann, in that a small region of specialized tissue is able to re-pattern the surrounding neuroepithelium when transplanted.

Several important signaling molecules have been localized to this region and are now known to play a key role in these patterning activities, including *wnt1*, *engrailed (en1)*, and FGF8. A member of the *wnt* gene family, *wnt1*, is expressed in this region (Figure 2.15), and when this gene is deleted in mice, the animals lose most of the midbrain and cerebellum (McMahon and Bradley, 1990). One of the earliest observed defects in these animals is the loss of expression of a transcription factor, *engrailed-1 (or en1)*, which is normally expressed in the region of the mesencephalon-metencephalon boundary. The expression of *en1* in this region has also been shown to be critical for normal development of midbrain and hindbrain structures. Mice homozygous for a targeted deletion in the *en1* gene are missing most of the cerebellum and the midbrain similar to the *wnt1*-deficient mice (Wurst et al., 1994). *en1* and *wnt1* were first identified in *Drosophila* segmentation mutants; when either of these genes is defective in flies, the animals have defects in segmentation. Moreover, in *Drosophila* the homologous gene for *wnt1, wingless,* is required for maintaining the expression of the *Drosophila engrailed* gene at the segment boundaries. Thus, the midbrain–hindbrain boundary is another example where the same basic mechanisms as those used in segmentation in *Drosophila* create differences and boundaries in the brain. In addition to the *wnt* and *engrailed* patterning system, the midbrain–hindbrain junction also expresses another key signaling molecule, *FGF8,* a receptor tyrosine kinase ligand. *FGF8* is necessary for both setting up this boundary and maintaining it, since mice deficient in *fgf8* show defects in cerebellar and midbrain development similar to the *Wnt1* and *En1* knockout animals (e.g., Meyers et al., 1998). *fgf8, En1,* and *wnt1* seem to be in an interconnected network, since deleting any one of them affects the expression of the other two. *fgf8*'s role in patterning the tissue around the mes-met boundary was demonstrated in a remarkable experiment; Crossley et al. (1996) placed a bead coated with *fgf8* protein onto a more anterior region of the neural tube and found that this molecule was sufficient to induce the repatterning of these anterior tissues into midbrain and hindbrain structures (Figure 2.14) (Crossley et al., 1996). Thus, the *fgf8* produced by midbrain/hindbrain acts like an "organizer" for the midbrain and hindbrain.

The model of how the midbrain–hindbrain signaling center arises described above can thus be extended as follows (Figure 2.16). *Xiro* activates both *Otx2* and *Gbx2* in this region of the developing CNS. *Gbx2* and *Otx2* cross inhibit one another, and it is at this point of inhibition that *fgf8* is expressed (Glavic et al., 2002). The interaction between *Otx2* and *Gbx2* maintains *fgf8* expression, and *fgf8* induces *engrailed* in those cells that express both *Xiro* and *Otx2.* Through these cross-regulatory loops between cells, the border is initially set up and maintained through development (Rhinn and Brand, 2001). The *FGF8* produced by this region then goes on to regulate growth of the progenitor cells in this region to ultimately produce the brain structures of the midbrain and hindbrain, including the cerebellum and the superior colliculus.

The unique signaling characteristics of the midbrain–hindbrain boundary suggest that such localized organizing centers may be a basic mechanism of brain patterning. There is evidence that other key organizing regions may exist between the dorsal and ventral thalamus and at the anterior pole of the neural tube. Moreover, as development proceeds and the brain expands, new organizers and signaling centers appear to pattern the newly expanded regions. It may be that the appearance of new signaling centers coincides with the expansion of the neuroepithelium past the

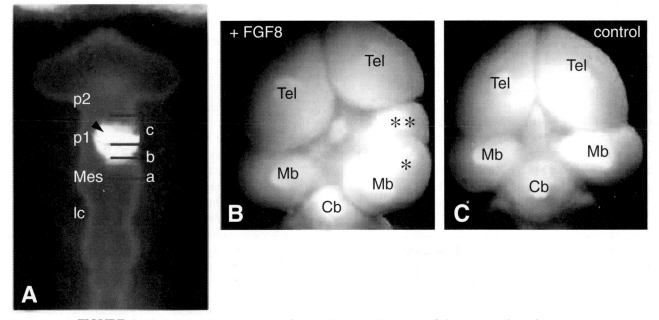

FIGURE 2.14 *FGF8* is a critical signal for the "organizer" activity of the mes-met boundary tissue. (A) Crossley et al. placed a bead (shown in arrow) of *FGF8* onto the telencephalon of the chick embryo and found that this caused a new mes-met boundary to form with a mirror duplicated midbrain (B), similar to the transplant experiment of Alvarado-Mallart. (C) Shows the control animal. (Modified from Martinez et al., 1999)

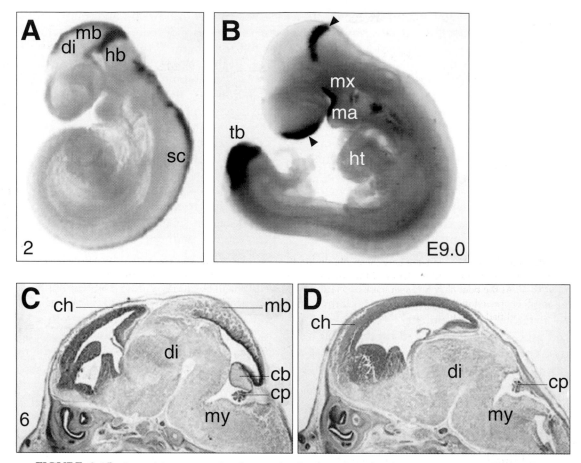

FIGURE 2.15 Several important signaling molecules have been localized to the midbrain-hindbrain boundary, a key signaling center in the brain. *wnt1* (A) *engrailed-1* and *fgf8* (B) form an interconnected network that specifies this boundary and is necessary for the growth of the midbrain and the cerebellum. Deletion of any of these molecules in mice results in a loss of the midbrain and reduction in cerebellar size. A section through the brain of a wild-type embryo is shown in C, while a *wnt1* knockout mouse brain is shown in D. Note the loss of the midbrain (mb) and cerebellum (cb) in the mutant brain. Other structures are normal (ch = cerebral cortex; cp = choroids plexus; di = diencephalon; my = myelencephalon). (A, B, D modified from Danielson et al., 1997; B modified from Crossley and Martin, 1995)

distance over which these molecules can signal. Once the number of cells exceeds the range over which the signal can act, the brain just makes a new signaling source. The same principle seems to be driving the construction of cell phone towers.

FOREBRAIN DEVELOPMENT, PROSOMERES, AND *PAX* GENES

To this point, we have explored how *Hox* genes control the specification of anterior-posterior position in the nervous system. However, *Hox* gene expression stops at the anterior boundary of the metencephalon. Are there similar transcription factors that control positional identity in the rest of the brain? Many other types

of homeodomain proteins are expressed in these more anterior regions of both vertebrate and invertebrate embryos, and they perform a role similar to that of *Hox* gene clusters in more caudal segments. Below, we explore the evidence that homeodomain proteins specify the structures that comprise the head and brain.

The most widely held view is that different parts of the brain are generated through the progressive subdivision of initially similar domains. The neural plate begins to show regional differences in the anterior-posterior direction at its formation. Embryologists at the beginning of the last century applied small amounts of dyes to specific parts of developing embryos and found that particular regions of the neural plate are already constrained to produce a particular part of the nervous system. Many embryologists have also used transplantation between species to define the contri-

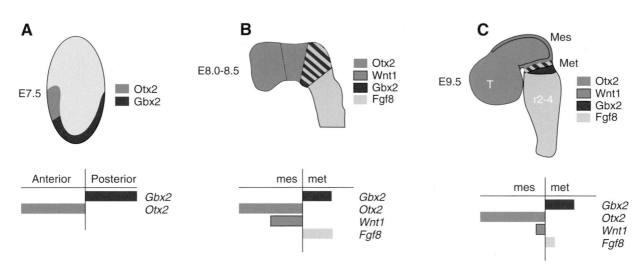

FIGURE 2.16 The model of how the midbrain-hindbrain signaling center arises. A. The initial distinction between the anterior and posterior of the embryonic nervous system is reflected in expression of *otx2* and *gbx2*. C. At the boundary between these two factors, the mes-met boundary forms, and *wnt1*, *en1*, and *FGF8* are all expressed in this region and act in a regulatory network to maintain their expression and this boundary. (Modified from Joyner et al., 2000)

butions to the mature brain of particular regions of the neural tube. One particularly useful interspecific transplantation paradigm that was developed by Nicole LeDouarin is to transplant tissues between chick embryos and quail embryos, as described in the previous section. Since these species are similar enough at early stages of development, the transplanted cells integrate with the host and continue developing along with them (Figure 2.17). The chick and quail cells can be later distinguished since the quail cells contain a more prominent nucleolus, which can be identified following histological sectioning and processing of the chimeric tissue. More recently, antibodies specific for quail cells have been generated, and these are also useful for identifying the transplanted cells. The combination of vital dyes, cell injections, and chick-quail transplant studies have produced a description or "fate map" of the ultimate fates of the various cells of the embryo. Figure 2.18 shows the fate maps for amphibian (Eagleson and Harris, 1990), avian, and mammalian neural tubes, for the basic forebrain regions that have been derived from these fate-mapping studies. The basic pattern has been elaborated upon to generate the wide diversity of brains that are found in vertebrates.

Although fate-mapping studies provide information about the fate of the different neural tube regions, embryologists have also investigated whether the fate of the cells is fixed or can be changed. The goal of these experiments, in general, is to provide a timetable for

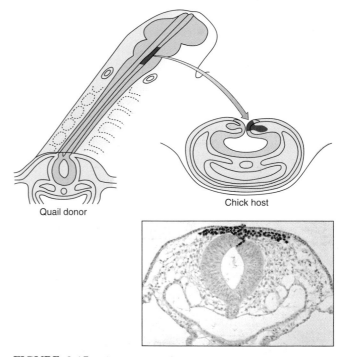

FIGURE 2.17 The interspecific transplantation paradigm was developed by Nicole LeDouarin using chick embryos and quail embryos. Tissue is dissected from quail embryos and then placed into specific regions of live chick embryos. In this case, the dorsal ridge of the neural tube, the region that will give rise to neural crest, is transplanted to a similar region in the chick. The chick and quail are similar enough to allow the quail to contribute to the chick embryo, and the quail cells can be specifically identified with an antibody raised against quail cells (bottom). (Modified from Le Douarin et al., 2004)

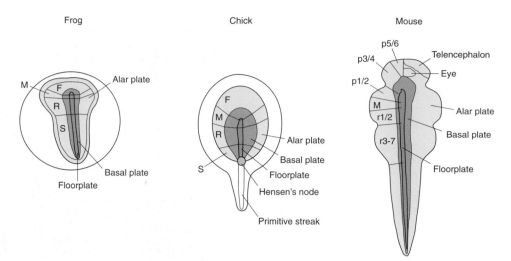

FIGURE 2.18 The fate maps for amphibian, avian, and mammalian neural tubes. The basic forebrain regions are common to all vertebrates; however, the basic pattern has been elaborated upon to generate the wide diversity of brains that are found in vertebrates. The rhombomeric and prosomeric organization of the mouse brain can already be recognized at this early stage by the pattern of expression of certain genes.

understanding the moments in development when molecular mechanisms are actively directing a specific region of the neural tube to its specific fate (i.e., is "specified"). To determine at what point in development this "specification" occurs, pieces of the neural plate are transplanted to ectopic locations in the embryo. If the transplanted cells give rise to a particular brain region, we say that it has already been specified. For example, a piece of the anterior neural plate, near the eye, is transplanted to the presumptive flank of another embryo. After sufficient developmental time has passed, the embryos are analyzed for the type of neural tissue that developed from the graft. In this case the finding is that, as early as late gastrula, a particular region of the neural plate will always give rise to anterior brain, including the eye. This occurs regardless of where the tissue is placed in the host animal. A number of embryologists carried out these types of experiments using various regions of the neural plate as the donor tissue, and the results consistently demonstrate that at some point in development, the cells of the neural plate take on a regional identity that cannot be changed by transplantation to some other place in the embryo. The fact that different regions of the neural plate are already committed to a particular fate has been extended in recent years by the observations that a number of genes are expressed in highly specific regions of the developing nervous system. In many cases the domain of expression of a particular transcription factor corresponds to that region of the neural tube that will ultimately give rise to one of the five brain vesicles, and the gene may con-

tinue to be expressed in that brain region throughout its development.

Many embryologists have taken advantage of the patterns in gene expression in the forebrain to gain insight into the basis of its organization. In what has become known as the prosomeric model of forebrain development, it is proposed that there are longitudinal and transverse patterns of gene expression that subdivide the neural tube into a grid of different regional identities (Puelles and Rubenstein, 1993). The expression of some of these genes is shown for the mouse embryo at two different stages of development (Figure 2.19). In many cases, the boundary of expression of a particular gene corresponds closely to the morphological distinctions between the prosomeres. For example, two genes of the *emx* class are expressed in the telencephalon, one in the anterior half of the cerebral hemispheres (*emx1*) and the other in the posterior half of the hemispheres (*emx2*). Thus, the telencephalic lobes can be divided into anterior and posterior segments on the basis of the pattern of expression of these two genes. Analysis of the expression patterns of additional genes has led to the conclusion that the prosencephalon can be subdivided into six prosomeres (Figure 2.19). They are numbered from caudal to rostral, and so prosomere 1 is adjacent to the mesencephalon. *P2* and *P3* subdivide what is traditionally known as the diencephalon, and *P4*, *P5*, and *P6* subdivide the telencephalon.

While the studies of regional expression of transcription factors present a model of brain organization and evolution, the functional analyses of homeo-

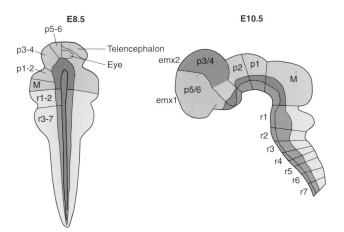

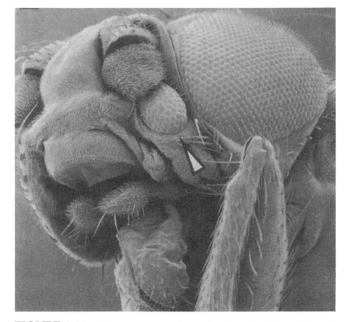

FIGURE 2.19 Prosomeric model of forebrain development; longitudinal and transverse patterns of gene expression that subdivide the neural tube into a grid of regional identities. The expression of some of these genes is shown for the mouse embryo at two different stages of development. Two genes of the emx class are expressed in the telencephalon, one in the anterior half of the cerebral hemispheres *(emx1)* and the other in the posterior half of the hemispheres *(emx2)*. Analysis of the expression patterns of additional genes has led to the conclusion that the prosencephalon can be subdivided into six prosomeres. They are numbered from caudal to rostral, and so prosomere 1 is adjacent to the mesencephalon, P2 and P3 subdivide what is traditionally known as the diencephalon, and P4, P5, and P6 subdivide the telencephalon.

FIGURE 2.20 Ectopic eyes are formed when the *Drosophila pax6* gene—*eyeless*—is misexpressed in other imaginal discs. Halder et al. (1995) misexpressed the *eyeless* gene in the leg disc in the developing fly and found that an ectopic eye was formed in the leg. This remarkable experiment argues for the concept that master control genes organize entire fields, or structures during embryogenesis, possibly by activating tissue specific cascades of transcription factors. (Courtesy of Walter Gehring)

domain factors have yielded remarkable evidence that these molecules are critically involved in defining the regional identity of the anterior brain. There are now many examples of regionally expressed transcription factors that have essential roles in brain development, but only a few will be mentioned. However, one principle that emerges is that several different classes of transcription factors are likely to be important in specifying the positional identity of cells in any particular region of the brain.

A key class of transcription factors that are critical for specifying regional differences in the nervous system are the *pax* genes. These genes have a homeodomain region, and they also have a second conserved domain known as the paired box (named for its sequence homology with the *Drosophila* segmentation gene, *paired*). There are nine different *pax* genes, and all but two, *pax1* and *pax9*, are expressed in the developing nervous system (Chalepakis et al., 1993). Several of these genes are also disrupted in naturally occurring mouse mutations and human congenital syndromes, and the defects observed in these conditions generally correspond to the areas of gene expression. *pax2*, for example, is expressed in the developing optic stalk and the otic vesicle of the embryo, and mutations in *pax2* in mice and humans cause optic nerve abnormalities, known as colobomas.

Perhaps the most striking example of *Pax* gene regulation of regional differentiation in the nervous system comes from the studies of *Pax6*. This gene is expressed early in the development of the eye, when this region of the neural plate is committed to giving rise to retinal tissue. Humans with a heterozygous disruption of this gene exhibit abnormalities in eye development, causing a condition known as aniridia (a lack of formation of the iris). In mice and humans with a homozygous disruption of this gene, the eyes fail to develop past the initial optic vesicle stage. A homologous gene has also been identified in *Drosophila* (as well as many other organisms), and mutations in this gene also disrupt eye formation in flies. And even more surprising, when this gene is misexpressed at inappropriate positions in the embryo, ectopic eyes are induced (Halder et al., 1995 (Figure 2.20). The ability of a single gene to direct the development of an entire sensory organ like the eye is striking, and while in flies the *Pax* genes act as if they are at the top of a hierarchy, and can be thought of as coordinating the signals and genes necessary to organize a "field" of the embryo's development, the situation in vertebrates is considerably more complex. The *Pax6* gene is one of several transcription factors that are expressed in the

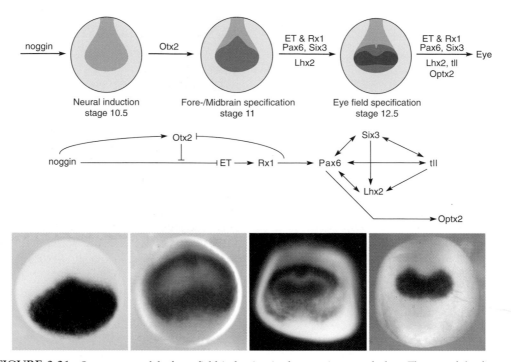

FIGURE 2.21 Summary model of eye field induction in the anterior neural plate. The top of the figure shows dorsal views of the neural plate of *Xenopus* embryos at successively later stages of development from left to right. Light blue indicates the neural plate, blue shows the area of *Otx2* expression, and dark blue represents the eye field. The diagram shows the complex relationships among the eye-determining transcription factors, including *pax6, Rx1, Lhx2, Six3, Otx2,* and *tll*. These genes act together to coordinate eye development in this specific region of the neural plate. The bottom panels show examples of *in situ* hybridizations for several eye transcription factors to show their specific patterns of expression in the presumptive eye-forming region of the embryo. (Modified from Zuber et al., 2003)

eye field, the region of neural plate fated to become the eye. Each of these factors is necessary for specification and growth of the eye. Mutations in any one of these transcription factors, including *pax6, Rx1, Lhx2, Six3, ET,* all have devastating effects on the development of the eye. Whereas overexpression of some of these factors on their own can cause the formation of ectopic eyes in *Xenopus* frogs, overexpression "cocktails" of several of the factors together have much more potency in inducing ectopic eyes (Zuber et al., 2003) (Figure 2.21). Thus, several different transcription factors may be necessary to control the expression of genes necessary for development of such a complex sensory organ as the eye.

DORSAL-VENTRAL POLARITY IN THE NEURAL TUBE

The early neural tube consists only of undifferentiated neural and glial progenitor cells. The neural tube is essentially a closed system, and the brain vesicles and developing spinal cord are fluid filled chambers. The surface of the tube, adjacent to the lumen, is known as the ventricular surface, since eventually the lumen of the neural tube goes on to form the ventricular system of the mature brain. The progenitor cells for neurons and glia of the CNS have a simple bipolar morphology and initially span the width of the neural tube. As these cells undergo mitotic divisions, they typically go through the M-phase of the cell cycle at the ventricular surface. The postmitotic immature neurons generated from the progenitor cells migrate away from the ventricular zone toward the margin of the spinal cord to form the mantle layer (see Chapter 3).

At the neural plate stage, several mechanisms are set in motion that will define the overall organization of the neural tube. First, the most ventral part of the neural tube becomes flattened into a distinct "floorplate." Second, the most dorsal aspect of the neural tube develops into a tissue known as the roofplate. Third, a distinct fissure, the *sulcus limitans*, forms between the dorsal and ventral parts of the neural tube along most of its length (Figure 2.22). These structures are an early sign that the neural tube is differentiating

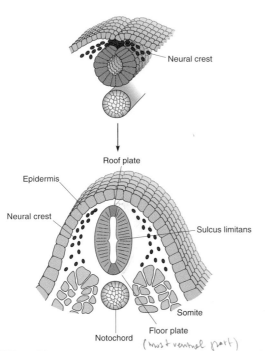

Neural crest

Roof plate

Epidermis

Neural crest

Sulcus limitans

Somite

Floor plate

Notochord *(most ventral part)*

FIGURE 2.22 The overall organization of the neural tube emerges soon after closure. The most ventral part of the neural tube becomes flattened into a distinct "floorplate." The most dorsal aspect of the neural tube develops into a tissue known as the roof plate. A distinct fissure, the *sulcus limitans*, forms between the dorsal and ventral parts of the neural tube along most of its length.

along the dorsal-ventral axis. Later, the neural tube will become even more polarized along this axis; in the ventral part of the tube, motor neurons will begin to arise, while in the dorsal part, the sensory neurons form. Experiments over many years have led to the conclusion that the distinct polarity of the neural tube arises largely because of the interaction between the surrounding nonneural tissue and the neural tube. Experiments during the early part of the twentieth century by Holtfreter demonstrated that the basic dorsal–ventral polarity of the neural tube was dependent on an adjacent, nonneural structure, called the notochord. Isolation of the neural tube from the surrounding tissues resulted in an undifferentiated tube, without obvious motoneuronal differentiation in the ventral tube. However, when he transplanted a new notochord to a more dorsal location, this induced a second floorplate (Figure 2.23) and motoneuron differentiation in the dorsal neural tube. Thus, the notochord is both necessary and sufficient for the development of the dorsal–ventral axis of the spinal cord.

The studies that led to identification of the signals that control dorsal-ventral polarity in the developing

spinal cord relied on the use of many molecular markers of cell identity that were obviously not around at the time Holtfreter was doing his experiments. These genes include the *pax* class of transcription factors discussed in the previous section, as well as a variety of other genes that are restricted to particular populations of both differentiated and/or undifferentiated cells within the spinal cord. The expression of some of the critical genes that define the dorsal-ventral polarity of the spinal cord are summarized in Figure 2.24. To track down the polarity signal released by mesoderm, a cell culture system was devised in which the notochord and the neural tube were co-cultured in collagen gels. The signal was first shown to be diffusible, since pieces of notochord could induce floorplate without touching the neural tube. In addition, the expression of motoneuron-specific genes, such as choline acetyltransferase, was also shown to depend on the notochord. A clue to the identity of the factor was uncovered in a rather roundabout manner. A crucial clue about the identity of the notochord signal would, again, come from *Drosophila*. During a large screen for developmental mutants in the fruit fly (Nusslein-Volhard and Wieschaus, 1980), a severely deformed mutant was found, named *hedgehog* for its truncated appearance. Subsequent cloning of the gene showed that this molecule resembled a secreted protein.

The link between hedgehog and the notochord-signaling molecule began with the identification of the mammalian homolog, called *Sonic hedgehog* (*Shh*). *Shh* is expressed initially in the notochord at the time when the dorsal–ventral axis of the neural tube is being specified (Roelink et al., 1994). Shortly after this time, the expression of sonic hedgehog begins in the differentiating ventral neural tube, leading to floorplate. This expression pattern is consistent with the transplantation experiments of Harrison and more recently of Tom Jessell and co-workers; both found that initially the ventralizing signal arises from the notochord, but soon after it is also found in the floorplate. To determine whether *Shh* was indeed the inducer of dorso-ventral polarity in the spinal cord, a small aggregate of *Shh* expressing Cos cells was placed next to the neural tube. The *Shh* released from these cells was sufficient to induce a second floorplate, as well as other genes normally expressed in the ventral neural tube. In further experiments, simply adding recombinant *Shh* protein to explants of neural tube was sufficient to induce them to differentiate as ventral neural tissues, including floorplate and motor neurons (Figure 2.25). These experiments thus show that *Shh* is sufficient to ventralize the neural tube during development. Two additional results show that *Shh* is required during normal development to specify the dorsal–ventral axis

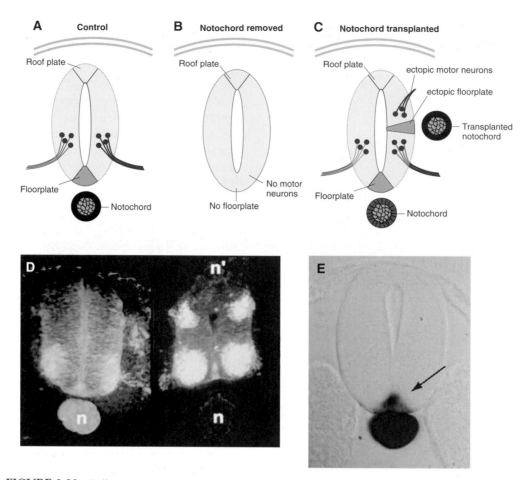

FIGURE 2.23 Differentiation in the neural tube is dependent on factors derived from adjacent, nonneural tissues. The diagrams at the top of the figure show that if the notochord, a mesodermally derived structure, is removed prior to neural tube closure, the neural tube fails to display characteristics of ventral differentiation, such as the development of the floorplate (blue) and the spinal motoneurons (red). This shows that the notochord is necessary for the development of ventral neural tube fates. If an additional notochord is transplanted to the lateral part of the neural tube at this same time in embryogenesis, a new floorplate is induced adjacent to the transplanted notochord. New motoneurons are also induced to form adjacent to the ectopic floorplate. Thus, the notochord is sufficient to specify ventral cell fates. In the lower panels, the experiment diagrammed at the top, the transplantation of an extra notochord, is shown next to a normal neural tube labeled with a marker for motorneurons. The extra notochord is labeled as *n'*. In the lower right, the expression of *sonic hedgehog* in the notochord and floorplate (arrow) of the neural tube is shown. (B, C, and D courtesy of Henk Roelink)

of the neural tube. First, antibodies raised against *Shh* will block the differentiation of floorplate and motor neurons when added to neural tube explants. Second, targeted deletion of the *Shh* gene in mice results in the failure of the development of the ventral cell types in the spinal cord (see Chapter 4).

In addition to its role in the ventralization of the neural tube, *Shh* is also expressed in the more anterior regions of the body axis immediately subjacent to the neural tube, in what is known as the prechordal mesoderm. Here the function of *Shh* is similar to that of the notochord and floorplate: it serves to induce ventral differentiation in the forebrain. In the forebrain, the

growth of the different brain vesicles gives rise to complex anatomy, and so the induction of ventral forebrain is critical for a number of subsequent morphogenetic events. Consequently, the loss of *Shh* signaling in the prechordal mesoderm produces dramatic phenotypic changes in embryos and the resulting animals. One particularly striking phenotype that arises from the disruption of *Shh* in embryogenesis is cyclopia (Roessler et al., 1997). The eyes normally form from paired evaginations of the ventral diencephalon (see above). However, in the neural plate, the eye field is initially continuous across the midline and is split into two by the inhibition of eye-forming potential by *Shh*

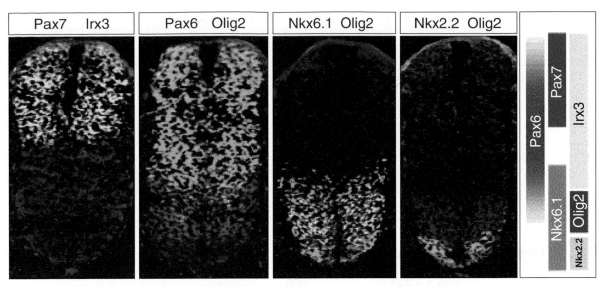

FIGURE 2.24 Several genes are expressed in restricted domains in the developing spinal cord; these have served as useful markers for positional identity of cells in this region of the nervous system. *Pax7, Irx3,* and *pax6* are all expressed in the intermediate and dorsal regions of the neural tube, while *nkx2.2, olig2,* and *nkx6.1* are all expressed in the ventral neural tube. Markers like these and others allowed Jessell and colleagues to dissect the signals controlling the identity of the different types of neurons in the spinal cord (see also Chapter 4). (From Wichtele et al., 2002)

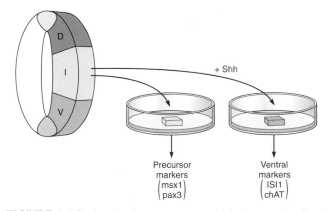

FIGURE 2.25 A cell culture system in which the notochord and the neural tube were co-cultured in collagen gels was used to find the polarity signal released by mesoderm. The signal was first shown to be diffusible since pieces of notochord could induce floorplate without touching the neural tube. Simply adding recombinant *sonic hedgehog* protein to explants of neural tube was sufficient to induce them to differentiate as ventral neural tissues, including floorplate and motor neurons. The dorsal-ventral polarity of the neural tube is controlled in part by a signaling molecule secreted by the mesodermally derived notochord. In the normal embryo, the notochord lies just ventral to the neural tube. The cells at the ventral-most part of the neural tube develop a distinct identity and morphology, and are known as the floorplate (blue). The from embryos at the time of neural plate formation in chick embryos, no floorplate develops, and the neural tube fails to develop motoneurons, or other features of its normal dorsal-ventral polarity. By contrast, if an additional notochord is transplanted to a more lateral position adjacent to the neural tube, an additional floorplate develops, and motoneurons are induced to form adjacent to the new floorplate.

from the prechordal mesoderm. *Shh* represses *Pax6* at the midline and induces *pax2*. *pax6* and *pax2* cross repress, creating a sharp border between developing retinal fields (*Pax6*) and optic stalk region (*Pax2*) that separate the developing retinas. When this signal is interrupted, the eye field remains continuous and a single eye forms in the midline. The subsequent elaboration of the forebrain depends on correct midline development, and so the lack of *Shh* disrupts later stages of brain development as well, leading to a condition known as holoprosencephaly.

The mechanism by which *Shh* induces ventral differentiation of the neural tube involves several interesting signaling steps. In both *Drosophila* and vertebrates, the hedgehog proteins undergo autoproteolysis to generate an amino-terminal fragment that is associated with the cell surface and a freely diffusible carboxyl-terminal fragment. The amino-terminal fragment is sufficient to elicit ventral differentiation as evidenced by floorplate and motoneuron differentiation. Since floorplate differentiation occurs at higher doses of recombinant *Shh* and motoneuron differentiation at lower doses of *Shh*, it has been proposed that a gradient of *Shh* from the notochord and floorplate patterns the neural tube into these two alternate fates. More will be said on this topic in Chapter 4.

Although the experiments described with *Shh* indicate that this molecule can have a profound effect in ventral patterning of the neural tube, more recent studies have shown that it is not the only factor with this capability. As we noted earlier in this section, retinoic acid is also secreted from the mesoderm and has effects in the neural tube, specifically in the developing hindbrain. Studies by Novitch et al. (2003) have found that RA, along with FGF, can almost completely replace the *Shh* signal and restore ventral development to tissue without any detectable *Shh*. This result, along with the finding that elimination of a downstream effector of *Shh* signaling, the *Gli* transcription factor, from mice can nearly completely rescue ventral development in the *Shh*-deficient mice (Litingtung and Chiang, 2000), indicates that *Shh* may be only one of several redundant molecular signals that pattern the ventral axis of the neural tube. As we saw for neural induction, a multiplicity of partly overlapping signals and transcription factors are responsible for the cellular diversity we know as pattern in the nervous system.

DORSAL NEURAL TUBE AND NEURAL CREST

The experiments of Harrison and others showed that removal of the notochord resulted in a neural tube without much dorso-ventral polarity. This implies that the dorsal neural tube is in some way the default condition, whereas the ventral structures require an additional signal to develop their fates. However, in the last few years it has become apparent that the dorsal neural tube also requires signals for its appropriate development. Before the neural tube closes, the future dorsal neural tube is continuous with the adjacent ectodermal cells. As the dorsal neural tube closes, the neural crest forms at the point of fusion of the neural tube margins. Thus, the neural crest is, in some sense, the most dorsal derivative of the neural tube, and has often been used as an indicator of dorsal differentiation. In addition, several genes specifically expressed in the dorsal neural tube at these early stages of development are critical for the specification of neural crest (e.g., *slug* and *snail*).

After extensive migration, the neural crest gives rise to an array of different tissues. In the trunk, the neural crest gives rise to the cells of the peripheral nervous system, including the neurons and glia of the sensory and autonomic ganglia, the Schwann cells surrounding all peripheral nerves, and the neurons of the gastric mucosal plexus. Several other cell types, including pigment cells, chromatophores, and smooth muscle cells, arise from the trunk neural crest. Neural crest also forms in the cranial regions, and here it contributes to most of the structures in the head. Most of the mesenchyme in the head, including that which forms the visceral skeleton and the bones of the skull, is derived from neural crest. The neurons and glia of several cranial ganglia, like the trigeminal sensory ganglia, the vestibulo-cochlear ganglia, and the autonomic ganglia in the head, are also derived largely from the progeny of the neural crest as well as from the cranial placodes. These placodes that give rise to the nose, the lens of the eye, the otic vesicle, and components of cranial sensory ganglia form a ring around the anterior edge of the neural plate and may be considered as a kind of anterior extension of the neural crest.

Because of the extensive migration of the neural crest cells, and the great diversity of the tissues and cell types to which neural crest cells can contribute, the neural crest has been studied extensively as a model for these aspects of nervous system development. In the next sections we will review what is known about the origin of the neural crest and the factors that control the initial aspects of its differentiation. Chapter 3 will detail additional studies of the factors that control neural crest migration, and Chapter 4 will deal with the cellular determination of various crest derivatives.

Classically, the neural crest has been thought to arise from the cells that form at the fusion of the neural folds when they become the neural tube. Vogt, using vital dyes to fate-map the different parts of the amphibian embryo, found that most of the neural crest forms from a narrow stripe of ectodermal cells at the junction between the neural plate and the epidermis. Subsequent studies using more sophisticated techniques have expanded this view. Le Douarin and her colleagues have extensively used the chick-quail chimera system described above to track the fate of the neural crest that arises from the different regions along the neuraxis to show the different types of tissues that are generated from different rostral-caudal regions (Figure 2.26). Bronner-Fraser and Fraser (1991) used single-cell injections to track the lineages of individual crest cells prior to their migration. The injected cells went on to divide, and they retained their lineage marker for several cell divisions. Many of the labeled cells went on to contribute to the tissues described above as the normal neural crest derivatives; however, some of the labeled cells that contributed to the neural crest also had progeny that populated the neural tube and the epidermis. Thus, although most of the cells in the neural crest field at the neural plate stage of development normally develop into neural crest, they are not restricted to this lineage. In addition, although in many embryos, the neural crest develops at the fusion of the neural folds, there are regions of the neuraxis in some species that do not form by the rolling of the neural plate. For example, in the fish, the neural tube

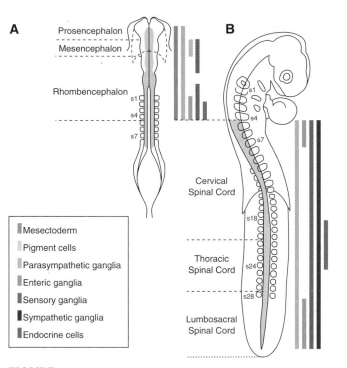

FIGURE 2.26 The fate map of neural crest in the chick embryo. Various types of tissues, including pigment cells, sensory ganglia, and endocrine cells, are derived from the neural crest. The cells migrating from the various positions along the neural tube give rise to different tissues; for example, the sympathetic ganglia arise from the neural crest of the trunk, but not from the head. Similarly, the parasympathetic ganglia arise from the neural crest of the head but not from the crest that migrates from most trunk regions. (Reproduced from Le Douarin et al., 2004)

forms first as a thickening of the neurectoderm, known as the neural keel, and tube formation occurs later by a process of cavitation, but the neural crest still forms from the lateral edges of the plate. Additional recent studies have also shown that although most of the neural crest normally arises from the lateral edges of the neural plate, there is a late-migrating population of crest cells that are derived from the neural tube.

The first experimental studies to indicate that the induction of the neural crest may involve some of the same factors as those responsible for neural induction were those of Raven and Kloos (1945). They found that neural crest was induced from ectoderm by lateral pieces of the archenteron roof, whereas neural tube was induced by medial pieces, such as the presumptive notochord. Similar results led Dalq (1941) to propose that a concentration gradient of a particular organizing substance originating in the midline tissue of the archenteron roof could set up medial-lateral distinctions across the neural plate—"the median strip of the archenteron roof, supposedly rich in organisine, would induce neural structures, while the more lateral

parts which elaborate it in smaller quantities, would induce neural crest." Since the cells that will ultimately develop into dorsal neural tube are initially immediately adjacent to the nonneural ectodermal cells, these could provide a signal for dorsal differentiation similar to the notochord-derived *Shh* for ventralization of the neural tube. This idea has been postulated for a number of years in various forms but has only recently been tested with perturbations of specific candidate-inducing molecules.

Several lines of evidence now support the hypothesis that the ectoderm provides the molecular signals to promote dorsal differentiation in the lateral regions of the spinal cord, and likely in the more anterior regions of the neuraxis. Moury and Jacobson (1990) first tested whether interactions between the neural plate and the surrounding ectoderm were responsible for the induction of neural crest by transplanting a small piece of the neural plate from a pigmented animal to the ventral surface of the embryo. When the embryo was allowed to develop further, the transplant rolled into a small tube and at the margins gave rise to neural crest cells, as evidenced by the pigmented melanocytes that migrated from the ectopic neural tissue. These results were extended by the similar experiments of Selleck and Bronner-Fraser (1995) in the chick embryo, and in addition, they used an explant culture system, in which neural plate and epidermis were co-cultured and analyzed for proteins and genes normally expressed by neural crest. They found that the neural crest was induced to form from the neural tube when placed adjacent to the epidermis. The initial steps toward identifying the crest inducer were made by Liem et al. (1995). BMPs, discussed in the previous chapter for their role in neural induction, also play important functions in specifying dorsal regional identity in the developing spinal cord. Liem et al. (1995) used a similar explant culture system as that used for the analysis of *Shh* effects on ventralization of the neural tube. The neural tube was dissected into a ventral piece, a dorsal piece, and an intermediate piece (Figure 2.27). They then analyzed the expression of genes normally restricted to either the dorsal neural tube or the ventral neural tube to determine whether these genes were specifically induced by co-culture with the ectoderm. They found that certain dorsally localized genes, such as *pax3* and *msx1*, are initially expressed throughout the neural tube and are progressively restricted from the ventral neural tube by *Shh* from the notochord and floorplate. However, co-culture with the ectoderm was necessary to induce the expression of other, more definitive, dorsal markers, such as *HNK1* and *slug*. BMPs were found to effectively replace the ectodermally derived signal,

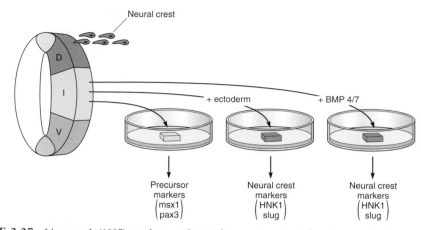

FIGURE 2.27 Liem et al. (1995) used an explant culture system to define the signals that specify dorsal cell fates. The neural tube was dissected into a ventral piece, a dorsal piece, and an intermediate piece, and the expression of genes normally restricted to either the dorsal neural tube or the ventral neural tube was used to determine whether these genes were specifically induced by co-culture with the ectoderm. They found that certain dorsally localized genes, like *pax3* and *msx1*, are initially expressed throughout the neural tube and are progressively restricted from the ventral neural tube by *Shh* from the notochord and floorplate; however, co-culture with the ectoderm was necessary to induce the expression of other, more definitive, dorsal markers, like *HNK1* and *slug*. BMPs were found to effectively replace the ectodermally derived signal, since these could also activate *HNK1* and *slug*, even from ventral explants.

since these could also activate *HNK1* and *slug*, even from ventral explants. Thus, there appears to be an antagonism between *Shh* from the ventral neural tube and BMPs from the dorsal neural tube; when BMP is added along with *Shh* to the explants, the *Shh*-induced motoneuron differentiation is suppressed.

In addition to the BMP signal that defines the border of the neural tube, there is evidence that the *wnt* signaling pathway plays a critical function in the specification of the neural crest fate (Deardorff et al., 2001). Treatment of neural plate explants with *wnt*, like those described for BMPs, is also sufficient to induce neural crest markers in the cells (Garcia-Castro et al., 2002), while blocking *wnt* signaling perturbs neural crest development. Several *wnt* genes are expressed in the developing ectoderm, adjacent to the point of origin of the crest, including *wnt8* and *wnt6*. Using a transgenic zebrafish line with a heat-inducible inhibitor of *wnt* signaling, Lewis et al. (2004) were able to precisely define the time in development when cells require the signal to become crest. They found a critical period when inhibiting *wnt* signaling was able to prevent neural crest development without affecting development of neurons in the spinal cord.

The model of dorsal-ventral polarity in the spinal cord that has emerged from these studies is as follows: BMPs and *wnt*s, expressed at the margin of the neural plate, induce the development of neural crest at the boundary of the neural plate and the ectoderm (Figure 2.28). BMPs and *wnt*s are also important for the development of the dorsal fates within the neural tube. *Shh*,

expressed first in the notochord and later in the floorplate, induces ventral differentiation in the neural tube. The *Shh* and BMP/*wnt* signals antagonize one another, and through this mutual antagonism they set up opposing gradients that control both the polarity of spinal cord differentiation and the amount of spinal cord tissue that differentiates into dorsal, ventral, and intermediate cell fates. Much more will be said about the later stages of development of spinal cord cells in Chapter 4.

PATTERNING THE CEREBRAL CORTEX

The cerebral cortex, the largest region of the human brain by far, is not a homogeneous structure, but rather has many distinct regions, each of which has a dedicated function. It has been known for over one hundred years that there are significant variations in the cellular structure (cytoarchitecture) of the cortex from region to region. The different regions of the cerebral cortex were exhaustively classified into approximately 50 distinct areas by Brodmann (1909). Although all neocortical areas have six layers, the relative number of cells in each layer and the size of the cells are quite variable and specialized to the specific function of that area. For example, the visual cortex, a primary sensory area, has many cells in layer IV, the input layer, whereas the motor cortex has very large neurons in layer V, the output layer.

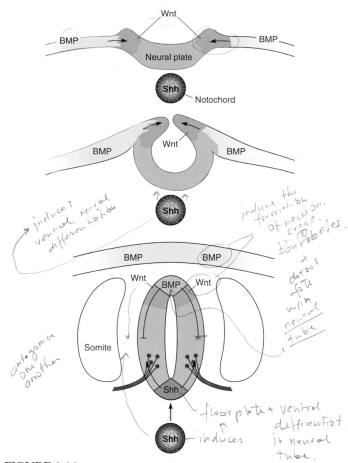

(handwritten annotations: "induce ventral differentiation", "antagonize one another", "induce the formation of neuron. crossboundaries + dorsal fate with neural tube", "floor plate + ventral differentiot, induces in neural tube.")

FIGURE 2.28 *Shh* is expressed first in the notochord and later in the floorplate and induces ventral differentiation in the neural tube. BMPs are expressed in the ectoderm overlying the neural tube and then in the dorsal neural tube cells later in development. These two signals antagonize one another, and through this mutual antagonism they set up opposing gradients that control both the polarity of spinal cord differentiation and the amount of spinal cord tissue that differentiates into dorsal, ventral, and intermediate cell fates.

Although for many years it has been thought that these different specializations occur later in development, as a consequence of the specific connections with other brain regions, more recent data indicates that the different areas have distinct identities much earlier in development, and these identities are not altered by changes in innervation (see Grove and Fukuchi-Shimogori, 2003). Like the other brain regions we have been discussing, the cerebral cortex arises from a layer of progenitors that comprises the early neural tube. In the specific case of the cerebral cortex, the anterior-most part of the neural tube, the telencephalon, is the source of these progenitors (Figure 2.1). The regional identities of the cortical areas can be monitored through the analysis of transcription factor expression. Two transcription factors that appear to have a role in the specification of regional identities in cortex are *pax6* (which we have already encountered for its role in eye development) and *emx2*. (Bishop et al., 2000; Mallamaci et al., 2000; Muzio et al., 2002) These two genes are expressed in opposing gradients across the cortical surface (Figure 2.29). *Emx2* is expressed most highly in the caudo-medial pole, while *pax6* is expressed highest at the rostral-lateral pole. Mutations in *pax6* cause an expansion of *emx2*'s domain of expression and ultimately an expansion of the areas normally derived from the caudal medial cortex, such as the visual cortex. Mutations in *emx2*, by contrast, cause the *pax6*-expressing domain to expand, and ultimately result in an expansion of the frontal and motor cortical regions.

The graded patterns of expression of *emx2* and *pax6*, along with the many examples of signaling centers we have already encountered in other regions of the developing nervous system, have led many investiga-

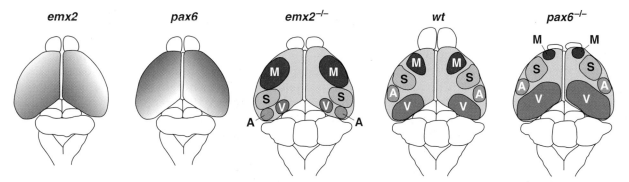

FIGURE 2.29 Two transcription factors critical for the specification of regional identities in the cortex are *pax6* and *emx2*. *Emx2* is expressed primarily in the posterior cerebral cortex and then gradually diminishes in expression toward the rostral cortical pole; *pax6* has the complementary pattern of expression. Loss of either the *pax6* gene or the *emx2* gene affects the cerebral cortical pattern of development. In the wild-type (wt) animal, the motor cortex (M) is primarily located in the rostral cortex, and the other sensory areas for somatosensation (S), auditory sensation (A), and visual perception (V) are located in the middle and posterior cortex, respectively. In the *emx2*-deficient mice, the pattern is shifted caudally, and a greater area is occupied by the motor cortex; by contrast, in the *pax6*-deficient mice, the visual cortex is expanded and the motor cortex is severely reduced. (Modified from Muzio and Mallamaci, 2003)

tors to postulate that similar signaling centers adjacent to the cortex regulate the regional expression of these transcription factors. We have already encountered the two most well-studied cortical patterning signals, FGF and retinoic acid. The most dramatic results have come from the studies of *fgf8*. *fgf8*, along with related *FGFs*, *fgf17*, and *fgf18*, are all expressed at the anterior pole of the developing telencephalon. To analyze the role of the *FGFs* in specifying cortical areal identity, Grove and her colleagues have misexpressed *fgf8* in different positions within the developing cortex (Grove and Fukuchi-Shimogori, 2003). These studies have monitored the identity of cortical regions both using the expression of region-specific transcription factors, like *pax6* and *emx2*, as well as analyzing later-developed properties of a region, like the barrel fields of the somatosensory map. Increasing the amount of *fgf8* in

the anterior pole causes a downregulation of *emx2* and a caudal shift in the cortical regions, with an expansion of the rostral regions (Figure 2.30). Blocking the endogenous *fgf8* signal, by expressing a nonfunctional FGF receptor to bind up all the available *fgf8*, causes the opposite result, a rostral-wards shift in the cortical regional identities. Most dramatically, placing a source of *fgf8* in the caudal cortex causes the formation of a duplicated, mirror image of cortical regions.

The graded pattern of expression of *emx2* and *pax6*, in part regulated by *fgf8* and other *FGFs* from the anterior pole, appears to represent an early stage in the process by which areas of the cerebral cortex become specialized for different functions. As we saw for the segmentation of the fly embryo at the beginning of the chapter, patterning is often accomplished by an initial gradient of expression that becomes further subdi-

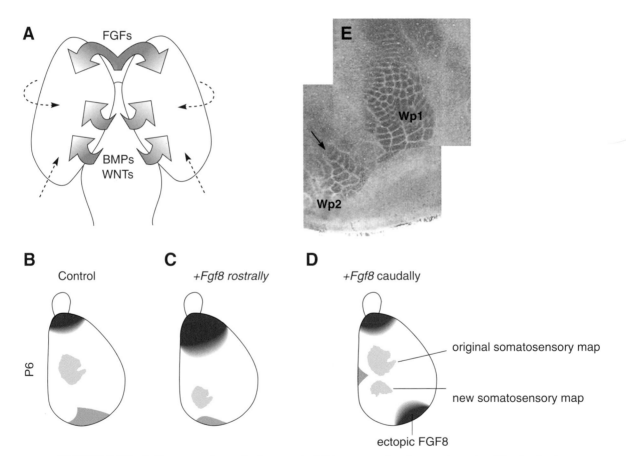

FIGURE 2.30 *Fgf8* patterns the cerebral cortex. A. *fgf8* is expressed at the anterior pole of the developing telencephalon, while BMPs and *wnt* genes are expressed in the posterior pole. Grove and her colleagues have misexpressed *fgf8* in different positions within the developing cortex. B. In the normal mouse, the barrel fields of the somatosensory map (yellow) are located near the middle of the cerebral cortex while *fgf8* (red) is expressed anteriorly and BMP (blue) is expressed posteriorly. C. Increasing the amount of *fgf8* (red) in the anterior pole causes a caudal shift in the cortical regions, including the somatosensory map. D. Placing a bead of *fgf8* in the caudal cortex causes the formation of a duplicated, mirror image of the somatosensory map. E. Micrograph of duplicated somatosory maps after the addition of an ectopic *fgf8* bead. Wp1 is the original map and Wp2 is the new map. (Modified from Grove and Fukuchi-Shimogori, 2003)

vided into finer and finer regions over time. The drive toward specialization seems to be fundamental to biology at all levels, from cells, tissues, organisms, and biological communities, and the cerebral cortex, arguably the basis for human preeminence, is no exception.

SUMMARY

The understanding of how the basic pattern of the nervous system is established has been put on a solid molecular ground in the past decade. One of the basic principles that has emerged from this work is that graded concentrations of antagonizing diffusible molecules are critically involved in setting up these patterns. These diffusible signaling molecules act to restrict the expression of specific transcription factors, which go on to regulate the expression of downstream target genes specific for the regional identity of part of the nervous system. One particularly well-conserved class of transcription factors, the *Hox* genes, is important in establishing and maintaining the regional identity of cells and tissues along the anterior-posterior axis of vertebrates throughout the hindbrain and likely the spinal cord. This conceptual framework holds true for vertebrates and invertebrates, and indeed, many of the molecular systems for generating specific parts of the nervous system have been highly conserved over the millions of years of evolution and considerable morphological diversity of animals.

Genesis and Migration

The human brain is made up of an enormous number of neurons and glial cells. The sources of all these neurons and glia are the cells of the neural tube, described in the previous chapters. Neurogenesis and gliogenesis, the generation of neurons and glia during development, occurs in many different ways in the various regions of the embryo. Part of the complex process of making the brain includes the proper migration of neurons and glia from their site of origin to their final position in the adult brain. Precisely orchestrated cell movements or migrations are an integral part of what is collectively known as *histogenesis* in the brain. This chapter describes the cellular and molecular principles by which the appropriate numbers of neurons and glia are generated from the neural precursors, and gives an overview of some of the complex cellular migration processes involved in the construction of the brain.

The number of cells generated in the developing nervous system is likely regulated at several levels. In some cases, the production of neurons or glia may be regulated by apparent intrinsic limits to the number of progenitor cell divisions essentially, a "cellular clock." The level of proliferation and ultimately the number of cells generated can also be controlled by extracellular signals, acting as mitogens, promoting progenitor cells to reenter the cell cycle and mitotic inhibitors that induce progenitor cells to exit from the cell cycle. However, it must also be remembered that the number of neurons and glia in the mature nervous system is a function not only of cell proliferation, but also of cell death. There are many examples of neuronal overproduction and subsequent attrition through programmed cell death; this process will be described in Chapter 7.

In the many small invertebrates, like the nematode, the lineages of the cells directly predict their numbers.

Since most of the mitotic divisions are asymmetric, the final number of cells that are produced during embryogenesis depends on the particular pattern of cell divisions and the number of cells that die through programmed cell death. The regulation of these divisions does not appear to depend on interactions with surrounding cells, but rather it is an intrinsic property of the lineage. The lineage of these cells also predicts the particular types of neurons that are generated from a particular precursor, and it appears that the information to define a given type of cell resides largely in factors derived directly from the precursors.

In the *Drosophila* central nervous system, neuronal number is also highly stereotypic. The neuroblasts of the insect CNS delaminate from the ventral-lateral ectoderm neurogenic region in successive waves (see Chapter 1). In *Drosophila*, about 25 neuroblasts delaminate in each segment, and they are organized in four columns and six rows. The pattern is basically the same for other insects and other arthropods, but the number of neuroblasts is dependent on the species (Doe and Smouse, 1990). Once the neuroblast segregates from the ectoderm, it then undergoes several asymmetric divisions, giving rise to approximately five smaller ganglion mother cells. Each ganglion mother cell then divides to generate a pair of neurons. These neurons make up the segmental ganglia of the ventral nerve cord and have stereotypic numbers and types of neurons.

In the vertebrate, the situation gets considerably more complex. The neural tube of most vertebrates is initially a single layer thick. As neurogenesis proceeds, the progenitor cells undergo a considerable number of cell divisions to produce a much thicker tube, with several layers. A section through the developing spinal cord is shown in Figure 3.1, and this basic structure is present throughout the developing central nervous

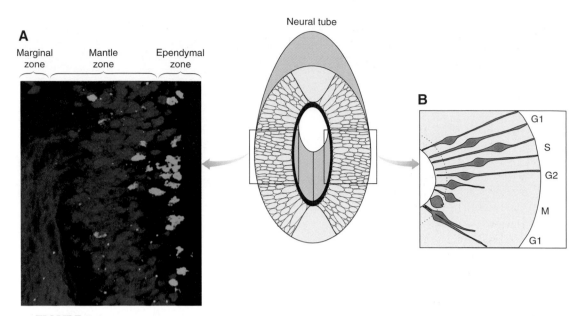

FIGURE 3.1 The process of neurogenesis in the neural tube. Section through the neural tube soon after it has formed shows a section through the chick embryo neural tube to the left and a schematic version of the progenitor cells to the right. The progenitor cells take up *BrdU* in the central region of the neural tube, indicating that they are in S-phase at this location. The cells then undergo mitosis at the ventricular surface, and this can be visualized with an antibody against a phosphorylated form of histoneH3, which is present only in late G2- and M-phases of the cell cycle. After the cells have withdrawn from the cell cycle and begun their differentiation as neurons, they migrate radially from the ventricular zone to the mantle zone, and these can now be visualized with an antibody against neurofilament protein. The movements of the progenitor cells are diagrammed to the left. The progenitor cells span the thickness of the neural tube, and their nuclei translocate to the mantle zone during the S-phase of the cell cycle. The nuclei return to the ventricular surface during G2 and the M-phase of the cell cycle always occurs at the ventricular surface. (A is courtesy of Branden Nelson)

system. The German developmental biologist Wilhelm His (an early proponent of the theory that the nervous system is composed of individual cells) divided the layers of the neural tube into the ependymal zone (near the ventricle), the mantle zone, and the marginal zone (containing postmitotic neurons). The ependymal zone, the innermost zone of the tube, is where the mitotic figures are located, and His thought these were actually different cells from the more elongate mantle cells (see Jacobson, 1991). Through the work of Sauer and others, we know that the nuclei of the progenitor cells undergo a constant up-and-down migration through the stages of the cell cycle, from the ventricular surface during M-phase to the mantle zone during S-phase (Figure 3.1). This phenomenon is known as *interkinetic nuclear migration*.

One of the clearest demonstrations of the constant motion of the nuclei of these cells came from the use of 3H-thymidine to label cells in the S-phase of the cell cycle during the active phase of DNA replication. Figure 3.2 diagrams this type of experiment in a section through the neural tube. If an injection of 3H-thymidine is made into an embryo and the tissue is

removed within an hour, the labeled cells are all found in the outer part of the ventricular zone, away from the ventricle. If the embryo is allowed to survive for 4 hours after the injection of thymidine, the labeled cells are all at the ventricular surface undergoing M-phase and can be seen as metaphase nuclei. If the embryo is allowed to survive for 8 hours, the labeled cells are in the outer half of the ventricular zone, and by 12 hours they are back at the ventricular surface. The thymidine labeling also shows a progressive increase in the number of labeled cells, as the cells divide. However, because the thymidine was only available for incorporation into cells for the first hour after injection (before it is cleared from the circulation), the cells labeled in subsequent divisions progressively dilute their label and appear more lightly labeled with each successive division. Although the function of interkinetic nuclear migration is unknown, it may be necessary for the progenitor cells to receive specific signals at different times in the cell cycle. Modern terminology now combines the ependymal and mantle zones into the term *ventricular zone*, the layer of the neural tube closest to the ventricle, where cells are in the cell cycle (Sauer, 1935).

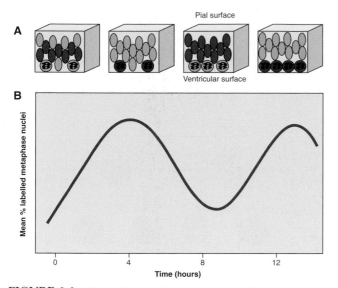

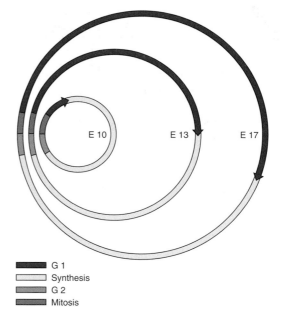

FIGURE 3.2 Thymidine labeling reveals the cell-cycle length of neural progenitor cells. A. After an injection of 3H-thymidine into a developing embryo, the S-phase cells incorporate the label into their DNA (red). When the labeled cells proceed to M-phase, they are recognizable as mitotic figures. A plot of the number of labeled mitotic figures (vertical axis) against the time after the thymidine injection shows that their number increases as the S-phase cells reach M-phase. As these cells complete M-phase and proceed into G1 and the next S-phase, they are no longer counted as mitotic figures, and their number drops. As the first cells finish their second S-phase and reenter mitosis, the number of labeled mitotic figures once again increases and we see a second peak. B. The length of time between the first and second peak is the time taken for the labeled cells to go from one mitosis to another, the cell-cycle length.

FIGURE 3.3 The overall length of the progenitor cell cycle increases during embryogenesis. The cell cycles of progenitor cells from the mouse cerebral cortex are plotted as circles of increasing size from E10 to E17. The increase in the cell-cycle period is largely due to an increase in the G1 phase, which nearly triples in length (shown in red).

The cells of the ventricular zone are the precursors of the differentiated neurons and glia of the central nervous system. These cells undergo from one to two cell cycles per day. In the early neural tube, many of the cells undergo symmetric cell divisions, producing two progenitor cells as daughters; however, some of the divisions produce asymmetric daughters: one daughter continues to divide and the other becomes a postmitotic neuron. In the spinal cord and in most other areas of the developing neural tube, the postmitotic neurons migrate from the ventricular zone to the marginal zone, where they continue their differentiation.

Thymidine labeling also allows one to determine the cell cycle length of a population of mitotically active cells. Since the thymidine is cleared from the circulation of mammalian embryos within one hour after an injection, a relatively small cohort of S-phase cells incorporate the label into their DNA. As described above, these cells then proceed through G2, and in M-phase they are recognizable as mitotic figures. If one counts the number of the mitotic figures at progressively longer intervals after the thymidine injection,

their number will continue to increase as a greater proportion of the S-phase cells reach M-phase (Figure 3.2). However, as these cells complete M-phase and proceed into G1 and the next S-phase, they are no longer counted as mitotic figures and their number drops. As the first cells finish their second S-phase and reenter mitosis, the number of labeled mitotic figures once again increases and we see a second peak. The length of time between the first and second peak is therefore the time taken for the cohort of labeled cells to go from one mitosis to another, the average cell cycle length.

In the vertebrate CNS, these types of experiments have been carried out for many regions of the brain, and again, some general principles emerge. First, the overall length of the cell cycle increases progressively during embryogenesis. Progenitor cells from the chick optic tectum, for example, have an overall cell cycle time of 8 hours on embryonic day 3, but this increases to 15 hours by embryonic day 6. A similar increase in cell-cycle period occurs in the mammal, as rat cortical progenitor cells increase their cell-cycle time from 11 hours on embryonic day 12 to 19 hours at embryonic day 18. The second generality that can be made is that the increase in the cell-cycle period is largely due to increases in the G1 phase. As shown in Figure 3.3, the M and G2 phases of the cell cycle change little from embryonic day 10 to embryonic day 19 in mouse cerebral cortex progenitor cells; however, the

G1 phase nearly triples in length. The lengthening of the G1 period likely reflects some regulatory process that restricts or slows reentry of the progenitor cells into the S-phase from G1, consistent with the idea that a limiting supply of growth factor controls this step (see next section).

The number of cells generated by a precursor cell in the ventricular zone depends on the stage of development and the region of the nervous system where the progenitor is located. Progenitor cells from the early embryonic CNS must generate a considerably greater number of progeny than those from animals nearing the end of neurogenesis. Similarly, progenitor cells from very large regions of the brain must have given rise to many more cells than those from the spinal cord. The thymidine studies described above showed that there were differences in the cell-cycle periods of ventricular zone cells of early and late embryos; thymidine labeling and, more recently, BrdU labeling of the mitotically active ventricular zone cells have given some information as to the number of progeny produced by the entire population of these cells as development proceeds. In the early embryonic cerebral cortex, for example, the number of cells more than doubles each day. Since it takes approximately a half day for a progenitor cell to generate two daughters, more than half of the progeny must continue to divide; that is, many of the cell divisions must produce two mitotically active daughters. During this early "expansion phase" of the progenitor cells, most of the cell divisions are symmetric, generating two additional progenitor cells (Figure 3.4); however, birthdating studies (see below) also show that some neurons are born during these early periods as well, and so some of the divisions must be asymmetric, generating a mitotically active daughter and a postmitotic neuron. As development proceeds, the cell-cycle time becomes progressively longer, and the number of new cells generated per day declines. Fewer cell divisions are symmetric and result in two progenitor cells at later stages of development, compared to the early stages of embryogenesis. Instead, in the later stages of neurogenesis, a greater proportion of the progenitor cells differentiate into neurons and glia (Caviness, 1996). By the end of neurogenesis, nearly all of the cells leave the cell cycle, and very few remain to generate new neurons.

The results from the thymidine and *BrdU* labeling studies have been nicely complemented by retroviral labeling of individual progenitor cells (Figure 3.5). The retroviral labeling technique takes advantage of the fact that retroviruses will only successfully infect and integrate their genes into cells that are going through the cell cycle. The genome of these viruses can be modified to contain genes that code for proteins not

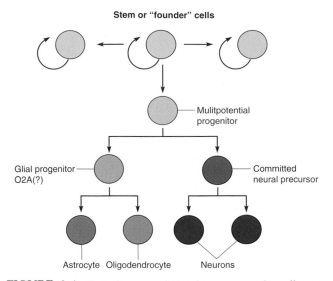

FIGURE 3.4 Basic lineage relationships among the cell types of the central nervous system of vertebrates. Through a variety of cell cultures and in vivo studies, the relationships among the various cell classes within the nervous system have been established. The early cells of the neural tube have the potential to generate an enormous number of progeny, and as a result are sometimes called founder cells or stem cells, which undergo symmetric cell divisions to produce additional founder cells as well as progenitor cells. (The term *stem cells* is also used to describe the persistent progenitors found in adult animals.) It is thought that the early founder cells also generate progenitor cells that are capable of a more limited number of cell divisions, and this is the reason that clones of progenitor cells labeled late in embryogenesis have fewer progeny. Nevertheless, the late progenitor cells are capable of generating both neurons and all macroglia, the oligodendrocytes, and the astrocytes. Although in vitro studies of certain regions of the nervous system, particularly the optic nerve, have shown that the lineages of astrocytes and oligodendrocytes share a common progenitor, known as the O2A glial progenitor, in the spinal cord, motoneurons and oligodendrocytes share a common progenitor. Thus, the lineage relationships shown may vary depending on the region of the CNS.

normally present in the nervous system but can be detected easily, such as green fluorescent protein (GFP) or alkaline phosphatase. Once the virus infects a cell, and the viral genome is integrated into the cell's DNA, the viral genes are inherited in all the daughter cells of the originally infected cell. Another important feature of this technique is that the virus is typically modified so that it is incapable of making more virus in the infected cells and spreading the infection to other cells. This means that only the daughter cells of the *originally infected* progenitors will express the viral genes. If a retrovirus that contains DNA coding for alkaline phosphatase is then injected into the developing brain and infects some of the proliferating progenitor cells, the progeny of the infected cells will still express the alkaline phosphatase gene even in adult animals. Labeling individual progenitor cells with retroviruses at differ-

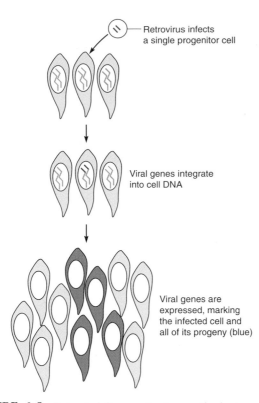

FIGURE 3.5 Retroviral lineage tracing methods are used to determine the relationships among the cell types. A retrovirus can infect and incorporate its genome into a cell only in the S-phase of the cell cycle. Thus, by engineering retroviruses to contain the gene for an enzyme not normally expressed in animal cells, like the bacterial beta-galactosidase gene or the jellyfish green fluorescent protein gene, and infecting the cells of the developing brain, the only cells that will contain the foreign genes are those that were proliferating at the time of the infection—that is, the progenitor cells. If only a relatively few viral particles are used for the infection, only a few progenitor cells will be labeled, and in this way all the progeny of a single infection can be tracked (blue cells). In practice, viral particles are diluted until only a few clones in each brain are observed.

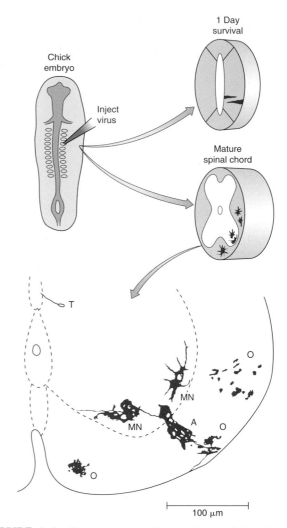

FIGURE 3.6 Clonal analysis of progenitor cells in the chick neural tube. Injections of a retrovirus with a reporter gene are made into the chick embryo neural tube. After either short or long postinjection survival periods, the spinal cord is sectioned and analyzed for the labeled progeny of the few infected progenitor cells. In the case shown, a single progenitor cell has been infected at this level of the spinal cord, and it has gone through a single-cell division to give rise to two daughter cells after one day. If the embryo is allowed to survive to a point where the spinal cord is relatively mature, and neurons and glial cells can be identified, the labeled cells can be assigned to specific cell classes. In the case shown, the progeny of the infected cell include motoneurons (MN), astrocytes (A), and oligodendrocytes (O) (Leber et al., 1990).

ent stages of brain development has confirmed that the number of progeny generated by a ventricular zone cell declines over the periods of neurogenesis. For example, retroviral infection of the progenitor cells in the early embryonic brain results in very large clones of labeled cells, but retroviral infection of progenitor cells in the brains of late staged embryos gives much smaller clones (Figure 3.5). The retroviral labeling technique also allows the discrimination of the contributions of neuronal and glial production to the total cell number. In the cerebral cortex, for example, the number of neurons per clone is typically less than 5, while the number of astrocytes and oligodendrocytes may be between 10 and 30 in the same aged embryo. In the chick spinal cord, the distribution of clones has been analyzed shortly after injections as well as in the more mature spinal cord. Clonally related cells were

typically found in radially oriented arrays when analyzed shortly after the viral infections; however, when similar clones were analyzed in the mature cord, the cells were typically more dispersed. Many of the clones contain both motoneurons and glial cells in the white matter. In the example shown in Figure 3.6, both astrocytes and oligodendrocytes are derived from the same progenitor cell that gave rise to the motoneurons (Leber et al., 1990).

In addition to the information on cell cycle that has been obtained from the use of H3-thymidine, it is also possible to use this isotope to determine the precise time points during embryogenesis when the neurons and glia are generated—that is, become postmitotic. While the progenitor cells are actively dividing, they are synthesizing DNA and incorporate the ^{3}H-thymidine. However, as noted above, since the thymidine is available for only a few hours, progenitor cells that continue to divide dilute their label over time. By contrast, those cells that withdraw from the cycle and become postmitotic soon after the ^{3}H-thymidine was administered will remain heavily labeled with the radioactive nucleotide. Thus, the postmitotic neurons generated, or "born," within a day after the ^{3}H-thymidine injection will have heavily labeled nuclei, and neurons generated later in development will be more lightly labeled. Unlabeled cells are those that withdrew from the cell cycle before the ^{3}H-thymidine injection. This technique, pioneered by Richard Sidman (1960), is known as thymidine "birthdating" and has been applied to virtually all of the areas of the developing mammalian nervous system (see Altman and Bayer, 1985, for review).

The thymidine birthdating studies revealed that the process of neurogenesis is remarkably well ordered. In many areas of the developing brain, there are spatial and temporal gradients of neuron production (Figure 3.7). The generation of the cerebral cortex, for example, proceeds from medial to lateral, and the retina is generated in a central to peripheral direction. In general, there are well-conserved and orderly sequences of generation of different types of neurons and glia. For example, in the cerebral cortex, the neurons are arranged in layers or lamina. During neurogenesis, the different laminae are generated in a sequence that is conserved from rats to monkeys. Similarly, in the retina, there are six main types of neurons and one type of glial cell, the Muller cell. Thymidine birthdating of a wide variety of species has shown that the various types of neurons are generated in a well-conserved sequence, even though the period of retinal histogenesis of a monkey takes place for more than a month while that of a frog occurs in less than two days.

Several additional generalizations can also be derived from the large number of thymidine birthdating studies that have been carried out in the different regions of the vertebrate CNS. As noted above, in many areas of the developing CNS, distinct types of neurons originate in a fairly invariant timetable. Often, the entire population of one type of neuron, like the spinal motoneurons, becomes postmitotic within a relatively short period of development. In general, large neurons

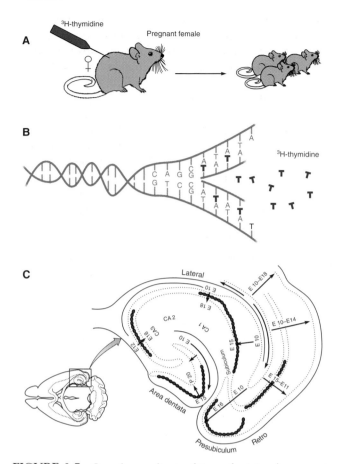

FIGURE 3.7 Complex gradients of time of origin of neurons in the hippocampus. Detailed thymidine birthdating analysis of mouse hippocampal formation shows the complex gradients in neuronal production typical of the central nervous system. Pregnant mice were given thymidine injections on the various days indicated (*E10–E18*), and the brains of the pups born from these dams were analyzed for labeled cells. Arrows indicate gradients in time. (Modified from Angevine, 1970)

are generated before small neurons in the same region. For example, pyramidal cells become postmitotic before granule cells in the hippocampus, cerebral cortex, and olfactory bulb, and in the cerebellum, Purkinje cells are generated prior to granule cells (Jacobson, 1977). Interestingly, it appears that the patterns of neuronal generation are also consistent with the hypothesis that phylogenetically older parts of the brain develop before the more recently evolved structures.

CELL-CYCLE GENES CONTROL THE NUMBER OF NEURONS GENERATED DURING DEVELOPMENT

The factors that control the number of divisions in both the vertebrate and invertebrate nervous systems are beginning to be understood. Many of the same molecular mechanisms that control the proliferation of progenitors in the nervous system are also important for the control of cell division in other tissues. Through the analysis of mutations in yeast cells that disrupt normal cell cycle, a number of the components of the molecular machinery controlling cell cycle have been identified. Remarkable progress has been made in recent years in understanding the proteins that control the mitotic cell cycle. An intricate sequence of protein interactions controls and coordinates the progress of a cell through the stages of cell replication. There has been a considerable amount of conservation of this molecular mechanism over the millions of years of evolution from the simplest eukaryotic cells, like yeast, to more complex animals and plants (see BOX).

Cyclins are a group of proteins that show dramatic changes in their expression levels that correlate with specific stages of the cell cycle. The association of cyclins with another class of proteins, called the *cyclin-dependent kinases* (*Cdk*), causes the activation of these kinases and the subsequent phosphorylation of substrate proteins necessary for progression to the next phase of the cell cycle. Different *cyclin/cdk* pairs are required at different stages of the cell cycle. For example, the binding of *cyclinB* to *Cdc2* forms an active complex that causes a cell to progress through the M-phase of the cycle, while the association of *cyclinD* and *cdk4* or *cdk6* causes these kinases to phosphorylate proteins necessary for progression from G1- to S-phase. One of the more dramatic examples of the importance of cyclins to the development of the nervous system is shown in Figure 3.8 (Geng et al., 2001). In this example, the *cyclinD1* gene was eliminated in mice using homologous recombination; the retinas of these mice are much smaller than those of normal mice because the progenitor cells of the *cyclinD1* –/– mice fail to proliferate at the normal rate.

One way in which the *cyclinD1/Cdk4* complex causes cells to enter S-phase is by phosphorylating a protein called retinoblastoma or *Rb* (see Cell Cycle Box). This phosphorylation causes the *Rb* protein to release another transcription factor, *E2F*, and allows the *E2F* protein to activate many genes that push the cell into S-phase. The *Rb* protein received its name from a childhood tumor of the retina, retinoblastoma, since defects in this gene cause uncontrolled retinal progenitor proliferation. In fact, the *Rb* gene was the first of a class of genes called tumor suppressors to be identified. Children who inherit a mutant copy of the *Rb* gene develop retinoblastoma when the second allele of this gene is mutated in a progenitor cell in the retina. *E2F* is then free to activate the genes that cause the progenitor to progress through the cell cycle, and

there is no active *Rb* around to stop the process. It is clear from this example that regulation of progenitor proliferation is critical both for making a normal retina and for preventing the uncontrolled cell proliferation that leads to cancer.

Other critical negative regulators of the cyclin-dependent kinases, *p27* and *p21*, are also expressed in the nervous system, and they are expressed in the final mitotic cycle of a progenitor, causing it to exit the cell cycle and differentiate into neurons or glia. The *p27* gene, for example, codes for a protein that interacts with the *Cdk* proteins but instead of activating the *Cdk* proteins, the *p27* protein inhibits the function of the *Cdk* it binds, and is therefore called a *Cdk-I* (for *Cdk* inhibitor). Deletion of the *p27* gene in mice causes the opposite phenotype as that of the *cyclinD1* knockout; that is, there is an overproduction of cells in the brain and other regions of the body in animals deficient in the *p27* gene. Figure 3.8 shows an example of continued proliferation in the retina of a mouse with the *p27* gene eliminated. Although the retina is not twice the size of a normal mouse, overall the mice are larger, and there are additional cells in the outer nuclear layer of the *p27*-deficient mice (Figure 3.8, arrows). What if both *p27* and *cyclinD1* are knocked out in mice? The surprising result is that the retina is now relatively normal (Figure 3.8 D). This implies that while these are key regulators of the cell cycle in normal development, when both the positive regulator and the negative regulator are removed, the system reaches a new balance. Overall, studies of cell-cycle genes in the CNS have led to the conclusions that these are critical regulators of neurogenesis and gliogenesis, but much more needs to be learned about their specific functions and interactions.

CELL INTERACTIONS CONTROL THE NUMBER OF CELLS MADE BY PROGENITORS

In many tissues in the body, secreted signaling factors have been identified that stimulate or inhibit the progress of mitotically active cells through the cell cycle. The signals that stimulate the proliferation of the mitotic cells are called growth factors or mitogens and were named for the tissue or cell type where they were first found to have mitogenic effects. For example, fibroblast growth factor (FGF) was first found to promote the proliferation of fibroblasts in cell cultures, whereas epidermal growth factor (EGF) was discovered as a mitogen for epidermal cells in vitro. These growth factors most commonly act to control the progression from G1- to S-phase of the cell cycle, and

CELL CYCLE

The core of the cell-cycle control machine are the cyclin-dependent kinases (Cdk) and their regulatory subunits, the cyclins. The cyclins were originally discovered as proteins that change dramatically in their levels of expression during the stages of the cell cycle. The transcription, translation, and destruction of each of these proteins are tightly tied to a particular stage of the cell cycle. The figure shows the period in the cell cycle when each of the cyclins is expressed.

The cyclins form complexes with specific CDKs, thereby activating the CDK to phosphorylate substrate proteins and consequently drive the cell through the next stage of the cycle. The best characterized of the cyclin/CDK pairs is that of *cyclinB* and CDK2, which control the passage of the cell through the M-phase of the cycle. In the late 1980s it was found that a cell-free extract of proteins could cause a cell to progress through the M-phase of the cell cycle. This activity was called MPF, for mitosis promoting factor. *CyclinB* and *cdk2* were discovered to be the active components of this activity. This complex of two proteins is a key regulator of this transition. Since that time there has been a considerable amount of study of these proteins and the other cyclins and their paired Cdk.

Other proteins regulate the cyclin/CDK activity as well. For a CDK to be activated it must be phosphorylated at a particular threonine residue, 161, and at the same time be dephosphorylated on residues threonine 14 and tyrosine 15. The proteins that effect these phosphorylation events are therefore important regulators of the CDK activity in a cell. These regulatory proteins include CAKs (CDK-activating kinases) and phosphatases (like string). In addition, there are a number of CDK inhibitors, small proteins that act to inhibit CDK activity by binding to

them and blocking their catalytic activity. The INK group, *p15*, *p16*, *p18*, and *p19*, all interact with *CDK4* and *6*, the CDK important in the transition from G1 to S-phase. A second class of these proteins includes *p21*, *p27*, and *p57* and can bind to and inhibit all known CDKs.

A particularly important cell-cycle transition is the entry in to the S-phase, which is the first step in the mitotic cycle. One of the most important regulators of this step is the *cyclinD* complex with either *CDK4* or *CDK6*. When a cell receives a stimulus to enter the cycle, *cyclinD* is increased in its expression (see below) and forms a complex with *CDK4/6*. When this complex is phosphorylated by CAK, it can catalyze the phosphorylation of several key substrate proteins necessary for the S-phase of the cycle. One of the most important of these is called the retinoblastoma protein, or Rb, which was originally identified as linked to a childhood retinal tumor. This protein is also one of a family of genes, known collectively as the pocket proteins, that act as tumor inhibitors; when they are absent or inactivated by a mutation, cells undergo unrestricted proliferation. In G1 the Rb protein is unphosphorylated, and this keeps the cell from entering the S-phase; when the *cyclinD*/CDK complex phosphorylates this protein, the cell can then express the proteins necessary to move into the S-phase.

The cell-cycle control machine shown in the figure is a highly coordinated complex sequence of protein interactions. For the most part, once the sequence is set in motion, it proceeds in an autonomous manner. However, there are a few points where extracellular signals can regulate the progression from one phase of the cycle to another. In mammalian cells, the major checkpoint is in the transition from the G1 phase of the cycle to the S-phase. This was first discovered by inducing tissue-cultured cells to cease their mitotic activity by withdrawing critical serum components from their medium. By adding back the serum for brief periods, it was found that once the cycle was started again, the serum was not needed, until after the cell had completed an entire cycle. In general, it is thought that when growth factors regulate cellular proliferation, they do so by acting at the G1 to S transition, either by promoting cyclinD expression/activity to stimulate entry into the next cycle, or alternatively by increasing the expression of a cyclin inhibitor, like *p27*, to block entry into the next cycle. In *Drosophila* cells, the G2 progression is also frequently the target of control. Cells in the developing eye imaginal disc are held in G2 until they receive an extracellular signal, hedgehog, to trigger their progression through the M-phase.

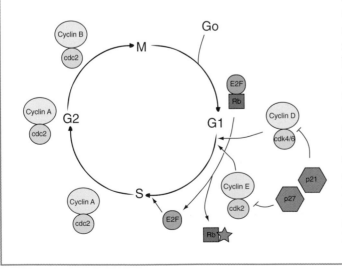

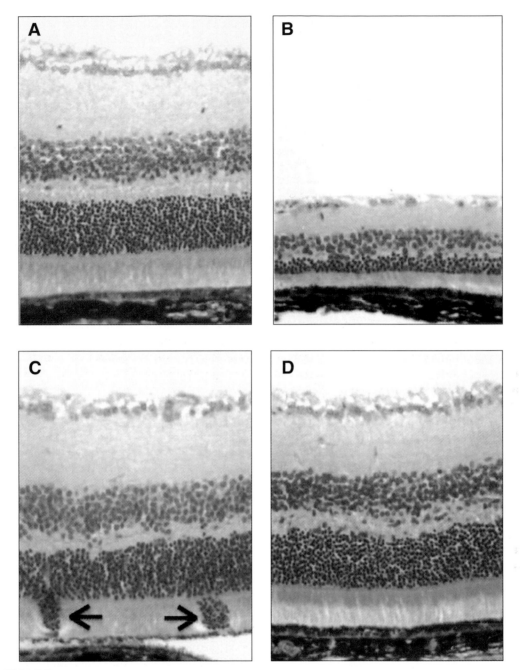

FIGURE 3.8 *Cyclin D1*-deficient mice have reduced cell proliferation in the nervous system. A. The normal retina is a laminated structure consisting of alternating layers of cells (stained purple) and dendrites (lighter stained layers). B. In mice that have had the *cyclinD1* gene deleted by homologous recombination, the layers are much thinner, and there are considerably fewer cells in the retina. This is due to the reduced amount of proliferation of the progenitor cells. On the other hand, if the gene coding for the negative regulator of the cell cycle, *p27kip*, is deleted in mice (C), the progenitor cells keep proliferating past the normal time when they would cease mitosis, and produce extra cells that abnormally spill out of their normal layer (arrows). D. When both *cyclinD1* and *p27kip* are deleted in mice, however, the retina returns to a remarkably normal appearance, indicating that the balance in proliferation has been restored when both the "gas" and the "brakes" of the cycle have been removed. Clearly, other regulators must take over when the *cyclinD1/p27kip* system is removed, but it is not clear yet what these are. (From Geng et al., 2001)

BOX

NEURAL CREST MIGRATION

The neural crest exemplifies the migratory process in embryology; no other tissue undergoes such extensive migration during development as the cells of the neural crest. As noted previously, the neural crest is a collection of cells that emerges from the dorsal margin of the neural tube, where it intersects with the ectoderm. Although the neural crest was first described by His in the chick embryo in 1868, the migration of the cells of the neural crest was first demonstrated by Detwiler (1937) by labeling the pre-migratory cells with vital dye. The neural crest from the trunk takes two basic routes (Weston, 1963) from the neural tube: the ventral stream, along which the cells that will form the sensory, enteric and autonomic ganglia follow, and the dorsal or lateral stream, in which the cells that will form the pigment cells in the epidermis predominate (Figure). The route that cells take is to some degree determined by the environment in which they find themselves. There are differences in the types of cells that differentiate from the neural crest at different points along the anterior-posterior axis of the embryo. For example, neural crest from the most anterior part of the developing spinal cord migrates into the gut to form the enteric nervous system, while neural crest from somewhat more caudal levels of the spinal cord never migrates in to the gut, but instead collects near the aorta and forms the sympathetic ganglion chain. Transplantation of neural crest cells from anterior (enteric ganglion forming) levels of the embryo to more posterior regions results in the anterior crest cells following the posterior pathways and making sympathetic neurons instead of enteric neurons (LeDouarin et al., 1975; LeDouarin, 2004).

What guides these cells to their proper locations in the embryo? Both permissive and repulsive cues, similar to those that guide growing axons (see Chapter 5), direct the neural crest through these two main routes. Several large extracellular glycoproteins and sulfated proteoglycans have been shown to be critical for the migration of many different types of cells. Two of these molecules, laminin and fibronectin, are known to support the migration of neural crest cells when the cells are dissociated from the embryo and plated onto tissue culture dishes. The receptors for these extracellular matrix molecules, heterodimers of integrin proteins, are expressed by the migrating neural crest cells, and perturbation of these receptors also inhibits neural crest migration. If either β1-integrin, or its heterodimeric partner, alpha4-integrin are blocked with specific antibodies, neural crest migration is blocked (Lallier et al., 1994; Kil et al., 1998). These results

and others have shown that the neural crest of the trunk primarily interacts with the extracellular matrix as the cells migrate to their various destinations.

Another characteristic feature of neural crest cells emerging from the hindbrain and trunk is that they migrate in a segmented manner. Trunk neural crest cells migrate through the rostral half of each somite but avoid the caudal half. What molecules are responsible for this restriction of their migratory routes? Another family of proteins appears to be necessary for this pattern of migration: the Ephrin receptors and ligands. These molecules were first identified for their roles in repulsive guidance of axonal growth (see Chapter 5). Two of the ligands, *Lerk2* and *HtkL*, are expressed by the caudal halves of the somites, while the receptor, *EphB3*, is expressed by the migrating neural crest cells (Krull et al., 1997; Wang and Anderson, 1997). If the neural crest cells are given a choice between fibronectin or the ephrin ligand, they avoid the ephrin. Moreover, if the soluble ligand is added to explants of trunk, the normal migratory pattern is disturbed, and the crest cells migrate on both halves of the somite (Krull et al., 1997).

The neural crest that migrates from the cranial regions of the neural tube has many unique features (Noden, 1975). As noted in Chapter 2, the neural tube has a considerable amount of pattern in the head very early in development. The regions of the brain are dependent on *Hox* and *Pax* gene expression. The neural crest that migrates from the cranial regions of the neural tube also has positional identity, and this is also dependent on the *Hox* code. The figure shows the migration of the neural crest from the rhombomeres. The cranial crest contributes many cells to three tissue bulges known as branchial arches. The neural crest that migrates into these arches will give rise to most of the skeletal and cartilage of the skull and face. Thus, although normally we think of the neural crest as "neural" in the head the bulk of the neural crest cells will develop into nonneural tissues. The unique contribution of the different regions of cranial neural crest has provided an opportunity to test for the specification of these cells and their migratory patterns. The crest cells from rhombomeres *r1* and *r2* migrate into the first (mandibular) arch, the crest from *r4* into the second (hyoid) arch, and the crest from *r6* and *r7* into the third arch (Kontges and Lumsden, 1996). The crest in each of these arches differentiates into specific skeletal elements of the face or jaw (Figure). The neural crest from each rhombomere continues to express the same pattern of *Hox*

BOX (*cont'd*)

NEURAL CREST MIGRATION

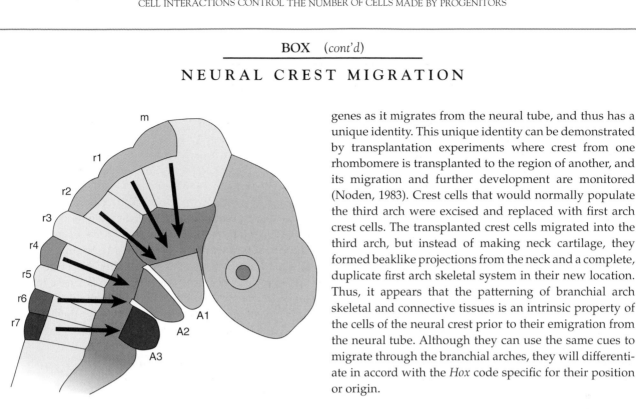

genes as it migrates from the neural tube, and thus has a unique identity. This unique identity can be demonstrated by transplantation experiments where crest from one rhombomere is transplanted to the region of another, and its migration and further development are monitored (Noden, 1983). Crest cells that would normally populate the third arch were excised and replaced with first arch crest cells. The transplanted crest cells migrated into the third arch, but instead of making neck cartilage, they formed beaklike projections from the neck and a complete, duplicate first arch skeletal system in their new location. Thus, it appears that the patterning of branchial arch skeletal and connective tissues is an intrinsic property of the cells of the neural crest prior to their emigration from the neural tube. Although they can use the same cues to migrate through the branchial arches, they will differentiate in accord with the *Hox* code specific for their position or origin.

---

therefore it is possible that the gradual lengthening of the G1-phase of the cell cycle in neural progenitor cells within the CNS (above) is due to an increasing dependence on these factors for progression through the cell cycle as development proceeds. The factors that have been shown to act as mitogens for the mitotically active cells in the progenitor cells of the vertebrate CNS are primarily those peptides that act on receptor tyrosine kinases, including FGFs, TGF-alpha, EGF, and insulin-like growth factors. However, there are many other types of signaling molecules that act on progenitor cells in the nervous system and also play a role in their proliferation. *Sonic hedgehog* and members of the *Wnt* protein family are examples of molecules that were involved in patterning the nervous system (reviewed in Chapter 2), but are also critical for the regulation of progenitor proliferation at later stages of brain development.

The identification of mitogenic factors for nervous system progenitor cells has relied on cell culture studies. Neural progenitor cells can be isolated from the developing brain by dissociating the cells with enzymes and putting them in culture dishes. The progenitor cells in the culture dishes will divide and make both additional progenitor cells and differentiated neurons (Figure 3.9). Some of the early studies tested many of the growth factors that stimulated proliferation of cells in other tissues and found that some of the

best mitogens for the progenitor cells of the nervous system were those that stimulated fibroblast cells (FGF) or skin cells (EGF). Progenitor cells express receptors for the various mitogenic factors, and depending on their location and stage of development, they are more responsive to one mitogen or another. Mitogenic factors like EGF and FGF bind to receptors of the tyrosine kinase signaling pathway to stimulate cell division by the upregulation of the S-phase cyclins, such as *cyclinD*. Since *cyclinD* expression will stimulate the cells to enter the S-phase, the regulation of *cyclinD* is a direct mechanism to control cell-cycle entry by a growth factor. In this way, the extracellular signals are integrated with the intrinsic cell cycle regulation machinery.

A factor first encountered in the context of patterning the nervous system (Chapter 2), *sonic hedgehog*, is also a key mitogen for nervous system progenitors. The *Shh* mitogenic pathway is important in many regions of the brain, particularly those of the dorsal brain (Figure 3.10). The way in which *Shh* acts in neurogenesis demonstrates the way in which differentiated neurons can feed back on the progenitors to maintain their proliferation and ensure that the correct number of neurons is generated during development. Although the *Shh* mitogenic pathway is used in many areas of the developing brain, the genesis of granule neurons in the cerebellum is a good example (Wechsler-Reya and Scott, 1999). The cerebellum is

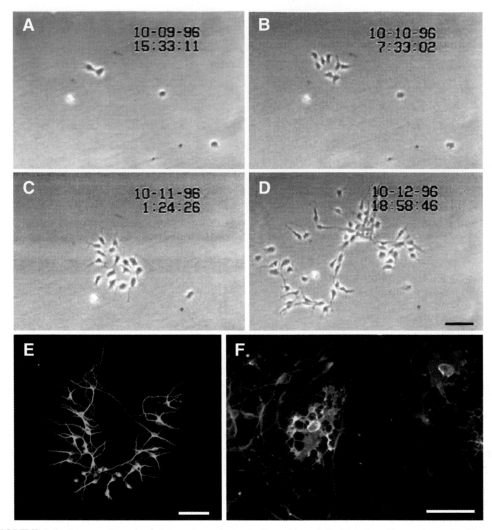

FIGURE 3.9 The proliferation of neural progenitor cells can be studied in vitro. Progenitor cells from the developing CNS can be studied in cell culture by dissociating them into single cells, diluting them to only a few cells in each well of a tissue culture dish, and then examining them daily for increases in their numbers. These micrographs, taken daily, document the proliferation of progenitors from the first division (A) to over 30 cells three days later B–D. Labeling the culture with antibodies against a neural specific protein shows that several of the new cells have developed into neurons, while others express antigenic markers of either oligodendrocytes or astrocytes. (From Qian et al., 1998, courtesy of S. Temple)

the large, highly convoluted structure at the back of the brain that is critical for smooth movements. The main output neurons from the cerebellar cortex are the Purkinje neurons. The Purkinje neurons are generated early in cerebellar development and form a layer one cell thick throughout the cerebellum. One type of neuron that connects with the Purkinje cell, the granule cell, is made in huge numbers by a nearby progenitor zone, the external granule layer. The Purkinje cells produce the mitogen, *Shh*, while the granule cell progenitors express the *Shh* receptors, *patched* and *smoothened* (named for *Drosophila* mutants defective in the homologous genes). The *Shh* released from the Purkinje cells

stimulates the granule cell progenitors to make more granule cells. If the *Shh* pathway is experimentally blocked, fewer granule cells are produced. If the *Shh* pathway is stimulated, granule cell production is increased. In this way, *Shh* is utilized to mediate the cell interactions between the differentiated Purkinje neurons and the neural progenitors. This pathway also provides another example of how a childhood tumor can result from a misregulation of neurogenesis. Children with mutations in the *Shh* receptor, *patched*, that mediates *Shh* signaling will develop a tumor called medulloblastoma, in which granule cell production is fatally uncontrolled (Goodrich et al., 1997).

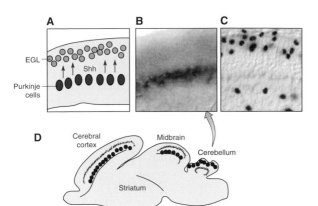

FIGURE 3.10 *Sonic hedgehog* secreted by neurons stimulates progenitor proliferation. The *Shh* mitogenic pathway is important in many regions of the brain, particularly those of the dorsal brain, like the cerebellum, the cerebral cortex, and the midbrain. *Shh* is produced by the differentiated neurons (red) to feed back on the progenitors to maintain their proliferation and ensure that the correct number of neurons is generated during development. In the cerebellum, for example, the *Shh* released from the Purkinje cells stimulates the granule cell progenitors to make more granule cells. (Modified from Ruiz i Altaba, 2002)

THE GENERATION OF NEURONS AND GLIA

The nervous system contains both neurons and glia, and both basic types of cells are produced in highly stereotypic ratios. What factors control the relative ratios of neurons and glia in the brain? Early in the development of the CNS, many, if not all, of the progenitor cells have the capacity to generate both neurons and glia. Retroviral lineage studies have shown that, for many regions of the nervous system, neurons, astrocytes, and oligodendrocytes can arise from a single infected progenitor cell. Davis and Temple (1994) have isolated progenitor cells from the embryonic cerebral cortex and cultured them as individual cells; they found that neurons, astrocytes, and oligodendrocytes can arise from a single progenitor cell in vitro. At later stages of development, the lineages of these cell classes can become separate. When cerebral cortical progenitor cells are labeled relatively late in development, the progeny of an infected cell may be restricted to only astrocytes or only neurons (Parnevales et al., 1991; Luskin et al., 1993). However, regardless of whether neurons and glia are made from a single division of a progenitor, as in some regions of the nervous system, or through separate lineages, the question of how these two very different cell types arise is an important one (Jacobson, 1977).

Cell culture studies have suggested that extracellular signaling factors, like those that control cell prolif-

eration of the progenitor, might also direct the progenitor cells to either a neuronal or glial lineage. In cell cultures, one can add defined factors and assay the effects on the production of either neurons or glia from the progenitor cells. These kinds of studies have led to some general principles, but also to many conflicting results that appear to depend on the region of the nervous system and the age of the embryo from which the cultures were derived. For example, *FGF2* and *Neurotrophin3* promote progenitor cells isolated from brain to develop primarily as neurons in most studies (Qian et al., 1998; Ghosh and Greenberg, 1995), and adding EGF (Kilpatrick and Bartlett, 1995), or CNTF (ciliary neuronotrophic factor; Bonni et al., 1997) to CNS cultures causes the cells to develop as astrocytes. However, members of the TGF-beta superfamily of molecules, like *BMP2* and *BMP4*, have also been shown to have effects on the multipotent progenitor cells, causing them to develop as neurons under some culture conditions (Loturco, 1997) and as astrocytes under others (Gross et al., 1997). Along these same lines, PDGF (platelet-derived growth factor) promotes oligodendroglial development in some assays (Raff et al., 1988) and neurons in other assays (Williams et al., 1997). The various factors that control the relative ratios of neurons and glia in the nervous system must eventually get translated to the nucleus and activate either the neuronal or glial program of gene expression, and recent evidence indicates that CNTF and *BMP2* act synergistically to activate the promoter of a critical astrocyte gene, GFAP; STAT binds directly to the GFAP promoter and activates the gene. Thus, this provides a direct transcriptional connection between the signaling molecule and a glial-specific gene. Together, while these studies highlight the complexity of the process of neurogenesis, the primary molecular pathways that promote neurogenesis and gliogenesis can be summarized as shown in Figure 3.11. However, it should also be noted that most of this work has been done in vitro, and so it is not known whether all these factors will have the same effects in vivo, in the intact ventricular zone. Nevertheless, studies of the effects of EGF and its receptor activation have shown that, even in vivo, this factor acts primarily to promote the production of astrocytes (Kuhn et al., 1997; Burrows et al., 1997), and biases cells away from neuronal differentiation.

One part of the central nervous system that has been particularly well characterized for its potential to form glia is the optic nerve. Raff and his colleagues have taken advantage of the fact that neurons do not develop in the optic nerve to carefully study the glial lineages in restricted glial progenitors (Figure 3.12). The nerve contains both astrocytes and oligodendrocytes, and in vitro studies have shown that a particular

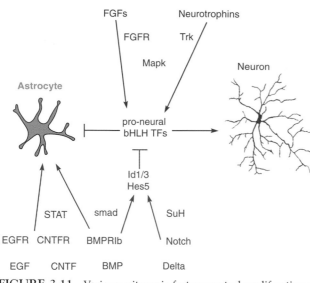

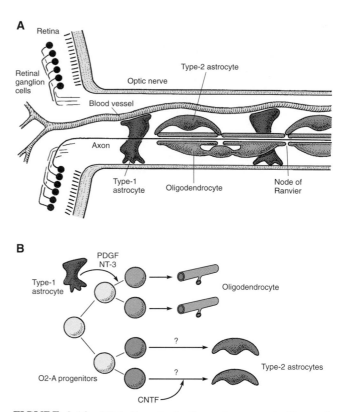

FIGURE 3.11 Various mitogenic factors control proliferation of the different types of progenitors in the nervous system. Neurogenesis and gliogenesis are regulated by many growth factors, and these are summarized in the figure. *FGF2* and *Neurotrophin3* promote progenitor cells isolated from brain to develop primarily as neurons, likely through the increase in expression of proneural *bHLH* genes, such as *NeuroD1*, EGF, and CNTF, which cause the progenitor cells to develop as astrocytes, and at least for CNTF this is known to work through the activation of the STAT transcription factor, which binds to the promoter of the glial-specific gene, GFAP. BMPs can synergize with CNTF to promote glial development, partly through the STAT pathway and partly through a direct inhibition of the proneural genes via the *Hes* pathway. Notch activation also activates the *Hes* pathway to promote gliogenesis and inhibit neurogenesis.

FIGURE 3.12 Glial diversity in the optic nerve. A. The optic nerve has three different types of glia, type 1 and type 2 astrocytes and oligodendrocytes. B. Culture studies show that type 1 astrocytes secrete PDGF and *NT-3*, which causes *O2-A* progenitors to divide. After a certain number of divisions, *O2-A* progenitors are timed to differentiate as oligodendrocytes or type 2 astrocytes if they are exposed to CNTF and other (?) factors. (Adapted from Harris and Hartenstein, 1999)

type of cell, the *O2A* progenitor (for **O**ligodendrocyte, and type **2 A**strocyte), can produce either astrocytes or oligodendrocytes, depending on the culture conditions (Raff et al., 1983). These *O2A* progenitor cells thus depend on signals to direct their differentiation. Multiple signals cause these cells to proliferate and develop as oligodendrocytes, including many of the same factors mentioned above: PDGF, NT3, and IGF-1. These factors are all produced by the astrocytes of the optic nerve, and PDGF is also produced by the retinal ganglion cells and present in their axons. In addition to these growth factors, the electrical activity of the axons in the optic nerve is also important for oligo progenitor cell proliferation; blocking electrical activity results in a decline in the number of oligodendrocytes in the nerve (see Barres and Raff, 1994, for review). Tethering the production of oligodendrocytes to the axons in the nerve may provide a way to ensure that sufficient oligodendrocytes are produced to properly myelinate all the axons.

Another critical genetic pathway that can control glial versus neuronal fate is the Notch pathway. The Notch pathway regulates lateral inhibition of cell fate

during the very early stages of brain development (Chapter 1). At the early stages of neural development, activation of the Notch pathway prevents epidermal cells from becoming neuroblasts in *Drosophila* by down-regulating the expression of the proneural genes *achaete* and *scute*. Similarly, overexpression of the proneural genes, either in flies or in frog embryos, causes ectopic neurons to form from the epidermis. Studies over the past 10 years have also demonstrated that the proneural genes continue to be expressed in the vertebrate nervous system throughout development. The neural progenitor cells express several proneural genes, including *NeuroD1*, *Neurogenin*, and *Mash1* (Figure 3.13). The expression of these genes in part determines whether a progenitor will produce a neuron or a glial cell. If the Notch pathway is activated, the progenitors primarily generate astrocytes; however, if the proneural genes are overexpressed in the progenitor cells, they are biased to generate more neurons than they normally would (Figure 3.14). Thus, the proneural genes function in both the initial forma-

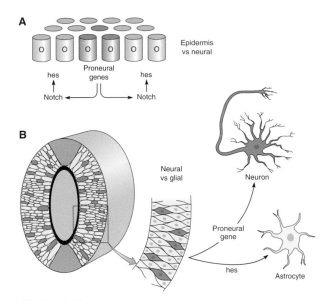

FIGURE 3.13 The proneural genes regulate neurogenesis at early and late stages of development. A. As described in Chapter 1, the proneural genes are important in the initial segregation of neural tissue from the epidermis in both *Drosophila* and vertebrates. B. Proneural genes are also important in the decision of a progenitor in the neural tube to generate either a neuron or a glial cell. The progenitors express one of several proneural genes, and this allows them to generate neurons. If the Notch pathway is activated in the progenitor cell, the proneural gene expression in the progenitor is inhibited by the antagonist *Hairy/enhancer of split* (*hes*) gene expression and the progenitors primarily generate astrocytes.

tion of the nervous system and in the control of the progenitor cells later in development to direct them to a neural fate.

CEREBRAL CORTEX HISTOGENESIS

In the next section, the histogenesis of two specific regions of the CNS—the cerebral cortex and the cerebellum—will be highlighted; histogenesis of these structures has been the subject of intense study for many years. As noted in the previous chapter, the cerebral hemispheres develop from the wall of the telencephalic vesicle. The neuroepithelial cells initially span the thickness of the wall, and as they continue to undergo cell division, the area of the hemispheres expands. At this early stage of development, the progenitor cells are thought to undergo primarily symmetric cell divisions, and their progeny both remain in the cell cycle. Soon, however, a few cells withdraw from the cycle to develop as the first cortical neurons. These neurons migrate a short distance to form a distinct layer, just beneath the pial surface, known as the preplate (Figure 3.15). The preplate consists of two dis-

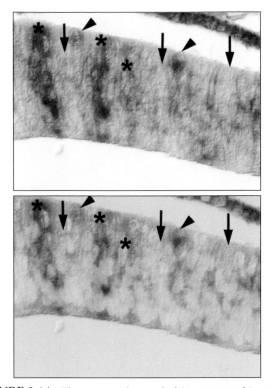

FIGURE 3.14 The proneural gene *Cash1* is expressed in progenitor cells. Combined in situ hybridization for the proneural gene chicken *achaete scute* homolog 1 (*Cash1*) and immunohistochemistry for *BrdU*, a marker for mitotically active progenitor cells shows that many of the *BrdU*-labeled cells also express the proneural gene *Cash1* in this section through the neuroepithelium of the embryonic chick retina. There are many other *BrdU*-labeled cells that do not express the *Cash1* gene, and presumably these express a different proneural gene. (Image thanks to Dr. Branden Nelson)

tinct cell types: a more superficial marginal zone, containing a group of large, stellate-shaped cells, known as Cajal-Retzius cells, and a deeper zone of cells called the subplate cells (Marin-Padilla, 1988; Allendoerfer and Shatz, 1994). The next stage of cortical development is characterized by a large accumulation of newly postmitotic neurons within the preplate (Marin-Padilla, 1988). These new neurons form the cortical plate. The cortical plate divides the preplate into the superficial marginal zone, composed primarily of the Cajal-Retzius cells, and the intermediate zone, composed of the subplate cells and increasing numbers of incoming axons. The developing cortex is thus described as having four layers: the ventricular zone, the intermediate zone, the cortical plate, and the marginal zone.

At the very earliest stages of cortical development, the processes of the progenitor cells span the entire thickness of the cortex. The first cortical neurons that are generated use the predominantly radial orientation of their neighboring progenitor cells to guide their

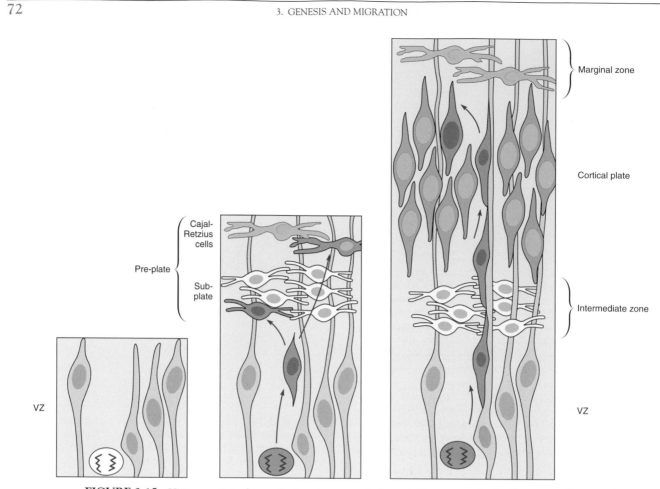

FIGURE 3.15 Histogenesis in the cerebral cortex proceeds through three stages. In the first stage of histogenesis, the wall of the cerebral cortex is made up of the progenitor cells, which occupy the ventricular zone (VZ). In the next stage of development, the first neurons exit the cell cycle (red) and accumulate in the preplate, adjacent to the pial surface. The neurons of the preplate can be divided into the more superficial Cajal-Retzius cells and the subplate cells. In the next stage of cortical histogenesis, newly generated neurons (red) migrate along radial glial fibers to form a layer between the Cajal-Retzius cells and the subplate. This layer is called the cortical plate, and the majority of the neurons in the cerebral cortex accumulate in this layer.

migration. However, the accumulation of neurons within the cortical plate results in a marked increase in cortical thickness. As a result, the processes of progenitor cells no longer are able to extend to the external surface of the cortex. Nevertheless, the newly generated cortical neurons still migrate primarily in a radial direction. How is this accomplished? To guide the newly generated cortical neurons to their destinations, a remarkable set of cells, known as *radial glia*, provide a scaffold. These glial cells have long processes that extend from the ventricular zone all the way to the pial surface. They form a scaffold that neurons migrate along. Serial section electron microscopic studies by Pasko Rakic first clearly demonstrated the close association of migrating neurons with the radial glial cells in the cerebral cortex (Figure 3.16). The migrating neurons wrap around the radial glial processes like a person climbing a pole. In recent years, it has been

possible to directly observe the process of neuronal migration in vitro using dissociated cell cultures (Edmonson and Hatten, 1987) or cortical slices (O'Rourke et al., 1995). These studies have confirmed that newly generated neurons migrate along the glial cells and that the process is saltatory, with migrating neurons frequently starting and stopping along the way.

The next phase of cortical histogenesis is characterized by the gradual appearance of defined layers within the cortical plate. As increasing numbers of newly generated neurons migrate from the ventricular zone into the cortical plate, they settle in progressively more peripheral zones. Meanwhile, the earlier-generated neurons are differentiating. Thus, later-generated neurons migrate past those generated earlier. This results in an inside-out development of cortical layers (Figure 3.17). Richard Sidman (Angevine and Sidman, 1961) used the 3H-thymidine birthdating technique

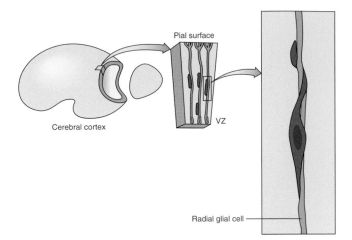

FIGURE 3.16 Migration of neurons along radial glia. The radial glial fibers extend from the ventricular zone to the pial surface of the cerebral cortex. A section through the cerebral cortex at an intermediate stage of histogenesis shows the relationship of the radial glia and the migrating neurons. The postmitotic neurons wrap around the radial glia on their migration from the ventricular zone to their settling point in the cortical plate. (Modified from Rakic, 1972)

described above to first demonstrate the inside-out pattern of cerebral cortical histogenesis. The neurons labeled in the cortex of pups born from pregnant female rats injected with thymidine on the 13th day of gestation were located in the deeper layers of cortex, whereas the neurons labeled after a thymidine injection on the 15th day of gestation were found more superficially (Figure 3.17). This inside-out pattern of cortical neurogenesis is conserved across mammalian species. Figure 3.18 shows the results of similar thymidine birthdating experiments in the monkey, where the process of neurogenesis is much more prolonged than in the rat. Thymidine injections at progressively later stages of gestation result in progressively more superficial layers of cerebral cortical neurons being labeled. Each cortical layer has a relatively restricted period of developmental time over which it is normally generated (Figure 3.18).

While the crawling of the neuroblast along the radial glial scaffold has been well recognized for over 30 years, recent direct visualization of the process in developing mouse cerebral cortex has yielded some surprises. In these studies, the ventricular zone was labeled using a dye that marked a subpopulation of the newly generated neuroblasts. As these cells left the ventricular zone, their leading processes were visible. Time-lapse imaging of dye-labeled neuroblasts shows clearly that many of the neuroblasts migrate just as predicted from the EM reconstructions of Rakic. However, direct visualization of the migration process also revealed that many of the neuroblasts move via a very different process, a process termed *somal translocation*.

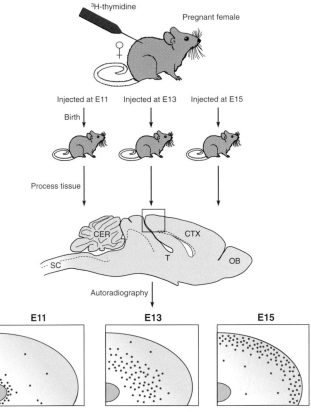

FIGURE 3.17 Birthdating studies demonstrate the inside-out pattern of cerebral cortical histogenesis. Pregnant female rats are given injections of 3H-thymidine at progressively later stages of gestation. When the pups are born, they are allowed to survive to maturity, and then their brains are processed to reveal the labeled cells. Neurons that have become postmitotic on embryonic day 11 are found primarily in the subplate (now in the subcortical white matter), while neurons "born" on day E13 are found in deep cortical layers, that is, V and VI, and neurons generated on E15 are found in more superficial cortical layers, that is, IV, III, and II. The most superficial layer, layer I, contains only the remnants of the preplate neurons (not shown). (Modified from Angevine and Sidman, 1961)

The migrating cell has a leading process that extends to the pial surface, while the cell body is still near the ventricular zone at this early stage of cerebral cortical development. Then, as the process gets progressively shorter, it draws the cell soma to the pial layer as if it were doing pull-ups on a bar (Nadarajah et al., 2001). At the same time, other neuroblasts are migrating with relatively constant leading processes, and presumably these are more like those described from the EM reconstructions. Some neuroblasts show both modes of migration at different points in their path.

The direct visualization of neuronal migration gave rise to another surprise. The relationship between the radial glia and the progenitor cells has long been thought to be separate. The glia were thought to be

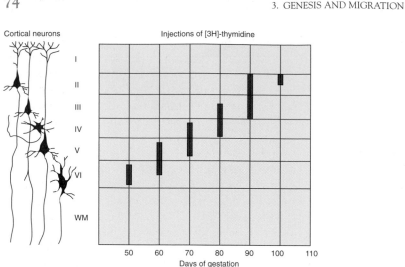

FIGURE 3.18 Birthdating studies in monkey further demonstrate the inside-out pattern of cerebral cortical histogenesis. In the monkey, where the histogenesis of the cerebral cortex is more protracted than in the rat, the production of the cortical neurons takes place over a 50-day time period. By labeling the pregnant female with 3H-thymidine at progressively later gestational ages, it is possible to determine the period of embryonic development when each specific cortical layer is generated.

FIGURE 3.19 Live imaging of GFP labeled radial glia shows that radial glia are the same cells as the progenitors. Noctor et al. (2002) used a retrovirus to label small numbers of cortical progenitor cells in slice cultures of the cerebral cortex of mice. In this example, they found that the radial glia (arrowhead) has undergone several cell divisions, and the progeny are migrating immature neurons (arrow). The neurons migrate along the radial glia that generated them. (From Noctor et al., 2001)

generated early in CNS development, and then the progenitor cells for the various neurons and glia coexisted, side by side with the radial glia. However, Noctor et al. (2002) used a retrovirus to label small numbers of cortical progenitor cells in slice cultures of the cerebral cortex of mice to directly visualize their genesis and migration (Figure 3.19). They found, to their surprise, that the radial glia themselves were the neuronal progenitors! Figure 3.19 shows an example of one of the clones they found. When the slice is viewed on the first day, the labeled cell is a single radial glial cell, with a process extending the entire width of the cerebral cortex; however, as they continue to analyze the clone on subsequent days, they find that the radial glia undergoes several cell divisions, and the progeny are not additional radial glia but migrating immature neurons. These neurons migrate along the radial glia that generated them. In addition to having the morphology of neurons, these migrating neurons label for neuron-specific markers, while the radial glial cell that generated them expresses proteins typical of radial glia. These results have been confirmed by cell culture studies of radial glia, as well as more sophisticated genetic studies of mice in vivo. Thus, our current view of the neuronal progenitor and the radial glia is that they are the same cell.

Most neurons are generated from cell division within the ventricular zone and migrate radially, either by crawling along the radial glia or by somal translocation. In addition to this predominantly medial migration of the newly generated neurons, however, it has been consistently noted that some populations of cortical neurons migrate tangential to the cortical surface. Lineage tracing studies give some indication as to the degree of this dispersion. The progeny of a progenitor cell labeled with a retrovirus can be widely dispersed within the cortex. In addition, chimeric animals expressing reporter genes show that a substantial fraction of the cortical neurons are not associated with nearby radial clusters of similar genotype. Observation of labeled neurons in cortical slice cultures has directly demonstrated this tangential migration of a subpopulation of the cells migrating out of the ventricular zone (O'Rourke et al., 1995).

In addition to the intrinsically generated, tangentially migrating neuronal population, at least some tangentially migrating cells are not derived from the cortical ventricular zone at all, but instead migrate all the way from the ventricular zone in a subcortical forebrain region, the lateral ganglionic eminence. Most of the neurons of the cerebral cortex are pyramidal in shape and use the neurotransmitter glutamate. However, there are other populations of neurons in the cerebral cortex, such as a class of stellate-shaped, GABA-containing inhibitory interneurons. In a study of chimeric mice, where single embryonic stem cells with *lacZ* reporter genes were injected into wild-type host embryos, clones of labeled progenitor cells could be clearly identified in the cerebral cortex from very early stages of development. Two patterns of clones were identified: those that were made up primarily of pyramidal neurons, arranged in radial columns or clusters (Figure 3.20), and those that were more widely dispersed and were GABA-containing stellate cells (Tan et al., 1998). Anderson et al. (1997) found that

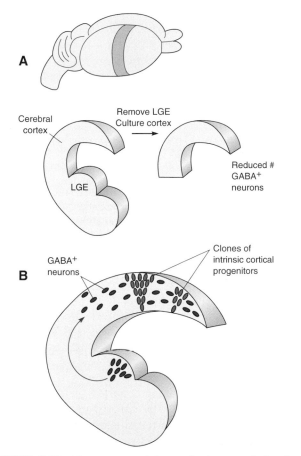

FIGURE 3.20 The neurons of the cerebral cortex derive from both intrinsic and extrinsic sources. Most of the neurons in the cortex are derived from the ventricular zone cells immediately below their adult location. Clones of cells in the cortex show only a limited amount of radial dispersion when labeled from very early stages of development using chimeric mice. However, the GABA containing stellate cells in the cortex arise from the lateral ganglionic eminence, a subcortical structure long thought to generate the neurons of the basal ganglia. If the cortex is isolated from the LGE early in its development and the cortex is allowed to develop further in vitro, the number of GABA containing neurons in the cortex is greatly reduced.

when cortical slices were cultured with the lateral ganglion eminences attached, these GABA neurons developed in the cortex. However, when the cortex was isolated from this subcortical region, the number of GABA-containing neurons was greatly reduced (Figure 3.20). Thus, it seems clear that the precursors of most GABA-containing interneurons in the cerebral cortex migrate all the way from the subcortical germinal ridges. Moreover, they could directly visualize this migration by labeling the premigratory population in the ganglionic eminence with the dye *DiI*, and track the migration of the GABA-containing neurons to the cerebral cortex. Clearly, the tangential migrating populations of young neurons within the cortex do not interact with radial glial cells in the same way as that described for radially migrating cortical neurons. However, at this time, much more is known about the factors that guide neurons along radial glia (see below) than for these tangentially migrating populations.

THE SUBVENTRICULAR ZONE: A SECONDARY ZONE OF NEUROGENESIS

As we have seen, neurons can at times migrate considerable distances from their point of generation in the ventricular zone. Although the cases described so far involve the migration of postmitotic neurons, there are also regions of the brain where the progenitor cells themselves migrate from the ventricular zone and continue to generate neurons in what are known as secondary zones of neurogenesis. There are three well-defined secondary zones of neurogenesis in the mammalian brain: the external granule layer (to be described in the section on cerebellar histogenesis), the subventricular zone, and the hippocampal granule cell precursors. The subventricular zone is a specialized region of the anterior lateral wall of the lateral ventricle (Figure 3.21). The SVZ forms as a secondary neurogenic zone in the late embryonic period in rodents, and, although most thoroughly studied in mice and rats, is present in all mammals examined to date (Jacobson, 1977). In rats, from *E11* to *E14*, mitoses occur exclusively at the ventricular surface; however, at *E16*, the SVZ forms, subadjacent to the VZ, and the SVZ continues to generate neurons and glia long after the VZ has ceased cell division at *E19*. The majority of the glia of the forebrain are thought to be derived from the SVZ. In recent years, the SVZ has become the subject of intense investigation owing to its maintenance in the adult brain, and more will be said about this later in this chapter. The progenitor cells of the granule

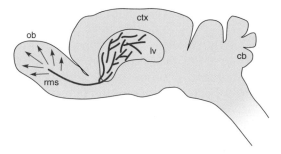

FIGURE 3.21 The cells generated by the subventricular zone in mature rodents migrate to the olfactory bulb in the rostral migratory stream. The cells migrate in chains, along extended astrocyte networks that lie adjacent to the lateral ventricles of the cerebral cortex. These networks are complex but in general have rostral-caudal orientation. (From Doetsch and Alvarez-Ruylla, 1996)

neurons of a region of the hippocampus known as the dentate gyrus originate from the ventricular zone for relatively short periods of embryonic development in the rat prior to *E14*. These progenitor cells then migrate to the growing dentate gyrus, and they, too, produce additional neurons for the rest of embryonic development, and even into adult life (see below).

Are radial glial cells also important for the migration of neurons and progenitors of neurons and progenitors in the secondary zones of histogenesis? Studies of the SVZ have been particularly illuminating. After the initial burst of gliogenesis (see above), most of the cells generated in the SVZ during the postnatal period and in mature rodents migrate to the olfactory bulb, in what is known as the rostral migratory stream (Lois et al., 1996). The cells migrate in chains, along extended astrocyte networks. These networks are complex (see Figure 3.21) but in general have rostral-caudal orientation. One might imagine that the association of migrating SVZ cells is analogous to the migration of cortical neurons along radial glia; however, the SVZ cells do not appear to require the glia. The migration of SVZ cells has been termed *chain migration* and is distinct from the migration of neurons along radial glia. The SVZ cells form a chain in vitro, even in cultures devoid of glial cells, and migrate by sliding along one another. Figure 3.22 shows this process of leapfrogging SVZ cells in a time-lapse series. Thus, glia might help to orient SVZ migration but are not essential for it.

CEREBELLAR CORTEX HISTOGENESIS

As noted above, the cerebellum is a large, highly convoluted part of the brain that is critical for control of our movements, particularly our balance. Cerebellar function is particularly susceptible to ethanol; the weaving motion of alcoholics is likely due to the cere-

bellar effects. The mature cerebellum is made up of several distinct cell types, each repeated in an almost crystalline array (Figure 3.23). The two most distinctive of these cell types are the giant Purkinje cells and the very small granule cells. Purkinje cells are the principal neurons of the cerebellar cortex, sending axons out of the cortex to the deep cerebellar nuclei. The cerebellar granule neurons are much more numerous than the Purkinje cells. In fact, the cerebellar granule cells are the most numerous type of neuron in the brain. In the mature cerebellum, they form a layer deep to the Purkinje cells, and their axons extend past the Purkinje cell layer into the molecular layer. The axons of the granule cells bifurcate in the molecular layer, into a T-shape, and these axons extend in the molecular layer for a considerable distance, synapsing on the Purkinje cell dendrites. One can think of the Purkinje cells as telephone poles and the granule cell axons as the telephone wires.

The generation of the intricate cerebellar architecture is a complex process. The large Purkinje neurons are generated from a ventricular zone near the fourth ventricle of the brainstem, in a manner similar to the way in which the neurons of the cerebral cortex are produced (as described in the previous section). Once they have finished their final mitotic division, the Purkinje cells migrate a short distance radially to accumulate as an irregular layer, known as the cerebellar plate. As the cerebellum expands, these cells become aligned to form a single, regularly spaced layer. The Purkinje cells then grow their elaborate dendrites. In addition to the Purkinje cells, the ventricular zone generates several other cerebellar interneurons, such as the stellate and basket cells.

In contrast to the somewhat standard pattern of neurogenesis of the Purkinje cells and the stellate and basket cells, the granule cells arise from a completely separate progenitor zone, known as the rhombic lip (Figure 3.24). The granule cell precursors are initially generated near the rim of the fourth ventricle but then migrate away from the ventricular zone, over the top of the developing Purkinje cells to form a secondary zone of neurogenesis, called the *external granule layer*. The cells in this layer continue to actively proliferate, generating an enormous number of granule cell progeny, thus increasing the thickness of the external granule layer considerably. The external granular layer persists for a considerable time after birth in most mammals and continues to generate new granule neurons. There are still granule neurons migrating from the external granule layer as late as two years after birth in humans (Jacobson, 1978).

Although the granule neurons are generated superficially in the cerebellar cortex, they come to lie deep to the Purkinje cells in the mature cerebellum. The developing granule neurons must therefore migrate past the

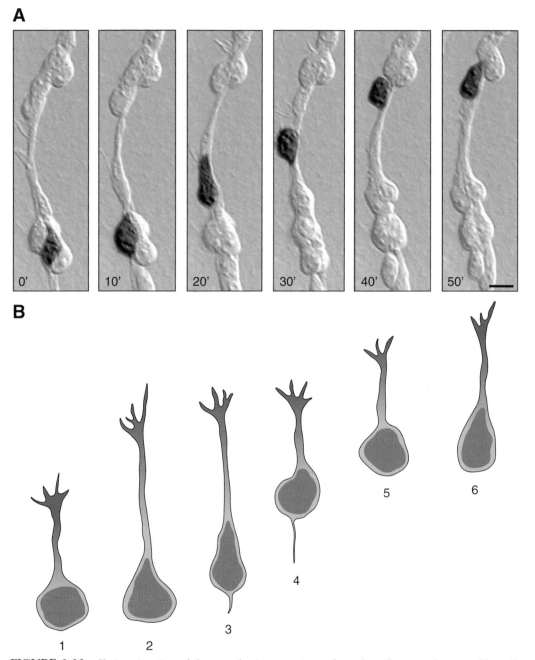

FIGURE 3.22 Chain migration of the rostral migratory stream from the subventricular zone. The cells normally migrate along complex astrocytic networks that are generally oriented in a rostral-caudal direction. However, the time-lapse series shows how these cells migrate in chains, by sliding along one another, and they can do this without any glia present in the cultures. (From Garcia-Verdugo et al., 1998)

Purkinje cells. Figure 3.25 shows what this process looks like, as originally described by Ramon y Cajal. Soon after their generation, after their final mitotic division, the granule cells change from a very round cell to take on a more horizontal-oriented shape as they begin to extend axons tangential to the cortical surface. Next, the cell body extends a large process at right angles to the axon. As this descending process grows deep into the cerebellum, the cell body and nucleus follow, leaving a thin connection to the axon. Meanwhile, the axons have been extending tangentially, and so the cell assumes a T-shape. The cell body eventually migrates past the Purkinje cell layer and then begins to sprout dendrites in the granule cell layer.

The migration of the granule cells is another example of the importance of radial glia in CNS histogenesis. As they migrate, a specialized type of radial glia, known as the Bergmann glia, guides the granule

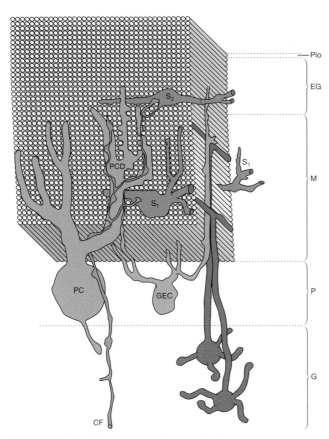

FIGURE 3.23 The neurons of the cerebellar cortex are arranged in a highly ordered fashion. In the mature cerebellum, the very large Purkinje cells (PC) lie in a single layer (P) and have an extensive dendritic elaboration that lie in a single plane. The granule cells (red) lie below the Purkinje cells in the granule cell layer (G) and have a T-shaped axon that runs orthogonal to the plane of the Purkinje cell dendrites, like phone wires strung on the Purkinje cell dendritic "poles" in the molecular layer (M). In addition to these distinctive cell types, the cerebellar cortex also contains other cell classes, the stellate cells (S) and the Golgi epithelial cells (GECs). (Modified from Rakic, 1971)

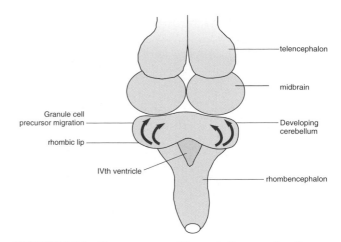

FIGURE 3.24 The precursors of the cerebellar granule cells come from a region of the rhombencephalon known as the rhombic lip. The rhombic lip is a region of the hindbrain that lies adjacent to the fourth ventricle. Cells from this region migrate over the surface of the cerebellum to accumulate in a multicellular layer—the external granule cell layer. This dorsal view of the developing brain shows the migratory path of the granule cell precursors from the rhombic lip of the rhombencephalon to the surface of the cerebellum (red arrows).

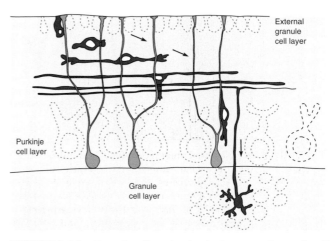

FIGURE 3.25 Granule cell production in the external granule cell layer is followed by the migration of these cells to ultimately lie deep to the Purkinje cell layer. Arrows show the migratory path a single neuron would take from its birth to the granule cell layer. The Bergmann glial cells are shown in red and function as guides for the migrating neurons. The migration of a granule cell is thought to take place along a single gila fiber, but in the diagram the migrating neuron is shown to be associated with several glial cells for clarity. (Modified from Ramon y Cajal, 1952)

MOLECULAR MECHANISMS OF NEURONAL MIGRATION

cells. EM studies, similar to those described for the cerebral cortex, first demonstrated the relationship between the migrating granule cells and the Bergmann glia (Rakic, 1971). Throughout the migration of the granule cells, they are closely apposed to the Bergmann glial processes. Hatten and her colleagues have been able to demonstrate directly the migration of granule cells on Bergmann glia using a dissociated culture system. When the external granule cell layer is removed from the cerebellum and the cells are cultured along with cerebellar glia, the granule glial cells migrate along the extended glial fibers in vitro. Time-lapse video recordings have captured the granule cell migration in action (Figure 3.26). The in vitro systems have also provided a way to explore the molecular basis of neuronal migration on glial cells in the CNS.

Several questions about the molecular mechanisms of neuronal migration have been explored in recent years using the in vitro cerebellar microculture system developed by Hatten and her colleagues. A particular

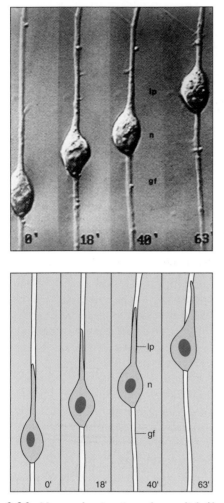

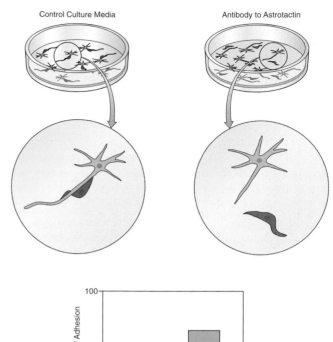

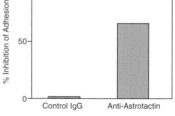

FIGURE 3.26 Neuronal migration along glial fibers can be observed directly. In cerebellar neuron-glial co-cultures, the neurons associate with the glial processes and slowly move along them. Using time-lapse microscopy, this neuronal migration can be directly observed and quantified. This has provided an excellent assay for investigating the molecular basis for neuronal migration. (Courtesy of M.E. Hatten)

FIGURE 3.27 In vitro assay demonstrates the importance of astrotactin in granule cell migration. When cerebellar glial cells are co-cultured with granule cells, the glial cells adopt an elongate morphology, and the neurons (red) crawl along the glial processes. When astrotactin antiserum is added to these cultures, the neurons no longer associate with the glial processes, and the long processes of the glial cells retract. This in vitro assay has also been used to analyze the effects of other molecules, like CAMs and extracellular matrix molecules, on granule cell migration.

class of cell surface proteins, known as cell adhesion molecules, or CAMs, are known to mediate the adhesion between cells and the migration of cells in many tissues in the embryo. Are these proteins necessary for the migration of granule cells along radial glial fibers? Under control conditions, the granule cells migrate on the radial glial cells in the cultures at about 33 μm/hr. To test whether the most abundant CAMs in the nervous system, NCAM, N-cadherin, and L1, are important for granule cell migration, Hatten's group added antibodies that specifically block the function of these molecules to the cerebellar microcultures (Stitt and Hatten, 1990). They found that none of these antibodies interfered with the migration of the granule cells. However, when they added an antiserum that

they had raised against cerebellar cells to the cultures, they found a significant inhibition of the granule cell adhesion (Figure 3.27). They concluded that some type of adhesion molecule was necessary for the migration of these cells but that it was not one of the previously known CAMs. To find this new CAM, they used a clever approach: they absorbed the antigranule cell antiserum with other types of neural cells and cell lines, thus getting rid of those antibodies from the serum that recognize common CAMs. What they were left with was an antiserum that recognized a single protein of approximately 100 kD, which they named astrotactin. Subsequent studies showed that astrotactin is a protein that resembles other CAMs but is in a distinct family. It is expressed in the migrating granule cells, and antibodies raised against astrotactin block the migration of the granule neurons along the glial fibers.

In addition to the use of in vitro systems for studying migration, advances in our understanding of the molecular mechanisms of cell migration have come about by analysis of naturally occurring mouse mutations that disrupt the normal migration of neurons. One important function of the cerebellum is to maintain an animal's balance. Lesions to the cerebellum in humans frequently produce a syndrome that includes unsteady walking, known as ataxia. Genetic disruptions of the cerebellum in mice produce a similar syndrome, and therefore they can be identified and studied. By screening large numbers of mice for motor abnormalities, several naturally occurring mutations have been identified that disrupt cerebellar development (Caviness and Rakic, 1978). Because of the nature of the symptoms, these mutant mouse strains have names like *reeler*, *weaver*, and *staggerer*. The mutant genes that underlie these phenotypes have been identified, and one of these mutants, *reeler*, has been particularly informative in understanding neuronal migration.

The *reeler* mutant mouse has ataxia and a tremor. Histological examination of individually labeled neurons in *reeler* mutant cerebral and cerebellar cortex revealed gross malpositioning of the cells. In the cerebellar cortex, the Purkinje cells are reduced in number and, instead of forming a single layer, are frequently present as aggregates. There are fewer granule cells, and most of them have failed to migrate below the Purkinje cells (Figure 3.28). Thus, they lie external to the Purkinje cells. In addition, many other regions of the CNS, including the cerebral cortex, show similar disruptions in normal cellular relationships. In the cerebral cortex of the *reeler* mouse, instead of the normal inside-out pattern, later generated neurons fail

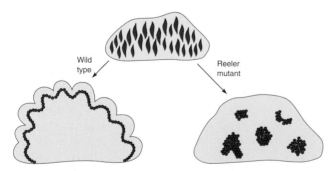

FIGURE 3.28 The *reeler* mutation in mice disrupts cerebellar development. In normal mice, the Purkinje cells form a single layer in the cortex. In *reeler* mice, the Purkinje cells do not migrate to the cortex but remain in large aggregates. The other cerebellar neurons are also malpositioned in the cerebellar cortex of the *reeler* mice. As a result, the normal function of the cerebellum in controlling balance is disrupted in the *reeler* mice, and so they "reel."

to migrate past those generated earlier. Thus, these mice have an outside-in organization.

The defective molecule underlying the *reeler* phenotype has been identified. It is a large glycoprotein, named *reelin*, containing over 3000 amino acids, and it bears similarities to some extracellular matrix proteins (D'Arcangelo et al., 1995). The reelin protein is expressed by the granule cells in the external granule cell layer from the very earliest stages of development. In the cerebral cortex reelin is expressed by the most superficial neurons, the Cajal-Retzius cells. Several additional mutant mice have been found to have *reeler*-like disruptions in their cortical lamination and act in the same molecular pathway as reelin. Mutations in the genes coding for a tyrosine kinase called disabled, LDLR8 and VLDLR (low and very low density lipoprotein receptors), ApoE (apolipoprotein E, an important factor in lipid transport), and cdk5 (a cyclin-dependent protein kinase—see Cell-Cycle Box) all cause defects in cerebral cortical neuroblast migration similar to those found in *reeler* mice (Jossin et al., 2003). In the *reeler* mice, the neuroblasts do not migrate past the subplate cells, and they appear to migrate at oblique angles rather than to track along the glial scaffold. The LDLRs, along with ApoE, form a receptor complex that phosphorylates the disabled protein upon reelin binding. Once phosphorylated, disabled can recruit other second messengers in the tyrosine kinase pathway and activate a host of cellular responses. The LDLRs and ApoE receptors are expressed in the migrating neuroblasts and in the radial glia themselves, while reelin is made by the Cajal-Retzius cells at the cortical surface. The observed cellular expression pattern of reelin and its receptors has led to two basic classes of hypotheses for its function during cortical development: (1) reelin might be a chemoattractant in cerebral cortex, causing the migrating neuroblasts to move toward the source of reelin in the Cajal Retzius cells in the superficial layers of the cortex; (2) alternatively, reelin could act as a stop signal at the cortical surface, telling the neuroblasts to "get off the track" and form a new cortical layer. In support of the second hypothesis, Dulabon et al. (2000) have found that adding reelin to migrating neuroblasts in cell culture causes them to stop their migration. However, this result is not inconsistent with reelin playing a role as a chemoattractant, since adding reelin to cell culture surrounds the migrating neuroblasts with the potential attractant and thus causes them to be attracted equally in all directions and to stop moving. To distinguish between these two possibilities, Curran's group generated a transgenic mouse that has reelin expressed under the control of the nestin promoter, and therefore expressed in the ventricular zone itself (Magdeleno et al., 2002). This misexpression experiment basically generates a reelin sandwich. Reelin is still expressed by the Cajal-Retzius cells, at the superficial surface of the cerebral cortex, but in these mice it is also expressed in the ventricular zone. If reelin is a stop signal or an attractant, the migrating neuroblasts should never leave the ventricular zone. However, they found that the nestin-reelin mice were essentially normal. The migrating neuroblasts still left the ventricular zone on schedule and in fact made basically normal layers (Figure 3.29). Therefore, it is

unlikely that reelin acts as either a chemoattractant or a stop signal. What then does this molecule do in the process of cell migration? These authors found a clue in a modification of this experiment. They mated the nestin-reelin mice with *reeler* mice. These mice only have the reelin in the ventricular zone and no longer express any reelin in the Cajal-Retzius cells. Although the cortical lamination was not perfect, it was improved over that of the reeler mouse. As noted above, reelin is also important for cerebellar Purkinje cell migration and lamina formation, and the nestin-reelin was even more effective in rescuing the cerebellar phenotype of the *reeler* mouse. Therefore, it looks as if it is less important where the reelin is localized in the developing cortex and cerebellum, as long as there is some reelin around.

In addition to reelin and astrotactin, a large number of molecules have been implicated in neuroblast migration in the cerebral and cerebellar cortices. Integrins are cell adhesion molecules that allow many different types of cells to attach to the proteins in the extracellular matrix. Since these adhesion receptors are necessary for the pial extracellular matrix formation, it is not surprising that they are required for the appropriate formation of the glial scaffold and hence the migration of the neuroblasts and correct positioning of the cerebral cortical neurons. It is as if you were trying to stand a ladder up without a wall to lean it against. Another class of molecules, the neuregulins and their receptors, likely has a very different role. Neuregulin, or glial growth factor, activates receptor tyrosine kinases called ErbBs (1–4) on the glial cell surfaces and promotes the appropriate differentiation and/or survival of the glial cells. Without the glial cells adopting their elongate morphology, the neuroblast migration is abnormal. Again comparing this with a ladder, it is as if you were trying to climb a ladder made of rubber.

In sum, many cellular and molecular interactions are necessary for proper arrangement of the neurons in the complex neuronal structures that make up the mature brain. Nearly all the neurons in the brain end up some distance from where they were generated in the ventricular zone, and the mature neuronal circuitry depends on cells getting to the right place at the right time. Mice with mutations in genes critical for neuronal migration have motor deficits, but it is likely that more subtle deficits are caused by less dramatic changes in neuronal migration. Several inherited mental retardation syndromes in humans are now known to be caused by defective migration of cortical neuroblasts. The beautiful choreography of neuronal migration is clearly an essential part of building a working nervous system.

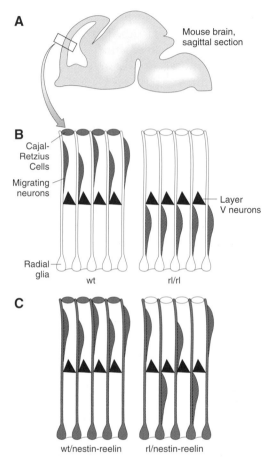

FIGURE 3.29 Is reelin a "stop" signal or a "go" signal? A transgenic mouse with reelin expressed in the wrong place might help sort this out. A. Sagittal section through a mouse brain to show where cortical sections are taken from. B. Mice with reelin expression under the control of the nestin promoter, and therefore expressed in the ventricular zone, generate a reelin "sandwich" for the migrating neurons. If reelin is a stop signal or an attractant, the migrating neuroblasts might never leave the ventricular zone. Surprisingly, the nestin-reelin mice had remarkably normal cortical lamination. C. Mating nestin-reelin mice with *reeler*-deficient mice leaves reelin only in the ventricular zone and no longer express in the Cajal-Retzius cells. This actually improved the lamination over that observed in the reelin-deficient mice. Therefore, it appears that it is less important where the reelin is localized, as long as there is some reelin around.

POSTEMBRYONIC AND ADULT NEUROGENESIS

The process of neurogenesis ceases in most regions of the nervous system in most animals. Neurons themselves are terminally differentiated cells, and there are no well-documented examples of functional neurons reentering the mitotic cycle. However, it has long been appreciated that in most species some new neurons are generated throughout life. There is considerable remodeling of the nervous system of insects during metamorphosis. Much of this remodeling occurs through cell death, but new neurons are also produced.

Many amphibians also go through a larval stage. Frogs and toads have tadpole stages where a considerable amount of body growth takes place prior to metamorphosis into the adult form. During larval stages, many regions of the frog nervous system continue to undergo neurogenesis similar to that in embryonic stages. One of the most well-studied examples of larval frog neurogenesis is in the retinotectal system. The eye of the tadpole, like that of the fish, increases dramatically in size after embryonic development is complete. During this period, the animal uses its visual system to catch prey and avoid predators. The growth of the retina, however, does not occur throughout its full extent, but rather is confined to the periphery (Figure 3.30). Similar to the way in which a tree grows, the retina adds new rings of cells at the pre-existing edge of the retina. This provides a way for new cell addition to go on at the same time the central retina functions normally. As the new retinal cells are added, they are integrated into the circuitry of the previously differentiated retina, into a seamless structure. At the same time that new cells are added to the peripheral retina, the optic tectum also adds new neurons. The coordination between neurogenesis in these two regions likely involves their interaction via the retinal ganglion cell projection of the tectum. The growth of the optic tectum, the brain center to which the retina sends its axons, occurs at its caudal margin, so the axons of the ganglion cell must shift caudally during this time. Fish retina also has an additional means of growth. As the retina grows, the sensitivity to light declines as retinal stretch causes a reduction in the number of rod photoreceptors. The fish maintains a constant sensitivity by adding new rods throughout the retina, not just at the peripheral edge. The new rods are generated by a specialized cell, the rod progenitor, which under normal circumstances generates new rod photoreceptors (Raymond and Rivlin, 1987). This specialized progenitor may not be entirely restricted in its potential, however, since, following damage to the

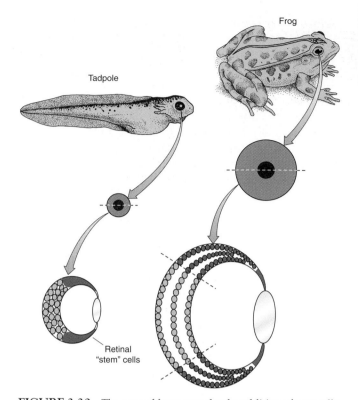

FIGURE 3.30 The eyes of frogs grow by the addition of new cells to the margin. The neural retina of the frog tadpole is derived from the neural tube, as described in a previous chapter. The initial retinal neurons are generated during embryogenesis. However, as the eye grows, the neural retina grows by means of a specialized ring of retinal stem cells at the peripheral margin of the eye (red). The retinal stem cells generate all the different types of retinal neurons to produce new retina that is indistinguishable from the retina generated in the embryo, and thoroughly integrated with it. In the newly post-metamorphic *Rana pipiens* frog, nearly 90% of the retina has been generated during the larval stages; all this time the retina has been fully functional. This process continues even after metamorphosis but much more slowly.

retina, these cells are stimulated to generate other retinal cell types as well.

One of the most well-studied examples of neurogenesis in mature animals comes from studies of song birds. In 1980, Fernando Nottebom reported that there was a seasonal change in the size of one of the brain nuclei important for song production in adult male canaries. In song birds, specific nuclei in the telencephalon of the brain are critical for the production of the song. The HVC nucleus is of particular importance for both song learning and song production (see Chapter 10). The HVC is almost twice as large in the spring, when male canaries are generating normal adult song, than in the fall, when they no longer sing. Nottebom initially proposed that this change in size might be due to seasonal changes in the numbers of synapses. In further studies of the HVC in male and female canaries, Nottebom also noticed that it was larger in males, which learn complex songs, than in female birds, which do not sing. Moreover, if adult females were given testosterone injections, the HVC nucleus grew by 90% and the female birds acquired male song (Nottebom, 1985).

To determine whether new neurons were added to the nucleus in response to the testosterone, female birds were injected with 3H-thymidine as well as testosterone, and the animals were sacrificed for analysis five months later. The researchers found that in both the testosterone-treated and control birds there were many thymidine-labeled cells, and many of these had morphological characteristics of neurons. They also analyzed birds immediately after the injections and found that the new neurons were not produced in the HVC itself, but rather were generated in the ventricular zone of the telencephalon and migrated to the nucleus, analogous to the way in which the nucleus is initially generated during embryogenesis. Subsequent studies have shown that the newly produced neurons migrate along radially arranged glial processes from the ventricular zone to the HVC (Garcia-Verdugo et al., 1998). Thus, ventricular zone neurogenesis is a normally occurring phenomenon in adult canaries (Figure 3.31). The progeny of the cells produced in the SVZ migrate to the HVC soon after their generation. There they differentiate into neurons, about half of which differentiate into local interneurons and half into projection neurons, which send axons out of the nucleus to connect with other neurons in the brain and form part of the functional circuit for song learning.

Thus, there appears to be a seasonally regulated turnover of neurons in the HVC and in other song control nuclei in the brain of the adult songbird. The turnover of neurons may correlate with periods of plasticity in song learning. Canaries modify their songs each year; each spring breeding season, they incorporate new syllables into the basic pattern, and then in the late summer and fall they sing much less frequently. Combining thymidine injections with measures of cell death and overall neuronal number in the HVC over a year, one can see two distinct periods of cell death, and each one is followed by a burst in the number of new neurons in the nucleus. Both of these periods of high neuronal turnover correlate with peaks in the production of new syllables added to the song. The neurogenesis is balanced by cell death, and during periods of new song learning the nucleus adds cells, while during periods when no song is generated, the song-related nuclei undergo regression. Is the rate of neurogenesis in the ventricular zone controlled by the seasonal changes in testosterone in the male birds? When the number of labeled cells in the ventricular

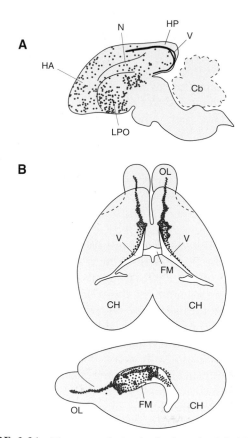

FIGURE 3.31 Neurogenesis in the brains of adult birds and rodents. A. Sagittal section of an adult canary brain showing the widespread pattern of mitotically active cells, many of which will go on to differentiate as neurons. B. Dorsal and sagittal views of an adult mouse brain showing thymidine labeling of proliferating subventricular zone cells and the rostral migratory stream (Modified from Smart, 1961; see also Jacobson, 1991).

zone is compared in testosterone-treated and untreated female birds, there are no differences—indicating that the rate of neurogenesis does not change in response to the hormone. However, it appears instead that the survival of the neurons in the HVC is seasonally regulated—neurons generated in the spring have a much shorter average lifespan than those generated in the fall. Thus, the seasonal changes in neuron number in the songbird HVC are not dependent on changes in the number of newly added cells, but rather relate to seasonally and hormonally regulated differences in the survival of the newly produced neurons.

Neurogenesis also occurs in the mature mammalian brain. Although for many years this view was regarded as somewhat heretical, it has become well accepted in recent years. The thymidine birthdating studies of Altman and Bayer, described at the beginning of this chapter, thoroughly documented the time and place of origin of neurons and glia of many regions of the rodent brain. It was found that many brain neurons are generated after birth in rodents. They next extended the labeling period to the second and third postnatal weeks and found that in one particular region, the olfactory bulb, thymidine-labeled cells were still found up to four weeks postnatally. These cells were generated in the subventricular zone in the forebrain and then migrated to the olfactory bulb (Figure 3.31). In recent years, it has been recognized that this neurogenesis also occurs in the adult hippocampus, and several studies have now shown that these cells are multipotent, like progenitor cells in the embryonic brain (Lois and Alvarez-Buylla, 1993; Luskin, 1993; Reynolds and Weiss, 1992).

In the last few years, Fred Gage and his colleagues have shown that the new neurons generated in the hippocampus are functionally integrated into the circuitry (Van Praag et al., 2002). To assay the function of the newly generated neurons, they used retroviral labeling in adult rats, similar to that which was described in the beginning of the chapter for labeling progenitors in the developing brain. Since a retrovirus will only infect and integrate into mitotically active cells, they were able to label the mitotically active hippocampal precursors with a retrovirus expressing the green fluorescent protein. When the authors examined the GFP-labeled cells after only 48 hours, the cells had a very immature morphology and resembled progenitors, like those found in the developing brain. However, when the animals were allowed to survive for four weeks, many of the GFP-labeled cells now expressed markers of differentiated neurons. Over the next three months these neurons continue to mature. To what extent are the new GFP cells functionally integrated into the hippocampal circuitry? The hippocampus can be sliced into thin sections while still functionally active, and the electrophysiological activity of the neurons monitored with microelectrodes (Figure 3.32). The newly generated granule cells had electrophysiological properties similar to those found in mature granule neurons, and they receive inputs from the major afferent pathway. Thus, newly generated neurons in the adult hippocampus integrate into the existing circuitry and function like those neurons generated during embryogenesis.

Why do mammals generate new neurons in these regions? Frogs and fish have eyes that grow, birds learn a new song. What is the advantage to the mammal? Although there are no studies of a change in the animal's abilities that correlate with this cellular addition in mammals, several possibilities may be entertained. Since both the olfactory and the hippocampus are involved in the formation of olfactory memories, the neuronal turnover in these regions could be important in a seasonal change in nests or mates. Altman observed that the neurogenesis of the brain proceeds in two basic phases. The large projection neurons (or macroneurons) are generated early in embryonic

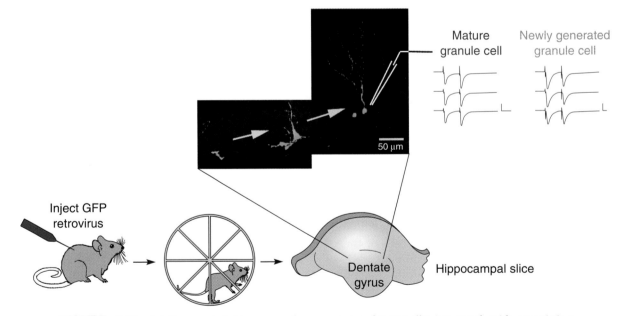

FIGURE 3.32 Adult-generated hippocampal neurons are functionally integrated with preexisting neurons. Van Praag et al. labeled proliferating cells in the hippocampus with a GFP-expressing retrovirus let the animals run on wheels to increase their production of new neurons, and then recorded from the GFP-labeled cells in hippocampal slices. The adult-generated neurons integrated into the hippocampal circuit and showed electrophysiological responses similar to their mature granule cell neighbors. (Modified from Reh, 2002)

development, while the smaller neurons (or microneurons) are generated later in development, through the postnatal period and even into adulthood. These later generated microneurons are then integrated into the framework provided by the macroneurons. Altman pictured this second stage of neurogenesis as a way for environmental influences to regulate the neurogenesis and produce a brain ideally suited to its environment. Although it has been difficult to prove that persistent neurogenesis in a particular region of the brain is necessary for behavioral plasticity in that brain region, recent studies are consistent with Altman's hypothesis. In one of the last reviews of his work, Altman (Altman and Das, 1965) summed up his hypothesis: "We postulate that this hierarchic construction process endows the brain with stability and rigidity as well as plasticity and flexibility." It is possible that if you are going to remember anything you have just read in this Chapter you have to make new neurons in your hippocampus!

SUMMARY

The enormous numbers of neurons and glia in the brain are generated by progenitor cells of the neural tube and brain vesicles. The progenitor cells from the early embryonic nervous system undergo many symmetric cell divisions to make more progenitor cells, while the progenitor cells in the late embryo are more likely to undergo an asymmetric division to generate neurons and glia. The production of both neurons and glia from the progenitor cells is under tight molecular control, and this allows the proper numbers of both neurons and glia to be produced for the proper functioning of the brain. Interactions between the neurons and the progenitor cells regulate their proliferation in both positive and inhibitory ways. Overall, a remarkable coordination takes place to regulate proliferation in the nervous system during development, and mutations in specific genetic pathways involved in neurogenesis can lead to childhood tumors and gliomas in adults. Once the developmental period of neurogenesis is complete, most areas of the brain do not generate new neurons, even after damage. This has led to the concept that you are born with all the neurons that you are going to ever have. However, in recent years, it has become clear that certain regions of the brain, the hippocampus and the olfactory bulb, continue to add new neurons throughout life. This continual addition of neurons in these regions may allow for greater plasticity in these specific brain circuits.

4

Determination and Differentiation

The nervous system is a coral reef of the body where evolution and development have collaborated to produce an extraordinary diversity of cell types. Neurons show enormous variety in cellular anatomy, physiological function, neurochemistry, and connectivity. For example, granule cells of the cerebellum are tiny, and have simple dendrites and bifurcated axons that release the excitatory transmitter glutamate, whereas cerebellar Purkinje cells are huge, have an impressively complex and electrically active dendritic tree, a single long spiking axon, and release the inhibitory neurotransmitter GABA. The differences between neurons can be much more subtle. All the motor neurons of the spinal cord share a common morphology, chemistry, physiology, and circuitry, yet they are distinctly specified molecularly so that they connect with particular presynaptic partners and postsynaptic muscles.

The fates of some neurons, particularly those of invertebrates, are the products of particular lineages. The fates of others, particularly those in vertebrates, appear to depend more on the local environment. Sydney Brenner suggested that neurons are either European or American. A neuron is European if its fate is largely the result of who its parents were. For American neurons, it is more about the neighborhood where they grew up. When one looks closely, however, it turns out that fate is not strictly controlled by either lineage or environment alone. Usually, it is the mixture of the two that is essential; the adoption of a particular fate is a multistep sequential process that involves both intrinsic and extrinsic influences. A progenitor cell may be externally influenced to take a step along a particular fate pathway, and so the unborn daughter of that cell has also, in a sense, taken the same step. A signal from the environment may act upon this daughter cell to refine its fate further, and the response of the daughter cell to the signal is to express an intrinsic factor consistent with its limited fates.

The environment in which neural progenitors divide and give rise to neurons is rich with diffusible molecules, cell surface proteins, and extracellular matrix factors. These extrinsic signals influence the genes that developing neurons express, which direct neuronal shape, axonal pathways, connectivity, and chemistry. The number of genes used to carry out this task of specification throughout the nervous system is impressive. It has been estimated that half of an organism's genes are expressed exclusively in the nervous system. Most of these are involved in various aspects of neuronal differentiation.

Some of the basic techniques that are used in approaching these questions are shown in Figure 4.1. Transplantation is a good technique for finding out whether a cell's fate has been intrinsically specified. For example, a progenitor from a donor animal is transplanted to a different part of a host animal. If the fate of the cell is unaltered by putting it in this new environment, then the cell is "autonomously determined" at the time of transplantation. If, however, the cell adopts a new fate, consistent with the position to which it was transplanted, then the fate at the time of transplantation is still flexible and can be "determined nonautonomously." Putting cells into tissue culture is another valuable technique. By isolating a cell from the embryo entirely, it is possible to assay the state of determination of a cell in the absence of all interactions. An advantage of this experimental system is that the culture medium and substrate can be controlled. In this way, potential extrinsic cues can be added and assayed for their effect on fate choice.

A very informative approach for studying the processes that lead neurons down particular differentiation pathways, at least in terms of identifying the factors

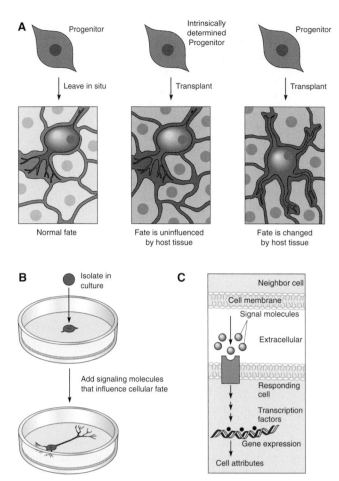

FIGURE 4.1   A. Testing fate by transplantation. On the left, a neural progenitor left in its normal environment turns into a particular type of neuron. In the middle, an intrinsically determined progenitor's fate is unchanged by transplantation to a different environment. On the right is an example of a progenitor whose fate is determined extrinsically and so is changed by transplantation to a different environment. B. An undifferentiated progenitor is placed into a culture dish, and signaling molecules are tested, which may influence the fate that the cell takes as it differentiates into a neuron. C. An extracellular signal originating from one cell can influence the fate of nearby cells by causing the responding cell to change its gene expression pattern.

that influence determination, is genetic manipulations such as mutational and transgenic analyses. Mutations in particular genes can alter the fate of certain types of neurons. With a genetic approach it is possible not only to show where and when the normal fate decisions are made but also to identify the gene product in question. Forward genetics uses random mutagenesis to define new genes that have effects on neural differentiation, while reverse genetics uses molecular engineering to knock out or overexpress particular genes (see Chapter 2) that are candidates for roles in neuronal fate determination or differentiation. Genetics combined with trans-

plantation or culture can reveal whether neuronal phenotypes are extrinsically regulated by the gene in question, as in the case of a gene that codes for a secreted factor, or intrinsically regulated, as in the case of a gene that codes for the receptor to such a factor.

This chapter examines the several facets of cell fate determination and differentiation, which have been investigated using such techniques. Each aspect is brought to light in a different system, and it is only through looking at several systems that one can begin to appreciate the full range of cellular and molecular mechanisms that lead a set of relatively simple looking progenitor cells to take on thousands of different neuronal fates.

TRANSCRIPTIONAL HIERARCHIES IN INVARIANT LINEAGES

As we discussed in Chapter 1, time-lapse studies of the development of the nervous system of the nematode *C. elegans* show that every neuron arises from an almost invariant lineage (Figure 4.2). In this system, the progenitors are uniquely identifiable by their position and characteristic patterns of division. Ablation of one of these progenitors usually leads to the loss of all the neurons in the adult animal that arise from that progenitor, indicating neighboring cells cannot fill in the missing fates. This is called mosaic development. To understand how these different precursors generate specific neurons, a genetic approach has been used, and mutants have been found that interfere with the development of particular neurons. These mutants are then used to dissect the mechanisms of neuronal fate. In this system, acquisition of neural identity is largely the result of a multistep, lineage-dependent, process of determination.

One of the best examples of such an analysis is that of the specialized mechanosensory cells in nematodes studied by Martin Chalfie and his colleagues (Chalfie and Sulston, 1981; Chalfie and Au, 1989; Chalfie, 1993; Ernstrom and Chalfie, 2002). Most nematodes wiggle forward when touched lightly on the rear and backward when touched on the front. By prodding mutagenized nematodes with an eyelash hair attached to the end of a stick, Chalfie and colleagues were able to find mutants that had lost the ability to respond to touch. Many touch insensitive worms have mutations in a group of genes involved in the specification of the mechanosensory cells. Mutations in the gene *unc-86* result in the failure of the mechanosensory neurons to form. *Unc-86* encodes a transcription factor that is expressed transiently in many neural precursors and particularly in the lineage

Zygote

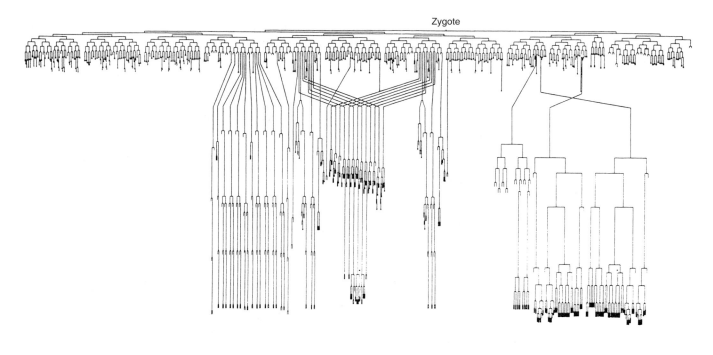

FIGURE 4.2   Complete lineage of *C. elegans* hermaphrodite. (Based on Sulston et al., 1983)

produced by a cell called Q. In wild-type animals, Q divides into two daughter cells, Ql.a and Ql.p (Figure 4.3A). Both of these cells divide once more, but only Ql.p produces a touch cell. The *unc-86* gene is turned on only in the Ql.p. and a mutation in *unc-86* (Figure 4.3B) results in the "transformation" of Ql.p into a cell that behaves like its mother, Q. We call this cell Q'. This transformed cell continues to divide, producing Q1.a' and Q1.p', but in the continued absence of *unc-86* function, the Q1.p' transforms into Q", which continues to behave like its mother Q' and its grandmother Q. Thus, mutations in *unc-86* affect the lineage of touch cells; mechanosensory neurons are never born in these mutants.

Another gene uncovered in Chalfie's screen of touch mutants is named *mec-3*. In these mutants (Figure 4.3C), the cells that would be touch sensitive are born, but they do not differentiate into mechanosensory neurons. Instead, they turn into interneurons. Thus, the *mec-3* mutation affects neural subtype determination. The *mec-3* gene codes for a transcription factor that is a member of the LIM-homeodomain family. Interestingly, the transcription of the *mec-3* transcription factor is directly regulated by the *unc-86* transcription factor. Cells fated to become touch cells all express *unc-86* at

first, and this transcription factor binds to the regulatory sequence of DNA that controls the transcription of the *mec-3* gene. Thus, *unc-86* mutants do not express *mec-3*. However, in normal animals, the protein UNC-86 leads to the expression of MEC-3, and when these two proteins are expressed in the same cell (the cell that will become a mechanosensory neuron), they physically interact to make a heterodimeric transcription factor with new specificity that activates genes that neither MEC-3 nor UNC-86 can activate on their own. Several of these are defined by mutations in other genes that cause touch insensitivity, such as the *mec-7*, *mec-12*, and *mec-17* genes. These three genes encode proteins that are used in the differentiation of the specialized touch cell cytoskeleton (Figure 4.3D). This system provides an excellent example of a simple hierarchical cascade of transcription factors, one regulating and interacting with the next, the end result of which is to turn on genes that the cell uses to realize its fate. Using a genomic approach, Chalfie and colleagues (Zhang et al., 2002) have tried to find even more genes involved in the touch cell pathway by looking for differences in profiles of all expressed genes in normal animals versus *mec-3* mutants. This approach identified up to 50 more genes in the pathway downstream

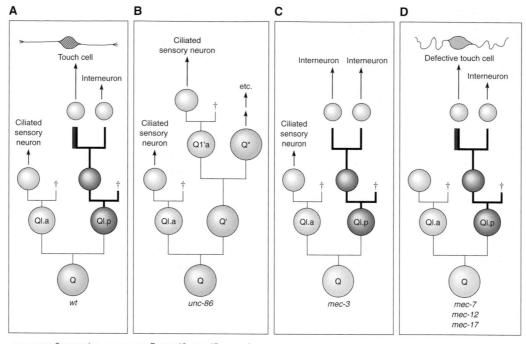

FIGURE 4.3 Intrinsic determinants at different steps on a neural lineage in the nematode. A. Normal lineage. B. If Q1.p cannot express *unc-86*, it becomes Q', a copy of its mother, Q. The result is a repeat of the previous division, which results only in Q progenitors and ciliated sensory neurons. C. *Mec-3* needs to be expressed in the touch cell. If it is not, as in *mec-3* mutants, this cell turns into an interneuron like its sister. D. In *mec-7*, *mec-12*, and *mec-17* mutants, the touch neuron is correctly specified but differentiates missing critical components of its morphology or function.

of *mec-3*, genes that are likely important for the function of the mechanosensory neurons.

Another set of *C. elegans* neurons that have been studied in detail are the hermaphrodite-specific egg-laying neurons (Desai et al., 1988; Desai and Horvitz, 1989). Mutants in genes involved in the determination or function of these neurons are unable to lay eggs. The result is that the self-fertilized eggs hatch inside the mother and begin to feed within their mother's uterus. The larvae proceed to devour their mother from the inside. Eventually, with only her epidermis intact, she becomes a bag of wriggling larval worms that in their hunger eventually eat through her cuticle into the world. Mutant lines in these genes thus seem to give their offspring a protected start in life characteristic of viviparous species but at what appears to be a mother's ultimate altruistic sacrifice. The "bag of worms" mutants have a phenotype that is easy to detect, and so a large collection of such mutants has been identified. Twenty or so genes have been found to define hierarchical transcriptional cascades affecting egg-laying neuron development. Surprisingly, only one of these is also nec-essary for the proper development of the touch cells. This is our friend *unc-86*. The role of *unc-86* in the determination of the egg-laying neurons, however, is quite different. Instead of controlling lineage as it does in mechanosensory cells, it regulates neurotransmitter expression and axon outgrowth in the egg-laying neurons.

The fact that there is surprisingly little overlap in the genes that are involved in these two systems suggests that the molecular cascades of neuronal determination must be complex and highly individualized. However, there are similarities that are worth emphasizing. In the case of both the egg-laying and the mechanosensory neurons, there is a hierarchical pathway, rich in transcription factors that operate through the specific lineages. These factors regulate other intrinsic transcription factors in a molecular cascade whereby the lineage, the specification, the differentiation, and finally the physiological properties of the neurons are established through a series of successive stages. This molecular strategy, we will see, is also used in the determination of neurons in most other species.

SPATIAL AND TEMPORAL COORDINATES OF DETERMINATION

The CNS of an insect develops from a set of individual neuroblasts that enlarge within the epithelium of the neurogenic region of the blastoderm and then delaminate to the inside, forming a neuroblast layer. All these neuroblasts, we learned in Chapter 1, express proneural transcription factors of the *Achaete-Scute* family that give them a common "neural" specification. However, each neuroblast is an individual and through successive divisions reproducibly gives rise to a unique set of neurons (Figure 4.4). How do these neuroblasts get their specific identities? They are arranged in reproducible columns and rows and so can be identified by their position. In Chapter 2, we discussed how the *Drosophila* embryo is finely subdivided in the anterior to posterior axis into stripes of expression of particular combinations of gap genes, pair-rule genes, *Hox* genes, and segment polarity genes. These genes provide neuroblasts with intrinsic positional information that reflects their location along the antero-posterior axis (Figure 4.5). *Hox* genes are expressed in the middle and posterior portions of the neural primordium, and the "head gap" genes are expressed more anteriorly in nested domains and provide positional information to the neuroblasts that give rise to particular brain regions or segments. Segment polarity genes control positional information within each individual segment (Bhat, 1999). These anterior-posterior

(AP) positional identity genes play important roles role in determining the identity of the neuroblasts as illustrated by the loss and/or duplications of particular sets of neurons. For example, *Wingless (wnt)* and *Hedgehog* proteins activate the expression of a gene called *huckebein* in some neuroblasts, and the transcription factors *Engrailed* and *Gooseberry* repress *huckebein* expression in other neuroblasts, thus establishing the precise pattern of huckebein protein in specific neuroblast lineages (McDonald and Doe, 1997).

Another set of genes divides the embryo and the nervous system along the dorsoventral axis. Three *homeobox* genes, *vnd*, *ind*, and *msh*, are expressed in longitudinal stripes within the neural ectoderm (Cornell and Ohlen, 2000). *vnd* is expressed in neuroblasts closest to the ventral midline, *msh* is expressed in the most dorsolateral stripe of the neurectoderm, and *ind* is expressed in an intermediate stripe between these two (Figure 4.5C). As is the case for the AP genes, mutations in these genes lead to loss of the neuroblast fates that normally express the mutated gene. The mechanism responsible for setting up these stripes involves responses of the promoters of these genes to threshold levels of the morphogen Dpp, which forms a gradient of expression from dorsal (high) to ventral (low). Once set up, the boundaries between the stripes of *vnd, ind,* and *msh* are sharpened by mutual repression.

A neuroblast in any position can thus be uniquely identified by expression of these spatial coordinate markers of latitude and longitude (Figure 4.5). These genes specifying position information along these two Cartesian axes collaborate to specify a positional identity for each central neuroblast in the developing organism. Once expressed in a neuroblast, the spatial coordinate genes are inherited by all the progeny of these cells, and act as intrinsic determinants of neuronal fate.

Each neuroblast divides asymmetrically to produce a copy of itself and a ganglion mother cell (GMC). The neuroblast divides several more times giving rise in ordered succession to a set of GMCs (see Chapter 1). Each GMC can thus be identified not only by the position of the neuroblast from which it arises but also from the order of its generation (e.g. whether it is the first, second, or third GMC to arise from a particular neuroblast). The first GMCs of a neuroblast lineage tend to lie deeper in the CNS and generate neurons with long axons, whereas the later arising GMCs stay closer to the edge of the CNS and generate neurons with short axons. In generating GMCs, neuroblasts go through a temporally conserved program of transcription factor expression (Figure 4.5 D and E). In the stages when the first GMCs are generated, most neuroblasts

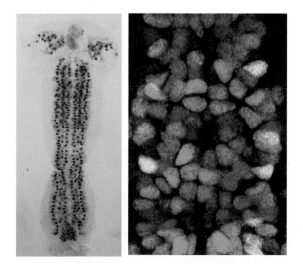

FIGURE 4.4 Neuroblasts of the *Drosophila* embryo. A. Shows the rows of neuroblasts labeled with an antibody to a late neuroblast-specific protein called Snail. B. Shows neuroblasts labeled with three different antibodies to the different neuroblast-specific proteins Hunchback, Eagle, and Castor. (Photos courtesy of Skeath and Doe)

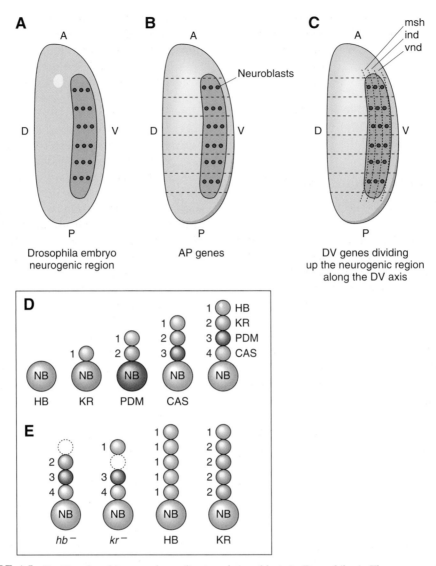

FIGURE 4.5 Positional and temporal coordinates of neuroblasts in *Drosophila*. A. The neurogenic region of a *Drosophila* embryo showing the rows and columns of neuroblasts. B. The embryo and the neurogenic region is divided segmentally into AP domains as described in Chapter 2. C. The neurogenic region is further divided into DV domains by the expression of transcription factors such as msh, ind, and vnd, creating a grid whereby each neuroblast has its own specific spatial coordinates. D. Neuroblasts transiently express temporally coordinated transcription factors *Hb*, then *Kr*, then *Pdm*, then *Cas*, but their progeny the GMCs maintain the transcription factor profile that was present at their birth. E. In *Hb* and *Kr* mutants or in animals that misexpress *Hb* or *Kr*, certain GMCs are missing or take on fates associated with the factors that the mother neuroblast expressed at the time when they were born.

express a transcription factor called hunchback (hb) and the GMCs generated at this time inherit this hb. Later, the same neuroblasts turn off *hb* and express a different transcription factor, Krueppel (Kr) instead. Now, all the GMCs generated at this stage inherit *Kr* expression, but not *hb* expression. If hb is eliminated from the neuroblasts when they are making GMCs, the neuroblasts generate early GMCs that cannot make early neuron fates. Similarly, if Kr is eliminated, then

later neural fates are missing. If instead, Hb is experimentally maintained in the neuroblasts at the stages when they would normally start expressing Kr, the neuroblasts keep making early GMCs (Figure 4.5) (Isshiki et al., 2001). The expression of the successive transcription factors is linked to the cell cycle, which functions as a kind of clock, since blocking the cell cycle blocks the succession and reactivating it reactivates the succession.

The expression of both spatial and temporal coordinate genes in neuroblasts is preserved in their progeny, the GMCs, and forms part of the increasingly rich inheritance of each developing neural progenitor. The ontogenetic roots of a neural progenitor can be read in the combination of transcription factors it expresses, and these factors in turn influence the cell's eventual phenotype.

ASYMMETRIC CELL DIVISIONS AND ASYMMETRIC FATE

Typically, the progeny of a cell division inherit the spatial and temporal coordinates expressed by the parental neuroblast at the time of birth. However, the parent cell often divides asymmetrically, giving intrinsic determinants to one daughter but not the other. As soon as they leave the neurectoderm behind, insect neuroblasts start dividing asymmetrically to produce two unequally sized daughter cells, a large second-order neuroblast remaining at the surface and the smaller GMC lying interiorly. How does a cell accomplish the partitioning of information selectively to one offspring and not the other? Two factors, Numb and Prospero (*Pro*), are expressed in most neuroblasts and are critical for asymmetric distribution of determinants of cell identity. At neuroblast division, these factors become localized to the smaller daughter, the GMC, where Prospero moves into the nucleus and positively influences GMC fate (Lu et al., 2000). Numb acts by inhibiting the Notch signaling pathway by binding to Notch and inactivating the transmission of a signal to the nucleus (Chapter 1). In the absence of Notch signaling, the GMC is free to move down the determination pathway.

Both Prospero and Numb are initially present throughout the entire neuroblast, so how do they get asymmetrically segregated? Figure 4.6 shows how this happens. Two proteins called Inscuteable (Insc) and Bazooka (Baz) form a complex, the Insc complex, that somehow recognizes and attaches to apical membrane of the neuroblast. The Insc complex orients the mitotic spindle along the apicobasal axis by anchoring the centrioles, which results in a vertical mitotic spindle. At the same time, the Insc complex, in conjunction with an actin-based cytoskeleton mechanism, drives the distribution of several key proteins along this vertical plane so they are inherited asymmetrically (Kaltschmidt and Brand, 2002). In particular, a cytoplasmic protein, Miranda (Mira) becomes enriched at the basal neuroblast pole such that when the cytokinesis separates the neuroblast's daughter cells, Mira is trapped in the GMC. It is Mira that binds the aforementioned determinants, Numb and Prospero, to the basal neuroblast pole and thus directs their localization to the GMC. The more apical cell does not differentiate into a GMC; rather it remains a neuroblast capable upon production of more Mira, Prospero, and Numb, to spit off another GMC at the next division. An example of the Numb protein segregating to a single daughter is shown in Figure 4.7.

An interesting example of a situation in which daughter cells adopt different fates due to asymmetric inheritance of Numb are the small sensory organs called sensilla, scattered over the body surface in *Drosophila*. The cells that compose each sensillum are usually clonal descendants of a single sensory organ precursor, SOP. The SOP cells are a bit like the neuroblasts that give rise to the CNS; they originally delaminate from the ectoderm during development in much the same way, dependent on proneural genes and Notch and Delta interactions, as described in Chapter

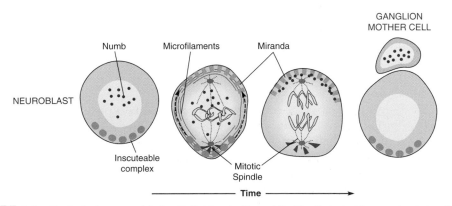

FIGURE 4.6 Control of asymmetrical cell division in *Drosophila*. The Inscuteable complex is localized to the apical pole of the neuroblast where it orients the mitotic spindle and causes the basal localization of asymmetrically localized determinants such as Miranda and Numb.

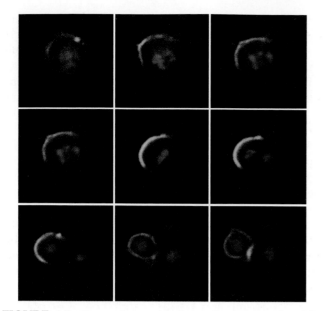

FIGURE 4.7 Frames from a time-lapse visualization of an SOP going through an asymmetrical division. The green label follows Numb, and the red label follows the chromosomes. In this sequence (courtesy of F. Schweisguth), it is easy to see Numb localized to one pole of the SOP and then inherited by a single daughter, spIIb.

1. Each specified SOP undergoes an invariant pattern of cell divisions. This division pattern has been investigated in detail for the external mechanosensilla (Guo et al., 1996; Schweisguth et al., 1996). The primary SOP (spI) for each macrochaete divides into an anterior daughter called spIIb and a posterior daughter called spIIa (Figure 4.8). SpIIa produces the outer two accessory cells: the socket cell and the shaft cell. The anterior daughter SpIIb divides into a neuron and a support cell, after first giving rise to a glial cell. These different fates arise through the reuse of the Notch signaling system during each of the cell divisions. When the SOP divides into spIIa and spIIb, these two cells interact with each other via Notch. The SpIIb fate is dominant, which is shown by the fact that if spIIb is ablated, spIIa will transform into spIIb. In Notch mutants, both cells become spIIb, and the result is no bristles or sockets, and when Notch is experimentally activated in both cells, they both turn into spIIa and there are no neurons or glia. Several intrinsic determinants, including the asymmetrically inherited Numb (see above), control the fate of spIIa versus spIIb (Figure 4.8). In this case, Numb is distributed to spIIb upon cell division. In the absence of Numb, the Notch pathway is active in spIIb, and it is transformed into spIIa; neither neurons nor support cells appear, but the sensilla form instead with double sockets and shafts. *Numb* mutants are thus insensitive to touch because the sensory bristles are uninnervated, hence the origin of the name.

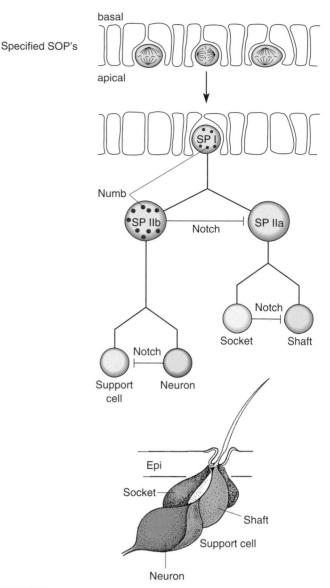

FIGURE 4.8 Lineage of a *Drosophila* external mechanosensory organ. From top to bottom: A Sensory organ precursor (SOP) enlarges and delaminates from the epithelium. It divides asymmetrically into spIIb, which inherits Numb, and spIIa, which does not. SpIIa divides again asymmetrically, as does spIIb slightly later. Notch signaling between daughters is involved in all these asymmetric divisions so that four daughter cells of the SOP have four different fates: support cell, sensory neuron, socket cell, and shaft cell.

GENERATING COMPLEXITY THROUGH CELLULAR INTERACTIONS

The compound eye of an insect is composed of a large number of identical unit eyes, called ommatidia, each with its own lens and array of cell types. This system provides a rather different example of how specific cells get their fates. Each of the 800 ommatidia in a *Drosophila* eye possesses 8 photoreceptors and 12 accessory cells. The 8 photoreceptors (R1–R8) are specialized sensory neurons (Figure. 4.9). Among the accessory cells, there are cone cells that form the lens of each ommatidium and pigment cells that surround the photoreceptors optically shielding one ommatidium from light that enters its neighbors. What is clear in this system is that lineage is not involved in the specification of the different cell types. Experimental results have shown that there is no clear clonal relationship among the cells of the ommatidia (Ready et al., 1976). In the developing *Drosophila* retina, cell–cell

interactions between the postmitotic photoreceptors and accessory cells are primarily responsible for specifying cell fate (Banerjee and Zipursky, 1990). Even after a cell has become postmitotic, it remains temporarily uncommitted to any particular differentiated fate. The mechanism controlling retinal cell fate diversification depends on the fact that the cells do not differentiate all at once, but follow a precise, reproducible temporal sequence of interactions. Thus, during late larval life, a wave of differentiation passes over the eye disc in a posterior to anterior direction (Figure 4.9). The wave front, or the position at which ommatidial differentiation begins, is morphologically visible as a morphogenetic furrow (MF), a narrow groove formed by apical constriction of the eye disc cells. As the furrow advances, cells in its wake aggregate into "rosettes" that foreshadow the regular ommatidial pattern. One cell is then singled out in each rosette. This becomes the R8 photoreceptor. The bHLH proneural gene *atonal* is turned on by the signaling protein Hh (a homolog of vertebrate *Shh*), which is expressed at the posterior tip of the eye disc (Kumar and Moses, 2000). Initially, *atonal* comes on in a continuous band of cells within the morphogenetic furrow. And this initiates a lateral inhibition mechanism that involves Notch signaling so that *atonal* expression becomes restricted to a mosaic of regularly spaced cells, which subsequently differentiate as R8. The set of well-spaced R8 cells continues to emit Hh, which signals across the MF toward the more anterior cells of the eye disc to induce the next set of R8 cells. This Hh-mediated feedback mechanism drives the morphogenetic furrow across the eye disc.

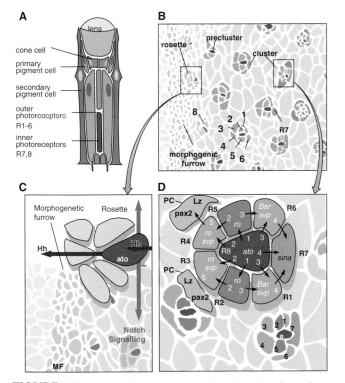

The first cells that join each R8 cell shortly after its determination become R2 and R5. It is believed that a signal emanating from R8 instructs the next cells to join the cluster, R2 and R5. These cells acting in combination with R8 help give the next cells that join the cluster R3 and R4 and then R1 and R6 their fates (Ready, 1989). The last photoreceptor to join the cluster is R7. Thus, ommatidial clusters incorporate cells in a manner analogous to how growing crystals incorporate molecules. That is, as new cells are added, they become neighbors of cells that are already incorporated and have particular fates. In this way, determination of a specific fate moves as a wave of crystallization across the eye disc, and so the developing *Drosophila* eye has been called a neurocrystalline array.

Each type of ommatidial cell expresses a unique set of intrinsic determinants. For example, *rough* (*ro*) is expressed in R2, R5, R3; R4, *Bar* appears in R1 and R6; *Seven-up* (*svp*) in R1, R6, R3, R4; *Prospero* (*pros*) in R7 and cone cells, *lozenge* (*lz*) in R1, R6, R7, and cone cells, and *Pax2* in cone cells only (Figure 4.9). Each of these

FIGURE 4.9 A. A schematic longitudinal section through an ommatidium depicting the different cell types. B. Diagram showing a surface view of part of the eye disc at a stage when photoreceptor clusters become assembled. C. A precluster in which the R8 precursor expresses *ato* in response to a previously generated hedgehog (Hh) signal. The R8 cell produces its own Hh and by this relay starts the next R8. R8s spread themselves out through the Notch lateral inhibition pathway. D. Cascade of photoreceptor determination in a developing ommatidial cluster. R8, the first cell to be determined expresses *atonal* (*ato*) and signals neighboring cells to become R2 and R5, which then express *rough* (*ro*). R2 and R5 in combination with R8 then signal the next set of neighboring cells to join the cluster of "fire" and become R3 and R4, which express *seven-up* (*svp*). On the other side, a cluster of seven cells is formed when R1 and R6 are induced to express *Bar*, *svp*, and *lozenge* (*lz*) by R2, R5, and R8. Finally R8, in combination with R1 and R6, induce the final photoreceptor R7 expressing *sevenless in absentia* (*sina*) and *prospero* (*pros*) to join. After the photoreceptors have joined, pigment cells expressing *lz* and *pax-2* are induced to join the cluster.

factors is linked to the normal differentiation of the respective cells in which it is expressed, as revealed by the fact that a particular cell type fails to develop in an eye disc that lacks the corresponding gene.

Shortly following its own determination, R8 puts out signals that activate two different signaling pathways, the Notch pathway and the Ras pathway (Freeman, 1997; Brennan and Moses, 2000). The Notch pathway, as we know, is activated by Delta. The Ras pathway is a highly conserved biochemical cascade of cytoplasmic kinases (Ras, Raf, MPK), which in this case are activated by the epidermal growth factor receptor (EGFR). R8 emits an EGF-like molecule, Spitz (Spi) that activates this receptor. Activation of these signaling cascades spreads concentrically from R8 to R2, R3, R4, and R5 and then the remainder of the ommatidial cells. The precise, temporally regulated activation of the EGFR and Notch signaling pathways assigns distinct phenotypes to the cells that join the ommatidial clusters.

The determination of the R7 cell deserves special mention in view of the pivotal role it has played in opening up the molecular-genetic study of signaling pathways. One of the first mutant screens in the field of cell determination took advantage of the fact that only R7 is sensitive to UV light. Thus, mutagenizing flies and screening for offspring that are blind to UV light yielded a fly line that lacked the R7 cell in every ommatidium and was therefore aptly called *sevenless* (*sev*) (Harris et al., 1976), (Figure 4.10). Lack of the receptor

causes the cell that would normally become R7 to develop as a cone cell instead. In several followup screens, many signaling molecules in the Ras pathway activated by the Sev receptor were identified (Rubin, 1991; Hafen et al., 1994) (Figure 4.11). Among them were the *Drosophila* homologs of Ras, Raf, MEK, MAPK, and Gap1, parts of the ras signaling pathway, as well as new genes in the signaling pathway such as *Son of sevenless* (*Sos*) and *Daughter of sevenless* (*Dos*). This pathway regulates the activity of transcription factors yan and pointed (pnt) that also were identified in such screens, and these factors control the expression of genes involved in the differentiation of R7. One of the most satisfying discoveries from the screens for the sevenless phenotype was the signal called Bride of sevenless (Boss), which binds to Sev. Boss is expressed specifically in R8 cells, so when any cell expressing Sev touches R8, the Ras signaling pathway fires in this cell and it becomes R7 (Reinke and Zipursky, 1988). These experiments provide an impressive demonstration of the power of genetic screens.

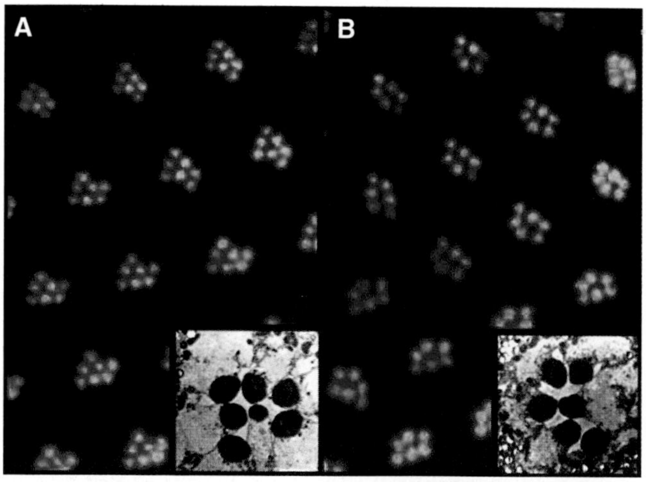

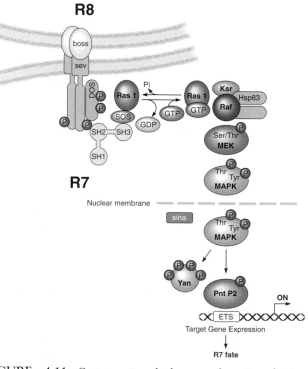

FIGURE 4.11 Components of the sevenless transduction pathway, include the seven transmembrane signaling molecule bride of sevenless (boss) which is expressed by R8, the sevenless receptor molecule (sev), and a number of downstream components of the signaling cascade such as daughter of sevenless and (Dos) and son of sevenless (Sos) and various members of the Ras Raf pathway ending in the activation of the yan and pointed genes. (Courtesy of E. Hafen)

FIGURE 4.10 Photoreceptors in the eye of normal flies (A) and sevenless mutants (B). The red images show light that is piped up through the clusters of receptors in each facet. The inserts are electron micrographs through a single facet and show cross sections of the photoreceptor array. Notice that the seventh central photoreceptor is missing in each facet of the mutant eye.

SPECIFICATION AND DIFFERENTIATION THROUGH CELLULAR INTERACTIONS AND INTERACTIONS WITH THE LOCAL ENVIRONMENT

The vertebrate neural crest is a transient stem cell population that arises along the lateral edges of the neural plate induced by the convergence of secreted signals (notably Wnts, BMPs, and FGFs), at the juxtaposition of neural plate, lateral epidermis, and subjacent paraxial mesoderm (see Chapters 1 and 2 for details of crest induction). As described in Chapter 3, these cells migrate from their site of origin at the dorsal-most part of the neural tube, along stereotypic pathways through the rest of the embryo. In this section, we discuss how the neural crest cells become specified toward different fates as they migrate through different environments. The neural crest progenitors continue to divide as they migrate until they coalesce at their destinations. Crest cells generate a variety of cell types. Not only does the crest produce the entire peripheral nervous system, including the autonomic and sensory ganglia, and the peripheral glia (Schwann cells), but it also produces endocrine chromaffin cells of the adrenal medulla, smooth muscle cells of the aorta, melanocytes, cranial cartilage and teeth, and a variety of other nonneural components. Because of its variety of descendants, crest has been a popular model for testing the mechanisms that generate cell diversity (Le Douarin, 1982).

To test whether premigratory crest cells are committed to a particular fate, Le Douarin and colleagues transplanted the crest between different anterior-posterior positions (Figure 4.12). These experiments took advantage of the chick–quail chimeric system described in Chapter 2. The results show that crest cells acquire instructions to differentiate during their migration, as well as when they arrive at their final destination. For example, crest cells from the trunk normally give rise to adrenergic cells of the sympathetic nervous system, whereas the more anterior crest cells from the vagal region give rise to cholinergic parasympathetic neurons that innervate the gut. When vagal crest cells from quail embryos were transplanted into the trunk region of chicken embryos, the transplanted vagal crest migrated along the trunk pathways and differentiated into adrenergic neurons in sympathetic ganglia. Similarly, trunk crest cells that were transplanted to the vagal region gave rise to cholinergic neurons of the gut. Similar experiments have been done to test the competence of crest cells to form a variety of different cell types, and the general conclu-

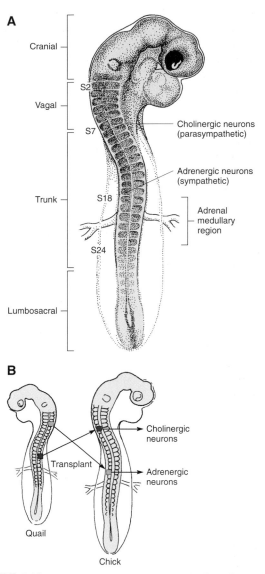

FIGURE 4.12 The environment influences the fate of neural crest cells. A. Crest cells from the trunk normally give rise to adrenergic cells of the sympathetic nervous system, whereas the more anterior crest cells give rise to cholinergic parasympathetic neurons that innervate the gut. B. When anterior crest cells from quail embryos are transplanted into the trunk region of chicken embryos, they differentiate into adrenergic neurons. Similarly, trunk crest cells that are transplanted anteriorly give rise to cholinergic neurons. (After Le Douarin et al., 1975)

sion is that crest cells display great flexibility in responding to local environmental cues. It could be that each region of the crest contains a complement of specified progenitors, only some of which survive in each location, but it seems more likely, given the evidence below, that commitment to a particular fate is a multistep process of determination.

Migrating crest cells become exposed to a sequence of instructive environments, each with a unique set of

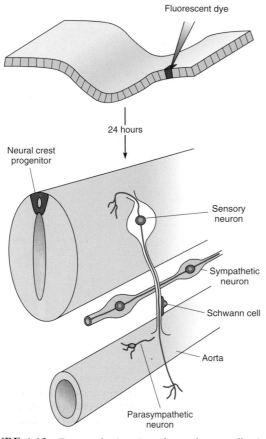

FIGURE 4.13 Fates and migration of neural crest cells. A single progenitor cell is injected with a lineage tracer, and its progeny are followed as they migrate out of the neural tube. Some may become sensory neurons, while others become Schwann cells or neurons of the autonomic nervous system. Environments these cells pass through on their migration routes influence their fate choice. (After Bronner-Fraser and Fraser, 1991)

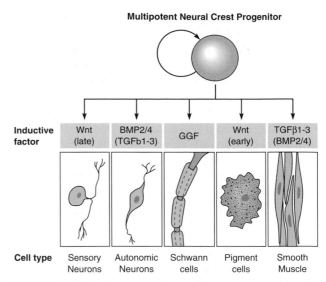

FIGURE 4.14 Different neural crest fates are promoted by a distinct set of extracellular signaling molecules. (After Dorsky et al., 2000)

factors, and the migrating cells respond to these factors and each other in a way that limits their potential. Initially, neural crest cells are multipotent, and labeling of single progenitors at the earliest stages of migration shows that these cells can give rise to a wide variety of derivatives (Bronner-Fraser and Fraser, 1988) (Figure 4.13). But as the cells migrate along particular routes, they segregate into several classes of more specialized progenitors. Thus, as development proceeds, they become more restricted. In the trunk region, an early decision separates postmigratory crest cells that will become the sensory progenitors, which remain in the somitic mesodermal region and express the proneural *bHLH* transcription factor Neurogenin (Nrgn2), from the autonomic progenitors, which do not (Lo et al., 2002). Transplantation studies with these two types of progenitors shows they can no longer make the full array of cell types. Nrgn2 is already expressed in premigratory crest cells and appears to bias but not deter-

mine them toward a sensory fate. It appears that the sensory cells of the dorsal root ganglia (DRG) inhibit other crest cells from assuming this fate. If the cells that normally make the DRGs are ablated, then later migrating crest cells will differentiate into sensory neurons (Zirlinger et al., 2002).

Experiments with purified populations of neural crest cells in culture suggest that different factors are involved in these restrictions (Figure 4.14) (Groves and Anderson, 1996; Anderson et al., 1997; Dorsky et al., 2000). It turns out that where cells encounter such factors on their migration route is very important in shaping appropriate destinies. BMPs are expressed in the dorsal aorta, where sympathetic ganglia form. BMPs turns on a program of neurogenesis in symphatho-adrenal (SA) cells (i.e. cells that are either part of the sympathetic nervous system or cells that are part of the adrenal gland) by inducing the expression of the paired domain transcription factor *Phox2b* which is required for the development of all of the autonomic nervous system and the proneural *bHLH* gene *MASH1*, which is required for the expression of neuronal markers in autonomic neurons (Pattyn et al., 1999; Schneider et al., 1999). However, further cues are needed if these cells are to become mature neurons. SA progenitors plated on a laminin-containing substrate in the absence of any growth factor form short, neuron-like processes. These processes become more extensive when the growth factor FGF is added to the medium, whereas the neurotrophic factor NGF, which is needed for the survival of sympathetic neurons (see Chapter 7) has no effect at this stage. The SA progenitors are initially unresponsive to nerve growth factor (NGF) because they do not express the NGF receptor. One of the effects of FGF is to induce the NGF receptor gene, thereby making the SA cells responsive to NGF, which stimulates their differentiation and survival as neurons (Anderson, 1993).

Sympatho-adrenal (SA) progenitors can also be isolated from the adrenal gland primordium of embryonic mammals and raised in culture; they give rise to two very different types of cells, the adrenergic sympathetic neurons and the endocrine chromaffin cells. Prior to differentiation, all SA progenitors express markers for both cell types. When SA progenitors are exposed to glucocorticoid hormone in vitro, which normally is produced in the adrenal gland, they develop as chromaffin cells. Glucocorticoids are steroid hormones that act on cytoplasmic receptors. After binding to the hormone (ligand) the receptor–ligand complex is transported to the nucleus where it acts as a transcription factor, binding to DNA and activating or repressing certain genes. In the case of the SA progenitor, glucocorticoids suppress the transcription of neuron-specific genes and activate the transcription of chromaffin cell-specific genes.

All sympathetic neurons start life producing the neurotransmitter noradrenalin. They receive the signal to be adrenergic. The Phox2b and MASH1 transcription factors induced by BMPs secreted by the aorta (Reissmann et al., 1996; Pattyn et al., 1999) (Figure 4.15) appear to be responsible for controlling the expression of tyrosine hydroxylase, a key member of the synthetic pathway for this transmitter. Many of these neurons send out axons to smooth muscle targets; these sympathetic neurons remain adrenergic throughout life. However, a few sympathetic neurons, for example, those that innervate sweat glands, switch their neurotransmitter phenotype late in development and become cholinergic; that is, they secrete the neurotransmitter acetylcholine (ACh). Neurotransmitter choice in these cells is a late aspect of cell fate that is regulated by the target (Francis and Landis, 1999). The fibers innervating sweat glands begin to turn off tyrosine hydroxylase and other adrenergic enzymes and begin to make choline acetyltransferase, the synthetic enzyme for ACh production (Figure 4.16). Evidence for the role of the sweat glands themselves in inducing the switch in phenotypes comes from transplantation experiments. Transplanting foot pad tissue, rich in sweat glands, to areas of the body that usually receive adrenergic sympathetic innervation leads to the induction of cholinergic function in the sympathetic axons that innervate the transplanted glands. Factors such as interleukin-6 are capable of causing an adrenergic-to-cholinergic switch in phenotype and have been purified from

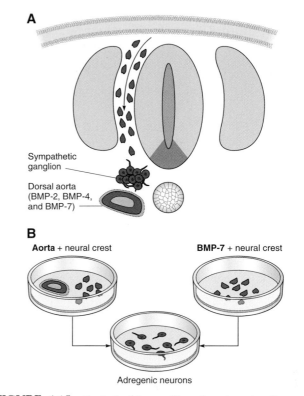

FIGURE 4.15 Control of transmitter phenotype by the aorta. A. Neural crest cells that migrate close to the aorta often become sympathetic ganglia with adrenergic neurons. The dorsal aorta is a source of BMPs. B. When neural crest cells are cultured with aorta or BMP-7, they turn on Phox2b, which activates the transcription of tyrosine hydroxylase and the cells become adrenergic neurons.

culture media, but the actual factor that operates in sweat glands to produce this effect in vivo has not yet been definitively identified. Nevertheless, these experiments make it clear that targets can retrogradely determine that transmitter type of the innervating neurons.

In the vertebrate peripheral nervous system, all glia, whether they are Schwann cells or glial support cells in the sensory and autonomic ganglia, arise from the neural crest and express *Sox10*. But the decision to express *Sox10* and commit to a glial fate happens late in the crest decision hierarchy, after the decision to be sensory or autonomic. A secreted protein called Neuregulin-1 (Nrg-1) induces crest cells to adopt glial fates (Britsch et al., 2001; Leimeroth et al., 2002) (Figure 4.17). When migrating crest cells are cultured in the absence of added Nrg-1, the majority of clones contain both neurons and glial cells, but if Nrg-1 is applied, most clones develop as pure glia. Neural crest cells express Nrg-1 only after they have migrated peripherally and coalesced into distinct masses as in the dorsal root or sympathetic ganglia. In fact, Nrg-1 is expressed only in those cells that have already started to exhibit a neuronal phenotype. The Nrg-1 receptor is expressed

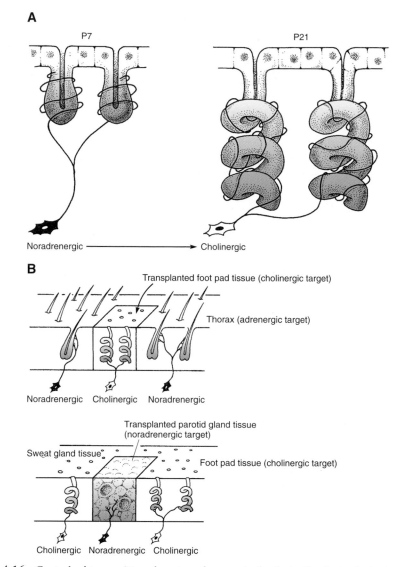

FIGURE 4.16 Control of transmitter phenotype by sweat glands in the footpad. A. A noradrenergic neuron begins to innervate developing sweat glands. As it does so, it switches and becomes cholinergic. B. Neurons that innervate hair follicles are noradrenergic, yet when a piece of footpad containing sweat glands is transplanted into hairy skin, the local neurons that innervate the transplanted sweat glands become cholinergic. Conversely, when a piece of parotid gland tissue, which is normally innervated by adrenergic neurons, is transplanted to the footpad, the local neurons that innervate it are noradrenergic. (After Landis, 1992)

by all migrating neural crest cells, so cells are sensitive to Nrg-1 as soon as they arrive at their destination but only the late cells to migrate in, those that do so after many neurons are already differentiating, see substantial amounts of Nrg-1. This is one reason why glial cells differentiate later than neurons. Another, as in the retina, has to do which Notch. Notch activation drives glial determination in crest-derived cells. As neurons begin to differentiate, they express Delta, which through Notch activation turns off proneural gene activity in neighboring cells, and thus

forces them down a non-neural pathway (Morrison, 2001).

COMPETENCE AND HISTOGENESIS

The cells of the cerebral cortex are generated near the ventricular surface of the closed neural tube. As described in Chapter 3, the cell divisions of cortical neuroblasts give rise to postmitotic neurons that

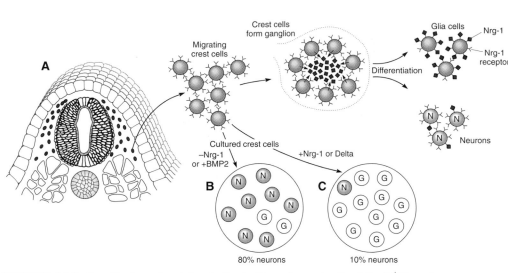

FIGURE 4.17 Specification of peripheral glia. A. Normally crest cells of the DRG produce neurons and gia (see text) and if placed in culture, crest cells give rise to both neurons (N) and glia (G). B. Removal of Nrg-1 or addition of BMP2 increase the number of neurons. C. Addition of the secreted signal Neuregulin to the medium or expression of Delta enhances the proportions of glial cells developing from the culture.

migrate along radial glia to settle in one of six different layers (Figure 4.18). In Chapter 3 we discussed the mechanisms that control the migration of these cells to their different layers; but what makes the cells of each layer distinct? The cells in the different layers of the cortex have layer-specific projection patterns. Thus, in the visual cortex, Layers 2 and 3 neurons are pyramidal cells that project to other cortical areas; Layer 4 cells are small stellate local interneurons that receive input from the thalamus; and Layers 5 and 6 cells are the largest pyramidal neurons and project to other parts of the brain, and even the spinal cord. How do the cells of a particular layer acquire their specific identities, such that they project to the appropriate targets?

Studies by McConnell and her colleagues have asked whether the progenitors of the cells destined for the deeper layers are somehow intrinsically different from progenitors of the cells destined for the upper layers, using cell transplantation in ferrets. In the ferret, Layer 6 cells are born in utero at embryonic day 29 (E29). Weeks later, cells born at P1 just after birth, are fated for Layers 2 and 3, and must migrate through Layers 6, 5, and 4 which have already formed. To test the idea that cells acquire laminar fate soon after they are born and before they migrate, cells generated in the ventricular zone of E29 ferrets were transplanted into older P1 hosts (Figure 4.19). Although their time of birth would have fated them for a deep layer, the experiments showed that many of the transplanted cells switched their fates and ended up in Layers 2 and

3, suggesting that these young cortical neurons are flexible with regard to fate (McConnell, 1988). Further studies showed that cell interactions are involved in this commitment since if the E29 cells are removed and cultured with other E29 cells for a number of hours, their deep layer fate is fixed even when challenged by transplantation to an older environment (McConnell, 1995). What about the reciprocal experiment, the transplantation of P1 precursors into E29 hosts? When P1 precursor cells were transplanted into younger brains, these cells all differentiated into Layer 2 and 3 cells, even though the host neurons around them were differentiating as Layer 6 neurons. Thus, these P1 cells seemed to have lost the competence to differentiate as deep layer cells (Figure 4.19).

This loss of competence appears to be a sequential process, as illustrated by the transplantation of progenitors in the middle stages of cortical development. When the progenitors of Layer 4 neurons born at E36 are transplanted into older P1 brains, in which Layers 2 and 3 were being generated, they also generate Layer 2 and 3 neurons. However, when E36 progenitors are transplanted into a younger E29 hosts, they show a restricted potential ending up in Layers 4 and sometimes 5, but not Layer 6 (Desai and McConnell, 2000) (Figure 4.19). These results suggest that environmental cues can influence precursors to produce neurons of different cell types, but that the competence of these precursors becomes increasingly restricted over time. Thus, they can respond to an older environment but not a younger one.

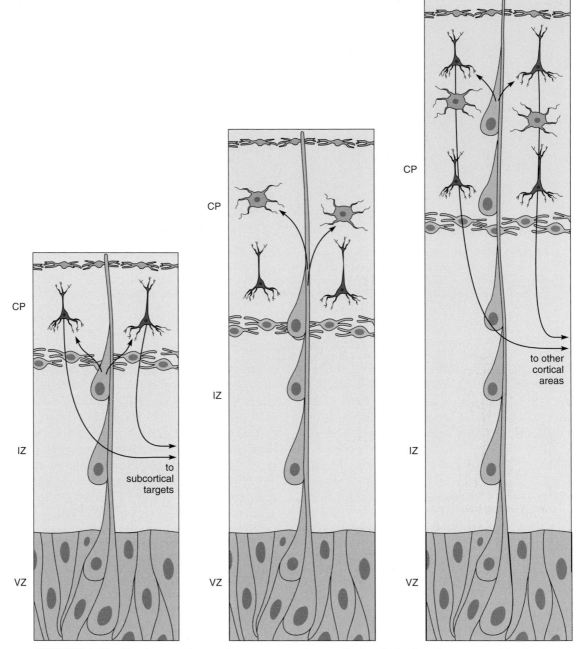

FIGURE 4.18 This figure picks up where Figure 3.15 leaves off, after the birth and migration of the Cajal-Retzius cells and the Subplate cells. A. The next neurons to be generated in the cortex are the pyramidal neurons of the deep layers, V and VI, whose axons project to subcortical targets. B. The next neurons to be born are the local interneurons in Layer IV of the cortex. C. Finally, the pyramidal cells of the upper layers, II and III, are generated. They send axons to other cortical areas.

THE INTERPLAY OF INTRINSIC AND EXTRINSIC INFLUENCES IN HISTOGENESIS

The vertebrate retina, like the mammalian cortex, is organized into layers, and there is a clear histogenetic pattern of cellular birth. To understand cell determination in this system, it is necessary to know the basic cell types. In addition to photoreceptors, which include rods and cones, the vertebrate retina contains interneurons, including horizontal cells, bipolar cells, amacrine cells, and projection neurons called retinal ganglion cells (RGCs). The RGCs are the output

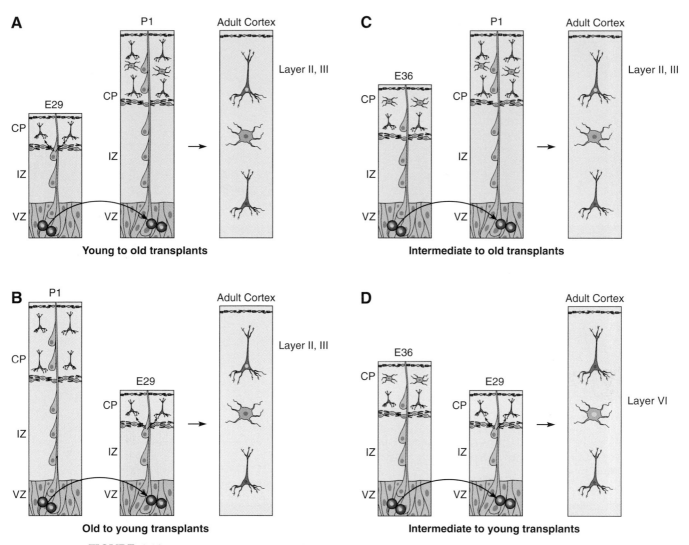

FIGURE 4.19 Progressive restrictions in fate determination in the cerebral cortex. A. Transplantation of cells from the VZ of an E29 ferret to a P1 ferret leads these cells to change from a deep layer (early) to a superficial layer (late) fate. B. But when P1 cells are transplanted to E29 hosts, they retain their superficial layer fates. C. Intermediate E36 generated cells normally destined for Layer IV can assume later fates when transplanted into an older host. D. But they cannot assume younger fates when transplanted into younger hosts.

neurons of the retina, and their axons form the optic nerve that relays the visual image to the brain. Thus, there are just six basic types of neuron in the retina and one type of intrinsic glial cell, called the Müller cell. Vertebrate retinal cells develop from a population of neuroepithelial progenitors, which produce this diversity of neurons and glia. Injection and infection of single retinoblasts with heritable markers produces clones of mixed, and seemingly random, cellular composition (Turner and Cepko, 1987; Holt et al., 1988; Wetts and Fraser, 1988), indicating that, as in the *Drosophila* retina, lineage is not the dominant mechanism of fate determination in these cells (Figure 4.20).

As in the cortex, it seems that the competence of retinoblasts becomes more restricted as development proceeds (Adler and Hatlee, 1989; Reh and Kljavin, 1989; Livesey and Cepko, 2001). During normal development, RGCs are born and differentiate first, next are amacrines, horizontal cells, and cones, and the later born neurons are rods and bipolar cells (Sidman, 1961; Cepko et al., 1996). The final cell type to be born is the Müller glial cell (Figure 4.21). This pattern is largely, though not entirely, preserved among vertebrates. To test the idea of changing competence, progenitors at various stages of retinal development are forced to differentiate in culture. If progenitor cells are isolated at a time when RGCs are normally born, they tend to turn

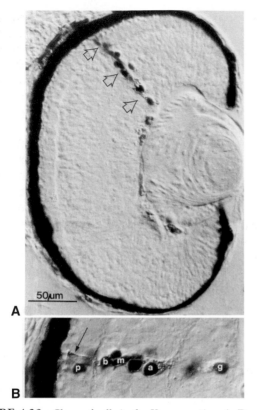

FIGURE 4.20 Clone of cells in the *Xenopus* retina. A. Daughters of a single retinal progenitor injected with horseradish peroxidase form a column that spans the retinal layers and contributes many distinct cell types. B. p, photoreceptor; b, bipolar cell; m, Müller cell; a, amacrine cell; g, ganglion cell.

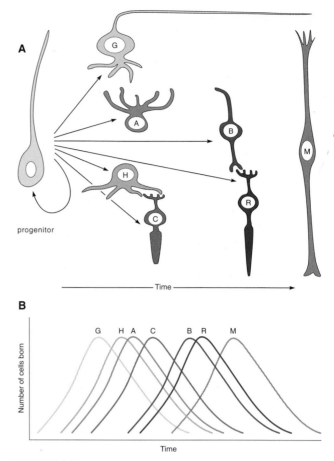

FIGURE 4.21 Determination in a vertebrate retina. A. A progenitor cell in the neuroepithelium divides several times and gives rise to clones of cells that contain all the major cell types of the retina, including Ganglion cells G, Amacrine cells A, Horizontal cells H, Bipolar cells B, Rods R, Cones C, and Müller cells M. These cells tend to be born at different developmental times, indicating a rough histogenetic order that is shown in (B).

into RGCs in culture. Progenitors isolated at a later stage tend to become rods (Watanabe and Raff, 1990) (Figure 4.22). And while it is not impossible, it is certainly more difficult for older progenitors to assume early fates than it is for early progenitors to assume late fates. This is shown by heterochronic experiments. If retinal cells that are born at the stage when RGCs are normally being born are labeled and mixed together into an aggregate and cultured in vitro, the labeled cells still differentiate into RGCs and few become rods. If, however, they are mixed with an excess of retinal cells that are several days older, the same cells have a higher probability of becoming rods (Watanabe and Raff, 1990). The other direction is much harder; that is, few late progenitors choose RGC fates, even when mixed with an excess of cells from earlier retinas (James et al., 2003).

The bHLH and homeobox transcription factors look as though they are intrinsic factors that help specify retinal cell fate. Among the homeobox factors, Chx10 is involved in bipolar fate, Prox1 in horizontal cell fate, and BarH1 in RGC fate. Among the *bHLH* genes, *Ath5*

is absolutely critical RGC fate. The example of *Ath5* is particularly interesting. *Ath5* is turned on transiently in retinal precursor development. In animals where *Ath5* is never expressed, the RGCs simply do not arise. Yet other retinal cells are formed in *Ath5* knockouts. This suggests that when this factor is downregulated, retinal precursors are no longer competent to make RGCs. Moreover, when retinal progenitors are forced out of the cell cycle by the expression of cell-cycle inhibitors, at a time when they express *Xath5*, they have an increased tendency to make RGCs, suggesting that the retinal progenitors are like the neuroblasts of the *Drosophila* CNS, which go through a succession of transcription factors. The external factors that influence competence may then turn out to be factors that influence cells to exit the cell cycle and differentiate rather than factors that influence the expression of specific *homeobox* or *bHLH* genes directly.

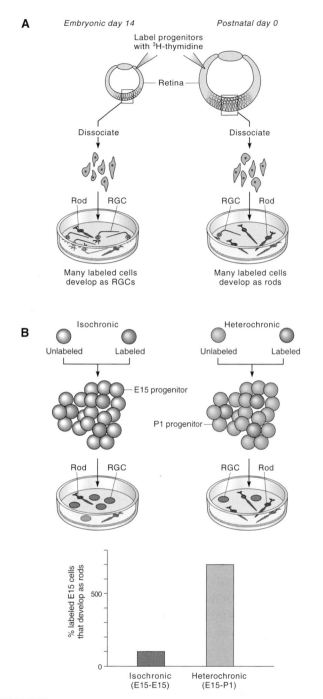

FIGURE 4.22 A. Birthdate and cell fate. Cells born on E14, even if dissociated into tissue culture, tend to differentiate into Retinal Ganglion Cells (RGCs), while cells born on P0, when dissociated, tend to differentiate into Rods. B. Early cell fate in the vertebrate retina is flexible and influenced by extrinsic factors. E15 cells labeled with thymadine and mixed with other E15 cells (isochronic) will not tend to differentiate as rods, while the same cells, if mixed with P0 cells (heterochronic) will. (After Watanabe and Raff, 1990)

Although the intrinsic transcription factors like Ath5 are critical for directing progenitors to specific fates, as noted above, cellular interactions can control the process to some extent. As we have already seen in examples from *Drosophila*, Notch signaling is an important mediator of cellular interactions that allow cells to choose different fates. Notch signaling is also very important in helping specify fate in the developing vertebrate retina. When Notch is overactive over an extended period, cell differentiation is delayed, and as a result, most retinal cells take late fates such as Müller glia (see above). Alternatively, if Notch signaling is blocked through the misexpression of dominant negative mutant genes, most retinal cells take on early fates such as RGCs. The idea proposed for Notch function in the vertebrate retina is that it allows only a certain number of cells to differentiate at any one time. This is critical when cells are making decisions based on the cues they receive in a rapidly changing environment. Thus, fates appear that are appropriate for the time at which they differentiate (Dorsky et al., 1997; Henrique et al., 1997). If all the cells were permitted to differentiate at the same instant, they might do so in more or less the same environment, and they might all choose the same fate!

Findings such as the above provide strong insights into the relationship between histogenesis and cell fate, as cells that pull out of the cell cycle at times when they express specific transcription factors may tend to produce fates based on these factors. A variety of external growth factors influence cell-cycle progression by affecting the expression of cell-cycle components (see Chapter 3). In the retina, as in many other parts of the CNS, glia are the last cells to pull out of the cell cycle. In the *Xenopus* retina, this is by virtue of an accumulation of a cell-cycle inhibitor called p27Xic1. If a retinal precursor expresses a proneural gene, like the atonal homolog Ath5, it may leave the cell cycle and differentiate as a neuron, but Notch activation in the precursor antagonizes the expression of the proneural genes like *Ath5* and allows precursor to build up p27Xic1. Interestingly, p27Xic1 is a bifunctional molecule. It has a cyclin kinase inhibitor domain to take the cells out of the cell cycle, and it has a separate Müller glial determination domain. The p27Xic1 forces the last dividing retinoblasts, those that have not been determined to become neurons, both to leave the cell cycle and to become Müller cells (Ohnuma et al., 1999). Thus, the neurogenic signaling pathway and the factors that regulate the cycle in the vertebrate retina are basic regulatory mechanisms that can be used to generate neuronal diversity by affecting the timing of differentiation in the changing external environment, linking histogenesis with determination.

INTERPRETING GRADIENTS AND THE SPATIAL ORGANIZATION OF CELL TYPES

The vertebrate spinal cord is composed of a variety of cell types, including motor neurons, local interneurons, and projection neurons. The embryonic spinal cord even contains a set of sensory neurons, called Rohon–Beard cells. These cells die by adult stages, and sensory input to the spinal cord is supplied by dorsal root ganglion neurons. The spinal cord primordium begins as a rectangular sheet of neural plate epithelium centered above the notochord. Lineage tracing experiments show that cells in the lateral edges of plate tend to give rise to Rohon–Beard cells and dorsal interneurons, while cells in the medial plate tend to give rise to motor neurons and ventral interneurons (Hartenstein, 1989) (Figure 4.23). Occasional clones are composed of different cell types, so it is thought that local position, rather than lineage, is involved in the generation of neuronal diversity in the spinal cord.

As we saw in Chapter 2, tissues outside the nervous system often provide critical signals that influence development within the CNS. In that chapter, we also reviewed the evidence that the notochord played a key role in establishing the dorsal-ventral axis of the neural tube, by providing a source of Shh. In this section, we will see how the same signal from the notochord is critical for spinal neuron differentiation acting as an organizing center for the induction of cell fate (Jessell et al., 1989). As we saw in the previous chapter, Shh secreted by the notochord induces the neural plate cells that are directly above to become the floorplate of the spinal cord. Once the floorplate has been induced in the ventral spinal cord, Shh signaling is relayed by the floorplate into the ventral neural tube. The ventro-lateral region of the tube that gives rise to motoneurons sees a fairly high dose of Shh, while further dorsally, where interneurons develop, the dose of Shh is lower. In response to this single gradient of Shh, several different neuronal types are generated. The most ventral neurons require the highest doses of Shh, and successively more dorsal ones require correspondingly less. When Shh is missing as in a knockout or is antagonized with an antibody, there is no floorplate, nor any ventral neuronal type in the spinal cord. When intermediate regions of the cord are exposed to higher levels of Shh, cells there take on more ventral fates, such as motor neurons.

How do cells at different dorso-ventral levels interpret their exposure to different levels of Shh to acquire different fates? Jessell and colleagues (Jessell, 2000) have shown that particular threshold levels of Shh turn on some *homeodomain* genes (Class II) and turn off

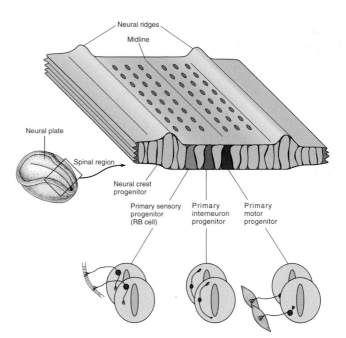

FIGURE 4.23 Rows of primary neurons in the neural plate. The most lateral row gives rise to sensory neurons (Rohon Beard Cells), the middle row gives rise to interneurons, and the most medial row gives rise to primary motor neurons. (After Hartenstein, 1989)

others (Class I). Thus, the read-out of the Shh level is first registered in neural tube neurons by expression of specific homeodomain proteins, which are either turned on or off at particular Shh thresholds (Figure 4.24). In this way, the ventral boundaries of Class I expression and the dorsal boundaries of Class II expression set up unique domains. The boundaries between these domains are sharpened through cross-repression of the two classes of genes. For example if the ventral border of the Class I *Pax-6* gene overlaps the dorsal boarder of the Class II *Nkx2.2* gene, cross-repression sets in, so that only one of these genes is expressed in any particular domain. By this process, specific domains uniquely express particular combinations of Class I and Class II homeodomain transcription factors. This results in spatial coordinates along the DV axis that are very similar to those used in setting up the mediolateral coordinate genes in *Drosophila*. Indeed, the *Drosophila* mediolateral coordinate genes share homologies with these vertebrate dorso-ventral neural tube genes, suggesting a conserved coordinate system.

Motor neurons arise from the domain that uniquely expresses Nkx6.1 but not Irx3 and Nkx2.2. Nkx6.1, unhindered by the repressive activities of these other factors, turns on OLIG2, a bHLH transcription factor that is required for motor neuron differentiation. OLIG2 in turn activates the expression of motor neuron specific transcription factors such as Mnr2, Hb9, Lim3, and Isl1/2. Once expressed, Mnr2 can regulate its own

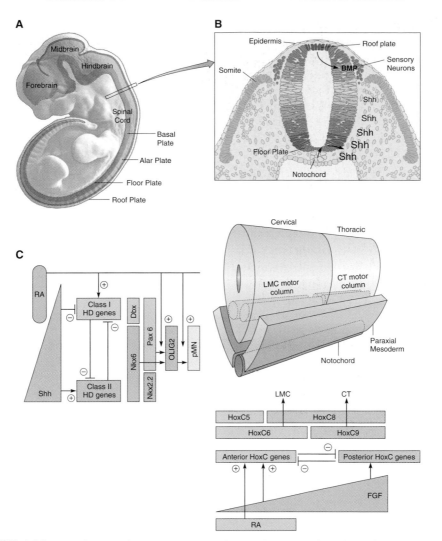

FIGURE 4.24 Specification of motor neurons in the vertebrate spinal cord. A. The neural tube, shown here for a mouse, is subdivided into four longitudinal domains: the floorplate, basal plate, alar plate, and roof plate. Motor neurons are derived from the basal plate. B. Schematic cross section of the neural tube. The notochord, which is located underneath the floorplate, releases *Sonic hedgehog (Shh)* which induces the floorplate to release its own *Shh*. This forms a gradient in the neural tube with high concentrations ventrally and low concentrations dorsally. BMP molecules released from the dorsal epidermis and roof plate form an opposing gradient. C. (Left) A gradient of *Shh* emanates from the notochord and floorplate; threshold levels of *Shh* turn on Class II HD genes. Retinoic acid (RA) expressed by the paraxial mesoderm induces the expression of Class I HD genes that are turned off more ventrally by threshold levels of *Shh*. Class I and Class II HD transcription factors cross-repress each other, creating sharp definitive boundaries at different dorso-ventral levels in the cord. Thus the boundary between Dbx and Nkx6 is more dorsal than the boundary between Pax6 and Nkx2.2. Between these boundaries, the OLIG2 *bHLH* transcription factor necessary for motor neuron specification is turned on by the concerted action of RA, Nkx6 and Pax6. OLIG2 and RA are necessary for the expression of motor neuron differentiation genes of the pMN family. (Below) A gradient of *FGF8* emanates from the mesoderm. High levels of *FGF8* turn on more caudal *HoxC* genes, whereas RA and low levels of *FGF8* turn on rostral *HoxC* genes. Rostral and caudal *HoxC* transcription factors cross-repress each other, creating sharp definitive boundaries at different rostrocaudal levels in the cord. The boundary between *Hoxc6* and *Hoxc9* establishes the boundary between the LMC of the cervical cord and the CT of the thoracic cord.

expression and is sufficient to drive spinal progenitor cells down a motor neuron pathway. This is shown by experiments in which motor neurons arise dorsally when Mnr2 is expressed ectopically in dorsal progenitors.

The motor neurons in the spinal cord are organized into functional columns that project to different muscle groups in the mature animal along the anterior to posterior axis of the body. Thus, in the cervical region is the Lateral Motor Column (LMC) that innervates forelimb muscles and in the mid-thoracic region is the Column of Terni (CT) that innervates the sympathetic chain. The anterior to posterior patterning of motor neurons into motor columns is accomplished in response to a gradient of FGF8 (high FGF8 posteriorly to low FGF8 anteriorly) secreted by the paraxial mesoderm. This gradient establishes domains of different *Hox* genes (Dasen et al., 2003). The anterior *Hox* genes inhibit the expression of the posterior *Hox* genes, and vice versa, so that sharp borders are established. The boundaries between these domains establish the boundaries of the different motor columns. We can appreciate that this logic is strikingly similar to that governing the positioning of the motor neurons in the ventral region of the spinal cord. Thus, exposure to gradients in both axes leads to the differential expression of *homeobox* genes that through cross-repression establish sharp borders and different motor columns in the ventral spinal cord (Figure 4.24).

These motor columns can be further divided into pools that innervate specific muscles. In zebrafish, each spinal segment has just three primary motor neurons: RoP, MiP, and CaP (for rostral, middle, and caudal primary, respectively; Figure 4.25). CaP innervates ventral muscle, RoP innervates lateral muscle, and MiP innervates dorsal muscle. If these motor neurons are transplanted to different positions a few hours before they begin axonogenesis, they seem to switch fate: for example, CaP transplanted into the RoP position can innervate lateral instead of ventral muscle (Eisen, 1991). These results suggest that the position of the cell soma specifies the axonal projection of the different primary neurons.

Motor pools that innervate individual muscles are distinguished by the expression of distinct members of

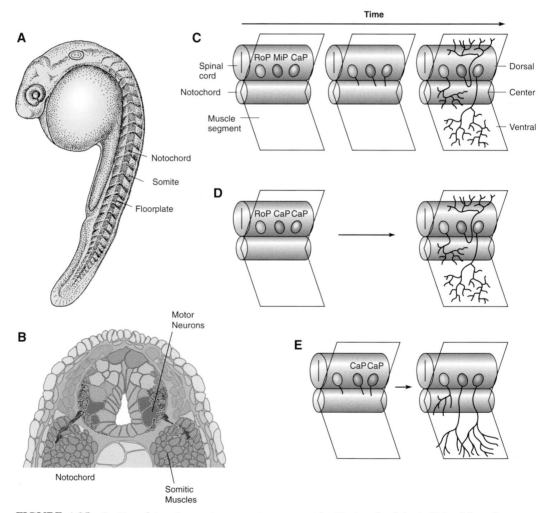

FIGURE 4.25 Position determines primary motor neuron identity in zebrafish. A. Zebrafish embryo at about 1 day old. B. Schematic cross section showing the location of the primary motor neuron and the somites that give rise to the axial musculature. C. The rostral (RoP), middle (MiP), and caudal (CaP) primary motor neurons of a single segment develop over a course of about 24 hours. D. If CaP is transplanted to the MiP position before axonogenesis begins, it develops a MiP axonal projection. E. However, if the transplant is done several hours later after the axons have begun to grow, the axonal fates are fixed.

the ETS family of transcription factors. For example, the *ETS* gene *ER81* is expressed in the motor neurons that innervate the limb adductor muscle in chicks, while the iliotrochanter motor neurons express the *PEA3 ETS* gene. Recent work suggests that these *ETS* genes regulate the expression of cell adhesion and axon guidance factors that help motor neurons recognize their targets (Price et al., 2002). Interestingly, the sensory afferents that innervate the stretch receptors in particular muscles express the same *ETS* gene as the motor neurons that innervate those muscles (Lin et al., 1998), and this helps the sensory neuron axons find the dendrites of these motor neurons completing the monosynaptic stretch reflex (Chen et al., 2003). Limb ablation studies show that signals from the periphery, perhaps from the muscles themselves, help establish the pattern of *ETS* gene expression in the sensory neurons that innervate particular muscles (Figure 4.26).

As described in Chapter 3, there are two main types of glial cells in the brain, the astrocytes and the oligodendrocytes. The oligodendrocytes are the cells of the vertebrate CNS that produce myelin sheaths around axons. While early views of CNS development proposed that these cells could arise throughout the CNS, we now know that these cells arise from relatively restricted domains in the ventral regions of the brain, and then they migrate to the axon tracts throughout the brain (Figure 4.27). Several years ago, it was discovered that a specific growth factor, PDGF, is a mitogen for the oligodendrocyte precursors in the spinal cord. Subsequent work demonstrated that the receptor for this mitogen, the PDGFα, is specifically expressed in a restricted domain of the neural tube that gives rise to the oligodendrocytes. It was therefore not too surprising that when several groups reported cloning oligodendrocyte-specific transcription factors, Olig1 and Olig2, these factors were expressed in the same ventral domain as the PDGF receptor. But what was very surprising was that the domain of Olig1/2 expression was the same domain from where motoneurons were arising! How do the same progenitors make both motoneurons and oligodendrocytes? The answer lies again in the changing competence of the progenitors. The cells that express Olig1 initially also express neurogenin2 (Nrgn2), a transcription factor that is similar to the other proneural *bHLH* transcription factors (Kessaris et al., 2001; Zhou et al., 2001). During the time when the cells in this region of the spinal cord express both Nrgn2 and Olig1, these progenitors generate motoneurons. During the same time that these cells are making motoneurons, the zone of olig1 is dorsal to the zone of Nkx2.2 expression. Later in development, the zone of Nkx2.2 moves dorsally to overlap with the Olig1/2 domain, while the Nrgn2 expression domain moves further dorsally and now no

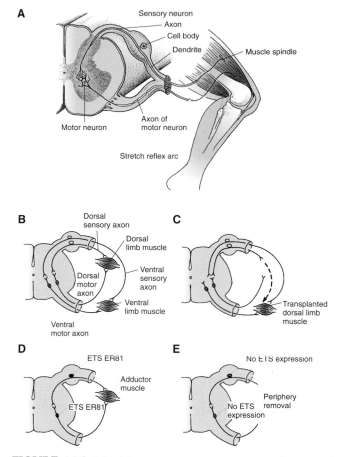

FIGURE 4.26 Matching sensory motor connectivity determined by muscles. A and B. Spindle afferents terminate on homonymous motor neurons. C. If the sensory fibers that normally innervate ventral muscles are forced to innervate dorsal muscles, they switch synaptic partners to the motor neurons that normally innervate dorsal muscles. D. Different muscles seem to induce the particular ETS molecules on both the motor and sensory neurons that innervate it. Thus, the motor and sensory neurons that innervate the Adductor muscle express the ETS ER81 molecule. E. If the peripheral muscles are removed, the motor and sensory neurons no longer express ETS molecules. (After Lin et al., 1998)

longer overlaps with the Olig1/2 expression domain. At this point in time, the progenitors that express both olig1/2 and nkx2.2 now start making oligodendrocytes. Experimentally misexpressing olig2 along with ngn2 causes the progenitors to produce motoneurons, whereas misexpressing olig2 and nkx2.2 causes the progenitors to produce oligodendrocytes. In olig2 knockout mice, neither oligodendrocytes nor motor neurons develop.

SUMMARY

In this chapter we have looked at the cellular determination in several different systems and have seen

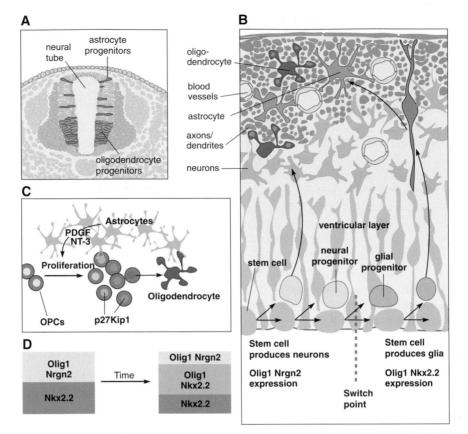

FIGURE 4.27 Glial cell development in vertebrates. A. Two main types of glial cells, oligodendrocytes and astrocytes, are formed in the neural tube. Astrocyte progenitors are distributed at all levels, whereas oligodendrocyte progenitors derive from the ventral neural tube. B. Oligodendrocytes form processes that wrap around axons and give rise to the myelin sheath. Astrocyte processes connect to capillaries and neurites. Glial progenitors and neural progenitors are derived from the same pool of stem cells that divide in the ventricular layer of the developing neural tube (bottom of B). At early stages, a stem cell generates neuroblasts. Later it undergoes a specific asymmetric division (the "switch point"), at which it changes from making neurons to making glia. C. In a culture system, optic nerve-derived oligodendrocyte progenitors (OPCs) depend on secreted signals from astrocytes. PDGF and NT-3 maintain OPC proliferation. In the absence of these factors, OPCs stop dividing and differentiate. The internal clock that determines when an OPC stops dividing depends on the level of the p27Kip1 protein, a cell cycle inhibitor that builds up over time and finally drives the cells to exit the cell cycle. D. The dorsally shifting expression of Nkx2.2 causes a change in gene expression in the progenitor cells driving the switch point (see text).

common themes such as successive restrictions in potency and potential as progenitor cells develop and divide. There is immense variation in the role of lineage versus environment in neuronal determination, with the general rule that invertebrates are more dominated by lineage mechanisms, while vertebrates are more dominated by diffusible signals and cellular interactions. Each determination pathway, however, usually brings its own mix of lineage-dependent and lineage-independent mechanisms. Of the transcription factors, bHLH factors of the proneural class help tell cells to become neurons and are antagonized by the Notch pathway which favors late differentiation of glia. Homeobox and paired domain transcription factors are often used to restrict neurons to certain broad classes linked to their position or coordinates of origin. Finally, POU, LIM, and ETS domain transcription factors may restrict cellular phenotypes even further. Of the signaling molecules, we find a particularly important role for the Notch pathway, but also important roles for BMPs, FGFs, and Hedgehog proteins. The last phases of determination involve each neuron interacting with its synaptic targets, which may provide the final differentiative signals for the maturing neuron. At the end of this process, the neuron becomes an individual cell with its own biochemical and morphological properties and its unique set of synaptic inputs and outputs.

5

Axon Growth and Guidance

Newborn neurons send out processes, threadlike axons that carry information to target cells and dendritic processes that receive inputs from other neurons. Some neurons, local interneurons, have short axons and make connections to cells in their immediate vicinity, while others, projection neurons, send long axons to distant targets. There is tremendous divergence and convergence of wiring. On the sensory side, peripheral neurons send axons into the CNS where they usually diverge to project to several distinct targets. Each of these targets contains neurons that also diverge to various targets of their own, and so on. Tracing pathways from the motor side backwards yields a similar complexity in convergence, with each motor neuron being innervated by many presynaptic neurons, and each of these having its own multitude of inputs. Thus, with thousands of target nuclei and billions of axons, the interweaving of axonal pathways is a remarkably complex tapestry. When looking at an adult brain, it is difficult to imagine how the precise patterns of connections were ever made. However, by looking at early embryonic brains, there is the possibility of seeing the very first axons on their way (Wilson et al., 1990; Ross et al., 1992; Easter et al., 1994) (Figure 5.1). These pioneer axons navigate in a simpler environment. But as the brain matures, more axons are added, and the weave becomes more intricate. As these later axons navigate, they are aided by pathways laid down by earlier axons. In this way, the rich tapestry of the brain wiring is accomplished by successive addition of new fibers that add complexity in a stepwise fashion.

An illustrative example of pioneer axonal navigation is that of the sensory neurons that arise in the distal part of a grasshopper leg (Keshishian and Bentley, 1983a, b, c). These cells, called Ti's, pioneer the tract that later developing sensory axons will follow to their targets in the central nervous system. If these Ti pioneers are ablated with a laser, then the later axons cannot find their way into the CNS. But how do the Ti's find their way in the first place? Part of the answer is that the Ti's use local cues on their journey. Spaced at short distances from one another are a series of well-spaced "guidepost" cells for which the Ti growth cones show a particular affinity (Caudy and Bentley, 1986). The distances between guideposts are small enough that a growth cone can reach out to a new guidepost while still contacting the previous one. Ti axons thus use these guidepost cells as stepping-stones into the CNS (Figure 5.2). Some of the guideposts are critical for pathfinding because when they are obliterated with a laser microbeam, the Ti axons get stuck and are unable to make it from one segment to the next (Bentley and Caudy, 1983).

An important insight into axonal navigation came from a simple experiment performed by Hibbard (Hibbard, 1965). He rotated a piece of the embryonic salamander hindbrain. In this tissue there are a pair of giant neurons called Mauthner cells that are responsible for rapid escape response. These neurons send large diameter axons, easily visible with silver staining, caudally down the spinal cord. In the rotated piece of hindbrain, Hibbard saw the axons of Mauthner cells initially grew rostrally instead of caudally, as though they were guided by local cues within the transplant. However, when the axons of these rotated neurons reached the rostral boundary of the transplant and entered unrotated neural tissue, they made dramatic U-turns and headed caudally down toward the spinal cord. This proved that axon navigation relies on cues provided by the external environment and is not just an intrinsic program of directions (Figure 5.3).

The axon of a projection neuron makes a journey to connect to its distant target much like a driver makes a journey from a particular address in one populated

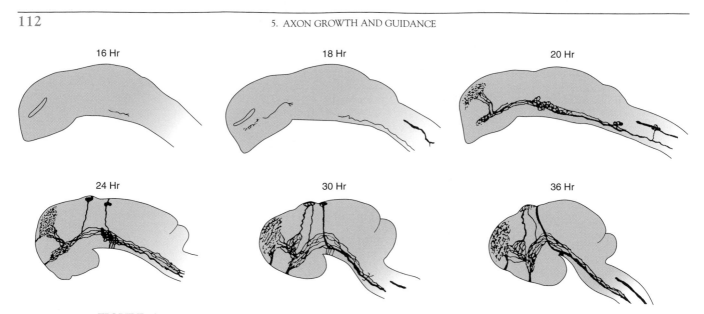

FIGURE 5.1 The increasing complexity of fiber tracts in the developing vertebrate brain. Antibodies against axons in the embryonic zebrafish brain at successive stages of development over the course of just 20 hours reveal that a variety of new axons are added at each stage. (After Wilson et al., 1990; Ross et al., 1992)

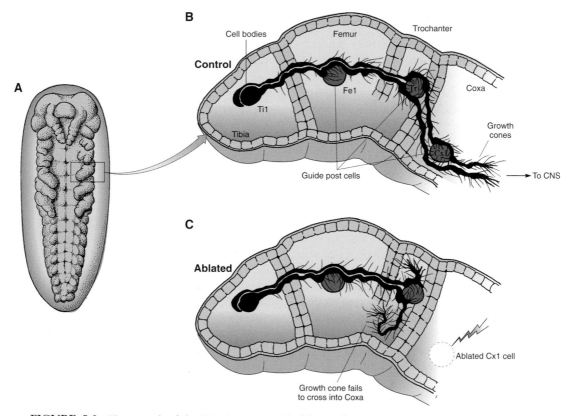

FIGURE 5.2 The growth of the Ti1 pioneers is aided by guidepost or stepping stone cells. A. A grasshopper embryo showing the developing legs. B. The Ti1 pioneers in black reach from one of the guidepost neurons to the next, successively contacting Fe1, Tr1, and Cx1 on their way to the CNS. C. When the Cx1 cells are ablated, the Ti1 cells lose their way and do not cross into the coxal segment of the embryonic leg. (After Bentley and Caudy, 1983)

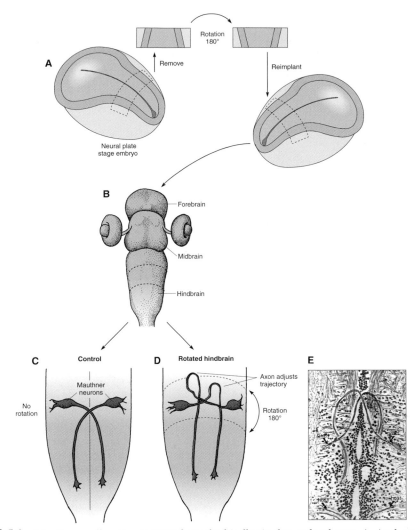

FIGURE 5.3 Mauthner cells grow posteriorly in the hindbrain due to local cues. A. At the neural plate stage, a segment of the hindbrain region of a salamander embryo is removed, rotated 180°, and reimplanted. B. shows a dorsal view of a larval brain of such an animal. C. The bilaterally symmetric giant Mauthner neurons in the normal unoperated larval brain. D. The trajectory of Mauthner axons in an experimental animal in which a segment of the hindbrain containing the Mauthner primordia is rotated. E. Photo of host and graft Mauthner cell axons in same animal. (After Hibbard, 1965)

city to another address in another city (Figure 5.4). As he or she pulls out from home on a particular street, the driver knows which turns to make, which roads to get on, the signs to follow, what to avoid, which exits to take, and how to recognize and stop at the correct destination. Unlike humans, most axons make these long journeys without errors. Growing axons are able to recognize various molecules on the surfaces of other axons and cells, and to use these molecules as cues to navigate the sometimes circuitous pathways to their particular destinations. They can also respond to diffusible molecules such as morphogens (see Chapter 2) that percolate through the embryonic brain and provide cues about overall orientation. They also need

motor abilities, as they must be able to move forward, make turns, and put on the brakes when they reach their target. They may also need to integrate information, for cues that have a particular significance during an early phase of their journey may have a different significance later on in the context of other signals, or they may need to adapt their responsiveness so that they remain sensitive as background levels of particular guidance cues change. These functions, the sensory, the motor, the integrative and the adaptive, are all contained within the specialized tip of a growing axon, called the *growth cone*. That growth cones are capable of all these functions is demonstrated in experiments in which growth cones surgically isolated from their

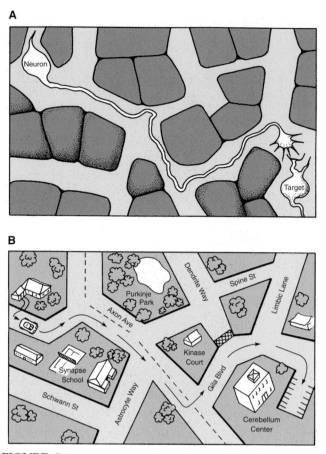

FIGURE 5.4 An axon growing to its target (A) is like a driver navigating through city streets (B). See text for details.

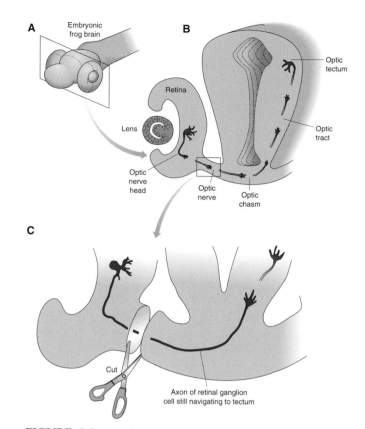

FIGURE 5.5 Axons grow from the retina to the tectum using their growth cones to guide them. A. A dorso-lateral view of the embryonic frog brain. B. Images of single retinal ganglion cell axons through the plane of section indicated in (A) shows that they as they grow to the tectum, they are always tipped by active growth cones. C. When the axon is separated from the cell body by cutting the optic stalk, time-lapse imaging shows that isolated growth cones still grow along the correct pathway. (After Harris et al., 1987)

cell bodies continue to grow (Bray et al., 1978) and even navigate correctly (Harris et al., 1987) (Figure 5.5).

THE GROWTH CONE

Growth cones were recognized more than a hundred years ago by the famous Spanish neuro-anatomist, Ramon y Cajal, as expansions at the tips of axons in fixed embryonic material. He imagined the growth cone as a sort of soft battering ram that extend-ing axons used to force their way through the packed cells of the embryonic brain (Ramon y Cajal, 1890) (Figure 5.6A). In 1910, Ross Harrison took pieces of embryonic neural tube and put them into tissue culture where he saw axons tipped with growth cones growing live across a microscope slide (Harrison, 1910). He was astounded to see the growth cones move and wiggle in real time. They seemed to feel their way along the surface, sending out long, thin filopodia and forming veils between the filopodia (Figure 5.6B). In

the 1930s, Speidel observed growth cones live in vivo at the ends of growing sensory axons in the trans-parent growing tail fin of a frog (Speidel, 1941). He followed single-growth cones over days, even weeks, and he watched how they responded to obstacles and injuries. He was impressed with the ability of these growth cones to change directions, branch, and respond to stimulation. Speidel described the growth rates as high as 40 μm/hr (Figure 5.6C). Such growth cones in vivo advanced in an amoeboid fashion and showed a number of delicate transient processes, con-sistent with Harrison's earlier in vitro observations.

Growth cones assume several morphologies and may travel at different speeds as they navigate through different parts of their pathways (Tosney and Landmesser, 1985; Bovolenta and Mason, 1987). Pioneer axons that are growing straight ahead have more traditional growth cones with several active filopodia and a few lamellipodia. The growth cones of follower axons that grow along earlier pioneer axons

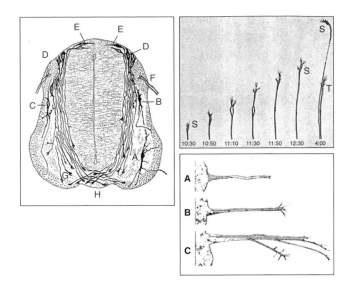

FIGURE 5.6 Early observations of growth cones. A. In the late 1800s, Ramon y Cajal saw expansions of axons near the ventral midline of the chick neural tube. B. In the 1930s, Speidel observed growing nerve fibers tipped with axons in the frog tail. C. In the early 1900s, Harrison grew neural explants in culture and watched them extend axons tipped with motile growth cones. (From Ramon y Cajal, 1890; Harrison, 1910; Speidel, 1941)

are generally simple and more bullet-shaped with few filopodia (Figure 5.7A). Growth cones get particularly complex, when they arrive choice points along the pathway (Figure 5.7B). Upon reaching an appropriate target of innervation, growth cones once again alter their shape. They display thin and highly branched terminals (Harris et al., 1987; Halpern et al., 1991) (Figure 5.7C). Static observations from fixed tissue provide neither the rate of growth nor the sampling strategy that growth cones employ as they make decisions. While Cajal's remarkable visual memory and artistry allowed him to produce accurate drawings of growth cones nearly a century ago, simple video microscopy systems now permit most laboratories to view and measure the growing axon.

Over the past 20 years, the standard compound microscope has been embellished with a number of technical wonders. These advances include low light-sensitive video cameras, scanning laser illumination, and specialized image processing software. Coupled with the recent introduction of fluorescent labels that are rapidly transported along living axons, one is now able to produce time-lapse movies of process outgrowth and innervation (Glover et al., 1986; Honig and Hume, 1986; Harris et al., 1987). Vital fluorescent dyes are commonly used to observe living processes. The lipophilic dyes, DiI and DiO, intercalate into the neuron membrane and diffuse rapidly down axonal processes; they are then visualized with epifluorescent

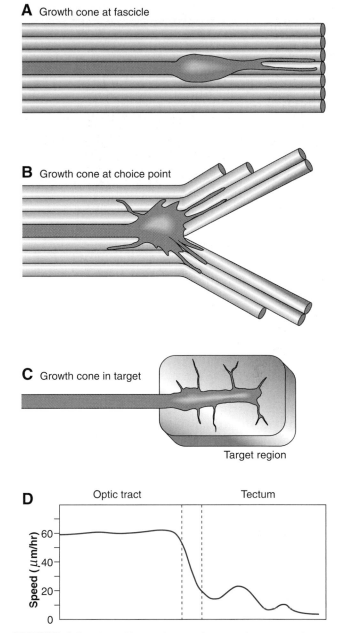

FIGURE 5.7 The different shapes of a growth cone. A. Growth cones that are fasciculating with other axons tend to be simple and have few filopodia. B. Growth cones at choice points are complex with filopodia and lamellipodia. C. Growth cones in the target region become even more complex and sprout backbranches along the axon shaft. D. Speed versus location plot for retinal axons in *Xenopus* embryos. They grow at about 60 μm/hr along the optic tract. When they arrive at the tectum, they slow down dramatically.

illumination. Fluorescent proteins of different colors such as Green Fluorescent Protein, (GFP), RFP (Red), and YFP (Yellow) are engineered variants of natural proteins found in certain types of jellyfish. When a growing neuron is transfected with such a GFP gene, it creates its own fluorescent protein. By genetically

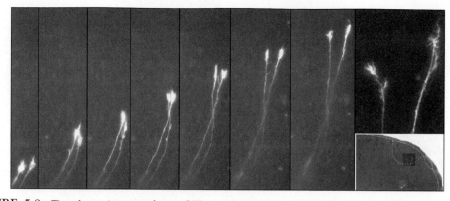

FIGURE 5.8   Time-lapse images of two GFP-expressing retinal axons growing in the optic tract of a *Xenopus* brain. The images on the left are successive time points spaced about 20 minutes apart showing the two elongating axons, tipped with growth cones, rearranging their relative positions in the optic tract. The image in the top right shows the initial branching of the same two axons in the tectum, the target structure for these axons. The image at the bottom right is a low-power view of the preparation at the beginning and at the end of the time-lapse. (Courtesy of Sonia Witte and Christine Holt)

engineering chimeric genes, combining GFP with a protein potentially involved in axon growth, one can test the effect of misexpressing the protein of interest in live axons. By putting the GFP gene under the control of a promoter that is active in a subset of developing nerve cells in a transgenic animal, it is possible to monitor a whole class of growing axons. One problem, often encountered in thick specimens, is the excessive level of out-of-focus fluorescence. Confocal microscopes overcome this obstacle and allow one to observe sharp fluorescent images in a limited depth of field (0.5–1.5 μm), called *optical sectioning*. This is accomplished by scanning a focused laser beam across the specimen in a point-by-point fashion. The emitted light from each "point" is acquired and used to reconstruct an image of the specimen. Using a combination of these techniques, it has become possible to watch axons grow out and innervate their target in the CNS of live embryos (Figure 5.8).

As growth cones crawl forward, they leave axons behind. This means that new material must be continually incorporated into the axon. In culture, when a particle is attached to a growing axon, it remains relatively stationary compared to the distal tip of the axon. This suggests that this new material is assembled distally, at the growth cone. Indeed, new glycoproteins are added preferentially at the distal tips of axons in the growth cone region (Hollenbeck and Bray, 1987). The incorporation of this new material is calcium dependent, suggesting that membrane is added by the calcium-mediated fusion of internal vesicles to the growth cone's surface. The addition of cytoskeletal components also takes place primarily at the tip of the growing process. This was shown by labeling neurons with fluorescent tubulin and actin, and then illuminating part of

the axon with a bright spot to bleach the fluorescence at a particular location (Figure 5.9). The result is that the bleached spot stays relatively still as the growth cone continues to advance, suggesting that these components are assembled distally. If one looks carefully at the middle of the axon, however, it is possible to see that there is also forward transport of some assembled microtubule fragments as well as cargo moving up and down the axon. In addition, if two beads are attached along the shaft of the axon, one notices a growing separation between these two markers, suggesting that some membrane is also added along the axon. Such, transport and interstitial growth is crucial as the brain or body enlarges after initial connections are made.

THE DYNAMIC CYTOSKELETON

If we want to know how a growth cones navigates, we have to understand how it moves, and this means looking into its dynamic cytoskeleton. The cytoskeleton of a growth cone is filled with molecules that are involved in cell movements (Heidemann, 1996; Letourneau, 1996; Dent and Gertler, 2003; Gordon-Weeks, 2004). The two most important elements are the microtubules that extend along the axon and splay out in the central domain, C-domain, of the growth cone, and the actin fibers, which are more prominent in the peripheral, P-domain (Figure 5.10). In the P-domain, actin forms the basic cytoskeleton of the lamellipodia, the veils between filopodia, and the filopodia themselves, which are filled with thick bundles of actin. Many other cytoskeletal-associated proteins in the growth cones do a variety of jobs such as anchoring

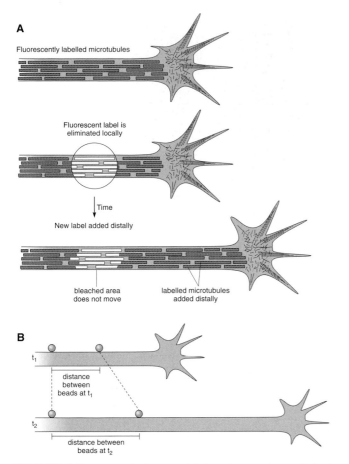

A Fluorescently labelled microtubules

Fluorescent label is eliminated locally

Time

New label added distally

bleached area does not move

labelled microtubules added distally

B

t_1

distance between beads at t_1

t_2

distance between beads at t_2

FIGURE 5.9 Microtubules are added at the growing end. A. A growing axon is labeled with fluorescent tubulin, and then some of this fluorescence is bleached by a beam of light (circle) focused on the axon near the growth cone. As the axon elongates distally, the bleached spot stays in approximately the same place (bottom panel), implying that the microtubules along the axon shaft do not move forward but rather that new microtubules are assembled at the distal tip. B. Two beads placed on an axon move further apart from each other as the axon grows, with the front bead moving further forward than the rear bead.

actin and microtubules to the cell membrane, to each other, and to other cytoskeletal components such as the molecular motors that generate force (Figure 5.11).

Drugs such as the actin-depolymerizing agent, cytochalasin, have been used to investigate the functions of actin in the growth cone. Treatment of growth cones with such drugs prevents filopodia formation, and such growth cones slow down dramatically, showing that actin-rich fibers are important for the forward progress of a growth cone. Cytochalasin-treated growth cones do not steer properly and usually lose their way in the developing organism. Thus, when the Ti1 pioneer neuron in the grasshopper limb is treated with cytochalasin, its axon meanders off course and often does not make the turns necessary to grow to its targets in the CNS (Bentley and Toroian-

Raymond, 1986) (Figure 5.12A). In the amphibian brain, retinal axons must make a posterior turn in the diencephalon to get to their targets in the midbrain. If these growing axons are treated with cytochalasin, they grow past the turning point (Chien et al., 1993) (Figure 5.12B). Thus, actin filaments are critical for growth cone navigation.

The actin filaments in filopodia are bundled and oriented so that their fast-growing barbed or plus ends are pointing away from the growth cone center out toward the periphery. Time-lapse observations in tissue culture show that, as the growth cone advances, the filopodia move backward from their base and shorten. The filopodia generate tension against the bulk of the growth cone, which they thus pull forward. This tension can be observed in culture, for when a single filopodium from a growth cone contacts another axon lying in its path, that filopodium can pull the axon toward the advancing growth cone like someone pulling the string on an arrow (Bray, 1979) (Figure 5.13). Similarly, a single filopodium that makes contact with a more adhesive substrate in tissue culture is able to steer the entire growth cone by pulling it toward the attachment point (Letourneau, 1996). The importance of single filopodia in directing growth cones in vivo has been demonstrated in the Ti1 pioneer axons of the grasshopper limb. Here it can be seen that when a single filopodium makes contact with a guidepost cell, then it attaches firmly while other filopodia retract (O'Connor et al., 1990) (Figure 5.13).

How is filopodial tension generated? The clutch hypothesis, formulated by Mitchison and Kirschner (1988) suggests that actin filaments become anchored to the substrate, engaging the clutch across the membrane though adhesion complexes. These actin filaments are then pulled centrally, probably by myosin molecules located at the base of filopodia (Figure 5.13). Indeed, when myosin function in growth cones is blocked, forward progress is slowed, while the filopodia themselves tend to lengthen as if a force that pulled them rearward into the growth cone was attenuated (Lin et al., 1996).

In culture, growth cones move straight ahead, and there are approximately equal numbers of filopodia on the left and right sides. Some conditions cause more actin polymerization or depolymerization on one side than the other, leading to an imbalance of filopodial number and resulting traction force, causing the growth cone to turn. In fact, this is suspected to be a main mechanism of growth cone reorientation. Indeed, if actin filaments are destabilized on one side of the growth by using a depolymerizing agent locally, the axon will turn in the other direction (Figure 5.14A) (Yuan et al., 2003). When a single filopodium is

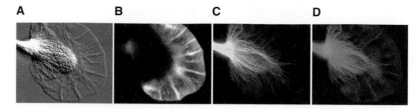

FIGURE 5.10 Views of an *Aplysia* growth cone. A. Namarski image showing the growth cone. The bulging central domain and the thin peripheral domain containing actin cables are visible. B. Labeling the actin filaments with a fluorescent probe reveals they are concentrated in the peripheral domain and the filopodia. C. Labeling the microtubules reveals that these structures are in the central domain. D. Pseudo-colored merged image of actin filaments (red) and microtubules (green). (From Paul Forscher)

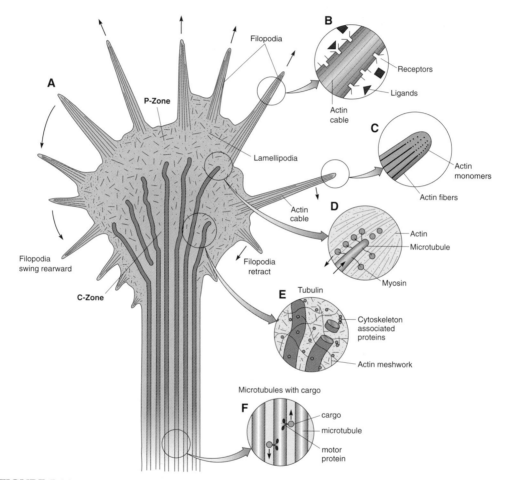

FIGURE 5.11 The structure of the growth cone. A. Actin bundles fill filopodia, that are bounded by membranes with cell adhesion molecules and various receptors, poke out at the advancing edge and are retracted at the trailing edge of the growth cone. Between the filopodia are sheets of lamellipodia that extend forward. They are filled with an actin meshwork that is continuous with that in the main body of the growth cone. Here also microtubules push forward and carry cargo to and from the cell body along the axon shaft as they enter the growth cone and fan out toward the filopodia. B–F. Close-ups of various regions show some of the molecular components of the cytoskeletal network that are localized in the growth cone.

detached from the surface of a culture dish using a fine glass needle, the growth cone snaps into a new direction that is consistent with the tension exerted by the remaining filopodia (Wessells and Nuttall, 1978).

Microtubules are the other main cytoskeletal elements of the growth cone. Pools of unassembled tubulin are concentrated in the growth cone, which is the most sensitive part of the axon to the effects of microtubule depolymerizing agents such as nocodozole (Brown et al., 1992). Natural microtubule stabilizing proteins, such as Tau, are also highly concentrated near the growth cone, suggesting that this is

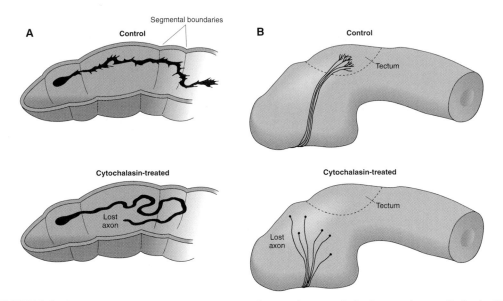

FIGURE 5.12 Actin filaments are necessary to guide growth cones. A. In the grasshopper limb, the Ti1 growth cones are hairy with active filopodia (top). If the growth cones are treated with the actin-depolymerizing agent, cytochalasin, the axon fails to navigate (bottom). B. In the vertebrate visual system, axons enter the brain from the optic nerve and grow toward the tectum by growing dorsally and turning posteriorly (top). When these axons are treated with cytochalasin, the axons fail to make the appropriate posterior turn, and most axons miss the tectum (bottom). (After Bentley and Toroian-Raymond, 1986; Chien et al., 1993)

where unpolymerized tubulin is fashioned into microtubules and stabilized. Microtubules run straight and parallel inside of the axon, but when they enter the base of the growth cone, they splay out and bend like soft spokes and sometimes appear as broken fragments (Figure 5.10). Like the actin filaments in the filopodia, the microtubules of the axons have a "plus" end where polymerization takes place, and this is positioned at the growing tip. The fact that axons treated with cytochalasin B continue to grow, albeit slowly, is probably due to the distal growth of microtubules, as depolymerization of microtubules inhibits axon elongation completely (Marsh and Letourneau, 1984). The control of microtubule assembly is partially controlled by post-translational modifications that affect their stability. A carboxyl terminal tyrosine is added to α-tubulin by the enzyme, tubulin tyrosine ligase, inside of the growth cone. The tyrosinated form of tubulin is quite dynamic and sensitive to depolymerizing agents. In contrast, tubulin loses the tyrosine group and becomes acetylated instead when it enters the axon, making axonal microtubules more stable (Brown et al., 1992). Dynamic microtubules in the growth cones can rapidly polymerize and extend transiently into the P-zone. Dynamic microtubules can also go through catastrophes, that is, rapid disassembly. This dynamic instability of the tyrosinated microtubules is critical for normal growth cone motility. If the dynamics are altered with a reagent like taxol which stabilizes microtubules growth cones advance more slowly

and become incapable of turning (Williamson et al., 1996).

Experiments on the local microtubule dynamics show that microtubules may be involved in growth cone turning in a similar way to actin (Figure 5.14). Thus, if the microtubule stabilizing agent taxol is delivered on one side of a growth cone, the growth cone will turn in that direction, while if a depolymerizing agent such as nocodazole is delivered to one side, the growth cone will turn the other way (Buck and Zheng, 2002). The similarity of the results with actin and microtubule destabilizers means that each of these cytoskeletal elements may be able to work independently to produce turning or that actin and microtubules interact in ways that are critical for proper growth cone navigation. The actin fibers at the periphery may restrict the dynamic microtubules from invading peripheral domains. It is thought that the myosin-driven retrograde flow of actin at the leading edges bends and breaks the dynamic microtubules and thus keeps many of them from successfully invading the P-domain. But dynamic microtubules continue to polymerize and probe the P-domain, as though they were trying to get a foothold. There is evidence that those microtubules that do successfully invade the peripheral domain become associated with actin bundles in a filopodium (Zhou and Cohan, 2004). Such interactions then lead to the stabilization of the both the microtubules and the actin bundles.

The keys to making the growth cone cytoskeleton dynamic in this way are the numerous proteins that are

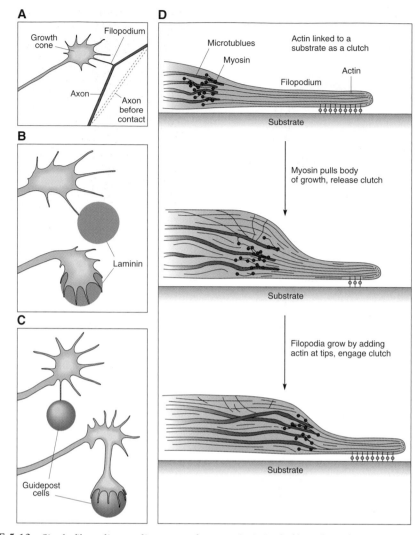

FIGURE 5.13 Single filopodia can direct growth cones. A. A single filopodium from a growth cone exerts tension and pulls on an axon it contacts in culture. B. A single filopodium touches a laminin-coated spot in a culture dish and reorients. C. A single filopodium of a Ti1 cell contacts a guidepost cell, and by the process of microtubule invasion becomes the new leading edge. D. In the hypothetical clutch mechanism, myosin at the base of filopodia, pulls on actin cables that are attached to the substrate through a transmembrane clutch and so pulls the main body of the growth cone forward. (After Mitchison and Kirschner, 1988)

associated with microtubules and actin. Several microtubule-associated proteins (MAPs) are found in growing neurites, and these proteins may regulate the growth process. Some MAPs seem to be involved in stabilizing microtubules and inducing them to form bundles. Different MAPs are differentially localized: MAP2 is mostly in dendrites, and Tau is in axons. Antisense mediated depletion or knockouts of many of these MAPs diminish neurite outgrowth (Letourneau, 1996). MAPs may also interact with actin fibers and cause bundling of F-actin, or cross linking of actin fibers to microtubules, thus leading to the stabilization of actin-microtubule complexes. Microtubule destabilizing proteins such as SCG10 and stathmin are also

important for growth cone function. For example, SCG10 overexpression in growth cones leads to enhanced dynamic instability of microtubules and increased neurite outgrowth (Grenningloh et al., 2004). There is also an array of actin-associated proteins (more than 50 have been identified) involved in growth cone function. Cofilin is an actin-binding protein found in the growth cone that increases the rate of actin dissociation, working a bit like SCG10 does on increasing dynamic instability, but on actin filaments rather than microtubules. Arps promote the branching of actin filaments, and the Ena/Vasp proteins that localize to filopodial tips act as anticapping agents, encouraging straight actin filaments to grow at the leading edge

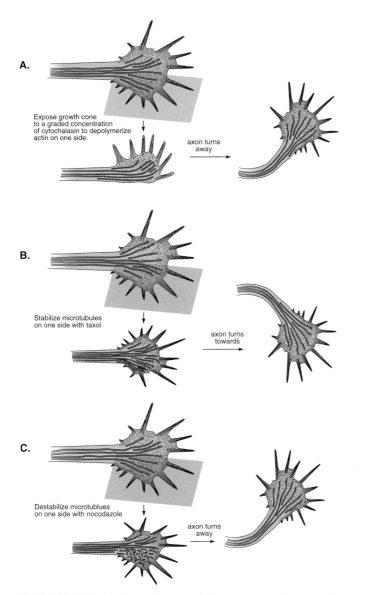

FIGURE 5.14 Actin and microtubules steer growth cones. A. Local depolymerization of actin on one side of a growth cone causes it to turn the other way. B. Local stabilization of microtubules on one side of a growth cone causes it to turn toward that side. C. Destabilization of microtubules on one side causes it to turn the other way. (After Buck and Zheng, 2002; Yuan et al., 2003)

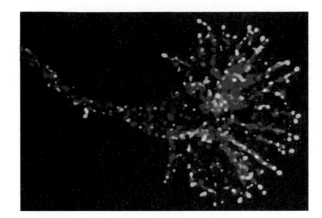

FIGURE 5.15 Image of a growth cone showing actin in green and Ena/Vasp in red located at the tips of the filopodia where these proteins act as anticapping agents, preventing the binding of actin capping proteins and thus encouraging plus end elongation. (From Lanier and Gertler, 2000)

WHAT DO GROWTH CONES GROW ON?

Growth cones have the ability to sense their environment and make choices based on extracellular information. These choices are made as the growth cone passes through a complex environment of chemical factors and physical terrain. On a dried cracked collagen surface, growing axons often follow the pattern of the stress fractures (Figure 5.16A). When tracts or commissures are wounded or experimentally severed, axons may be physically impeded from growing as they normally would. In such cases, it is sometimes possible to provide axons with an artificial mechanical pathway across the wound (Silver and Ogawa, 1983) (Figure 5.16B). Mechanical support is necessary and influential, but axon growth and guidance are based on molecular mechanisms. To discover these mechanisms, factors are often tested in tissue culture to see whether they influence the rate and direction of axonal growth. Many molecules have been identified in this way, and such studies show that a single type of growth cone can respond to a variety of different cues. For example, embryonic spinal neurons from frog embryos respond to at least a dozen different factors tested, and this implies that growth cones are highly attuned sensory beasts with rich arrays of receptor molecules to sample various aspects of their environment. Genetic approaches in *Drosophila* and nematodes have led to the identification of many genes for guidance cues and their receptors, which when disrupted lead to growth and pathfinding errors.

When neurons are plated on plain glass or tissue culture plastic, the cells often attach but rarely put out long neurites with active growth cones. However,

(Lanier and Gertler, 2000; Lebrand et al., 2004) (Figure 5.15). All these actin and microtubule-associated proteins and many others have to be regulated properly as the growth cone navigates, and there are a large and growing number of identified kinases and phosphatases that add and remove phosphate groups from these proteins, thus activating and inactivating them. These enzymes are in turn regulated by receptors on the growth cone surface that sense the substrates and guidance molecules that the growth cone encounters in its journey.

A

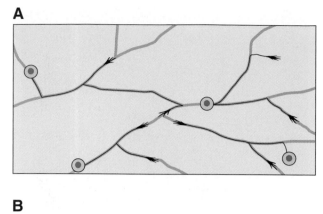

B

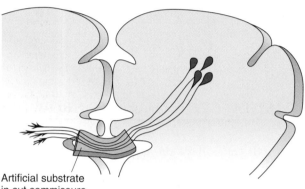

Artificial substrate
in cut commissure

FIGURE 5.16 Axons may follow mechanical pathways. A. The axons of neurons on a dried collagen matrix growing through the cracks. B. Axons of the corpus callosum can use an artificial sling to grow from one side of the brain to the other.

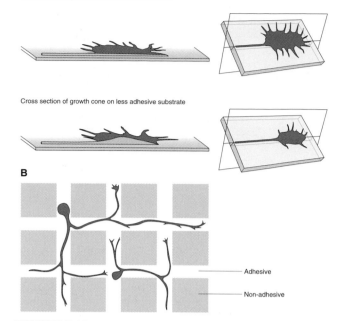

A

Cross section of growth cone on adhesive substrate

Cross section of growth cone on less adhesive substrate

B

Adhesive

Non-adhesive

FIGURE 5.17 Growth cones and adhesion. A. On a very adhesive substrate growth cones are flattened, have lots of filopodia, and do not move rapidly (top). On a less adhesive substrate, growth cones are more compact, rounded, have fewer processes, and often move more quickly. B. Neurites in culture given a choice between an adhesive and a nonadhesive substrate will tend to follow the adhesive trails.

when they are plated on polycationic substrates, such as polylysine, that stick well to negatively charged biological membranes, the neurons are much more likely to initiate axonal outgrowth. The growth cones of such neurons are flattened against the substrate as though they adhere very strongly to it (Figure 5.17). When axons are grown on a patterned dish that offers them the choice between a nonadhesive substrate, versus an adhesive substrate, the growth cones follow the adhesive trail (Letourneau, 1975; Hammarback et al., 1985) (Figure 5.17). These findings led to the idea that growth cones might simply follow gradients of adhesion in the developing organism. In fact, this may be the case for some neurons.

In the wing of the moth, sensory axons of the wing grow in the distal to proximal direction along a basal lamina of epithelial cells (Nardi and Vernon, 1990). Examination of this epithelium microscopically shows that it becomes increasingly complex toward the base. Transplantation of the epithelium suggests that this change corresponds to a gradient. Axons readily cross onto a transplant that has been moved in the proximal to distal direction, but avoid distal transplants that

have been moved proximally, suggesting that axons readily grow onto more adhesive membranes but will not onto less adhesive ones (Nardi, 1983).

To measure growth cone attachment to various cell adhesion molecules, a pipette can be positioned in a culture dish containing growing axons, and culture media can be squirted at the growth cones in an attempt to "blast" them off the substrate (Figure 5.18). The longer the growth cone stays attached to the surface, the stronger its adhesion must be. Neurite growth rate can then be measured on these same substrates for comparison (Lemmon et al., 1992). Moreover, when axons are given a choice between pairs of naturally occurring cell adhesion molecules, they do not necessarily grow preferentially on the more adhesive substrate. Interestingly, it turns out that the most adhesive substrates, such as the lectin concanavalin A, are not good supporters of axon outgrowth. In fact, growth cones tend to get stuck on such a surface. They seem unable to retract their filopodia efficiently and thus grow very slowly. Thus, the ability to detach is just as important as the ability to attach, and the molecules that have just the right amount of adhesion, not too little and not too much, are best able to support axon growth.

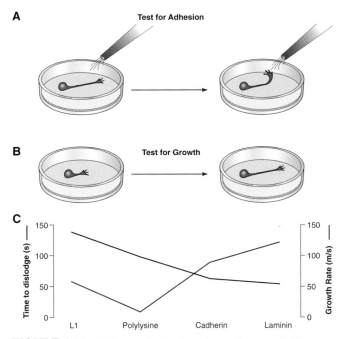

FIGURE 5.18 Differential adhesion of growth cones. A. To quantitate adhesivity, a measured blast of culture medium is directed at the growth cone. At a particular time, the growth cone becomes detached. B. Growth is quantified by axon length increase over an interval time. C. By using such tests, it can be shown that the class of gray neurons has a different adhesion profile than the class of red neurons. (After Lemmon et al., 1992)

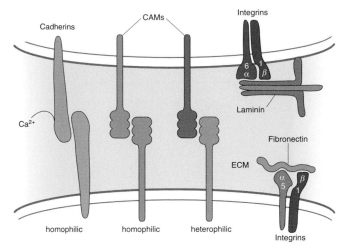

FIGURE 5.19 A few of the classes of adhesion molecules expressed on the growth cone. Cadherins are calcium-dependent homophilic adhesion molecules. Some members of the IgG super-family of cell adhesion molecules, CAMs, can bind homophilically; others are heterophilic. Integrins composed of various alpha and beta subunits bind to a variety of different extracellular matrix components with distinct affinity profiles.

Many molecules that are excellent supporters of axon growth have been isolated from the extracellular matrix (ECM) (Bixby and Harris, 1991). These factors were initially purified from culture media that was conditioned by cells known to support axonal outgrowth. Some of the most abundant proteins in ECM, such as laminin, fibronectin, vitronectin, and various forms of collagen, all promote axon outgrowth. Many of these ECM proteins are large and have many different functional domains for cell attachment, for collagen attachment, and for protein interactions. Different neurons seem to show a preference for particular ECM molecules. Vertebrate CNS cells grow particularly well on laminin, while peripheral neurons often seem to grow better on fibronectin. In experiments where retinal neurons are given a choice between laminin and fibronectin laid down in alternate stripes, retinal axons clearly prefer laminin, though they will grow on either substrate if given no choice.

Integrin is a receptor for many different extracellular matrix proteins, and it is composed of two subunits, α and β (Figure 5.19). The extracellular matrix molecule that an axon will respond to is largely a matter of which integrin molecules are found at the growth cone as the specificity of integrin for particular ECM proteins depends on the combination of α and β subunits that are expressed (McKerracher et al., 1996). There are 18 different α and 8 different β subunits, and it is clear that different tissues, including different neural tissues, use different subunit combinations. The $\alpha 5$ subunit is particularly good at binding to fibronectin, while the $\alpha 6$ subunit is better at binding to laminin. Over the course of development, axons may change which integrin subunits they express and thus change their sensitivity to a particular ECM molecule. For instance, chick retinal ganglion cells express $\alpha 6$ and grow well on laminin when they are young. As they mature, they stop expressing this subunit and lose their ability to respond to laminin just as their axons make contact with the tectum (Cohen and Johnson, 1991).

WHAT PROVIDES DIRECTIONAL INFORMATION TO GROWTH CONES?

It has been suggested that four types of molecular cues influence the direction in which growth cones will travel (Tessier-Lavigne and Goodman, 1996). These are divided into short-range and long-range cues, each of which may be either attractive or repulsive (Figure 5.20). (1) Contact adhesion: a sudden increase in the adhesivity of one cellular substrate compared to another may cause axons to switch pathways; (2)

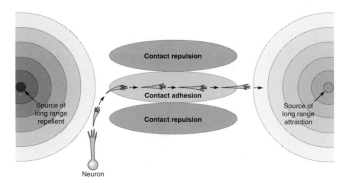

FIGURE 5.20 Short-range (or contact) and long-range attractant and repellents guide an axon. This axon is pushed from the left and pulled from the right by long-range cues, while hemmed in along a narrow pathway by contact adhesive and repulsive cues. (After Tessier-Lavigne and Goodman, 1996)

Contact repulsion: when growth cones bump into cells with repulsive membrane molecules, they may collapse or turn away and grow in a different direction; (3) Long-range attraction: growth cones may exhibit positive chemotaxis behavior to certain diffusible molecules originating at distant sources and so grow toward these chemoattractants; and (4) Long-range repulsion: growth cones may exhibit negative chemotaxis to diffusible molecules and so orient away from the sources of such factors. In the following sections, we will introduce many of the specific guidance cues that work according to these basic mechanisms. In many systems, it has become clear that growth cones respond to a combination of these cues, and the molecular directional signaling can become quite complex.

CELL ADHESION AND LABELED PATHWAYS

Growth cones, in addition to using ECM, make contact with the membranes of other cells and axons. In this, they are supported by a set of cell adhesion molecules (CAMs) (Walsh and Doherty, 1997). There are a host of such molecules, and they come in several classes (Figure 5.19). The most prominent class is the IgG superfamily. Members of this class have extracellular repeat domains similar to those found in antibodies, reflecting perhaps an ancient adhesive function for the IgG superfamily. Another class is the calcium-dependent cadherins. One property that many CAMs share is their ability to bind homophilically. Homophilic binding means that proteins of the same type bind to each other. Thus, if two cells express the same homophilic CAM on their surfaces, the CAM on the one cell will act as receptor for the CAM on the

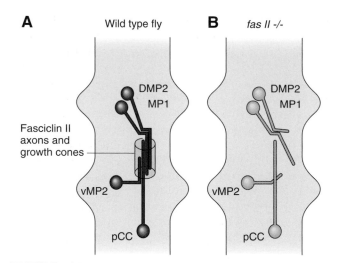

FIGURE 5.21 A homophilic mutant in flies. A. In normal *Drosophila* embryos, DMP2, MP1, vMP2, and pCC axons fasciculate in a longitudinal bundle in which all axons express fasII. B. In a *fasII* mutant, these axons do not fasciculate normally. (After Grenningloh et al., 1991)

other cell and vice versa, causing the two cells to adhere to each other. To test whether a particular CAM is homophilic, nonadherent cells are transfected with the CAM and assayed with respect to whether they then form aggregates. During outgrowth, axons expressing the same CAMs often fasciculate with one another. Thus, single pioneer axons may use a specific CAM to guide the follower axons that express the same CAM, and these axons can attract the fasciculation of still later axons of the same type. In this way, pioneer axons can become the founders for large axonal tracts within the CNS.

Monoclonal antibodies raised against axonal membranes have been used to search for cell adhesion molecules that are expressed on particular fascicles of fibers during neural pathway formation. Several molecules were discovered in this manner. Some CAMs show a particularly restricted expression and result in very specific defects in axon growth. For example, a *Drosophila* CAM, called fasciclinII (FasII), is expressed on a subset of longitudinal tracts or fascicles in the CNS (Bastiani et al., 1987). In FasII mutants, axons that normally express this gene defasciculate, and the longitudinal tracts in which they run become disorganized (Grenningloh et al., 1991) (Figure 5.21). When FasII is transgenically misexpressed on CNS neurons that would not normally express FasII, their axons tend to join together abnormally. Another CAM in *Drosophila*, called IrreC, is expressed in the optic lobes, and IrreC mutants show specific miswiring of the optic pathway (Ramos et al., 1993). The vertebrate homolog of IrreC has been implicated in the formation of specific tracts in the spinal cord and cerebellum. In mammals, limbic-associated membrane protein, LAMP, is an IgCAM expressed by neurons throughout the limbic system, a cortical and subcortical area of the brain that functions in emotion and memory (Horton and Levitt, 1988). Administration of LAMP antibodies to developing mouse brains results in abnormal growth of the fiber projections in the limbic system, suggesting that LAMP is an essential recognition molecule for the formation of limbic connections (Pimenta et al., 1995). LAMP has three IgG domains, and these domains appear to participate in different ways to enhance local wiring. LAMP enhances neurite outgrowth through homophilic interactions with other neurons in the limbic circuit using the first of its IgG domains, but inhibits neurite outgrowth of non-LAMP-expressing neurons using heterophilic interactions involving the second IgG domain (Eagleson et al., 2003). The function of the third domain is not known.

Neural cell adhesion molecule (NCAM), the earliest identified of the CAMs, appears to be expressed on all vertebrate neurons and glia (Edelman, 1984). NCAM exists in many different forms, some with an intracellular domain that may interact with the cytoskeleton and some without. The extracellular portion of NCAM can be highly modified by the addition of carbohydrates, particularly sialic acid residues. Nonsialylated forms are very adhesive compared to the sialylated forms. In the course of a neuron's development, there may be changes in the sialylation state of its NCAM, thus adjusting its adhesion (Walsh and Doherty, 1997). For example, developing motor neurons of the chick grow out of the spinal cord and enter a complicated plexus region, where they cross in many directions and eventually segregate into distinct nerve roots leading to their appropriate muscles. During the time when these axons are growing in the plexus, the NCAM they express is highly sialylated. This keeps the axons that are headed toward different muscles from fasciculating indiscriminately with one another. If the sialic acid is digested away with endoneuraminidase (Endo-N), errors in pathfinding occur in the plexus region, and motor axons exit into the wrong peripheral nerves (Tang et al., 1994) (Figure 5.22).

Growing axons respond to a variety of cues as they navigate, so the loss of just one of these may affect axon growth and navigation in a limited way. In many gene deletion studies in *Drosophila*, there is a significant increase in pathfinding errors, but the majority of axons usually grow along the correct pathway. For instance, in *fasII* mutants mentioned above, although longitudinal tracts are somewhat defasciculated, the axons are still able to grow in the correct directions. In some CAM knockouts, for instance the NCAM knock-

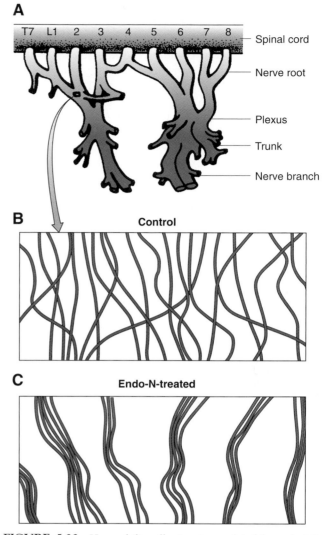

FIGURE 5.22 Homophilic adhesion is regulated by polysialic acid. A. The brachial plexus region in the chick where motor axons destined for particular muscles sort out into their correct nerve roots. B. Higher magnification of the plexus region showing fascicles breaking up and axons regrouping with other axons. C. After treatment with Endo-N to remove sialic acid residues from N-CAM, the axons do not defasciculate properly and stay in large fascicles. As a result, innervation errors are made. (After Tang et al., 1994)

out in the mouse, the phenotype is surprisingly subtle, suggesting that it is just one of a number of molecules that are used for this purpose. This idea of molecular redundancy for pathfinding in the developing nervous system is supported by the finding that sometimes doubly mutant embryos, which have deletions of more than one particular adhesion molecule, exhibit severe axonal growth defects, whereas each of the single mutants is relatively normal. For example, *Drosophila* that lack FasciclinI, a CAM that is expressed in commissural fascicles, have a normal looking CNS. To test whether FasI is part of a redundant system, a series of double mutants were made with *fasI* and several other putative guidance or growth cone function mutants, each of which also had no striking phenotype on its own. One of the double mutants tested showed pathfinding defects in combination with *fasI*. This was the *fasI/abl* double mutant (Elkins et al., 1990). *abl* codes for a tyrosine kinase, which probably functions in a signal transduction pathway in the growth cone. Since both genes had to be knocked out to cause axon disorientation, the two genes are probably part of two distinct molecular pathways, either of which may suffice for axon guidance. Similarly, in vertebrate tissue culture, motor axon growth over muscle fibers is not seriously impaired unless two or more adhesion molecules are simultaneously disabled (Tomaselli et al., 1986; Bixby et al., 1987). Axonal growth is such an important part of building an organism that such fail-safe molecular mechanisms often operate to help ensure that the nervous system is properly wired.

Early axon tracts can be thought of as subway lines: the Orange Line, the Red Line, and the Green Line. As the pioneer axons grow, they express particular CAMs on their surfaces, creating a labeled "line" that other growth cones can follow. Guidance along previously pioneered tracts by selective adhesion is referred to as the labeled pathways hypothesis (Goodman et al., 1983). When a new neuron sends out its axon, the growth cone is able to choose between specific pioneer axons because it expresses complementary CAMs. Evidence for labeled lines or pathways comes from the embryonic grasshopper central nervous system. Here, for example, the axon of the G-neuron extends across the posterior commissure along a pathway pioneered by the Q1- and Q2-neurons (Bastiani et al., 1984; Raper et al., 1984). Once it has crossed, the growth cone of the G-neuron pauses for a few hours while its filopodia seem to explore a number of different longitudinal fascicles in the near vicinity. In the electron microscope, it can be seen that that filopodia from the G-neuron growth cone preferentially stick to the P-axons of the A/P-fascicle. The G-growth cone then joins the A/P fascicle and follows it anteriorly to the brain. If the P-axons are ablated before the G-growth cone crosses the midline, the G-growth cone acts confused upon reaching the other side (Figure 5.23). It does not show a high affinity for any other longitudinal bundle or even the A-axons. As a result, it often stalls and does not turn at all. Thus, the P axons seem to have an important label on their surface that the G growth cone can recognize, possibly because of a specific receptor on the G-cell membrane. Examination of CAM expression with the electron microscope confirms that specific CAMS are distributed on the surface of axons in particular fascicles (Bastiani et al., 1987). In vertebrates,

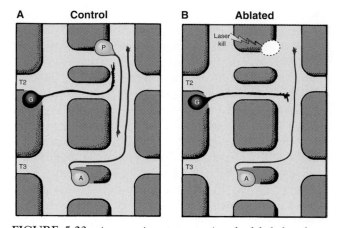

FIGURE 5.23 An experiment supporting the labeled pathway hypothesis. A. In a control embryo, the G growth cone, after crossing the midline, fasciculates with P-axons and not A-axons. B. When the P-neuron is ablated, the G growth cone stalls and does not fasciculate with the A-axons. (After Raper et al., 1984)

too, there is evidence that some axons use a labeled pathway mechanism. The tract of the postoptic commissure (TPOC), for example, is a pioneering tract for axons from the pineal. The pineal axons fasciculate with the TPOC as they turn posteriorly at the boundary between forebrain and midbrain. If the TPOC is ablated, pineal axons often fail to make the appropriate turn (Chitnis et al., 1992).

Just as you may have to change lines at a subway stop to reach your final destination, so axons may have to change pathways. To do so, an axon must change the CAM on its surface. Such a change in the expression of particular CAMs has now been seen in a number of systems at places where axons switch directions. For example, when axons in the central nervous system of *Drosophila* travel on longitudinal tract, they express FasII, but when they leave the longitudinal tract and turn onto a horizontal commissure, they stop expressing FasII and express FasI (Figure 5.24). In addition to following a scaffold of CAM-expressing axons, a new axon may also pioneer a new route during the last leg of its journey and add a new CAM to help future axons reach the same site. Thus, the simple scaffold of the first pioneers with a small number of CAMs becomes increasingly complex as more axons and more CAMs are added to the network.

Not all CAMs are homophilic. Some are involved in heterophilic interactions with other CAMs. For example, the CAM, TAG-1, which is expressed on commissural interneurons of the spinal cord, binds to a different CAM, called NrCAM which is expressed on glial cells in the floorplate (Stoeckli and Landmesser, 1995). Antibodies to NrCAM can be used to perturb the heterophilic interaction between these two CAMs,

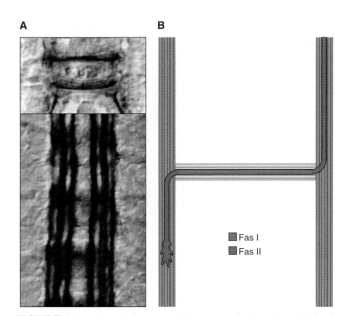

FIGURE 5.24 CAM changing. A. Two panels showing FasI (top) and FasII (bottom) distribution in the embryonic CNS of *Drosophila* as revealed with specific antibodies. B. Axons express different CAMs on different segments. A commissural axon in an embryonic *Drosophila* CNS. This axon expresses FasII in the longitudinal pathway to help it fasciculate with other FasII expressing axons in this pathway, switches to FasI while it is in the commissure and fasciculating with other FasI expressing axons, and then switches back again to FasII once it has reached the other side. (After Zinn et al., 1988; Lin et al., 1994; Goodman, 1996)

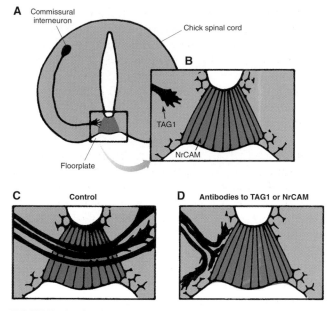

FIGURE 5.25 Crossing the midline of the vertebrate spinal cord. A. A commissural interneuron sends an axon toward the floorplate. B. At higher magnification and with antibodies, we see that the growth cone of the commissural neurons expresses the heterophilic CAM, TAG1, while the floorplate cells express a heterophilic CAM partner, NrCAM. C. In a normal animal, these commissural axons cross the midline, while in (D) if either TAG1 or NrCAM function is perturbed with antibodies, the axons do not cross. (After Stoeckli et al., 1997)

and the result is that commissural interneurons are much less likely to cross the floorplate. Instead, they remain ipsilateral (Stoeckli et al., 1997) (Figure 5.25).

REPULSIVE GUIDANCE

In addition to adhesion molecules and extracellular matrix molecules that promote axon growth, there are factors that do just the opposite. These are the inhibitory or repulsive factors (Kolodkin, 1996). Tissue culture experiments have indicated that neural tissue produces substances that repel some axons. In these sorts of experiments, the trajectories of axons are observed when they are cultured in the presence of tissues they normally avoid. For example, the axons of dorsal root ganglia (DRG) innervate the dorsal spinal cord and generally do not enter the ventral regions of the spinal cord. When these DRGs are grown in tissue culture alongside pieces of dorsal and ventral spinal cord, most DRG axons preferentially invade the dorsal cord and avoid the ventral cord even if they have to grow in a circuitous fashion around it (Peterson and Crain, 1981) (Figure 5.26A,B). Similar results are

obtained by co-culturing pieces of olfactory bulb with septum, a medial structure of the forebrain. Axons from the bulb run laterally in the forebrain appearing to grow away from the septum. When an explant of bulb tissue is placed near the explant of septal tissue, the olfactory axons emerge from the side of the explant opposite the septum (Pini, 1993) (Figure 5.26C,D). These data provide indications that a chemorepulsive mechanism could play a role in axon guidance.

When two explants are co-cultured, there is usually an intermingling of axons, but in the case of retinal and sympathetic co-cultures, it is apparent that the axons avoid one another. Time-lapse video films made of growth cones from one explant as they approach the axons of the other show that these growth cones collapse when they make contact with the foreign axon. They lose their filopodia, retract, and become temporarily paralyzed (Kapfhammer and Raper, 1987a) (Figure 5.27). Often, after a few minutes, a new growth cone is formed which advances once more until it again encounters the unlike axon in its path and again collapses. These studies demonstrate that just the briefest contact from a single filopodium is all that is necessary to elicit this aversive behavior, strongly suggesting that growth cones sense a repulsive signal on

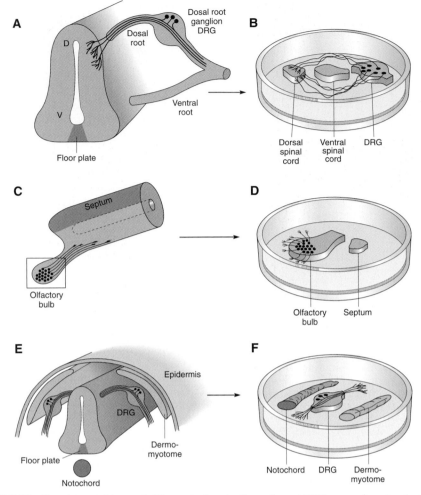

FIGURE 5.26 Repulsive guidance. A. The central projections of most DRG axons do not enter the ventral horn of the spinal cord, but rather make synapses in the dorsal horn. B. When cultured together, DRG neurons avoid ventral spinal cord explants to grow to dorsal targets. C. The telencephalon shows olfactory tract fibers originating from the olfactory bulb traveling in the lateral region, far away from the medial septum. D. When cultured together, olfactory bulb axons travel away from the septum indicating the existence of a diffusible chemorepellent. E. Surround repulsion. DRG axons outside the spinal cord elongate in a bipolar fashion between the dermomyotome and the ventral spinal cord and notochord. Many surrounding tissues, including the epidermis, the dermomyotome, the floorplate, and the notochord, secrete diffusible repellents. F. When placed in a collagen gel between a piece of notochord and dermomyotome, DRG axons extend in a bipolar fashion, similar to their pattern in vivo. (After Peterson and Crain, 1981; Pini, 1993; Keynes et al., 1997)

the surface of the other axon. By pairing different types of explants in such cultures and observing the growth cone interactions, a variety of different collapsing activities effective on different types of growth cones were discovered, showing that there is a rich heterogeneity of repulsive interactions between neurons (Kapfhammer and Raper, 1987b).

Attempts to purify collapsing factors biochemically were aided by a bioassay in which reconstituted membrane vesicles were added to cultures of axons growing on laminin substrates. When vesicles enriched in collapsing activity from the CNS were added to cultures, they caused the immediate collapse and paralysis of all the sensory ganglion cell growth cones on the plate. A 100kd glycoprotein that could cause growth cone collapse, initially called Collapsin, was eventually purified sufficiently to obtain a partial protein sequence (Luo et al., 1993). Through use of the sequence data, the gene was obtained. Collapsin turned out to be a member of a large molecular family, which was then renamed the semaphorin family owing to the role that these molecules have in signaling. Collapsin became known as Sema3A. The first member of the semaphorin family to be identified in grasshoppers, SemaI, (Kolod-

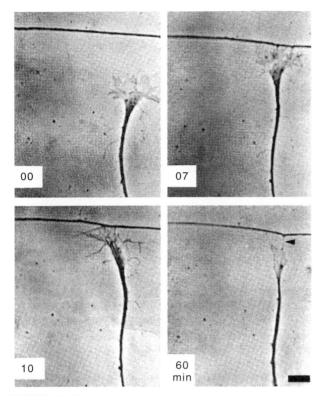

FIGURE 5.27 Growth cone collapse. A time-lapse series of a growth cone from a retinal ganglion cell encountering an axon of a sympathetic axon in culture. Upon first contact, the growth cone retracts and collapses. (From Kapfhammer and Raper, 1987a)

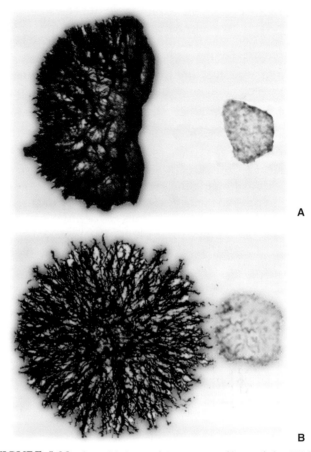

FIGURE 5.28 Sema3A is repulsive to pain fibers of the DRG. A. A DRG (left) is plated next to a small clump of COS cells (right) that are transfected with the *Sema3A* gene. NGF is added to the medium so that pain fibers grow out. There are few neurites, and most of these grow away from the COS cells. B. In a control experiment with untransfected COS cells, the DRG puts out neuritis in all directions. (After Messersmith et al., 1995)

kin et al., 1993), was originally named FasciclinIV because it is expressed on particular axon fascicles. Sema1 is also expressed in stripes near segment borders on the limb bud epithelium, and antibodies that neutralize Sema1 function in the limb allow the Ti1 pioneers to cross the segment border, suggesting that this molecule normally serves a repellent function.

Sema3A is responsible for the repulsive guidance of DRG axons in the ventral spinal cord, described above. Sensory neurons that carry information from pain receptors to the CNS grow into dorsal roots of the spinal cord and synapse locally with dorsal interneurons, avoiding the more ventral spinal cord. Both ventral spinal cord and COS cells made to express Sema3A are able to repel these pain neurons (Messersmith et al., 1995) (Figure 5.28). In Sema3A knockout mice, however, the axons of these neurons enter the ventral spinal cord. In contrast, afferents from stretch receptors dive ventrally to synapse on motor neurons in normal animals and are not repelled by Sema3A. These results suggest that the family of Semaphorin signals is specifically arranged in the CNS to repel specific axons, presumably those that have receptors to particular family members. The ectoderm, dermomy-

otome, and notochord, it turns out, also repel DRG axons (Figure 5.26E, F). If a DRG is placed between a dermomyotome and a piece of notochord in a collagen gel, mimicking its in situ position between these tissues, the result is bipolar axon growth that looks very much like the in vivo trajectory of these neurons and suggests that "surround repulsion" may be a key mechanism for shaping the process trajectories of the developing neurons (Keynes et al., 1997).

An interesting wrinkle concerning repulsive guidance is that the receptor for this cue must first bind the repellent molecule before turning away from it. Since the specificity is often high, this means that the affinity between the receptor and the repellent molecule is also strong. This may not be a problem if the repulsive cue is in the form of a diffusible gradient, but often repellent molecules are attached to cell membranes or the ECM. In such cases, one might imagine that if the affinity is high enough, the axon would attach rather than be repelled. Two mechanisms have been discovered that appear to resolve this problem. The first is extracellular protein clipping in which the ectodomains of activated receptors are attacked by metalloproteases and cleaved, breaking the attachment between the growth cone and its substrate (Hattori et al., 2000). The second is endocytosis in which the entire receptor-ligand complex is internalized into the growth cone (Zimmer et al., 2003). In each case, the molecular bonds holding the growth cone to the repellent surface are neutralized allowing the growth cone to retract.

The repulsive factor responsible for the guidance of olfactory tract axons away from the septum has been identified as the vertebrate homolog of a *Drosophila* protein called Slit, which is the ligand for Slit receptor, Robo (Li et al., 1999). The olfactory bulb axons express Robo, which enables them to sense the Slit. Motor neurons of the vertebrate spinal cord also express Robo and grow away from the ventral midline, which expresses Slit (Brose et al., 1999). The axons of motor neurons and olfactory bulb neurons also grow away from cells transfected with Slit in culture. Slit and Robo will become more important to us later in the chapter when we delve further into the issue of midline crossing.

CHEMOTAXIS, GRADIENTS, AND LOCAL INFORMATION

In a process termed *chemotaxis*, growth cones claw their way up concentration gradients of diffusible attractants to their source, or in the case of negative chemotaxis turn down concentration gradients of repellents. For chemotaxis to occur, the growth cone must be positioned in the gradient such that one side is exposed to a higher level of the factor than the other. Gradients of different molecules can be experimentally produced by ejecting solution from the tip of an electrode and allowing the concentration to dissipate as it spreads out into the tissue culture media. By using this method, it was demonstrated that chick dorsal root axons turn toward a source of Nerve Growth Factor (Figure 5.29) (Gundersen and Barrett, 1979; Zheng et al., 1994), although since then a large number of molecules have been shown to have activity in such pipette-based turning assays. These experiments indicate that growing processes have a mechanism for recognizing small concentration differences across a relatively small distance. Using a controlled gradient in a collagen gel, Rosoff et al. (2004) have been able to show that growth cones can sense a difference in con-

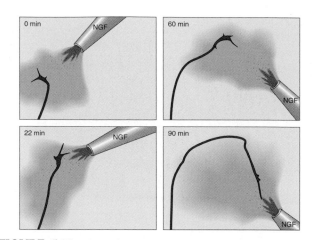

FIGURE 5.29 Growth cones can rely on chemotaxis to orient their growth. A sensory neuron turns toward a pipette that is ejecting nerve growth factor (NGF) and thus producing a diffusible gradient. Each time the pipette is moved, the axon reorients its growth. (After Gundersen and Barrett, 1979)

centration of one molecule in a thousand from one side of the growth cone to the other, making the growth cone the most sensitive reader of chemical gradients known in biology. In the case of pipette-based turning assays, growth cones grow forward at a reasonable pace and turn toward or away from the pipette without speeding up, suggesting that these guidance factors are not simply acting as a general growth-promoting or inhibiting substances but as a directional cues. Growth cones that turn toward attractants also send out more filopodia on the side where the concentration is higher and fewer filopodia on the other side, and the opposite is true for diffusible chemorepellents. This presumably creates differential traction forces toward the attractant or away from the repellent.

Do axons use similar gradient-based mechanisms of growth cone guidance in vivo? One of the first examples of an in vivo gradient attracting axons by chemotaxis was that of trigeminal ganglion sensory axons in the mouse growing to the maxillary pad epithelium at base of the whiskers. This is the most heavily innervated skin in the entire body. When the maxillary pad and trigeminal ganglion are removed and placed near each other in a three-dimensional collagen gel, the trigeminal axons preferentially grow toward the explant of whisker pad, even when there are competing target explants of neighboring pieces of epidermis even closer (Lumsden and Davies, 1986) (Figure 5.30). Thus, it was hypothesized that the maxillary pad emits a tropic agent for axon growth. This factor was nicknamed "max factor," which was later identified as the combination of BDNF and NT-3 (O'Connor and Tessier-Lavigne, 1999), two neurotrophins that will be discussed further in Chapter 7.

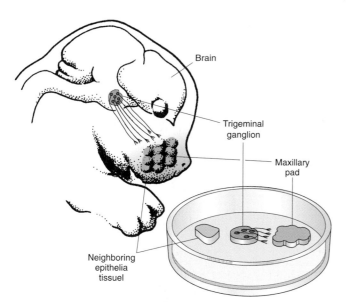

FIGURE 5.30 Chemotactic agents from target tissues. Sensory axons from the trigeminal ganglion heavily innervate the maxillary pad of the mouse face, the site of the whisker field. When the trigeminal ganglion is placed into a three-dimensional collagen gel with the maxillary pad tissue and another piece of epithelium, the axons leaving the ganglion grow toward their appropriate target, suggesting that it is releasing a chemotropic agent. (After Lumsden and Davies, 1986)

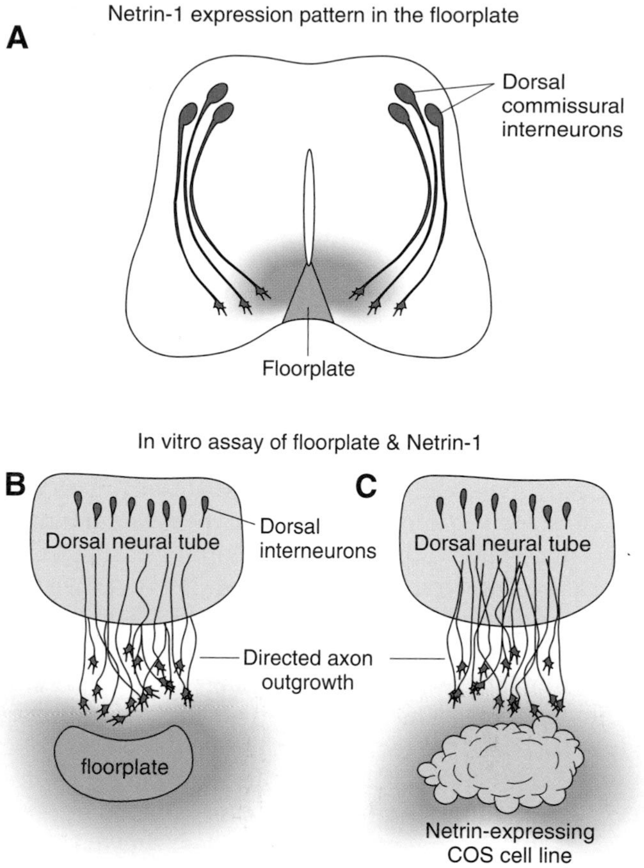

Netrin-1 expression pattern in the floorplate

A

Dorsal commissural interneurons

Floorplate

In vitro assay of floorplate & Netrin-1

B Dorsal neural tube

Dorsal interneurons

Directed axon outgrowth

floorplate

C Dorsal neural tube

Netrin-expressing COS cell line

FIGURE 5.31 Dorsal commissural interneurons are attracted by a gradient of netrin. A. Dorsal commissural interneurons grow directly to the ventral midline of the spinal cord along a gradient of netrin that is released by floorplate neurons. B. In collagen gels, dorsal interneurons are attracted at a distance to the floorplate. C. They are also attracted to netrin released from a pellet of COS cells which have been transfected with the netrin gene. (After Kennedy et al., 1994)

A suggestion for a diffusible guidance mechanism also came from work in nematodes. Here a gene called *unc-6* is expressed at the ventral midline, and *unc-6* mutants show the circumferential guidance defects of certain axons toward the ventral midline (Hedgecock et al., 1990; Culotti, 1994). Nematodes with a deletion of the *unc-40* gene also show defects in the migration of neuronal processes toward the ventral midline. Unlike *unc-6* mutants, the *unc-40* phenotype is cell autonomous, with *unc-40* being expressed in the navigating axons. This suggests that Unc-40 might be a receptor for Unc-6. The first diffusible axon attractant identified turned out to be a homolog of Unc-6. In the spinal cord of vertebrates, dorsal commissural interneurons grow to the ventral midline, cross the floorplate, and then turn 90° and grow in a longitudinal tract beside the floorplate. Two factors were biochemically purified from embryonic chick brains using a bioassay for directed outgrowth from commissural interneurons in explants of dorsal spinal cord. They were partially sequenced, and the genes encoding these proteins were pulled out of a cDNA library (Kennedy et al., 1994; Serafini et al., 1994). They were called netrin-1 and netrin-2 after the Sanskrit "netr," meaning "one who guides." In the spinal cord netrin-1 is expressed in the floor plate (Figure 5.31). The netrins are secreted molecules found largely attached

to cell membranes, but they are somewhat diffusible and can clearly reorient growing commissural axons toward a local source of netrin over a distance of hundreds of microns (Colamarino and Tessier-Lavigne, 1995) (Figure 5.31). In netrin-1 knockout mice, the dorsal commissural interneurons of the spinal cord do not grow all the way toward the ventral midline. Netrins have also been found in *Drosophila* where they serve a similar role in guiding commissural axons to the ventral midline. The sequence of the netrin gene showed it to be a homolog of the nematode gene *unc-6* (discussed below). In vertebrates, DCC binds netrin, is expressed in commissural interneurons, and is essential for the attraction of their axons to the floorplate (Keino-Masu et al., 1996). The DCC protein, which is a transmembrane receptor, is a homolog of the nematode Unc-40 (Chan et al., 1996). A mutation in the *Drosophila* netrin receptor homolog, which is called *frazzled*, shows similar defects in commissural guidance (Kolodziej et al., 1996). These results suggest a strongly conserved function of netrins and netrin receptors in chemoattraction toward the ventral midline.

Although guidance to a distant target through the target's release of a diffusible cue would appear to be a reasonable way to guide axons to their targets, it seems that this solution is used rather rarely in the nervous system, and target tissues do not generally put out long range attractants. Rather, as we shall see, axons tend to use a number of intermediate targets and several distinct local cues on the way to their destinations. The early neuroepithelium is like a patchwork quilt of various guidance cues that act as attractants and repellents and operate in context with particular ECM molecules and CAMs. This rich, detailed molecular array covers the entire developing brain. Axons read the terrain and respond appropriately. Consistent with this analogy are the results of various embryonic perturbations. If the tectum is removed, retinal axons grow toward the missing tissue, suggesting that optic axons use these local cues rather than a diffusible attractant from the tectum as they grow along the optic tract (Taylor, 1990). If a small piece of the optic tract neuroepithelium is rotated 90° before the axons enter it, then they become misoriented when they enter the rotated transplant (Harris, 1989) (Figure 5.32) and correct their course of growth when they exit. These results confirm that the neuroepithelium contains local information to which growing axons respond and that they are not simply following gradients of attractants released by their targets.

As a growth cone migrates through the embryonic nervous system, it encounters new molecular cues every 10 to 50 μms or so. How does such a molecularly complex terrain get established in the neuroepithelium? It appears that the early molecular events that pattern the embryo play a role in setting up the domains of these guidance cues. The *homeobox* genes that provide segmental identity (see Chapter 1) also direct the expression of various axon guidance cues, and pioneer axons are often found at the boundaries of brain territories that express different *homeobox* positional markers. Moreover, axonal tracts in embryonic brains exhibit unusual patterns when the expression of *homeobox* genes is perturbed (Wilson et al., 1997). Sometimes the molecules that are involved in the initial patterning of the nervous system are also used as cues in pathfinding. We learned in Chapter 2, for example, that *Wnts* are expressed in caudal high to rostral low gradient during gastrulation. The same *Wnt* gradient appears to be involved in giving cues to spinal axons about whether they are traveling toward

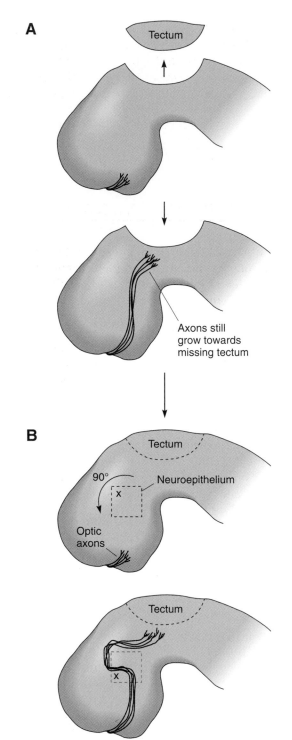

FIGURE 5.32 Retinal axons follow local guidance cues in the neuroepithelium. A. When the tectum is removed, the axons still grow correctly to the tectum, indicating that the tectum is not the source of a diffusible attractant. B. A piece of neuroepithelium in front of the retinal axons is rotated 90° (top). When the retinal axons enter the rotated piece, they are deflected in the direction of the rotation, but they correct their trajectories when they exit the rotated piece, showing that these axons pay attention to localized cues within the epithelium. (After Harris, 1989, and Taylor, 1990)

or away from the brain (Lyuksyutova et al., 2003) (Figure 5.33). *Shh*, which as we have seen participates in patterning the ventral neural tube, also acts as a guidance cue at the ventral midline of the spinal cord (Charron et al., 2003) and at the optic chiasm (Trousse et al., 2001). Conversely, BMPs that pattern the dorsal spinal cord are repulsive to the growth cones of dorsal interneurons that grow away from the dorsal midline (Augsburger et al., 1999).

THE OPTIC PATHWAY

The optic pathway from the retina to the tectum presents one of the most complete examples of the various different kinds of guidance cues that are used in guiding a single axon from its origin to its target (Dingwell et al., 2000; Johnson and Harris, 2000). Retinal axons travel centrifugally from a peripheral location on the retina toward the optic nerve. When they get to within about 50 µm of the optic nerve head, they encounter a high concentration of netrin, which acts as an attractant for RGC axons (Deiner et al., 1997). At the optic nerve head on the vitreal surface of the retina is a layer of the ECM molecule laminin. The combination of laminin and netrin is repulsive rather than attractive, and so these retinal axons turn away from the surface of the retina and dive into the optic nerve where they travel until they enter the brain near the chiasm (Hopker et al., 1999). Here they encounter

the repulsive guidance molecule Slit and the morphogen Shh that also acts as a repellent to RGC axons. Slit and Shh are expressed anterior and posterior to the chiasm, but not in the chiasm itself (Erskine et al., 2000; Trousse et al., 2001). Thus, these molecules corral the RGC axons into the chiasm proper. Zebrafish mutants called *astray* do not have a functional Slit receptor in their RGCs, and in these mutants fewer RGC axons find the chiasm, and many get lost when they enter the brain (Fricke et al., 2001; Hutson and Chien, 2002). At the chiasm itself, there is a high concentration of another repulsive guidance molecule called EphrinB. The RGCs from the ventrotemporal part of the mouse retina express EphB, the receptor for EphrinB, and so ventrotemporal retinal axons do not cross the chiasm but remain ipsilateral while the dorsal and nasal axons, insensitive to EphrinB, cross the chiasm to the other side of the brain (Nakagawa et al., 2000; Williams et al., 2003). Once in the optic tract, retinal axons are influenced by the repulsive guidance cue Sema3A and the ECM heparan sulfate so that they are guided toward the tectum (Walz et al., 1997; Campbell et al., 2001; Irie et al., 2002). At the front of the tectum, the RGC axons encounter a sudden drop of FGF, which signals that they have entered the target area (McFarlane et al., 1995). In the target area they encounter orthogonal gradients of EphrinA and EphrinB that signify tectal coordinates (see Chapter 6). In the frog embryo, retinal growth cones encounter all these cues and several others during their journey of no more than about 800 µm (Figure 5.34).

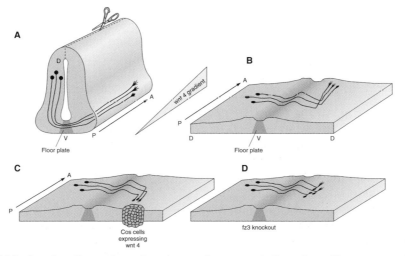

FIGURE 5.33 Local gradients of morphogens can orient axons. A. Commissural interneurons of the spinal cord, once they cross the ventral midline, grow anteriorly toward the brain, and up the *Wnt4* concentration gradient. B. These axons can be seen well in a filleted preparation grown in culture. The neural tube is sliced open at the dorsal midline and flattened out. Label is applied to the commissural interneurons. C. Commissural interneurons grow posteriorly if a ball of COS cells expressing *Wnt 4* is placed on the posterior side of such an explant. D. In a fz3 knockout, lacking the *Wnt4* receptor, commissural interneurons do not grow either anteriorly or posteriorly once they cross the midline. (After Lyuksyutova et al., 2003)

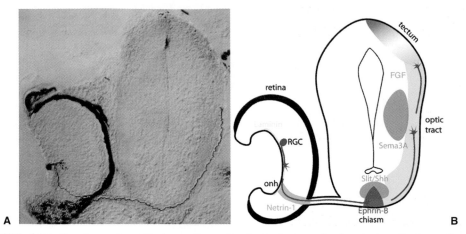

FIGURE 5.34 Local cues for retinal ganglion cells. A. A photograph of a single retinal ganglion cell growing toward the tectum. The cell has been filled with a dye for visualization. B. Various guidance cues that the growth cone of the RGC uses to orient toward its target in the tectum are artificially colored in, although in close correspondence to the known distribution of these guidance cues. (Courtesy of Christine Holt)

THE MIDLINE

A very interesting problem in axonal guidance is that commissural axons generally only cross the midline once; they do not return. This seems a bit odd if the midline has solely an attractive function. This point is brought home in a *Drosophila* mutant *roundabout* (*robo*). Commissural axons in *robo* mutants cross the midline and then recross it at the next commissure. They do this over and over, sometimes going in circles (Figure 5.35). Robo is the receptor for a midline repellent Slit, and *slit* mutants show a similar phenotype. Interestingly, there is another class of mutant in *Drosophila* called *commissureless* that has the opposite phenotype (Seeger et al., 1993; Tear et al., 1993). In these mutants, commissural neurons are unable to cross the midline and remain ipsilateral (Figure 5.35). Only axons that cross the midline express *commissureless*; those that do not, never cross the midline. At the midline, it appears that Comm binds Robo and removes it from the growth cone surface into internal vesicles so that the axons are not repelled by Slit (Keleman et al., 2002; Myat et al., 2002). Once they cross the midline, Comm protein levels are reduced and Robo returns to the surface so the axons do not recross.

In *Drosophila*, there are actually three different *Robo* genes, and each one has a different affinity for Slit. The Robo with the highest affinity is expressed on longitudinal fascicles that are farthest away from the ventral midline, the source of Slit, while the Robo with the lowest sensitivity to Slit is expressed on axons that grow right next to the midline. Thus, the particular

form of Robo which an axon expresses determines where it forms its longitudinal fascicle (Rajagopalan et al., 2000; Simpson et al., 2000). Later axons use a combination of this Robo code, plus the expression of a particular homophilic CAM or fasiclin to join their appropriate fascicle.

Vertebrate commissural interneurons also only cross the midline once. But here a different mechanism is at play. Instead of gaining sensitivity to a midline repellent, they lose sensitivity to the midline attractant, Netrin. Whole pathway preparations were placed into culture to show that once these axons had crossed the midline, they no longer responded to a source of netrin, and that if they were exposed to a source of netrin before they reached the midline, they were no longer attracted to the midline (Shirasaki et al., 1998). In addition, Slit at the vertebrate midline activates the Robo receptor on these axons. Once activated, the intracellular domain of the Robo receptor binds to the intracellular domain of the DCC receptor and blocks Netrin signaling, and so once these axons have crossed the midline they only sense the repellent activity of Slit and not the attractive activity of Netrin (Stein and Tessier-Lavigne, 2001).

ATTRACTION AND REPULSION: DESENSITIZATION AND ADAPTATION

The responses of growth cones to the same cues may change as axons progress toward their final destinations. We have just seen examples of this at the ventral midline, which can be considered an interme-

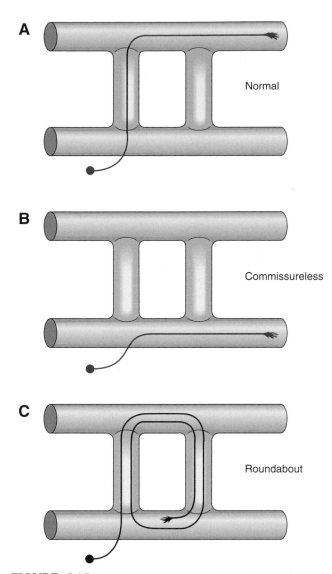

FIGURE 5.35 Midline crossing mutants in *Drosophila*. A. In normal flies, many neurons cross the midline once in a commissure and then travel in longitudinal fascicles on the other side. B. In *commissureless* mutants, the axons do not cross but travel in longitudinal tracts on the same side. C. In *roundabout* mutants, the longitudinal tracts do not form properly because the axons keep crossing back and forth. (After Seeger et al., 1993)

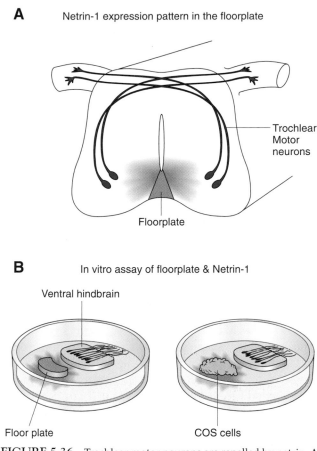

FIGURE 5.36 Trochlear motor neurons are repelled by netrin. A. Trochlear motor neurons arise in the ventral neural tube at the midbrain/hindbrain region. They grow away from the ventral midline to decussate and leave the brain dorsally. B. Trochlear neurons in a collagen gel explant culture grow away from the floorplate, and C. from COS cells expressing Netrin. (After Colamarino and Tessier-Lavigne, 1995)

diate target. These axons must not get stuck where the Netrin concentration is highest, but should move on and take the next portion of their journey. One way to do this is to make an attractant either unattractive or even downright repulsive. Netrin (or Unc-6) is attractive for axons that express DCC (or Unc-40). Mutants in DCC (or Unc-40) disrupt the migration of axons toward Netrin (or Unc-6). The *unc-5* mutant has the opposite phenotype. It disrupts dorsal migration (Culotti, 1994), but the *unc-5* phenotype is dependent on wild-type *unc-6* function. Interestingly, when neurons whose axons normally grow toward the Unc-

6 at the ventral midline, are made to misexpress Unc-5, these axons now grow dorsally instead, again depending on the normal expression of Unc-6 to do so. Unc-5, like Unc-40, codes for a transmembrane protein, possibly another Unc-6 receptor, but is involved in chemorepulsion rather than chemoattraction.

This means that Netrins can act as chemorepulsive agents for neurons that express Unc-5 type receptors and chemoattractive factors for neurons that express Unc-40 or DCC-type Netrin receptors. That Netrin can act as a long-range chemorepulsive factor in vertebrates has been demonstrated in trochlear motor neurons whose axons grow dorsally away from the ventral midline (Colamarino and Tessier-Lavigne, 1995) (Figure 5.36). In collagen gels, these axons grow away from explanted floorplate tissues or from cells transfected with a Netrin-1 expressing gene. The switch from attraction versus repulsion to Netrin can be mediated by the expression of a different receptor such as

Unc-5, but it can also be mediated by DCC alone, depending on the level of intracellular cAMP (Ming et al., 1997). When cAMP inside the growth cone is high, Netrin acts like an attractant to spinal and retinal axons, but when cAMP is experimentally lowered using drugs that block adenylate-cyclase, then the response to Netrin by the very same axons is one of avoidance and repulsion (Figure 5.37). Similarly, many other attractants can be switched to repellents by pharmacologically down-modulating cAMP in growth cones. One hypothesis for how this works is that the signal from the guidance receptor catalyses cytoskeletal polymerization/depolymerization reactions. In high cAMP this reaction favors polymerization, while in low cAMP it favors depolymerization (Ming et al., 1997; Song et al., 1997). Although some guidance cues are primarily modulated in this way by cAMP, others such as NGF and Sema3A are modulated by cGMP (Song et al., 1998).

How might this modulation happen in vivo? Earlier, it was noted that the combination of Netrin and laminin is repulsive to retinal axons and drives them from the retinal surface into the optic nerve. Laminin dramatically reduces the cAMP level in these growth cones, so that when these retinal axons are grown on high levels of laminin in vitro, the cAMP level in the growth cones is low and Netrin is repulsive (Hopker et al., 1999). If the cAMP level is artificially raised, retinal growth cones on laminin are once again attracted to Netrin. If the same neurons are grown on low levels of laminin or fibronectin, cAMP is high in the growth cone, and Netrin is attractive, but this can be switched by experimentally lowering cAMP. So it seems that environmental cues, such as which ECM molecules the growth cones are most exposed to, can modulate whether a guidance cue is attractive or repulsive.

As axons grow past intermediate targets, such as the optic nerve head, they are also getting older and maturing. This intrinsic aging process may also affect their internal cAMP levels and thus the way an axon responds to a guidance cue like Netrin. By the time retinal axons enter the brain, they are no longer responsive to Netrin (even if they are grown on low levels of laminin), and by the time they reach the tectum Netrin is repulsive. This change in responsiveness happens in a stage-dependent way even in isolated retina explants where the axons are not exposed to the optic pathway (Shewan et al., 2002). Another way that growth cones can change their sensitivity to a particular guidance cue is to make or degrade proteins that are critical for responding to the cue, such as a receptor. Retinal growth cones do not "see" Sema3A in their pathway until they grow along the optic tract. At early stages of pathway navigation, retinal axons do not express Neuropilin, the Sema3A receptor, and so Sema3A is neither attractive nor repulsive to them. However, by the time these axons have entered the optic tract, they express Neuropilin and are repelled by Sema3A (Campbell et al., 2001).

It was initially thought that all proteins in the growth cone were made in the cell body and shipped to the growth cone. It now appears that the growth cone is full of mRNAs, ribosomes, and other translational machinery, and can respond to guidance cues by making new proteins that are involved in growth and navigation (Campbell and Holt, 2001). Growth cones also have degradation machinery, including ubiquinating enzymes and proteosomes (Campbell and Holt, 2001). Both synthesis and degradation can be activated in growth cones by guidance cues, and if either process is experimentally inhibited, a number of guidance cues become completely ineffective, suggesting that growth cone navigation may depend on the local manufacture and degradation of proteins. mRNAs, such as beta-actin mRNA, which code for

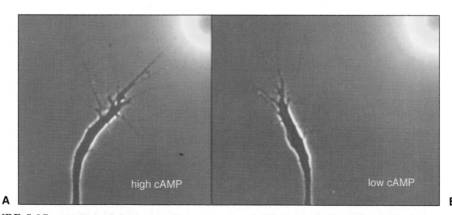

FIGURE 5.37 cAMP modulates growth cone turning. A. When internal cAMP is high, the growth cone of embryonic spinal neurons grows toward a source of Netrin, ejected by a pipette. B. When cAMP is pharmacologically lowered, the same neurons are repelled by Netrin. (After Ming et al., 1997)

proteins that form the growth cone filopodia, are selectively transported to growth cones (Bassell et al., 1998), a finding that suggests that the concentration of cytoskeletal components in the growth cone may be affected by guidance cues. Another way that local protein synthesis might be involved in guidance is in the manufacture of new receptors once growth cones have reached intermediate targets; this would allow them to become sensitive to new guidance cues on the next leg of the journey (Brittis et al., 2002).

Various studies have shown that growth cones adapt to guidance cues. For example, if a growth cone bumps into an axon with a repulsive guidance cue (as mentioned above), it will collapse and retract. When it regrows, it may run into the same axon again and repeat the collapse to regrowth cycle. However, after several such cycles, it has been observed that growth cones generally become desensitized to the collapse-inducing factor and are able to grow over repulsive axons (Kapfhammer et al., 1986). In another study, axons were tested to see how far they would crawl along a membrane on which there was an increasing gradient of the repulsive guidance factor, EphrinA. Axons were started on a platform of different levels of EphrinA, and those that started on higher concentrations ended up growing the farthest, suggesting that they had partially adapted to EphrinA (Rosentreter et al., 1998). Adaptation can be subdivided into two parts, desensitization followed by resensitization. Recent work suggests that desensitization involves the internalization and degradation of receptors to guidance factors (Piper et al., 2005), while resensitization involves making new proteins that counteract this process (Ming et al., 2002; Piper et al., 2005). Growth cones can thus adjust the levels of the proteins that are critical for axon navigation. When this is added to the ability of growth cones to regulate cyclic nucleotides that can rapidly mediate a switch between attraction and repulsion, the picture that emerges of the growth cone is that of a very autonomous machine capable of continually redefining itself as it navigates through the embryonic brain.

SIGNAL TRANSDUCTION

Signal transduction in growth cones is the process by which guidance cues exert their effects on the dynamic cytoskeleton. Some aspects of signal transduction such as the activation of kinases, phosphatases, cyclic nucleotide levels, and protein turnover have been mentioned in the above sections. CAMs, integrins, repulsive factors, attractive factors, and growth factors are received by receptors on the surface of the growth cone. Many such receptors have intracellular domains that have enzymatic activity when an extracellular ligand is bound and are thus able to amplify the signal (Strittmatter and Fishman, 1991). For example, the Robo receptor has one kind of intracellular domain that mediates a repulsive response to Slit, while the Frazzled receptor (the *Drosophila* homolog of DCC) has a different intracellular domain that mediates attraction to Netrin. The specific action of these intracellular domains is made clear in experiments where domains are switched, as in Robo-Frazzled fusion proteins (Bashaw and Goodman, 1999), where the intracellular domain of Frazzled is fused with the extracellular domain of Robo, the repulsive Slit receptor, which makes neurons attracted to Slit. Similarly, neurons expressing Frazzled-Robo fusion proteins are repelled by netrin.

Receptors that have intracellular tyrosine kinase (RTKs) or tyrosine phosphatase (RTPs) activity have been found in abundance on growth cones (Goldberg and Wu, 1996; Goodman, 1996). An RTK in *Drosophila* called *derailed* is expressed on particular fascicles in the nervous system, and *derailed* mutants exhibit striking pathway errors (Callahan et al., 1995). Derailed appears to be a receptor for a wingless protein that is expressed on the posterior commissure and acts as a repellent to Derailed expressing axons (Yoshikawa et al., 2003). Some receptors do not possess intracellular enzymatic domains on their own. However, they may be able to recruit other RTKs, such as the FGF receptor, to do the work. Thus, there is a CAM binding site on the FGF receptor, and NCAM binding can stimulate phosphorylation of growth cone proteins via FGF receptor activity (Williams et al., 1994). In addition to RTKs, some receptors are thought to signal through cytosolic tyrosine kinases and phosphatases. Among the nonreceptor tyrosine kinases (NRTK) are *src*, *yes*, and *fyn*. Neurons from single-gene mouse knockouts that have no *src* are specifically unable to grow on the CAM, L1, while neurons from single-gene knockout mice that have no *fyn* are unable to grow on NCAM (Beggs et al., 1994). A number of RTPs are found predominantly in growth cones, such as DLAR, which is expressed in *Drosophila* motor axons. Mutants in these genes can also lead to growth and pathfinding defects (Desai et al., 1996). It is not precisely known what the critical targets of each of these phosphorylation and dephosphorylation reactions are in the growth cone for any of these pathways, but it is likely that many of them act on proteins involved in cytoskeletal dynamics. Well known among these are the small GTPases of the Rho-family (Luo et al., 1997; Dickson, 2001). These exist in two states: an active GTP bound state and an inactive GDP bound state. Rho-family molecules are involved in various actin rearrangements, which are in turn regulated by a host of effector molecules like the Guanidine Exchange Factors (GEFs) that exchange GDP for GTP. The best known Rho-family members are RhoA, Rac1, and Cdc42. RhoA is involved in actin mediated neurite retraction, while Rac1 and Cdc42 promote filopodia and lamella formation and thus neurite extension.

Growth cone filopodia are long, motile, and covered with receptors, so they are able to sample and compare different parts of their local environment. They also have a very high surface-to-volume ratio, which can help convert membrane signals into large changes in intracellular messengers such as calcium. Filopodia can show localized transient elevations of intracellular calcium. These transients reduce filopodial motility. By stimulating Ca transients through uncaging in the filopodia on one side of a growth cone, the growth cone will turn (Gomez et al., 2001). Experiments in which calcium is uncaged on one side of a growth cone that is moving forward generally cause the growth cone to turn toward the side that has the elevated calcium (Zheng, 2000). Calcium may also be released from internal stores in response to a signal from the cell surface or enter the growth cone through calcium channels, and may stimulate or inhibit neurite outgrowth. Serotonin, which stops growth cone advancement in certain neurons of the snail *Heliosoma*, appears to work by increasing calcium levels locally, and stopping behavior can be elicited using calcium ionophores. In this preparation, it is possible to cut single filopodia off an active growth cone and study its behavior in isolation. Such isolated filopodia react to the application of serotonin by showing a marked increase in calcium along with a shortening response (Kater and Mills, 1991), giving an excellent insight into just how localized sensory and motor responses in growth cones are. Collapse responses to some factors are blocked by drugs that inhibit calcium elevations, whereas in other neurons, calcium appears to stimulate neurite outgrowth and growth cones will grow in the direction of agents that increase intracellular calcium on one side of the neuron over the other. It is not entirely clear how calcium triggers these responses; a likely possibility is that calcium stimulates the activity of certain cytoplasmic kinases, such as CAM kinase II or PKC, which then go on to affect the cytoskeleton.

GAP-43, a *g*rowth-*a*ssociated *p*rotein in axons, is highly enriched in growth cones of growing and regenerating axons (Skene et al., 1986). It is an internal protein that is associated with the cytoplasmic membrane and various cytoskeletal proteins. The function of GAP-43 has been tested by overexpressing it in cul-

tured neurons and in transgenic mice and by inhibiting its expression. The results show that extra GAP-43 enhances axon outgrowth, while inhibiting GAP-43 compromises growth and leads to stalling (Fishman, 1996). In GAP-43 knockout mice, retinal axons stall and then take random courses when they reach the optic chiasm (Strittmatter et al., 1995). When GAP-43 is overexpressed in transgenic mice, exuberant growth occurs and pathfinding errors are also made. These results suggest that regulated levels of GAP-43 are essential for the normal responses of axons to external cues. The activity of GAP-43 is regulated, in part, by phosphorylation through PKC and dephosphorylation through a phosphatase. The phosphorylated form of GAP-43 seems to stimulate its activity and promote outgrowth, while the dephosphorylated form is less active. It is not yet known how GAP-43 regulates axon growth, but evidence suggests that it directly links the cytoskeleton with the protrusive membrane of the growth cone, thus enhancing motility.

Local protein synthesis within the growth cone is possible because of the presence of mRNAs, which are targeted and transported to the growth cone, and the presence of a full complement of protein synthetic machinery. Similarly, protein endocytosis and degradation machinery, including the ubiquitinating enzymes that target proteins to the proteasome for degradation, are present in the growth cone (Campbell and Holt, 2001). When a guidance molecule stimulates protein synthesis or degradation, it does so through MAP kinase pathways. There are several different MAP kinases, some of which eventually activate a protein called Target of Rapamycin, mTOR (rapamycin being a natural toxin of stimulated protein synthesis in various cell types). TOR is a kinase that phosphorylates a protein called eIF4E-BP, a negative regulator of the translation initiation factor eIF4E (Campbell and Holt, 2001). Other MAP kinases activate specific ubiquitin ligases that attach ubiquitin molecules to particular proteins targeting them for degradation. In fact, it seems likely that there are a variety of downstream kinases that regulate different aspects of the growth cone response when stimulated by a single guidance cue.

SUMMARY

We began this chapter by comparing axonal navigation with human navigation. We mentioned the need for a motor, and we have seen that the growth cone by virtue of its dynamic cytoskeleton is able to locomote forward, turn, stop, and even retract. We mentioned the need for guidance cues, and we have seen a variety of cues attached to the extracellular matrix and cell surfaces, and as diffusive gradients of guidance molecules. Some of these factors promote growth and adhesion, and others inhibit growth and adhesion. We suggested that these various signals had to be integrated and communicated to the motor, and we have seen a variety of intracellular agents in the growth cone that can be regulated in response to external cues and communicated to the active cytoskeleton. We are, however, still a long way from understanding how axons grow to their targets. The molecules mentioned in this chapter are used only in some neurons, and it is fair to say that for even the best studied neurons, we understand only small parts of their navigation, but not their entire route. Many more guidance factors are known but are not mentioned in this chapter, and many others remain to be discovered. Our insights into how these cues regulate growth and guidance are still in their infancy but it is possible that these insights will bring us closer to understanding how to rejuvenate adult neurons so that they can regenerate.

BOX 1

AXON REGENERATION

In the adult mammalian CNS, axons fail to regenerate following injury (Ramony Cajal, 1928; Aguayo et al., 1990). Axons that are cut find it difficult or impossible to cross the lesion site, and many neuronal cells whose axons are cut die as the result of the injury. This means that injuries that break axons in the spinal cord of an adult human can lead to permanent paraplegia or quadriplegia. Work on axonal regeneration is therefore of intense medical interest. The inability of adult central axons to regenerate is in stark contrast to the situation in the peripheral nervous system where regeneration is possible and the situation in lower vertebrates such as fish and amphibia. In these animals, for example, retinal ganglion cells are fully capable of regeneration (Piatt, 1955), and severing the optic nerve in a salamander, an insult that would lead to permanent blindness in an adult human, is followed by the regrowth of these axons and the restoration of vision. The failure of regeneration in the adult mammalian nervous system is also in contrast to the ability of the developing nervous system to send out long axons, and the capacity of central axons to regenerate is lost during the early stages of mammalian development (Kalil and Reh, 1982). It is as though there were a connection between evolution and development in the ability of axons to regenerate central axons. Perhaps the key to central regeneration is to find a way of making the damaged tissue act more like it did during the time when it was developing.

For both intrinsic and extrinsic reasons central neurons are incapable of regeneration. Let us look at the extrinsic factors first, because more work has been done on this aspect of the problem. Several lines of evidence point to the importance of extrinsic factors. For instance, the axons that are able to regenerate following a pyramidal tract lesion in neonatal hamsters or cats grow *around* the lesion site and are not able to penetrate the injury site (Bregman and Goldberger, 1983). Thus, there is thought to be something inhibitory at the lesion site. The importance of extracellular cues in vivo is clearly illustrated by the ability of peripheral nerve grafts to support central axonal regrowth (Richardson et al., 1980; David and Aguayo, 1981; Aguayo et al., 1990). In a set of classic studies, it was shown that while transected central axons were unable to grow within the CNS, they could grow for many centimeters through a sheath of nonneuronal cells that ordinarily provide insulation to motor axons in the periphery (Figure 5.38). Indeed, while embryonic and peripheral

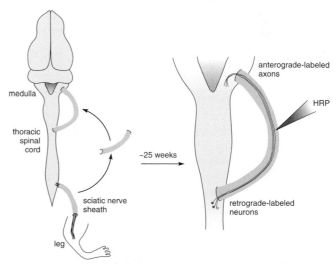

FIGURE 5.38 Central neurons in spinal tracts do not regrow long axons after they are transected, but if they are allowed to innervate a sheath of peripheral nerve, they can regrow over substantial distances. (After David and Aguayo, 1981)

BOX 1   (cont'd)

glial cells support neurite outgrowth, adult astrocytes and oligodendrocytes appear to inhibit neurite outgrowth. When a central nerve bundle is injured in a mammal, the axons are usually unable to regrow across the wound and thereby reestablish connections they had lost. Part of the problem, it appears, is the invasion of the wound site with various glia, which produce repulsive cues that the axons cannot navigate around. By X-irradiating mouse spinal cords during neonatal development, it was possible to create mice that are deficient in glial cells. In these animals, spinal axons regenerated past a transection point, a behavior that they never display in normal animals (Schwab and Bartholdi, 1996).

To find the molecular components involved in inhibiting central regeneration, the system was brought into culture where it was found that CNS neurons stop, and sometimes collapse, when they touch oligodendrocytes. Liposomes from these cells and preparations of myelin were used to identify an inhibitory factor that causes CNS growth cones to collapse. A monoclonal antibody to this factor, which is now called Nogo, was then made and tested in culture for its ability to block the collapsing activity. In the presence of antibody, axons grew over oligodendrocytes without stopping or collapsing. The antibody was then tested in vivo, using mice with partially severed spinal cords. In the presence of Nogo antibodies, many more axons were able to regenerate beyond the crush than in control animals, and there was considerable functional recovery suggesting that Nogo is a critical component of the failure of spinal regeneration (Schnell and Schwab, 1990; Bregman et al., 1995), although there has been some debate as to whether knocking out Nogo in mice leads to enhanced regeneration. The Nogo receptor was identified and found to be a receptor for other myelin-derived inhibitory factors such as Myelin Associate Glycoprotein (MAG), indicating that this receptor may provide an insight into where the signals that inhibit regeneration converge (Fournier et al., 2001).

Astrocytes often accumulate around CNS wounds, forming complex scars. These cells produce an extracellular matrix that is inhibitory to axon regeneration, and one of the key components of this inhibitory material may be chondroitin sulfate glycosaminoglycan chains found on many proteoglycans in the astroglial scar (Asher et al., 2001). Even when plated on the growth-promoting ECM component, laminin, spinal neurons stop growing when they confront a stripe of chondroitin sulfate. In culture, the inhibitory component can be digested away with chondroitinase, rendering the matrix more permissive to axon growth and regeneration. To see if chondroitinase could be used to treat models of CNS injury in vivo, rats whose spinal cords had been transected, were treated locally with the enzyme. Such treatment restored synaptic activity below the lesion after electrical stimulation of corticospinal neurons and promoted functional recovery of locomotor activity (Bradbury et al., 2002).

It is clear that progress is being made on the extrinsic factors that inhibit axon regeneration, but there is still the problem that older central axons simply do not regenerate very well even when the conditions are good. Adult axons can grow for a short distance ($<500\,\mu m$) in many central locations (Liu and Chambers, 1958; Raisman and Field, 1973). The growth of very young neurons does not appear to be so restricted in an adult nervous system. Human neuroblasts are able to form long axon pathways when transplanted into excitotoxin-lesioned adult rat striatum (Wictorin et al., 1990). Similarly, mouse embryonic retinal ganglion cells are able to grow long distances within the rostral midbrain of neonatal rats and selectively innervate some normal targets (Radel et al., 1990). Indeed, myelin inhibits regeneration from old but not young neurons. What do these young axons have that older axons do not? It was found that the levels of cAMP in young growth cones are much higher than in older axons (Cai et al., 2001). By increasing the cAMP levels, one can turn old neurons into neurons that behave more like young ones in terms of their regenerative potential (Qiu et al., 2002). Moreover, the recovery from spinal injury in neonatal rats is markedly reduced by lowering cAMP levels.

Another key intrinsic difference between old and young axons may have to do with protein synthesis. Young growth cones are full of protein synthetic machinery, but the axons of older neurons do with less of such machinery. There appears to be a good correlation between the ability of a growth cone to make new proteins and its ability to regenerate in vitro. The challenge now will be to find ways to crank up the protein synthetic machinery in the growth cones of damaged CNS neurons to see if this can aid recovery. When considering all these data, it seems that full recovery after an injury may require a strategy for dealing with both intrinsic and extrinsic factors.

BOX 2

TISSUE CULTURE

The tissue culture technique, a mainstay of all biological research in the last century, has continuously embraced innovative solutions to address neurobiological questions (Bunge, 1975; Banker and Goslin, 1991). In fact, Ross Harrison invented tissue culture to study axon outgrowth (Harrison, 1907, 1910). His original preparations consisted of pieces of tissue, now termed *organotypic cultures*. Such cultures may now be obtained from vibratome sections of neural tissue and grown under conditions that promote thinning to a monolayer, thus providing greater access and visibility of individual neurons (Gähwiler et al., 1991). Slices are attached to a coverglass and placed in rotating tissue culture tubes (hence the term *roller-tube culture*) such that the tissue culture media transiently washes over them. If one requires a slice of tissue with somewhat greater depth, then the cultures can be grown statically at the gas–liquid interface by using tissue culture plate inserts that provide a porous stage for the tissue and a reserve of media below (Stoppini et al., 1991). The relative simplicity of modern organotypic preparations has resulted in a wealth of data on the interaction between afferent and target populations, as described in the text.

In the arena of primary dissociated cell cultures, it has become feasible to isolate particular cell types. This may be performed by an immunoselection technique in which a cell-specific antibody is adsorbed to a plastic Petri dish, creating a surface on which one cell type will selectively attach. This approach has led to a 99% pure retinal ganglion cell preparation (Barres et al., 1988). A different means of separating cells relies upon selective pre-labeling with a fluorescent dye and subsequently performing fluorescent-activated cell sorting (FACS). When passed through such a device, single cells are sequentially monitored for fluorescence and then selectively diverted to a receiver tube if they are labeled. This approach led to the isolation of retrogradely labeled spinal motor and pre-ganglionic neurons (Calof and Reichardt, 1984; Clendening and Hume, 1990). Finally, it is possible to isolate large and small cell fractions following centrifugation on a Percoll density gradient, and then further enrich the cells with a short-duration plating step, which allows the more adhesive cells (e.g., astrocytes) to be retained on a treated surface. This approach led to the isolation of a >95% pure granule cell population from cerebellar tissue (Hatten, 1985). Once specific cell types have been isolated, they may be mixed together in known ratios, or plated on two surfaces that are subsequently grown opposite one another as a sandwich. This technique allows one to produce a "feeder layer" of astrocytes on one surface that promotes survival of low-density neuronal cultures. It may also allow the experimenter to discriminate between contact-dependent and contact-independent phenomena.

Having obtained the neurons and glia of interest, tissue culture offers the opportunity to perform insightful manipulations. For example, it is possible to produce a nonuniform distribution of growth substrates to test the role of specific molecules in axon guidance (Letourneau, 1975) or stripes of membranous material as assays for the identification and characterization of axon guidance factors (Walter et al., 1987a; Walter et al., 1987b). The technique has been extended to create gradients of laminin or neuronal membrane on a surface that subsequently serves as the tissue culture substrate (McKenna and Raper, 1988; Baier and Bonhoeffer, 1992). The gradients can be visualized and quantified by including a fluorescent or radioactive marker along with the intended substrate. Gradients of soluble molecules can be produced in vitro with repetitive pulsatile ejection of picoliter volumes from a micropipette tip into the tissue culture media (Lohof et al., 1992). The concentration gradient is quantified by ejecting a fluorescein-conjugated dextran and measuring the fluorescent signal at increasing distances from the pipette tip. Three-dimensional tissue culture is also possible by embedding neurons or explants in gelatinizing collagen or mixture of ECM material. Such gels not only provide a more "realistic" substrate for axons to grow through, they also allow for the formation of relatively stable gradients of soluble factors that can percolate through the gel undisturbed by flows and currents that happen when the experimenter moves the culture dish. It was in such gels that evidence for diffusible guidance factors released from targets such as Max Factor and Netrin was first obtained (Lumsden and Davies, 1986). In such gels, it is also possible to inject diffusible factors in known quantities to create designer gradients of particular concentrations and steepnesses. In such designer gradients, it was possible to show that a growth cone can sense a difference across its width of one molecule per thousand (Rosoff et al., 2004).

BOX 3

DENDRITE FORMATION

Dendrites, the neuritic processes that are the main receivers of synaptic input, are perhaps the most distinctive features of neuronal morphology. Dendritic trees differ from axonal arbors in a variety of ways; the most obvious being that dendrites have more postsynaptic specializations while axons have more presynaptic specializations. Dendrites also grow somewhat differently to axons (McAllister, 2000; Whitford et al., 2002). The cytoskeleton of a dendrite is also different from that of an axon, usually having a higher ratio of microtubules to actin filaments and more rough endoplasmic reticulum and polyribosomes. Axonal microtubules have their plus ends pointed distally, while the dendritic microtubules have a mixture of plus- and minus-ends leading. Microtubule associated proteins are also differentially distributed in axons and dendrites. For example, MAP2 is located in dendrites, while tau is located mainly in axons. Treatment of cultured neurons with antisense constructs that reduce MAP2 or tau expression have the expected specific effects on the formation of dendrites or axons, indicating that these proteins are particularly critical in the formation or stabilization of these structures (Liu et al., 1999; Yu et al., 2000). Certain membrane proteins are also differentially distributed among axons and dendrites; for instance, transmitter receptors are more common on dendrites, whereas certain cell adhesion molecules and GAP-43 are found mainly on axons (Craig and Banker, 1994). This polarity implies a sorting mechanism, but the molecular basis of sorting different proteins to different neuronal processes is not yet understood. The initial polarity of neuronal cells in terms of axon versus dendrite is not yet well understood, although current evidence suggests that this polarity may be controlled by the same cues, such as Par3, that specify polarity in epithelial cells and asymmetric cell divisions (Shi et al., 2003).

Hippocampal cells in culture initially put out several short neurites tipped with small growth cones (Figure 5.39). Initially, all of these processes are identical; for example, they all have GAP-43 at their tips. Soon, however, one of these processes begins to extend more rapidly than the others, and as it does so it gathers axonal specific markers so soon that only the axon has a GAP-43 tipped growth cone and the other processes begin to assume dendrite specific markers. Interestingly, if the emerging axon is selectively cut off, the longest of the short processes starts to grow faster than the others and it becomes the axon. Thus, neurons have an axon versus

dendrite polarity that is, to a certain extent, internally regulated through a feedback mechanism by which the axon inhibits the other neurites from assuming the axonal identity they would attain by default (Goslin and Banker, 1989). If at an early stage of polarization when all processes are equal, the actin depolymerizing agent cytochalasin is transiently applied locally to just one neurite, that neurite will become the axon. If, however, cytochalasin is applied uniformly to all the neurites, then surprisingly they all become axons (Bradke and Dotti, 1999). This suggests that actin instability, possibly allowing microtubule invasion, may be a key to the decision of a process to become an axon or a dendrite. This model is supported by direct manipulation of molecules that control microtubule stabilization, such as collapsin response mediator protein (CRMP-2, a MAP) and a GEF

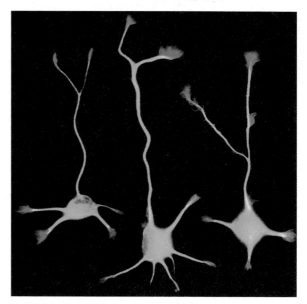

FIGURE 5.39 In tissue culture, a hippocampal neuron begins by putting out several minor processes that are basically equivalent. One of these, the future axon, then begins to grow faster than the other process and collects axon-specific components like GAP43 and tau. After the axon has elongated, dendrites begin to grow and express dendrite-specific components such as MAP2. This figure shows three young hippocampal neurons in culture stained for microtubules (red) and actin (green). At this stage, one process is elongating while the shorter processes are not yet definitive dendrites. If at this stage, the emerging axon is cut, then a minor process, which would have otherwise become a dendrite, begins to grow more rapidly and becomes the axon. (From Ruthel and Hollenbeck, 2003)

BOX 3 *(cont'd)*

called Tiam1 (Kunda et al., 2001; Fukata et al., 2002). When either of these proteins is overexpressed in a developing hippocampal cell, all the processes become axons; when their function is reduced, all the processes become dendrites. It is thought that these proteins help microtubules invade actin networks. External influences also affect the axonal versus dendritic decision perhaps by affecting these cytoskeletal-associated proteins. For instance, if a hippocampal cell is plated at the interface between a laminin and polylysine, the axon almost invariably grows on the laminin substrate.

Dendritic growth-promoting factors can also be found in the environment. Dendritic outgrowth from mouse cortical neurons was specifically enhanced by astrocytes derived from the forebrain (Le Roux and Reh, 1994). Similar results were obtained with glial conditioned medium. Superior cervical ganglion neurons grown in serum-free medium in the absence of nonneuronal cells were unipolar and only grew axons. When the same neurons were exposed to serum, they became multipolar and developed processes that could be categorized as dendrites by morphological and antigenic criteria (Bruckenstein and Higgins, 1988a,b). Thus, serum contains factors that stimulate dendritic extension. The bone morphogenetic proteins, BMP2 and BMP6, were subsequently found to selectively induce the formation of dendrites and the expression of microtubule-associated protein-2 (MAP2) in sympathetic neurons in a concentration-dependent manner (Lein et al., 1995; Guo et al., 1998). Dermatan sulfate also specifically enhances dendritic growth. Dendritic retraction occurs in many regions of the developing brain. Leukemia inhibitory factor (LIF) and ciliary neurotrophic factor (CNTF) specifically cause dendritic retraction in SCG cells (Guo et al., 1997; Guo et al., 1999). Axon growth is unaffected by these factors. Taken together, these results suggest the existence of separate but extensive molecular mechanisms for promoting and inhibiting dendrite outgrowth that parallel the growth-promoting and collapsing mechanisms known to be involved in axonogenesis.

It has been thought that mature dendrite formation is somehow dependent on the axon making proper connections to its target. The situation, however, may not be so one-sided as it seems that dendrites can also "search" for their inputs (Jan and Jan, 2003). In some cases, growing dendrites are tipped with dendritic growth cones that appear as miniature equivalents to axonal growth cones. Dendritic growth cones express receptors for several classes of guidance factors and indeed, factors such as Slits, Netrins, and Semaphorins can influence the growth of dendrites (Whitford et al., 2002). Interestingly the same factors that do one thing in axons can do a different thing in dendrites. For example, Sema3A can attract the apical dendrites of cortical neurons toward the pial surface, while the same factor can repel the axons of the same pyramidal cells.

Even though the final size, shape, and complexity of the dendritic tree are sensitive to innervation, the dendrite is able to develop in a largely autonomous fashion. One of the interesting examples of the independence of dendritic growth from innervation is the case of the Purkinje cells in weaver mutant mice. In these mice, the granule cells do not migrate properly into the cortex of the cerebellum and thus fail to make synapses on the Purkinje cells. The Purkinje cells nevertheless make a dendritic tree that, although smaller and less well formed than the trees of properly innervated Purkinje cells, are still characteristically complex (Bradley and Berry, 1978). The most dramatic demonstrations of the ability of neurons in the absence of synaptic input to produce dendritic trees come from culture experiments. Conditions have been worked out in which pyramidal neurons from the hippocampus, principal neurons of the SCG, and even cerebellar Purkinje cells are able to develop a characteristic dendritic tree in dissociated cell culture.

Each type of neuron has a characteristic dendritic tree. In some neurons, like the Purkinje cell, the dendritic tree is enormously complex and supports synaptic input from thousands of presynaptic fibers. In other neurons, like some sensory neurons, the dendritic tree is simple, consisting of a single postsynaptic process. In the central nervous system of a cockroach or a leech, the dendritic tree of each identified cell has a unique signature branching pattern recognizable from individual to individual. What drives these particular morphologies? The Rho family of GTPases is critical for proper dendritic morphology. These GTPases are key regulators of actin polymerization. Cdc42 is required for multiple aspects of dendritic morphogenesis (Luo, 2000, 2002). For example, in neurons that are mutant for Cdc42, dendrites are longer than normal, branch abnormally, and have a reduced number of spines. Extra activation of RhoA, via experimental expression of a constitutively active form of the molecule, leads to a dramatic simplification of dendritic branch patterns. Such experiments suggest that different Rho family members have distinct roles in regulating dendritic morphogenesis. Several transcription factors, identified in *Drosophila*, appear to dramatically regulate

BOX 3   (*cont'd*)

dendrites (Jan and Jan, 2003). The *hamlet* gene is expressed in externalsensory neurons that have very simple monopolar dendrites, and when these neurons are mutant for *hamlet*, their dendritic trees become highly branched so they look very much like another class of sensory neurons called multidendritic neurons, which normally do not express *hamlet*. Conversely, multidendritic neurons express high levels of another transcription factor called *cut*, which is expressed at low levels in the external sensory neurons. Overexpression of *cut* in the latter causes them to acquire multiple dendrites, while loss of *cut* function in multidendritic neurons causes them to lose their dendritic complexity. Active axonal inputs are important for dendrite growth and branching. The complexity of the dendritic tree is, in many systems, proportional to the amount of innervation. Thus, for example, the dendritic trees of the principal sympathetic neurons

of the superior cervical ganglion (SCG) are larger and more complex in larger mammals that have more inputs onto these cells. Moreover, if inputs to the SCG are reduced or silenced, this causes a concomitant decrease in the complexity of the dendritic tree that develops (Purves and Lichtman, 1985; Voyvodic, 1989). Tectal dendrites are innervated by the axons of RGCs. Time-lapse recordings of tectal dendrites show that dendritic branches are very dynamic and can appear or disappear within minutes. Active visually driven input on these dendrites enhances their growth in early development through the activation of NMDA receptors, CAM kinase II and Rho GTPases. Activity stabilizes dendritic branches at later stages (Cline, 2001). Many dendritic trees are dynamic throughout the life of the animal. The plasticity of these branches in response to experience and neuronal activity will be discussed in Chapter 9.

6

Target Selection

As a growth cone nears the end of its journey, it must find appropriate target cells with which to synapse. This has to be done with immense accuracy because a properly functioning nervous system depends on precise patterns of neural connectivity. The task seems daunting because the growth cone may be confronted with thousands or even millions of roughly similar cells in the general target area from which it will have to choose only one or a few as postsynaptic partners. The process of target selection can be broken down into a number of conceptual steps (Holt and Harris, 1998) (Figure 6.1). First, as they near the target area, axons defasciculate from the tracts or nerves that they are growing along. Next, they enter the target area, slow down, and begin searching for their postsynaptic partners who might also be searching, via dendritic growth, for the axons that will innervate them. As they search, the axon terminals begin to branch in the target area. Molecular barriers may be erected around the borders of the target area so that the incoming axons are corralled until they find the most suitable partners. In large target areas, there is often a topographic mapping strategy so that the axon and its postsynaptic partner can meet at particular molecular coordinates. Some targets are multilayered, and it is important for axons to find the appropriate synaptic layers within the correct topographic region. Having finally arrived at the correct location, the growing axons choose particular postsynaptic cells and perhaps particular dendrites or regions of these cells.

There is a final phase of target selection, which has to do with the refinement of connections and is often based on neuronal activity and dependent on synapse formation (see Chapter 8). As synapses are formed, the nervous system can begin to function and test out the wiring for connections that are misplaced. This func-

tional verification of target selection is used to refine the connections further so that the end product works in the real world. The refinement of connections is dealt with more thoroughly in Chapter 9. The consideration of target selection as the product of a series of discrete decisions makes it easier to appreciate how immense precision can be developmentally built into the nervous system.

DEFASICULATION

In order to enter a target area or to find a target cell with which to make synapses, it is often first necessary to exit from a tight bundle. Nerves, tracts, columns, bundles, and fasciculi often travel past a variety of potential targets. As they do so, specific axons or groups of axons peel off of these common pathways, so that they can enter the target. In the last chapter, we saw that homophilic adhesion molecules such as N-CAM and its homologs can cause similar axons to fasciculate together into nerves or tracts in the nervous system, and so one of the first questions in targeting is how do axons leave these tracts and nerves? How do they defasciculate? Changes in cell adhesion seem necessary. In *Drosophila*, there is a mutant called *beaten path* in which motor axons fail to defasciculate from the main nerve and as a result bypass their targets (Vactor et al., 1993). The protein encoded by this gene, Beat-1a, thus seems required for selective defasciculation of motor axons at these choice points. Indeed, Beat-1a appears to be an anti-adhesion factor that is secreted by growth cones (Fambrough and Goodman, 1996). There are other Beat-family members, such as Beat-Ic, which are not secreted but are membrane-bound proteins that have pro-adhesive functions (Pipes et al.,

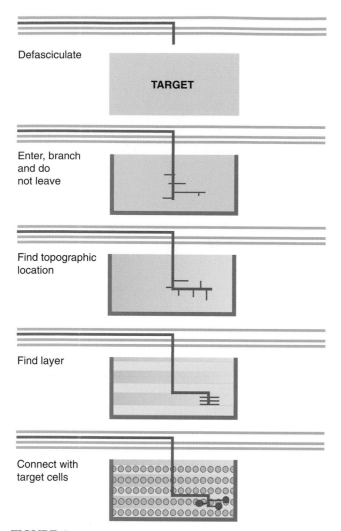

FIGURE 6.1 Conceptual stages of targeting. From top to bottom, an axon defasciculates in the region of the target. It enters the target and begins to branch, and is prevented from exiting by a repulsive border. The axon responds to a topographic gradient that promotes branching at the correct location. It then selects a particular layer and finally homes in on particular target cells. (After Holt and Harris, 1998)

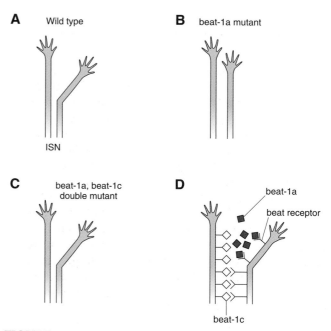

FIGURE 6.2 Defasciculation is regulated by Beat proteins. A. Two motor axons growing off the intersegmental nerve (ISN) are shown, one that branches off the nerve in the region of its target. B. In a *beat-1a* mutant, the motor axon at right does not defasciculate. C. The *beat-1a* mutant phenotype is rescued in a *beat-1c* mutant background. D. Model where the axons are fasciculated, integral membrane beat-1c protein binds the beat receptor. Where the axons defasciculate, soluble beat-1a binds the receptor, breaking the adhesion. (After Vactor et al., 1993; Fambrough and Goodman, 1996; Pipes et al., 2001)

2001). Interestingly, *beaten path* embryos are rescued by reducing the levels of cell adhesion molecules, like Beat-1c, and thus it is the balance between these pro- and anti-adhesive functions that regulates whether particular axons will take the appropriate exits or stay on the main road (Figure 6.2). Other factors may also serve to alter the balance, including repulsive guidance factors such as Semaphorin-1a, which can also cause axons to branch off the main pathways (Yu et al., 2000).

Branching patterns of nerves are mediated by defasciculation in vertebrates too. One way to reduce fasciculation is to secrete an anti-adhesive factor, like Beat-1a. Another is to post-translationally modify adhesion molecules before putting them into the mem-

branes. In the previous chapter, we looked at the role of polysialic acid (PSA) in decreasing the adhesivity of N-CAM in the nerve plexus, where defasciculation was necessary to get different motor neurons sorted into the correct mixed nerves. The relative PSA levels of L1 and N-CAM are also important in balancing axon–axon versus axon–muscle adhesion during target innervation (Landmesser et al., 1990). If PSA is removed, the result is increased axon fasciculation, reduced nerve branching, and reduced target innervation.

It seems likely that the targets themselves contribute to the defasciculation of the axons that will innervate them. In *Drosophila* mutants lacking mesoderm, the main motor nerves form, but motor axons fail to defasciculate from these bundles. Experiments by Landgraf et al. (1999) have shown that founder myoblasts are the source of defasciculation cue(s) and that a single founder myoblast can trigger the defasciculation of an entire nerve branch. This suggests that the separate targets, through the release of possibly different defasciculation factors at different locations, lead to the patterned branching of nerves.

TARGET RECOGNITION AND ENTRY

Targets can be internal organs, sensory cells, muscles, or other neuronal nuclei in CNS or PNS. What makes these tissues the targets for specific innervation? The sympathetic nervous system, with its diversity of end organs, provides an excellent opportunity to study this question. For sympathetic neurons, neurotrophins are the key to initial innervation. There are several different neurotrophins, originally named for their effects on neuronal growth and survival. In Chapter 7, we will see that the survival of sympathetic and other neurons is critically dependent on receiving enough of these target-derived trophic factors. However, the axons must first enter the targets to gain access to a supply of the neurotophins, and neurotrophins like NGF have roles in target entry that are distinct from their roles in survival (Glebova and Ginty, 2004). Many sympathetic neurons grow along vasculature to reach their various somatic targets. These blood vessels are a source of the neurotrophin, NT-3. In the absence of NT-3, sympathetic cells often fail to invade their targets (Kuruvilla et al., 2004). Take, for example, the epidermis of the external ear. It is innervated by sympathetic neurons, and in NT-3 knockout mice, sympathetic fibers fail to invade the external ear postnatally. Exogenously administered NT-3 into the ear rescues the sympathetic innervation of the mutant mice (ElShamy et al., 1996) (Figure 6.3).

When sympathetic fibers get right into the target region, they often switch the neurotrophin they are most interested in from NT-3 to NGF (Kuruvilla et al., 2004). The pancreas and other internal organs, which are invaded by different sets of sympathetic fibers, express NGF, which is critical for target invasion. If NGF is overexpressed under the control of a beta-cell specific promoter in the islets of the pancreas in transgenic mice, there is a dramatic increase in the innervation of the islets (Hoyle et al., 1993). The role of NGF in attracting innervation has medical implications. For example, pancreatic cancer is particularly invasive to neural tissue, and this may be because it attracts innervation (Schneider et al., 2001). Similarly, pancreatic transplants may benefit by the local application of NGF to help attract innervation (Reimer et al., 2003). A more interesting example of target recognition by sympathetic axons concerns another neurotrophin, glial-derived neurotrophic factor (GDNF) (Ledda et al., 2002). GDNF is recognized by the c-Ret receptor on the growth cones of innervating axons, and it is also recognized by a GPI-linked secreted receptor called GFRα1. The peripheral targets of the c-Ret expressing axons, such as the epidermis, secretes GFRα1, which

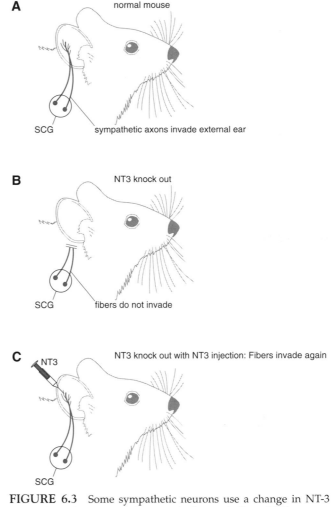

FIGURE 6.3 Some sympathetic neurons use a change in NT-3 expression to innervate their targets in the ear. A. Some SCG neurons project to and arborize in the pinna of a normal mouse. B. In NT-3 knockout mice, these fibers do not invade the pinna. C. Restoration of targeting by injection of NT3 into the ear. (Adapted from ElShamy et al., 1996)

captures circulating GDNF and holds it to the target sites, thus creating a very high level of GDNF right around the target, which attracts innervation from c-Ret expressing axons.

Neurotrophins are also involved in the innervation neuronal targets by nonsympathetic neurons. A particularly interesting example is the innervation of the inner ear, which includes the vestibular organs of balance and the cochlear organs of hearing (Ernfors et al., 1995), (Figure 6.4). Neurotrophins are first expressed in the otocyst during the time at which ganglion cells with neurotrophin receptors send their processes toward this structure. Indeed, BDNF may be expressed by the hair cells, which are the cellular targets of this innervation, well before the hair cells have fully differentiated (Hallbook and Fritzsch, 1997).

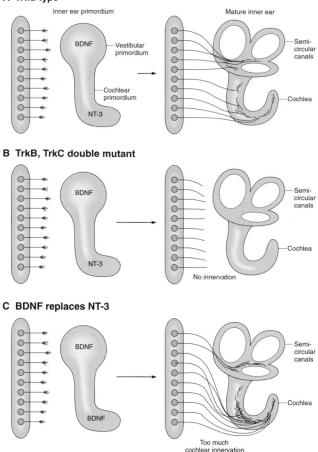

FIGURE 6.4 Innervation of the inner ear is regulated by BDNF and NT-3. A. In-the-wild-type animal, the vestibulo-cochlear ganglion, all of whose neurons express TrkB and TrkC, grow toward the developing inner ear, which has a vestibular and a cochlear primordium. As the system develops and the primordia develop into semicircular canals and a cochlea, the ingrowing axons innervate both parts of the inner ear. B. In BDNF, NT-3 or TrkB, TrkC double mutants, the inner ear remains uninnervated. C. In transgenic mice in which BDNF has been knocked into the NT-3 coding region, the cochlear region becomes innervated by the vestibular part of the ganglion. (After Ernfors et al., 1995; Fritzsch et al., 1997; Fritzsch et al., 2004; Tessarollo et al., 2004)

NT-3 is also expressed in the developing inner ear, and all the innervating fibres possess receptors for both neurotrophins: Trk-B for BDNF and Trk-C for NT-3. Knockout experiments of the ligands show that BDNF is necessary for the innervation of the vestibular hair cells, whereas NT-3 is more important in the innervation of the cochlear hair cells, for that is where these factors are most heavily expressed (Fritzsch et al., 1997). In mice that lack both BDNF and NT-3, or both TrkB and TrkC, there is a complete loss of innervation to the inner ear (Fritzsch et al., 1997). Interestingly, if the BDNF coding sequence is inserted into the NT-3

gene in a transgenic mouse, the result is the expression of BDNF throughout the inner ear, and all the fibers that normally innervate the NT-3 rich areas survive and innervate the cochlea as usual (Tessarollo et al., 2004). So the neurotrophic factors can substitute for each other in a way. However, these transgenic mice show excessive innervation of the cochlea from neurons that would normally innervate the vestibular regions, a miswiring that probably occurs because the changed spatiotemporal expression pattern of BDNF. This incorrect projection can be enhanced by knocking out the normal expression of BDNF in the vestibular region. These results suggest that correct temporal and spatial pattern of neurotrophin expression may be critical for the correct innervation of these inner-ear targets.

SLOWING DOWN AND BRANCHING

In the previous chapter, we discussed how the axons of retinal ganglion cells navigate to their targets in the optic tectum. In this chapter, we discuss how these axons find their postsynaptic targets within the tectum. The ability to look at these processes as they are happening has been important in establishing some aspects of targeting. For example, time-lapse observations of fluorescently labeled RGC axons in *Xenopus* embryos grow at a rate of about 60um/hr in the optic tract but slow to about 16um/hr when they enter the optic tectum (Harris et al., 1987) (Figure 6.5). Once within the optic tectum, these terminals may advance in a saltatory, stop and start, manner. Why do growth cones slow down when they reach tectum? Retinal axons grow toward their target on a pathway that is rich in FGF, and this molecule has been found to promote axonal growth in the tract and in vitro. As retinal axons enter the tectum, they encounter a sudden drop in external FGF because the tectum expresses very little of it (McFarlane et al., 1995). Therefore, one cue that decreases the growth rate of retinal axons at the target is a drop in FGF levels. If excess FGF is added, or if the retinal axons are made insensitive to FGF, the retinal axons do not respond to the target and often grow by it, so they may read the drop in FGF as a target entry signal (McFarlane et al., 1995; McFarlane et al., 1996; Webber et al., 2003).

As retinal growth cones slow down in the tectum, time-lapse images show that they also become much more complex. Branches begin to form, and many of these arise at some distance behind the axonal tip (Harris et al., 1987). Thus, axonal arbors are built in a way that is reminiscent of the way the arbor of a tree

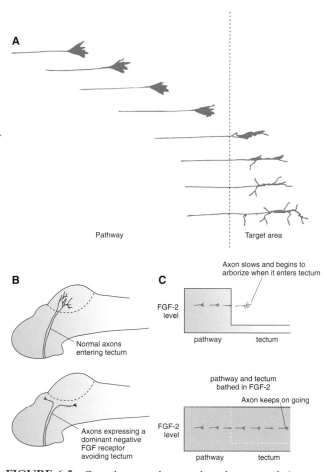

FIGURE 6.5 Growth cones change when they enter their target zones. A. Images from a time-lapse movie of a retinal ganglion cell growing in the optic tract and then crossing (at the dotted line) into the tectum. The simple growth cone becomes much more complex and slows down dramatically as it enters the target. B. Tectal innervation by control retinal axons (top) and tectal avoidance by retinal axons that misexpress a dominant negative FGF receptor. C. Retinal axons slow down and branch when they reach the tectum in control animals (top), but when the pathway is exposed to high levels of FGF-2, the axons keep going and do not innervate the tectum. (Adapted from Harris et al., 1987; McFarlane et al., 1995; McFarlane et al., 1996)

into the optic tectum of live *Xenopus* laevis tadpoles increases the branching of RGC terminals, whereas blocking BDNF reduces axon arborization (Cohen-Cory and Fraser, 1995). Altering levels of BDNF in the retina had no effect on RGC branching in the tectum, indicating that the branch-promoting effects of BDNF are local on axon terminals (Lom and Cohen-Cory, 1999; Lom et al., 2002).

The sensory neurons of the dorsal root ganglion (DRG) provide another example of branching. These axons enter the spinal cord through the dorsal root and then bifurcate in to grow in the anterior and posterior directions. Collaterals then sprout from these branches to innervate the gray matter of the cord (Ozaki and Snider, 1997). Using an in vitro assay, it was found that Slit2 promotes the formation and elongation of branches in DRG neurons (Wang and Scott, 1999). The identification of repulsive guidance molecules such as Slits and Semas, which can control growth cone guidance on the one hand and promote branching on the other, suggests that there may be a link between repulsion and branching. Indeed, in vitro observations have indicated that branches form behind the tip of the growth cone whenever the growth cone collapses in response to any collapse-inducing agent, even a mechanical one (Davenport et al., 1999). These observations are consistent with the finding that the growth cones of callosal axons in the cortex pause for many hours beneath their targets prior to the development of branches (Kalil et al., 2000). Imaging of dissociated living cortical neurons shows that wherever a growth cone undergoes a lengthy pause, and then advances again, filopodial remnants of the paused growth cone are left behind on the axon shaft (Szebenyi et al., 1998). Here, the cytoskeleton of the axon appears more splayed apart and fragmented, and it is from these regions that new branches form (Dent and Kalil, 2001) (Figure 6.6). Such results demonstrate that growth cone pausing is closely related to axon branching.

develops, with new branches arising along the stems of older branches. Many of these new branches are not tipped with growth cones themselves but appear like worms wriggling out of the parent branch. What causes axons to branch in this way? The tectum expresses the repulsive guidance cue Sema3A. Sema3A, when applied to retinal growth, cones in vitro, causes them to collapse, but this collapse is transient, and recovery from collapse is often associated with branching (Campbell et al., 2001). Thus, Sema3A may stimulate terminal branching in the tectum. In addition, the tectum is a source of BDNF, which also promotes branching of RGC axons. Injection of BDNF

BORDER PATROL AND PREVENTION OF INAPPROPRIATE TARGETING

Once they have recognized and entered a target area, slowed down, and started to branch within, axons may be prevented from exiting the target area by repulsive cues at the perimeters. Sema3a, which we discussed in the last chapter, repels the growth cones of cutaneous sensory neurons. Analysis of knockout mice supports a critical role for Sema3a as an exclusion factor confining the peripheral ends of these axons to the correct target areas of the skin (Taniguchi et al.,

A Pausing growth cone   **B** Developing branch   **C** Elongating branch

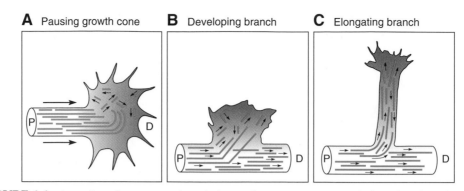

FIGURE 6.6   Axons branch at pause points. A. A growth cone pauses, microtubules splayed. B. The growth cone moves on but leaves behind it a zone where the cytoskeleton remains somewhat disorganized. C. A branch forms at this zone. (After Dent and Kalil, 2001)

1997). In these mice, axons that are normally restricted from innervating skin that expresses Sema3a now enter these territories. Sema3a is also expressed at the posterior boundary of the olfactory bulb, where it seems to act to restrict olfactory axons to the bulb, preventing them from entering the telencephalon (Kobayashi et al., 1997). The repulsive molecule Ephrin-A5, which we will discuss in more detail below with respect to topography, reaches its highest concentration just posterior to the superior colliculus or tectum, the target of retinal axons, suggesting that this ligand may also serve as a factor that confines these axons to the target. Recent studies show that retinal axons extend freely beyond the posterior border of the superior colliculus in Ephrin-A5 knockout mice (Frisen et al., 1998).

This raises the question of whether this type of mechanism is used to help segregate neural circuits that carry different kinds of information to different, but nearby, centers of the brain, thus preventing inappropriate targeting. Re-innervation and cross-innervation experiments show that when the normal targets of axons have been surgically removed, functional synapses can indeed be made on the wrong targets. Similarly, when the brain is injured, the normal targets of some axons may die and nearby regions may become denervated. In these cases, axons that originally innervated the injured areas may sprout new growth cones to invade denervated but inappropriate targets. To test how promiscuous axons are, and whom they will synapse with if given the chance, one can test a variety of foreign targets with different axonal populations. For example, to know how determined retinal ganglion cell axons are to invade a specific target, one of their normal targets, the visual thalamus, was left to degenerate by ablating the visual cortex and a neighboring nonvisual area, the somatosensory thalamus, was denervated (Metin and Frost, 1989). In this case, retinal axons innervated the somatosensory thalamus (Figure 6.7). In a similar experiment, retinal ganglion cell axons innervated the auditory thalamus (Roe et al., 1992). The thalamocortical connections have not been changed and are basically normal in these animals, giving rise to the weird condition that these animals process visual information in the somatosensory or auditory cortex, thus perhaps having the conscious sensation of feeling or hearing the visual world (Figure 6.7). Normally, of course, the nuclei of the thalamus have modality-specific innervation. The question is whether segregation is a result of molecular barriers that normally separate these brain areas. It is interesting to note, then, that high levels of Ephrin-A2 and Ephrin-A5 define a distinct border between the visual and auditory thalamus. If the normal input to the auditory thalamus is denervated and the visual thalamus is spared, retinal axons seem happy to remain in their uninjured normal targets. However, when this experiment is done in knockout mice that lack both Ephrin-A2 and Ephrin-A5, there is extensive rewiring and retinal axons invade and innervate the deafferented auditory thalamus (Lyckman et al., 2001) (Figure 6.7). These findings suggest that signals that induce innervation may compete with barriers, such as repulsive guidance molecules, that serve to contain axons within the normal targets.

Border patrol is not the only mechanism for maintaining appropriate targeting. In cross-innervation experiments in amphibians, in which extensor motor nerves are forced to innervate flexor muscles and flexor motor nerves are forced to innervate extensor muscles, the animals develop expected inappropriate motor behaviors after surgery. Interestingly, however, normal behavior usually recovers after a rather long period of time. This was first interpreted as the animals learning to use these muscles in a new way, but detailed anatomical investigations of these animals

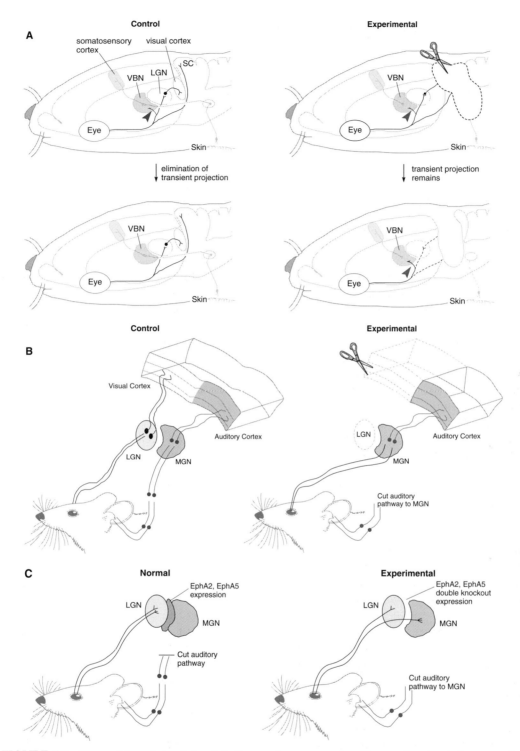

FIGURE 6.7 Cross wiring of visual signals to the somatosensory and auditory and cortex. A. In a normal ferret, the auditory input goes indirectly to the medial geniculate nucleus of the thalamus and from there to the auditory cortex. Retinal input is to the LGN and then on to the visual cortex. B. When the auditory pathway is cut and then is left to degenerate by ablation of the visual cortex, the visual input sprouts into the MGN, which projects as usual giving visual physiological properties to the auditory cortex. C. EphrinA2 and EphrinA5 are expressed at the border between the LGN and the MGN. Even though the auditory pathway to the MGN has been cut, the retinal fibers do not invade the MGN, but they do so in EphrinA2, EphrinA5 double knockouts. (After Metin and Frost, 1989; Roe et al., 1992; Lyckman et al., 2001)

showed that the crossed nerves, over the course of time, had managed to uncross themselves and find their original muscles again (Mark, 1969). Competition experiments between original and foreign nerves for the innervation of particular muscles show that the original nerves always have an advantage (Dennis and Yip, 1978). Thus axons, although they will innervate inappropriate denervated targets when their own targets are unavailable, seem to have a natural preference for their own original targets.

TOPOGRAPHIC MAPPING

In many targets there is a topographical relationship between the position of the innervating neuron and the position of its terminal arbor in the target field. A good example is the visual system. RGCs in a particular position in this retina are maximally stimulated from a region of the visual world, and these cells send their axons to a particular region in the tectum. Neurons in neighboring retinal positions send their projections to neighboring regions in the tectum. This orderly projection preserves visual topography in the brain and provides a neuroanatomical basis for the contiguity of perceived visual space. Similarly, most central auditory nuclei have a representation of the cochlea's frequency axis. Such maps may be referred to by the anatomical substrates that they preserve (e.g., retinotopic, cochleotopic) or by more perceptual terms (e.g., visuotopic or tonotopic). Even simple animals like the nematode, with only 301 neurons, have ordered arrays of sensory receptors that make somatopically organized central projections. These help them respond appropriately to stimuli that strike the animal from different directions. There is a second type of neural map, a computational map, which can be revealed by recording from neighboring single nerve cells in vivo. What is represented in such maps is not obvious from the anatomy of the connections, yet these maps may also display orderly representations of a physical parameter. For example, in the auditory system, we find nuclei that display topography of sound source location, even though the ear contains only a one-dimensional array of spiral ganglion cells representing sound frequency. Such maps are constructed from cells that extract information and are referred to by the functional characteristic that they encode (e.g., map of auditory space) rather than a piece of tissue. There are also stranger maps, such as maps of smell that we will discuss below.

What is the developmental basis for the establishing topographic projections in the nervous system? In the late 1800s John Langley discovered that superior cervical ganglion (SCG) neurons mediate reflexes in a topographic manner (Langley, 1897, 1985). When Langley stimulated the first or top thoracic root to the ganglion in the rat, this activated ganglion cells that caused dilation of the pupil. When he stimulated the fourth thoracic root to the ganglion, the blood vessels of the ear constricted. This suggested that there was some sort of topographical organization within the SCG. All reflexes were immediately lost when the preganglionic nerve to the SCG was cut, but the fibers reinnervated the SCG in several weeks, as peripheral nerves often do in mammals, and the autonomic reflexes recovered. The surprising discovery was that the connections reformed with such precision that all reflexes were reestablished accurately (Figure 6.8). This result suggested that individual SCG neurons have some mechanism that enables the regenerating preganglionic fibers to distinguish one SCG neuron from another.

The sympathetic chain ganglia provide a simple system in which to examine somatotopic specificity because each ganglion is selectively innervated by

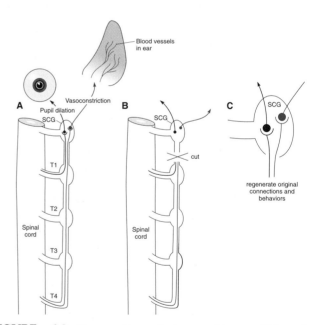

FIGURE 6.8 Regeneration of topographic specificity. A. Langley's classic study showed that stimulation of preganglionic root T4 relayed through ganglion cells in the SCG that caused vasoconstriction of the ear pinna vessels, whereas stimulation of root T1 excited other SCG neurons that caused dilation of the pupil. B. When Langley cut the sympathetic tract above T1, all these sympathetic reflexes were abolished, but with time they recovered. C. Shows the specificity associated with this regeneration, such that the axons that enter the chain at T4 reinnervate the SCG cells that cause ear vasoconstriction, and the axons that enter at T1 reinnervate the cells that cause pupil dilation. (After Langley, 1897, 1985)

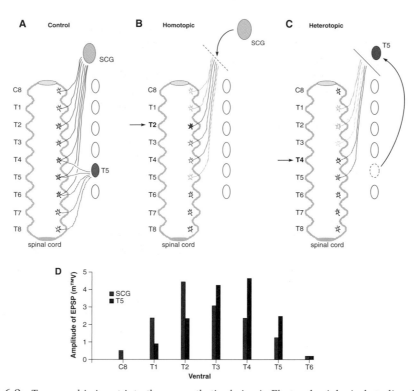

FIGURE 6.9 Topographic input into the sympathetic chain. A. Electrophysiological studies show that the SCG receives input from many roots but primarily the more anterior ones. The ganglion at T5 receives its primary inputs from more posterior roots. B. When the SCG is removed and replaced with another SCG, the axons that reinnervate it tend to be from more anterior roots. C. When a T5 ganglion is put in place of the SCG, its neurons still tend to get innervated by more posterior roots even though the ganglion is in an anterior position. D. This topographic specificity of reinnervation is reflected in the shape of the histogram of EPSP amplitudes as a function of nerve root stimulation for the homototopic and heterotopic transplants. (After Purves et al., 1981)

afferents from a limited number of spinal cord segments. Thus, the superior cervical ganglion (SCG) is primarily innervated by preganglionic afferents from thoracic segments T1-T4, whereas the more caudally located fifth thoracic ganglion (T5) is primarily innervated by afferents from T4-T7 (Nja and Purves, 1977). In one experiment, a T5 ganglion was transplanted to different locations along the sympathetic chain, exposing this target to afferents from a large range of spinal cord segments (Purves et al., 1981). Selective reinnervation was then assessed electrophysiologically. The sympathetic chain ganglia were dissected out along with the ventral nerve roots through which all preganglionic fibers course from the ventral spinal cord. Stimulating electrodes were then placed on the ventral roots from each spinal cord segment, and an intracellular recording was obtained from the reinnervated T5 ganglion. The spinal segments that innervate each T5 neuron were thus recorded. The results clearly indicated that T5 neurons were selectively reinnervated by their original spinal segments (Figure 6.9). This was not merely an artifact of the host transplantation site

because when the SCG was placed in the same location, it, too, became reinnervated by its original set of afferents. These experiments strongly suggest that axons from different rostrocaudal levels can distinguish individual sympathetic ganglion cells, which must also carry some label of their rostrocaudal origin.

CHEMOSPECIFICITY AND EPHRINS

In the early 1940s, Roger Sperry cut the optic nerve of a newt, rotated the detached eye 180° in its orbit, and assayed the visuomotor behavior of the animal after its nerve had regenerated. The newts, and in subsequent studies, frogs, behaved as if their visual world were back-to-front and upside-down: when a lure was presented in front of them, "they wheeled rapidly to the rear instead of striking forward" and when the lure was presented above "the animals struck downward in front of them and got a mouthful of mud and moss" (Sperry, 1943) (Figure 6.10). This led him to propose

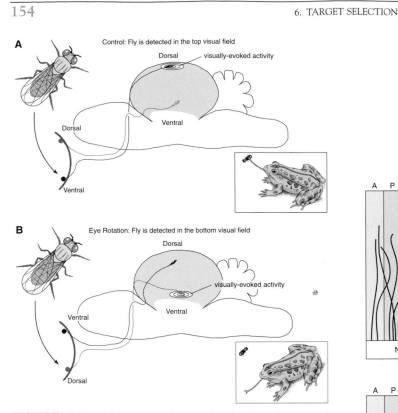

FIGURE 6.10 Maladaptive topography implies chemospecificity. A. A normal frog sees a fly above on its ventral retina, which projects to the dorsal tectum, leading to a snap in the appropriate upwards direction. B. A frog with a rotated eye sees the same fly on what used to be the dorsal retina, which projects as ever to the ventral tectum, leading to a snap in the wrong downward direction. (After Sperry, 1943)

that topographic nerve connections between the retina and its main central target, the optic tectum, were the result of anatomical rather than experiential features of the nervous system, as Sperry's unlucky frogs never did learn to snap in the correct direction. Sperry reasoned that the retinal fibers mapped onto the tectum according to original anatomical coordinates of the eye. The explanation he gave was the possible existence of biochemical tags across the retina and tectum. He postulated the existence of two or more cytochemical gradients "that spread across and through each other with their axes roughly perpendicular" (Sperry, 1963). These separate gradients successively superimposed on the retinal and tectal fields and surroundings would stamp each cell with its appropriate latitude and longitude expressed in a kind of chemical code with matching values between retinal and tectal maps."

The chemoaffinity hypothesis inspired many biologists and biochemists to try to find the molecules that were responsible for topographic targeting in the retinotectal system. Such studies often took an in vitro approach, and for over 20 years, not much progress was made. Friedrich Bonhoffer and colleagues made a

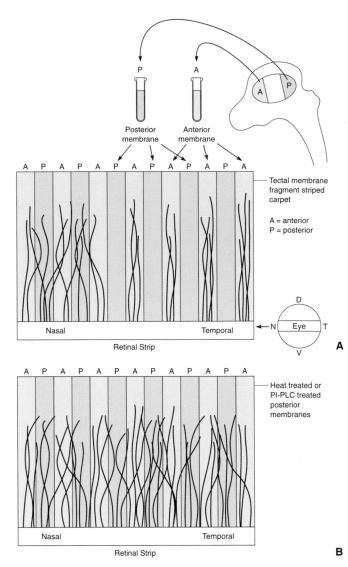

FIGURE 6.11 The striped carpet assay. A. An equatorial strip of retina spanning the nasal (N) temporal (T) extent is positioned on a striped carpet of alternating anterior (A) and posterior (P) tectal membranes. The nasal fibers from the retinal explant grow on both A and P tectal membranes, but the temporal fibers grow only on the A membranes. B. If the tectal membranes are denatured or treated with PI-PLC, which releases PI-linked membrane proteins, the temporal axons also grow on both types of membranes, suggesting that the P membranes normally have a PI-linked repulsive guidance molecule. (After Walter et al., 1987a; Walter et al., 1987b; Walter et al., 1990)

breakthrough when they used membranes from anterior and posterior parts of the tectum to make a striped carpet. When retinal tissue from the temporal retina was cultured on such striped carpets, they found that these retinal axons grow preferentially on anterior tectal membranes (Walter et al., 1987b) (Figure 6.11). Surprisingly, when the posterior membranes were heated, treated with formaldehyde, or exposed to an enzyme (PI-PLC) that removes PI-linked membrane molecules, temporal axons no longer showed such a preference (Walter et al., 1987a; 1990). This suggests that the relevant activity is a membrane-linked protein that is repulsive to temporal axons, and to which nasal axons are rather insensitive. By examining the choices that temporal axons make between membranes extracted from successive rostrocaudal sixths of the tectum, it became clear that this inhibitory activity is graded across the tectum, highest at the caudal pole and lowest at the rostral pole. Two repulsive factors, now called Ephrin-A5 and Ephrin-A2, were then identified as the inhibitory molecules involved by the Bonhoffer and Flanagan laboratories (Cheng et al., 1995; Drescher et al., 1995). Ephrins, of which several are now known, come in two subfamilies, a GPI-linked or A-type, and a transmembrane or B-type. Their receptors, known as Ephs, also divide into two A- and B-type families. The Ephrin-As generally activate Eph-As, while the Ephrin-Bs generally activate Eph-Bs (Flanagan and Vanderhaeghen, 1998). Both Ephrin-A5 and Ephrin-A2 are expressed in a posterior (high) to anterior (low) gradient in the tectum (Figure 6.12). The

retina, as expected, shows a gradient of Eph-As, the receptors for these ligands. Temporal axons that have high levels of Eph-A avoid the posterior pole of the tectum that has the highest level of the Ephrin-A ligands. When Ephrin-A2 is misexpressed by transfection across the entire tectum in chick embryos, temporal axons find it difficult even to enter the tectum. When membrane stripes are made from the transfected anterior tectal cells, temporal axons will not grow on them.

These results predict that when the Ephrins are knocked out, there will be mapping errors. In mutant mice, in which Ephrin-A5 is knocked out, temporal axons map more posteriorly (Frisen et al., 1998), but the mapping phenotype is even more striking in double knockouts of both Ephrin-A2 and Ephrin-A5 (Figure 6.13). In these mice, the anteroposterior order is largely, though not totally, lost. Temporal axons terminate all over the tectum and freely invade the posterior poles (Feldheim et al., 2000). The fact that there is still some order left in this projection suggests that there may be as yet other undiscovered chemospecifity factors that are involved. Knocking out Eph-As,

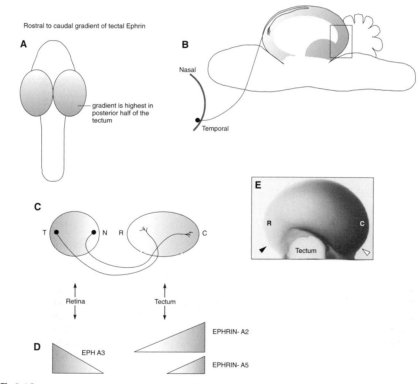

FIGURE 6.12 Nasotemporal to anteroposterior retinotopic guidance system. A. There is a gradient of Ephrins in the tectum, high in the posterior pole and low in the anterior pole. B. A retinal ganglion cell in the temporal retina expresses active receptors for these tectal Ephrins and avoids the posterior tectum. C and D. Show the opposing gradients of active Eph-A receptor expressed in the retina and the gradients of A-type Ephrins in the tectum. This system can at least partially account for topographic mapping in this axis. E. The Ephrin-A gradient is shown in the tectum of a chick that uses a soluble Eph-A receptor fused to alkaline phosphatase to reveal the distribution of the Ephrin-A ligands in the tectum. (After Cheng et al., 1995; Drescher et al., 1995)

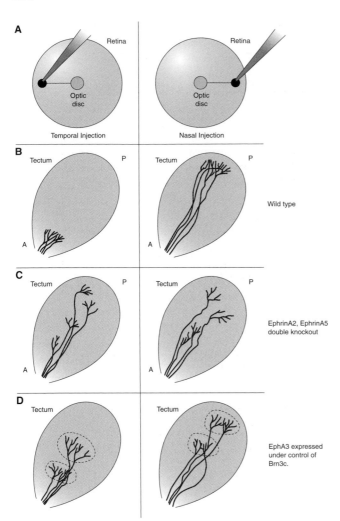

FIGURE 6.13 Topographic mapping in Ephrin-A2 and Ephrin-A5 double knockouts and with mosaic Eph-A3 misexpression. A. Label is injected into the temporal (left) or nasal (right) retinas. B. The result in the normal mice is a projection to the anterior (left) or posterior (right) colliculus. C. In the Ephrin-A2, Ephrin-A5 double knockouts, termination zones are all over the AP extent of the colliculus for both nasal and temporal injections. D. Two separate overlapping maps form from the RGCs that express Brn3c and thus extra Eph-A3 and those that express the normal amount of Eph-A3. (After Brown et al., 2000; Feldheim et al., 2000)

the receptors to the Ephrins, corroborates these findings (Feldheim et al., 2004). If Eph-A3 is knocked out in temporal axons in the chick, they project more posteriorly than they normally would within the tectum. A gene disruption of mouse Eph-A5 receptors caused similar map abnormalities. A very interesting experiment with Eph-A receptors involves gene targeting to elevate Eph-A receptor expression in just one subset of retinal ganglion cell in the mouse, while leaving the neighboring ganglion cells to express normal levels of Eph-A. The effect of this manipulation is to produce two intermingled ganglion cell populations, one that

expresses more receptor and thus should be more sensitive to Ephrin-A, and another that is normally sensitive. The results are that the two populations of ganglion cells from the same eye form separate shifted maps in the tectum, leading to a kind of double vision (Brown et al., 2000) (Figure 6.13). The RGCs that express higher levels of Eph-A map more rostrally than those that express normal levels. This finding clearly favors a model in which retinal growth cones, by the levels of Eph-A they express, read the levels of Ephrin-A in the tectum to establish a graded map.

The data described above, however, does not fully explain the problem of topographic mapping across the anterior to the posterior axis of the tectum. Since Ephrin-As are noted as axon repellents, one of the key questions that remains is why any axons bother to go to the posterior tectum, especially as all axons express at least some Eph-As and should prefer to map to the anterior tectum. Is there a counterbalancing attractive mechanism? One idea in this regard is that the tectum is a source of a limited supply of neurotrophin, for which retinal axons compete (Wilkinson, 2000). The nasal axons that have the least Eph-A find the competition less fierce in the posterior tectum, which is why they map there. This idea may explain the otherwise puzzling observation that removal of the Ephrin-As from the tectum not only causes a posterior shift for temporal axons, but also causes an anterior shift for nasal axons, as if the two populations were competing. There is, however, another, though not mutually exclusive, explanation that has to do with the finding that Ephrins are not always repulsive. A systematic in vitro analysis shows that Ephrin-A2, while capable of inhibiting the growth of temporal axons at high concentrations, actually promotes the growth of these axons at low concentrations (Hansen et al., 2004) (Figure 6.14). Moreover, the transition from growth inhibition to growth promotion varies across the retina according to how much Eph-A is expressed in RGCs; so nasal axons with low levels of Eph-As may actually be attracted to the posterior tectum.

The possibility that Ephrins are involved in attractive as well as repulsive signaling is much clearer in the other axis of retinotectal map formation, in which axons from the dorsal retina map to the lateral tectum while axons from the ventral retina map to the medial tectum. Eph-B receptors have been found in a ventral (high) to dorsal (low) gradient in the retina, whereas Ephrin-B ligands are found in a medial (high) to lateral (low) gradient in the tectum (Braisted et al., 1997; Holash et al., 1997) (Figure 6.15). Interestingly, Ephrin-Bs are also expressed in the retina, in a dorsal (high) to ventral (low) gradient, whereas Eph-Bs are expressed in the tectum in a lateral (high) to dorsal (low) gradient. These expression patterns suggest that attraction rather than

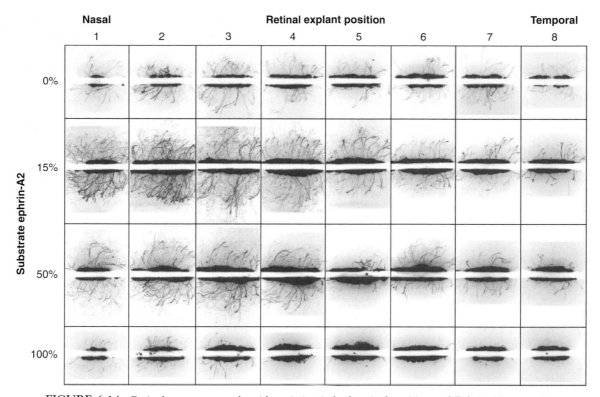

FIGURE 6.14 Retinal axon outgrowth, with variation in both retinal position and Ephrin-A2 concentration. Representative photographs showing outgrowth from the eight contiguous explant positions (numbered 1 through 8) across the nasal-temporal axis of the retina, grown on substrates containing different proportions of membranes from Ephrin-A2 DNA-transfected and untransfected cells. Outgrowth varies with both retinal position and Ephrin concentration. Responses to Ephrin-A2 membranes vary from total outgrowth inhibition (at higher concentrations and more temporal positions) to several-fold outgrowth promotion (at lower concentrations and nasal positions). (After Hansen et al., 2004)

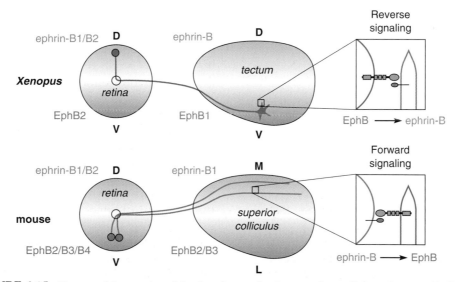

FIGURE 6.15 Topographic mapping of the dorsal ventral retina onto the mediolateral tectum. In *Xenopus*, Ephrin-B expressing retinal ganglion cells from the dorsal retina via reverse-signaling are attracted to the high levels of Eph-B in the lateral tectum, while in chicks Eph-B expressing retinal ganglion cells are attracted via forward signaling to the high levels of Ephrin-B in the medial tectum. Abbreviations: D, dorsal; V, ventral; M, medial; L, lateral. (From Pittman and Chien, 2002)

repulsion is the overriding mechanism at work in this dimension. Here Eph-B expressing axons of the ventral retina are attracted to Ephrin-Bs in the medial tectum, and Ephrin-B expressing RGC axons of the dorsal retina are attracted to Eph-B expressing cells in the lateral tectum via backwards signaling from the receptor to the ligand. In *Xenopus*, the prevention of all Ephrin-B/Eph-B interactions causes dorsal axons to project medially rather than laterally (Mann et al., 2002a). This effect seems to depend on Ephrin-B function in the axons because the same phenotype occurs if retinal axons express a dominant negative form of Ephrin-B that lacks an intracellular domain. Thus, reverse signaling seems to attract Ephrin-B expressing dorsal retinal axons to Eph-B-expressing cells in the lateral tectum (Figure 6.15). But then why does the ventral retina project to the medial tectum? When the Eph-B2 and Eph-B3 receptors are knocked out, there is an ectopic projection to the lateral tectum and the phenotype is even stronger, if the Eph-B receptors are replaced with receptors that are unable to signal (Hindges et al., 2002). This result suggests that forward signaling through the intracellular domain of the receptor is critical for ventral axons to map to the medial tectum (Figure 6.15).

These studies on Ephrin-As and Ephrin-Bs and their receptors strongly verify Sperry's chemospecificity ideas for retinotectal mapping by providing the molecular identities of at least some cytochemical tags of the kind that he proposed. The next question is whether the Ephs and Ephrins are involved in setting up topographic projections in other regions of the nervous system. Certainly, the fact that there are many of these ligands and receptors is consistent with such a possibility, as are the histological findings that the CNS is painted with a rich pattern of these ligands and receptors often in reciprocal graded arrangements of A-type ligands with A-type receptors and B-type ligands with B-type receptors (Zhang et al., 1996). Work in a number of systems has now established that this is the case. For example, there is evidence that Ephrin/Eph signaling is used in establishing the visuotopic projection from the retina to the visual thalamus (Feldheim et al., 1998), the somatotopic map of the body surface on the primary somatosensory area of the cortex (Vanderhaeghen et al., 2000), and the tonotopic projection from the cochlea onto the nucleus magnocellularis in the hindbrain (Person et al., 2004).

SHIFTING AND FINE TUNING OF CONNECTIONS

Branching can be topographic. In the frog and the fish, retinal axons make branched terminals in the correct topographic location in tectum (Harris et al., 1983; Stuermer and Raymond, 1989), but in the chick and the mouse, axons overshoot their termination points and subsequently make interstitial branches at the correct topographic position behind the growth cone (Nakamura and O'Leary, 1989; Simon and O'Leary, 1990, 1992) (Figure 6.16). This is a process of map refinement. The branches that form along the shaft of RGC do so with good topographic specificity, which is enhanced through the preferential arborization of appropriately positioned branches and elimination of ectopic branches, thus further refining the topography. Topographic refinement may occur throughout life. Consider the case of a goldfish. It hatches as a tiny 1 mg animal and over the course of its life may attain a weight of 1 kg or more. It has increased in volume a millionfold. As the animal grows, the retina grows in proportion by adding cells circumferentially at the rim or margin. The tectum grows as well but mostly at the caudal end. In order for the retinotectal map to remain evenly distributed, the retinal axons must continually retract anterior branches and send out new branches more posteriorly in the tectum (Gaze et al., 1979). For example, axons from the center of a large adult retina are from the oldest retinal ganglion cells that were born when the fish was just a small larva (Figure 6.17). These axons used to project to the center of the larval tectum whose cells remain at the anterior pole of the tectum as new tectal cells are added caudally, but now they project to cells in the middle of the large adult tectum, perhaps a millimeter or so away (Easter and Stuermer, 1984). These axons have continued to switch their preferred targets to more posterior cells throughout their lifetimes. A similar type of shifting reorganization of connections is evident when half of the retina or half of the tectum of a fish is ablated. When half the retina is ablated, the projection of the remaining half of the retina expands to cover the entire tectum. When half the tectum is removed, the retina's projection compresses to cover the remaining half (Schmidt and Easter, 1978; Schmidt and Coen, 1995). This sort of regulation is also observed in neonatal hamsters with a partially deleted superior colliculus (Figure 6.17). This form of topographic expansion or compression, like the natural shift that is a consequence of the asymmetric growth of the tectum, does not depend on the activity patterns in retinal fibers. The regulation can occur in the dark or even in the continuous presence of tetrodotoxin (TTX) which blocks action potentials (Meyer and Wolcott, 1987).

These shifting connections are part of larger developmental phenomenon whereby, once a topographic map is roughly established, it is adjusted, modified,

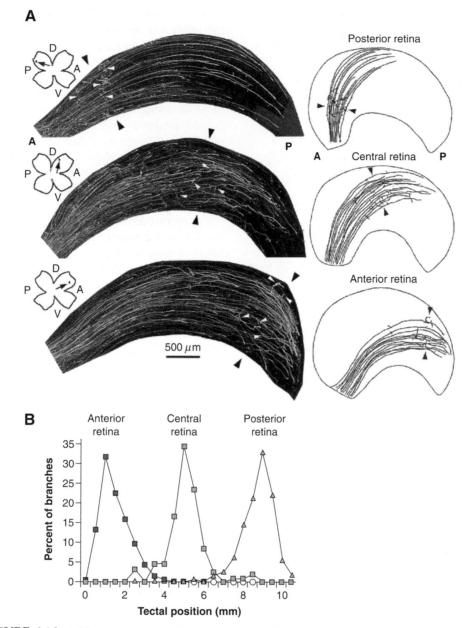

FIGURE 6.16 RGC axons overshoot their correct termination zone in a position-dependent, differential manner in chick embryos. A. RGC axons in tectal whole mounts labeled by a small focal DiI injection into peripheral temporal retina (*top*), central retina (*middle*), or peripheral nasal retina (*bottom*) on E11. The injection sites are shown in drawings of the retinal whole mounts and marked by *arrows*. The relative positioning of the labeled axons and branches within the tectum is shown to the *right*, with drawings of the outline of each tectum on which the labeled axons and branches are traced. Axons overshoot their topographically correct TZ (the predicted locations of the TZs are marked with *black arrowheads*), but the distribution of interstitial branches along the axon shafts (*white arrowheads*) is strongly biased for the location of the future TZ along the anterior (*A*)-posterior (*P*) tectal axis. Peripheral temporal axons exhibit the greatest overshoot and peripheral nasal axons the least. B. Distribution of interstitial branches along the axon shaft expressed in percentage. The A-P tectal axis was divided into 500 μm bins, and the number of branches in each bin is graphed as the percentage of total branches for each of the three groups of injections. (After Yates et al., 2001)

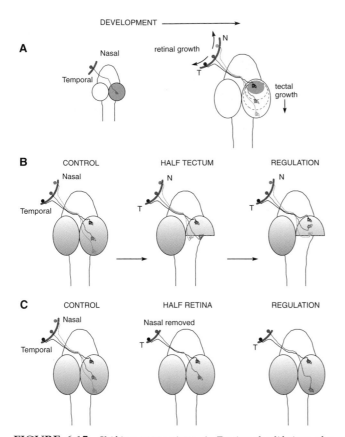

FIGURE 6.17 Shifting connections. A. During the lifetime of a frog or fish, its eye and brain continue to grow. The retina grows circumferentially like a tree, but the tectum grows in expanding posterior crescents. As a result, new retina that is added temporally must send axons to the anterior primordial tectum, while fibers from the central primordial retina must shift posteriorly, and new nasal fibers map to the new posterior tectum in order to keep the map in topographic order. B. If half the tectum is removed from a fish, after about a month the retinotopic map will regulate and compress, mapping out evenly over the remaining half tectum. C. Similar regulation occurs when half the retina is removed. The remaining projection eventually expands over the whole tectum. (After Schmidt and Easter, 1978; Gaze et al., 1979; Schmidt and Coen, 1995)

and fine-tuned. Part of this refinement may be based on growth patterns or injury, as above, but refinement also has an activity or experience-dependent aspect. Without impulse activity, the retinotectal map of the goldfish is topographic, but the sizes of the receptive fields recorded in the tectum are larger and less precise than normal. Analysis of individual retinal axonal arbors shows that they are up to four times as large as normal ones (Schmidt and Buzzard, 1990). Repeated examination of single retinal arbors over time shows the effects that activity has on branching and topography. When retinal activity is abolished by tetrodotoxin (TTX), the result is increased branch addition and elim-

ination, or in other words decreased branch stability (Schmidt and Buzzard, 1990).

The mechanisms by which activity may have such effects on the fine-tuning of connections are discussed in Chapter 9. Here, we would simply like to point out how activity may affect topographic maps in the nervous system. A very interesting example in this regard is the somatosensory system. The discovery of a somatosensory representation of the body, a homunculus in the case of humans, was discovered by the neurosurgeon, Wilder Penfield (Penfield, 1954a). While performing operations to remove epileptic foci in the brains of fully conscious patients, Penfield took the opportunity to study the organization of the cortex by locally stimulating different regions with an electrode. When he stimulated points in the postcentral gyrus, patients reported the sensation of touch in specific areas of their bodies. Stimulation of neighboring points caused the patients to experience sensations in neighboring parts of their body surface although there were occasional jumps, such as between the hand and the face. By mapping these sensations on the cortex of different patients, Penfield was able to come up with a consistent somatosensory homunculus and in the precentral gyrus a matching topographic homunculus where stimulation caused movements of specific body parts (Figure 6.18). One striking feature of the homunculus that Penfield noticed immediately is the relative magnification of parts of the map. This appears to be a consistent feature of many maps in the CNS. The largest features of the human homunculus are the lips, tongue, and tips of the fingers. In contrast, the representation of the upper back is quite small. In other animals, the somatosensory cortex has an expanded representation of different body parts: the hands of the raccoon, the snout of a star nose mole, and the whiskers of the mouse, for example, are particularly enlarged. The differential magnification of certain body parts in the cortical representation of the body is probably due to the density of peripheral innervation. Thus, in humans, each fingertip has almost as many sensory receptors as the whole of the upper back. In mice, the vibrissae are most heavily innervated.

Central representations of the somatosensory system are flexible and may depend on sensory stimulation, especially during early life. In the mouse, single cortical areas, called *barrels*, are devoted to each vibrissa. The barrel fields of the cortex are almost equal in size to the somatosensory cortex devoted to the rest of the body (Woolsey and Van der Loos, 1970). There are five rows of barrels that correspond to the five rows of vibrissa. When a bristle is destroyed by cauterization in early life, the cortical barrel that represents it

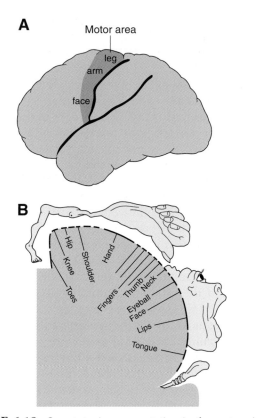

FIGURE 6.18 Somatotopic representation in the cortex. A. The motor area of the precentral gyrus of the cerebral cortex was stimulated electrically in human patients during neurosurgery. B. A "homunculus" of the body on the motor cortex illustrates the sequence of representation as well as the disproportionate representation given to the various muscles involved in skilled movements. (After Penfield, 1954b)

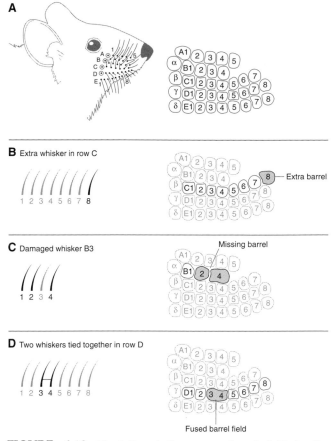

FIGURE 6.19 Plasticity of the mouse barrel field in the somatosensory cortex. A. The correspondence between bristles and barrel field in the cortex of a normal mouse. B. An extra whisker in row C leads to the formation of an extra barrel in the appropriate location in the cortex. C. Neonatal damage to the B3 whisker causes the shrinkage of this barrel and the expansion of neighboring ones. D. Tying two whiskers together causes their barrel field to coalesce. (After Woolsey and Van der Loos, 1970)

shrinks, while the neighboring barrels expand into the territory of the cortex originally devoted to the cauterized whisker (Dietrich et al., 1981; Simons et al., 1984) (Figure 6.19). When two whiskers are glued together, their cortical barrels fuse. Perhaps the most surprising finding is the case of a mouse that was born with an extra whisker, as sometimes happens. This mouse had an extra barrel in its cortex (Van der Loos et al., 1984). From these results, it is clear that the neural representation of the body surface has flexibility in its structure. The sensory fields themselves and their activity guide this flexibility.

When an adult, through accident or medical intervention, loses sensation in one area of the body as happens when a peripheral nerve is cut, the cortical representation of that area may be invaded by representation from neighboring parts. This is thought to be one reason why people who have lost a limb may report sensations in the phantom limb, especially

when a part of the body is touched whose cortical representation is adjacent to the missing limb (Ramachandran and Rogers-Ramachandran, 2000). A touch to the face in such a person can be experienced as a touch on the missing hand. The explanation is that nerve fibers that carry information about touch on the face invade the neighboring cortical area that used to receive such information from the lost limb. The rest of the brain, however, has not yet "learned" the change in the meaning of the input to this part of the cortex, and still interprets it as a touch to the hand (Figure 6.20). Experiments with monkeys, in which a cuff of TTX on the nerve temporarily paralyzes a single finger, have shown that there is a rapid reorganization of the somatosensory map in the cortex. Within days, the representation of the insensitive finger shrinks, and the

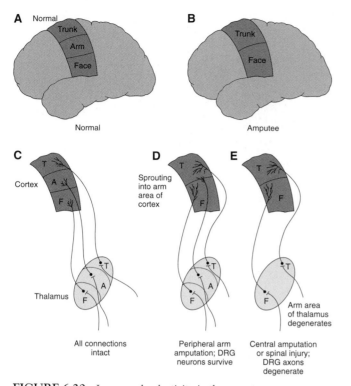

FIGURE 6.20 Large-scale plasticity in the somatosensory cortex. A. The normal organization of the cortical topography in the somatosensory system. B. Damage to the arm may result in large-scale reorganization of the cortical map. C. There is an isomorphic mapping of the somatosensory thalamus onto the primary cortex. D. When the arm is damaged peripherally and the sensory neurons in the DRG survive, the reorganization is cortical rather than thalamic. E. When the damage is more central causing the DRG cells to degenerate, both the thalamus and cortex get reorganized. (After Merzenich, 1998)

representation of the neighboring fingers expands (Merzenich and Jenkins, 1993). These cortical changes in the representation of somatotopy can be extremely large even in normal animals, as was revealed by an unusual experiment at the National Institutes of Health. Antivivisectionists stole a set of experimental monkeys after their somatosensory cortex was first mapped, and they were not recovered until about 10 years later. When the scientists remapped their somatosensory cortex, they found that the extent of the rearrangement was dramatic, a matter of tens of millimeters (Palca, 1991). Thus, minor reorganization of the cortical somatosensory map is happening throughout life, presumably influenced by experience and activity. Even normal use can change topographic representations in an impressive way. Monkeys trained to use just one finger to feel textural differences for a few hours a day over a period of months have a hugely expanded cortical representation of that finger (Recanzone et al., 1992).

THE THIRD DIMENSION, LAMINA-SPECIFIC TERMINATION

Many parts of the nervous system are layered structures like the tectum and the cortex, and innervating axons must not only map to their correct topographic position in two dimensions, but they must also find the appropriate layer in which to synapse, making targeting a three-dimensional problem. Lamina-specific targeting may involve a variety of different issues. In many cases, a laminated target is composed of layers of physiologically and molecularly distinct cells types, and the innervating axons must therefore choose between different cell types, possibly based on chemical differences. In some cases, the layers are composed of essentially similar cells, but the innervating axons, through an activity-based competition, segregate them into layers. This latter case can be considered an example of the refinement of synaptic connections and so will be dealt with in Chapter 9.

We have already encountered the first kind of the laminated structure in the central projections of DRG fibers in the spinal cord. These axons enter the spinal cord and make synapses in various laminae of the dorsal horn or ventral gray matter depending on their modality. For instance, stretch receptors make monosynaptic contact with motor neurons in the ventral horn. In contrast, pain and temperature sensory fibers innervate neurons in dorsal laminae of the spinal cord (Figure 6.21). The result of this laminar arrangement by different types of input is that somatosensory modalities sort out in the spinal cord and so make a multilayered registered map, such that a column of cells in the spinal cord represents one area of the body with different modalities at different depths. Multimodal, layered maps are used in several places in the nervous system. Why do only stretch receptors penetrate the more ventral layers of the spinal neuropil? In the previous chapter, we described the varying sensitivity of different classes of neurons to the repulsive effects of semaphorin3A, which is expressed in the ventral layers only. Semaphorin3A repulses pain receptors and thermoreceptors, which therefore map to dorsal layers, while stretch receptors ignore Semaphorin3A and map to ventral layers (Messersmith et al., 1995). In mice in which the Semaphorin3A gene is knocked out, the terminals of the pain and thermoreceptive axonal terminals appear to extend into the ventral regions of the spinal cord, similar to the stretch receptors and this layer-specific targeting is abolished (Taniguchi et al., 1997; Catalano et al., 1998) (Figure 6.21).

The cerebral cortex is another example of a highly laminated structure, composed of different cell types in

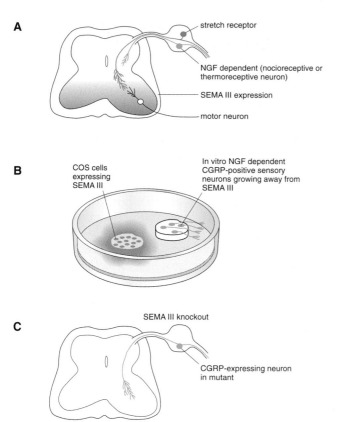

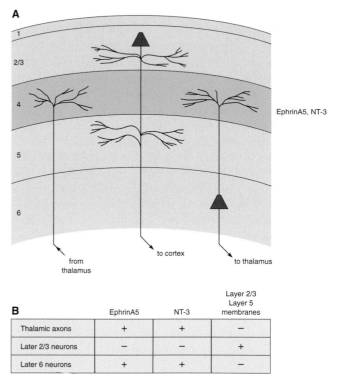

FIGURE 6.21 The role of Sema3A in keeping some axons out of a target region. A. Stretch receptors project their central axons into the ventral horn of the spinal cord where they synapse on motor neuron dendrites. Nocioreceptive and thermoreceptive axons, which are NGF dependent, however, terminate in the dorsal gray matter of the spinal cord. Sema3A is expressed only in the ventral cord. B. COS cells in culture repel nocioceptive and thermoreceptive axons from DRG cultures. C. In Sema3A knockout mice, these sensory neurons extend into the ventral horn. (After Messersmith et al., 1995)

FIGURE 6.22 Layer-specific targeting in the sensory cortex. A. Thalamic and Layer 6 axons invade the cortex and selectively branch in Layer 4, which expresses Ephrin-A5 and NT-3. The branches of Layer 2/3 neurons avoid Layer 4. B. In vitro studies show that Ephrin-A5 and NT-3 enhance growth and branching of thalamic and Layer 6 neurons, but inhibit the growth of Layer 2/3 neurons. There is also in vitro evidence to suggest that the membranes of Layer 2/3 and Layer 5 cells are specifically inhibitory to the growth of thalamic and Layer 6 axons.

different layers. The different layers of the cortex are innervated by different inputs. In vitro studies using membrane fractions of cells in the different layers suggest that the targeting cues are membrane-associated (Castellani and Bolz, 1997). Somatosensory thalamic neurons, for example, innervate layer 4 of the somatosensory cortex and cross through layer 5 without branching (Bolz et al., 1996) (Figure 6.22). Ephrin-A5 is expressed on the membranes of cells in layer 4 but not layer 5, and in vitro studies show that membrane-bound Ephrin-A5 increases the branching of thalamic axons and that there are as yet unidentified repulsive activities to these neurons on the membranes of layers 2/3 and 5 cells (Mann et al., 2002b). Layer 6 neurons in the cortex, like thalamic cells, also arborize in layer 4, and some results suggest that the axons of these neurons respond to Ephrin-A5, as do thalamic axons (Castellani et al., 1998). Unlike the thalamic cells and layer 6 cells, layer

2/3 cells in the cortex send out axons that do not branch in layer 4. For the axons of layer 2/3 cells, in vitro experiments show that Ephrin-A5 inhibits rather than promotes branching. Interestingly, NT-3, which is expressed in layer 4, also promotes axonal branching of layer 6 axons while it inhibits branching of layer 2/3 axons (Castellani and Bolz, 1999). Finally, it has been shown that blocking impulse activity in the cortex also impairs the selective branching of layer 6 axons in layer 4. In summary, these studies demonstrate that many familiar factors are at work, leading to laminated projections, and that some factors may have bifunctional roles in promoting the branching of some axons and inhibiting the branching of others.

Laminar-selective growth is even more impressive in the chick tectum where there are 16 layers that receive input from at least 10 different sources. The retinal ganglion cells contribute input to just three of these layers, and each retinal ganglion cell sends its terminals to just one of these three layers. These retinorecipient layers express a number of molecules including one unidentified factor known only because it binds a particular plant lectin, known as VVA, which labels all three retinorecipient layers, but none of the other layers (Yamagata et al., 1995). To study the components of lamina-specific termination in this system, sections of formaldehyde-fixed tectum were put into a culture dish with live retina. Amazingly, the retinal axons grow into the correct layers in this situation. Yet if VVA is added to this preparation, retinal axons become unable to map to the correct laminae. Several different cadherin molecules, such as N-cadherin, R-cadherin, and T-cadherin, are also expressed in different combinations of the tectal laminae, with N-cadherin being selectively present in the retinorecipient layers. Antibodies to N-Cadherin when added to these cultures also cause lamination errors, though of a different type, with some axons stopping in the retinorecipient layers but not extending in them. If BDNF is added to the in vitro preparation, it does not affect the appropriate targeting or retinal axons to the retinorecipient layers, but it does cause excessive growth and branching in these layers. These studies suggest that different molecules regulate different aspects of laminar-specific innervation, including recognition, innervation, and branching (Sanes and Yamagata, 1999).

The retina is a multilayered structure, and the layers where synapses occur, such as the inner plexiform layer, are refined into functionally specialized sublaminae, such as the ON sublaminae that contain the synaptic terminals of bipolar and amacrine cells that fire when light is turned on and transmit this signal to ON-type retinal ganglion cells whose dendrites are in the same sublamina, and the OFF sublamina that does the same for lights off. Sidekick 1 (Sdk-1) and Sidekick 2 (Sdk-2), have been identified as sublamina-specific molecules within the inner plexiform layer (Yamagata

et al., 2002). Sdks are homophilic CAMs of the IgG superfamily, and each Sdk is expressed by the presynaptic terminals of a subset of bipolar and amacrine cells and the postsynaptic dendrites of a subset of ganglion cells that project to a common sublamina. Ectopic expression of Sdk-1 in Sdk-negative cells redirects their processes to the Sdk-1 positive sublamina, and similar experiments with Sdk-2 show that it directs processes to the Sdk-2 sublamina (Figure 6.23). Retinal ganglion cells, although they express Sdks, are not absolutely critical for the formation of these sublaminae. As in the zebrafish mutant *lak*, ganglion cells are never born, and yet the ON and OFF sublaminae form, though in a delayed and slightly disarrayed way (Kay et al., 2004). That homophilic adhesion molecules bind axonal terminals to the dendritic processes of cells that are destined to synapse onto each other makes a good deal of sense. As we will see in the next section, this kind of process is used a great deal when we consider targeting at the cellular or synaptic level.

CELLULAR AND SYNAPTIC TARGETING

The final step in targeting comes when axonal terminals actually make contact with the specific cells or parts of cells with which they will synapse. Choosing a specific postsynaptic partner is aided by getting the terminal branches to the right topographic and laminar position, but the next stage involves the actual adhesion of specific terminals to specific postsynaptic target cells. A good example of how individual axons choose particular target cells is the neuromuscular system of the *Drosophila* larva. In each segment of the

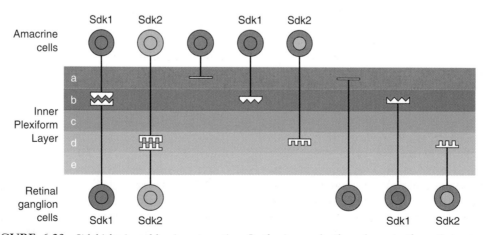

FIGURE 6.23 Sidekicks in sublaminar targeting. In the inner plexiform layer in the retina, neurons expressing the same sidekick send processes to the same sublamina, thereby establishing lamina-specific synaptic connections. Overexpression of a sidekick shifts connectivity. (After Yamagata et al., 2003)

larva, the growth cones of about 40 motor neurons touch about 30 different muscles before they select those one or two onto which they will synapse (Nose et al., 1992b; Broadie et al., 1993). The differences between the muscles are subtle since they are, by and large, similar to one another. Each muscle has a variety of molecules on its surface, many of which are the same as the molecules expressed on the membranes of all its neighbors. The difference is that an individual muscle cell also expresses cell surface molecules that are shared with only some of its neighbors, and it usually expresses different concentrations of the same molecules as its neighbors (Winberg et al., 1998).

Some motor neurons normally innervate muscles that express high levels of Netrin. In embryos in which Netrin is not expressed, the axons that would normally innervate some of these muscles go to inappropriate muscles (Mitchell et al., 1996). However, two other muscle cells that normally express Netrin remain innervated in these mutants, suggesting that additional recognition molecules must be involved in these muscles. Connectin is a second molecule that plays a role in nerve–muscle specificity in this system. It is a homophilic cell adhesion molecule that is expressed, under the direct control of a *homeotic* gene on the surface of a subset of motor neurons and the muscle cells that they innervate (Gould and White, 1992; Nose et al., 1992a; Meadows et al., 1994; Nose et al., 1994, 1997; 1997; Raghavan and White, 1997). There are few neuromuscular innervation defects in *connectin* mutants, however; when connectin is expressed ectopically on all muscles in transgenic flies, motor axons frequently make targeting errors and invade nontarget muscles adjacent to their normal targets. The defects seen with connectin overexpression may be attributed to increased adhesion between different muscles that do not normally adhere to each other, making it difficult for the axon to take its usual pathway through the muscle field.

A third factor that plays a role in this system is the homophilic adhesion molecule, FasII, which is also expressed differentially on subsets of muscle fibers (Schuster et al., 1996a, b). Since FasII is expressed on many muscle cells at different levels, a more subtle experiment was done, which was to use various cell-type specific promoters to switch the relative levels of FasII expressed on specific muscles. The result is that extra synapses form on muscles that express higher levels of FasII at the expense of synapses formed on neighboring muscles that do no not have increased FasII. This is true when the level of FasII on the less innervated muscles is high or low, so it is the relative and not the absolute level of FasII that is important. FasIII, another homophilic adhesion molecule, and

SemaII, a secreted growth cone repulsive factor, are also expressed on overlapping specific muscle subsets in *Drosophila*. As with connectin mutants, loss of function mutants in these molecules displays no serious effects on neuromuscular targeting (Winberg et al., 1998). But as for the other molecules described, misexpression of FasIII or SemaII in inappropriate muscles leads to dramatic targeting effects. The change in probability of particular motor neurons targeting particular muscles caused by experimentally changing the levels of the single-cell adhesion molecule is consistent with the idea that growth cones are able to distinguish targets by relative changes in the concentrations of a number of such molecules. Furthermore, targeting errors caused by the increase in an attractive or adhesive factor can be compensated by a simultaneous increase in a repulsive factor, showing that indeed the combination of amounts of various such factors is what counts. In summary, the results with FasII, FasIII, connectin, and SemaII, and the netrin suggest that cellular targeting in the *Drosophila* neuromuscular is based on a combinatorial code involving all these molecules and perhaps others (Figure 6.24).

In nematodes, a very intriguing case of cellular targeting is provided by the hermaphroditic specific motor neurons that innervate the vulva. In *syg-1* and *syg-2* mutants, vulval muscles remain uninnervated, and the neurons make ectopic synapses on inappropriate targets (Shen and Bargmann, 2003; Shen et al., 2004). The Syg-1 and Syg-2 proteins are adhesion molecules of the IgG superfamily. Syg-1 is expressed in the neuron and its binding partner Syg-2 is normally expressed transiently not on the postsynaptic targets but on a vulval epithelial guidepost cell. This interaction is critical for the maturation of the axonal terminal in preparation for synapse formation on the adjacent region of the neuron. Interestingly, if Syg-2 is expressed under the control of a promoter that causes it to be localized on other epithelial cells, the hermaphrodite-specific motor neurons begin to make presynaptic specializations at these ectopic sites. Thus, heterophilic binding between Syg-1 and Syg-2 is involved in setting up the formation of appropriate synapses.

SynCAMS and cadherins (especially protocadherins) are large families of homophilic adhesion molecules that may add another level of specificity by helping presynaptic terminals make contacts at the correct subcellular locations (Abbas, 2003; Yamagata et al., 2003). There are more than 50 different cadherins, and many of them are expressed at subsets of synapses. As many as 20 different genes of the cadherin superfamily genes are expressed in restricted patterns in the developing tectum (Miskevich et al., 1998), and recent speculation is that these types of molecules, like the

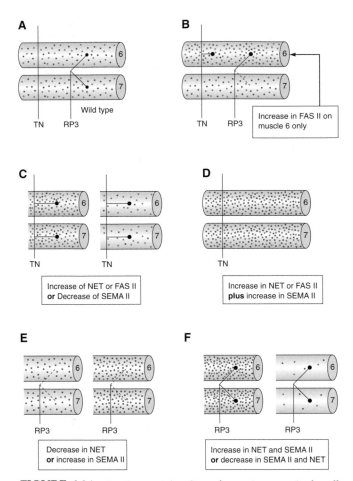

FIGURE 6.24 Combinatorial coding of targeting at a single-cell level. A. In wild-type *Drosophila* larva, the TN nerve does not innervate muscle fibers 6 and 7, whereas the RP3 nerve innervates both. B. When FasII is increased on muscle 6 only, both the TN nerve and the RP3 nerve innervate this muscle differentially. C. When Netrin expression or FasII expression is increased on both muscles, or when SemaII is decreased on both, the TN nerve innervates both 6 and 7. D. When Netrin or FasII is increased but SemaII is also increased simultaneously, then the TN nerve does not innervate 6 and 7. E. When there is a decrease in Netrin or an increase SemaII, the RP3 nerve does not innervate muscles 6 and 7. F. However, when there is either an increase or a decrease in both SemaII and Netrin, the RP3 nerve innervates as normal. (Adapted from Winberg et al., 1998)

Sdks and Sygs discussed above, are involved in a synaptic targeting code (Redies and Takeichi, 1996). As an example of how this might work, consider the N- and E-cadherins, which are distributed at synaptic junctions in a mutually exclusive pattern along the dendritic shafts of single pyramidal neurons (Fannon and Colman, 1996). Ultrastructural examination of double immunolabeled material revealed the existence of many unlabeled synapses on these cells as well as raising the possibility that synapses at these other synapses are linked by other cadherins or CAMs. In the

hypothalamus, it has been found that some protocadherins are found just at excitatory synapses (Phillips et al., 2003). Of course, these molecules are likely to do more than simply glue particular synapses together; they are also likely to be involved in the maturation, structural organization, and stabilization of synapses, topics that will be covered in more detail in Chapter 8.

SNIFFING OUT TARGETS

The way that the axons of olfactory receptor neurons find their particular targets in the olfactory bulb of the vertebrate brain has proven to be an extraordinary case of cell-specific targeting. Olfactory receptors are located in the olfactory sensory epithelium of the nose, and they send axons into the bulb where they make connections with second-order cells in synaptic complexes called *glomeruli*. Physiological studies reveal that distinct odorants cause activity in distinct glomeruli, and in the zebrafish, a careful anatomical study showed a reproducible pattern of about 80 glomeruli that have the same position and size from individual to individual (Baier and Korsching, 1994). Surprisingly, the receptors projecting to a single glomerulus are scattered all over the olfactory epithelium in a fairly random pattern, and there is no regionalization of odorant receptors on the sensory epithelium (Figure 6.25). So point-to-point mapping does not occur as it does in the visual or somatosensory systems. In the mammalian olfactory epithelium, about 1000 different genes code for the seven transmembrane receptors to odorants (Buck and Axel, 1991). By using in situ hybridization, it is possible to label all the olfactory sensory cells that express a particular receptor molecule. Odorant receptor genes are expressed in nonoverlapping subsets of sensory neurons, each neuron expressing only one of the 1000 odorant receptor genes. All the sensory neurons that express a particular odorant receptor are located within one of four zones, and neurons in each of these zones send axons to the glomeruli situated in matching zones of the olfactory bulb. So there is some topography, but again surprisingly within each zone, all the receptors that express a particular odorant receptor gene are dispersed widely and randomly throughout the zone, so there is no spatial topography with the zones. The amazing thing is that all the olfactory neurons that express the same olfactory receptor genes, though distributed over a wide area of the olfactory epithelium, nevertheless project their axons to single glomeruli in specific regions of the olfactory bulb (Vassar et al., 1994). And although there are more than 2500 choices of glomeruli, the olfactory neurons that express the

A

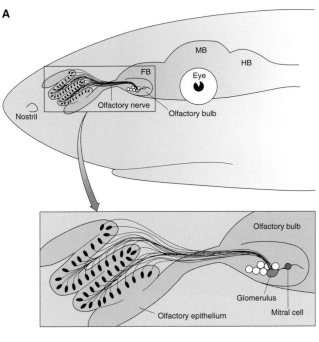

B

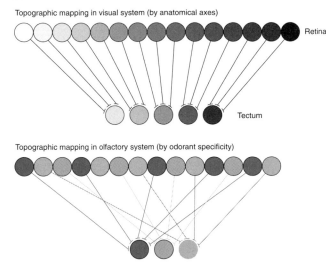

Topographic mapping in visual system (by anatomical axes)

Retina

Tectum

Topographic mapping in olfactory system (by odorant specificity)

FIGURE 6.25 Comparison of topographic mapping in the visual system, where neighboring cells project to neighboring targets creating a central representation of visual space; and the olfactory system where cells of the same type are intermingled and yet their axons sort out and converge forming an odor representation map.

same gene map to just one of two of these (Mombaerts, 1996).

The zone-to-zone mapping of the olfactory neurons onto the bulb uses, it seems, guidance cues and topography cues that we have already discussed in enough detail, such as cell adhesion molecules and repulsive guidance factors. But what concerns us in this case is the question of how the axons of a single class of receptor converge onto topographically fixed glomeruli so that a

consistent olfactory map is created in the brain (Vassar et al., 1994; Mombaerts, 1996). There are two major possibilities. The first is that the expression of a particular receptor gene is linked to the expression of particular set of guidance molecules. The other, more radical, possibility is that the odorant receptor molecules themselves are expressed on axonal growth cones and are involved in sniffing out the correct postsynaptic target area. Let us look at this more interesting possibility first. The first question is whether odorant receptor molecules are expressed in axons and growth cones. The answer is yes, both odorant receptor mRNA and protein are expressed in growth cones (Barnea et al., 2004) (Figure 6.26). The next question is what happens when a receptor is knocked out? Are the axons unable to find their targets? To answer this question, marker genes such as *lacZ* have been knocked into specific receptor loci. Thus, *lacZ* is expressed in the cells that would have expressed a particular receptor and by examining the distribution of *lacZ*, which is transported down the axons of these cells, one can see that the axons appear disoriented and do not converge on their targets (Wang et al., 1998). This suggests that the olfactory receptors are critical for accurate targeting in the bulb.

Swapping receptors is a powerful way to test this idea (Mombaerts et al., 1996; Wang et al., 1998). Thus, in another set of experiments, a specific odorant receptor gene was replaced by a fusion gene driving not only *lacZ*, but also the cDNA for a different receptor, M71, so that the axons misexpressing this receptor are easy to visualize (Figure 6.27). When olfactory neurons that target to distant regions of the bulb have their receptors swapped, they target neither to their normal glomeruli (P2 in this case) nor to the glomeruli typical of their new odorant receptor (M71). Instead, they map to a new specific glomerulus somewhere in-between, suggesting that, although odorant receptors do have some role in targeting, there must be other factors that guide these axons to their particular targets. In fact, there is accumulating evidence that this is so. Like the muscles of *Drosophila* larvae discovered above, different combinations of cell adhesion and guidance molecules appear to be expressed on various receptors. However, when receptors are swapped between sensory neurons that have nearby targets in the same region such as when the P3 receptor is expressed under the P2 promoter, the axons do target to the precise vicinity of P3, proving that the odorant receptors are very important for targeting to the exact right place.

Even a minor change in the coding region of an olfactory receptor gene causes a change in the target destination of the axons that express this gene (Feinstein and Mombaerts, 2004). The critical amino acid residues that affect guidance tend to be clustered in

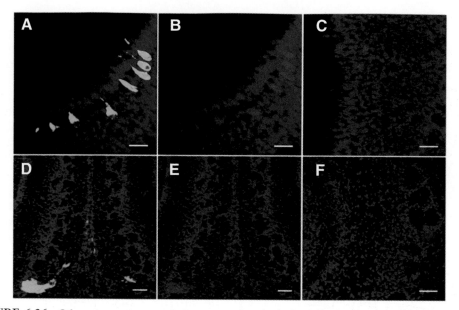

FIGURE 6.26 Odorant receptor protein is expressed on both dendrites and axons of olfactory sensory neurons. A.–C. Staining of mouse olfactory epithelium with A. an antibody to extracellular epitope on a particular odorant receptor, B. an antibody to the cytoplasmic epitope of the same receptor, and C. an antibody to a different receptor. Scale bars, 10 μm. D–F. Staining of mouse olfactory bulb with the same antibodies. (From Barnea et al., 2004)

the transmembrane domains. When these are interchanged between two parent receptors, the axons expressing the hybrid receptors may map to the same glomerulus as the axons that express one of the parent molecules they may map or to a different glomerulus altogether. Whatever they do, all the axons that express the chimeric proteins seem to behave the same way. These results suggest that odorant receptor molecules expressed on axons and growth cones may mediate homotypic adhesion between like axons. In agreement with this idea is the finding that the majority of axons that express the same receptor fasciculate with each other before they enter the glomerular neuropil (Potter et al., 2001). However, some axons always seem to follow more tortuous courses without fasciculating before they terminate in the target glomerulus. This indicates that homotypic adhesion cannot be the whole story and that there is also is a direct axon to glomerulus-specific interaction mediated by the olfactory receptor molecule. If the odorant receptor molecule is used in target recognition in this way, the next question is what these receptors on the growth cone are sensing and whether particular regions of the olfactory bulb "smell" different to these axons.

Interestingly, activity may play a critical role in the refinement of the target selection of odorant sensory axons. In the section on topographic mapping above, we discussed a role for such a mechanism in the refinement

of somatosensory axons in the mouse to specific barrel fields in the cortex. Could something similar be happening here? Certainly, it is true that axons; that are active at the same time tend to terminate together; this is part of the refinement process, and olfactory neurons with the same receptor molecule would respond the same way to odorants. However, the capacity of cells expressing a particular receptor to converge to discrete glomeruli in the olfactory bulb appears to be uncompromised or minimally affected in transgenic or mutant mice in which the activity patterns of olfactory neurons are altered, and elimination of all synaptic transmission from olfactory sensory neurons appears to leave the initial targeting process initially intact (Belluscio et al., 1998; Lin et al., 2000; Zheng et al., 2000). However, if olfactory sensory neurons express a mutation that makes them electrically unresponsive to odorants, though they initially map correctly to their target glomeruli, they then fail to compete with active axons expressing the same receptor and are eliminated (Zhao and Reed, 2001). Interestingly, these axons can be rescued by nose plugs, which cause the potentially active axons to be odorant deprived. These studies suggest that competitive activity is necessary to stabilize specific connections in the bulb. The idea of competition is strengthened by the discovery that very early in the innervation of the bulb, some glomeruli are innervated by more than one receptor type of axon, but this sorts out

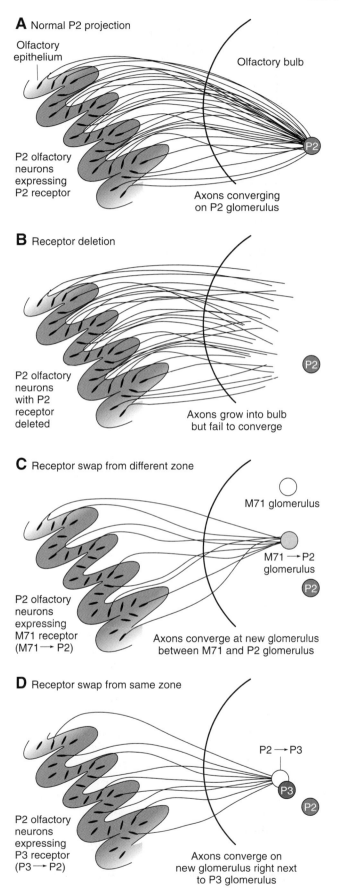

A Normal P2 projection

Olfactory epithelium

Olfactory bulb

P2 olfactory neurons expressing P2 receptor

Axons converging on P2 glomerulus

B Receptor deletion

P2 olfactory neurons with P2 receptor deleted

Axons grow into bulb but fail to converge

C Receptor swap from different zone

M71 glomerulus

M71 → P2 glomerulus

P2 olfactory neurons expressing M71 receptor (M71 → P2)

Axons converge at new glomerulus between M71 and P2 glomerulus

D Receptor swap from same zone

P2 → P3

P2 olfactory neurons expressing P3 receptor (P3 → P2)

Axons converge on new glomerulus right next to P3 glomerulus

postnatally such that eventually each glomerulus is innervated by only a single receptor type (Zou et al., 2004). The importance of successful physiological activity in mapping is also supported by experiments showing that inhibition of all spontaneous activity in a subset of olfactory sensory neurons that express a single odorant receptor causes a selective unraveling of the target glomerulus, with the axons that originally invaded this glomerulus now making inappropriate contacts in various regions of the bulb (Yu et al., 2004). There is a suggestion that activity is important for the continued expression of particular odorant receptors. Therefore, physiologically ineffective olfactory neurons, like those just mentioned, may express other receptors and cause the axons of these cells to begin to wander the bulb in an attempt to innervate other glomeruli.

Insects have olfactory sensory neurons in their antennae, and these also send axons into olfactory glomeruli in the brain, according to a "one receptor type one glomerulus" rule. However, in *Drosophila*, odorant receptor gene manipulations suggest that, unlike mammals, the receptor molecules themselves are not critically involved in targeting. However, specific olfactory receptor neuron classes require the cell surface protein Dscam (Down Syndrome Cell Adhesion Molecule) in order to target to the correct glomeruli (Hummel et al., 2003). In Dscam mutants, the axons of these cells often do not find the correct target glomerulus. Multiple forms of Dscam RNA are detected in the developing antenna, and Dscam protein is localized to developing ORN axons. Dscam is an IgG-superfamily member, and the gene encoding it has a huge number of potential splice variants. Many of these variants are expressed differentially in subsets of neurons, including subsets of olfactory sensory neurons. Amazingly, there are more than 38,000 isoforms of the Dscam protein (Schmucker et al., 2000) (Figure 6.28), and studies suggest that each isoform binds homophilically to itself but does not bind to other isoforms; even closely related isoforms exhibit isoform-specific homophilic binding (Wojtowicz et al., 2004). It has not escaped the attention of developmental neurobiologists that such diversity could contribute to the specificity of neuronal connectivity.

FIGURE 6.27 Olfactory receptors are involved in central targeting. A. Axons from olfactory neurons expressing the P2 receptor converge on the P2 glomerulus. B. If the P2 receptor is deleted, these axons do not converge on any glomerulus. C. If these neurons are made to express the M71 receptor instead of the P2 receptor, they converge on a glomerulus somewhere in-between P2 and M71. D. If they are forced to express the P3 receptor and the P3 glomerulus is in the same zone as the P2 glomerulus, they converge on a glomerulus right next to P3. (After Mombaerts, 1996; Wang et al., 1998)

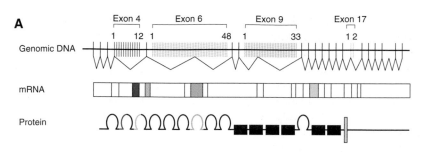

B Differential use of alternative exons generates enormous receptor diversity.

cDNAs	Exon 4	Exon 6	Exon 9	cDNAs	Exon 4	Exon 6	Exon 9
1	1	2	14	26	6	27	16
2	1	5	5	27	6	30	17
3	1	8	11	28	6	34	7
4	1	33	20	29	6	36	16
5	1	43	20	30	6	37	23
6	2	4	13	31	7	6	5
7	2	18	30	32	7	7	12
8	2	25	8	33	7	18	5
9	2	28	13	34	7	19	5
10	2	37	30	35	7	20	21
11	2	39	13	36	7	24	9
12	3	3	13	37	7	42	8
13	3	3	32	38	8	4	26
14	3	15	13	39	8	6	16
15	3	27	7	40	8	19	15
16	3	24	13	41	8	30	25
17	3	37	19	42	8	48	24
18	4	30	4	43	9	35	3
19	4	42	32	44	10	16	2
20	4	46	8	45	10	37	14
21	5	6	8	46	10	42	25
22	5	21	2	47	10	48	26
23	5	26	19	48	12	7	16
24	5	26	19	49	12	15	16
25	5	19	18	50	12	18	9

FIGURE 6.28 Multiple forms of Dscam are generated by alternative splicing. A. The *Dscam* gene spans 61.2 kb of genomic DNA. Dscam mRNA extends 7.8 kb and comprises 24 exons. Mutually exclusive alternative splicing occurs for exons 4, 6, 9, and 17. 1 of 12 exon 4 alternatives, 1 of 48 exon 6 alternatives, 1 of 33 exon 9 alternatives, and 1 of 2 exon 17 alternatives are retained in each mRNA, as deduced from cDNA sequence. Variable exons are shown in color: exon 4, red; exon 6, blue; exon 9, green; and exon 17, yellow. Constant exons are represented by gray lines in genomic DNA and white boxes in mRNA. The splicing pattern shown (4.1, 6.28, 9.9, 17.1) corresponds to that obtained in the initial cDNA clone. The alternatively spliced exons 4, 6, 9, and 17 encode the N-terminal half of Ig2 (red), the N-terminal half of Ig3 (blue), the entire Ig7 (green), and the transmembrane domain (yellow), respectively. B. Alternative exons are expressed. RT-PCR was performed on total RNA isolated from 12–24 hour embryos. Fifty individual cDNA clones (numbers 1–50) were isolated and sequenced across exons 4, 6, and 9. Alternative exons used in each cDNA are indicated. Color coding of exons in schematic corresponds to scheme in A. 49 of the 50 cDNAs contain unique combinations of alternative exons. (From Schmucker et al., 2000)

SUMMARY

Pathfinding to the vicinity of a correct target is only the first step in the process of selecting appropriate postsynaptic cells on which to synapse. Having come to the doorstep of the target population, axons use a variety of signals, such as relative changes in growth factors, to enter the target and begin to arborize.

Growing axons are often encouraged to enter the target at one site and discouraged from exiting the target at another site through repulsive barriers. There are a variety of molecules within the target zone, including gradients of Ephrins and CAMs, which conspire, often in combination, to encode different possible target cells along various axes and layers. The incoming afferents are distinguished from each other by the presence of different amounts of various receptors on their surfaces so that they respond differentially to the different target cells. These gradients of ligands and receptors are often the result of very early patterning events in the embryo, such as those that lay down rostrocaudal and dorsoventral patterns (see Chapter 2). Molecular cues, in the form of homophilic or heterophilic adhesion molecules, especially molecules that can show great diversity, are often involved in the final cellular and synaptic levels of targeting.

The molecular nature of target recognition leads one to appreciate how a nervous system can wire up to a fairly high degree of precision in the absence of function. As we have also seen, however, the fine-tuning of neural connections is heavily dependent on synaptic activity, and before the final connections are made, synapses are tested. In Chapter 9, we shall learn much more about the role of neural activity on synapse formation.

Naturally Occurring Neuron Death

Nervous system differentiation is accompanied by tremendous growth; neuron cell bodies and dendrites expand, glial cells and myelin are added, blood vessels arborize, and extracellular matrix is secreted. Even after the period of neurogenesis has largely ended, the human brain continues to increase in size from approximately 400 grams at birth to 1400 grams in adulthood (Dekaban and Sadowsky, 1978). Surprisingly, neurogenesis and this later period of growth overlap with a tremendous loss of neurons and glia, both of which die from "natural causes." Depending on the brain region, 20 to 80% of differentiated cells degenerate during development (Oppenheim, 1991; Oppenheim and Johnson, 2003).

Whether or not a neuron survives depends on many factors (Figure 7.1). Soluble survival factors may be supplied by the postsynaptic target, by neighboring nerve and glial cells, or by the circulatory system. Neurons also depend upon the synaptic contacts that they receive, and deafferentation leads to atrophy or death. These diverse signals are referred to as *trophic factors* because one cell is nourished or sustained by another. The first part of this chapter will describe the characteristics of cell death in the developing nervous system. Relatively little will be said of injury-evoked cell death. We then discuss the trophic factors and intracellular signals that regulate this process. Finally, we will learn that electrical activity and synaptic transmission can have an important influence on neuron survival.

WHAT DOES NEURON DEATH LOOK LIKE?

Naturally occurring neuron death was discovered over a century ago by John Beard (1896), who followed the fate of a very large, easily recognized neuron found at the surface of the skate spinal cord. He found that these Rohon-Beard cells were born in the neural crest and differentiated in the spinal cord, sending out processes to the ectoderm before degenerating. At first, it was difficult to accept the concept that neurons were born, only to die a short time later. Although there had been many reports of neuron death after their target had been removed (see below), it was not clear that postmitotic neurons were lost in any significant number during normal development (Clarke and Clarke, 1996).

We now understand that nerve cells participate actively in their own demise through gene transcription and protein synthesis, and this process is termed *apoptosis* or programmed cell death (PCD). To the trained eye, a dying neuron looks quite different from a healthy one (Figure 7.2). The chromatin becomes very condensed and aggregates at the nuclear membrane, in a process called *pyknosis*. The plasma membrane remains intact, but the neuron gradually shrinks and small protuberances, called *apoptotic* bodies, are pinched off from the cell body and are phagocytosed by *macrophages* (Figure 7.3). PCD can also follow a second course, called *autophagic cell death*, in which the cell contents are sequestered in autophagic bodies and destroyed by the cell's own lysosomes (Kinch et al., 2003).

As the large, crescent-shaped aggregates of nuclear material form, enzymes are activated that cleave the DNA, producing fragments of about 180 base pairs. Although this process is too small to see anatomically, it is possible to stain the broken ends of DNA strands with molecular markers. One technique, called *TUNEL* (for **T**erminal transferase **U**TP **N**ick **E**nd **L**abeling), employs an enzyme that attaches labeled nucleotides to the exposed ends of the DNA fragments (Figure 7.4).

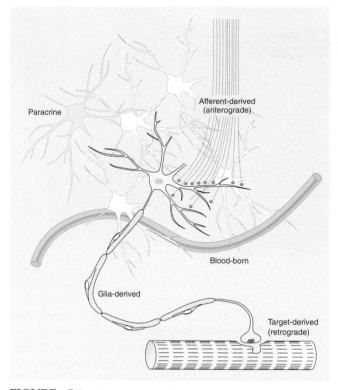

FIGURE 7.1 Five sources that influence neuron survival. Neurons can receive survival signals from the cells that they innervate (target-derived), from their synaptic inputs (afferent-derived), from neighboring neuron cell bodies (paracrine), from distant sources via the circulatory system (blood-born), and from nonneuronal cells (glia-derived).

This approach is useful when studying cell death in a large population of cells that has no clear boundaries, such as an area of cerebral cortex. However, it should not be considered a bullet-proof characterization of cell death. Unlabeled cells may, in fact, enter a cell death pathway in which chromatin breakdown and condensation are not featured (Oppenheim et al., 2001).

The degenerative changes that follow a traumatic injury are called *necrosis*, and they are usually distinct from apoptosis. Following injury, the mitochondria stop producing energy, and the neuron becomes unable to regulate ionic content and hydrostatic pressure. The neuron and its organelles begin to swell, lysosomal enzymes become activated, and cytoplasmic components are broken down. Finally, the soma bursts open (Figure 7.3). There is an important difference between a neuron that dies gracefully by budding off neat little packages of membrane (apoptosis), compared to one that dies violently by retching catabolic enzymes on its neighbors (necrosis). Clearly, a graceful death is unlikely to injure healthy neighboring neurons, and serves as an efficient means to eliminate these cells. For the most part, we will consider only the

mechanisms that accompany apoptosis in the absence of injury or insult to the nervous system.

EARLY ELIMINATION OF PROGENITOR CELLS

Although most of this chapter is devoted to the signals that mediate survival and death in differentiated neurons, there is a prominent period of apoptosis during neurogenesis (Blaschke et al., 1996; Blaschke et al., 1998; Rakic and Zecevic, 2000). During late embryonic development, cells within the proliferative zone of the mouse cerebral cortex display heavy labeling with a variant of the TUNEL technique, indicating a high level of DNA fragmentation (Figure 7.5A). Electron micrographs reveal cells with dark condensed nuclei and other anatomical hallmarks of apoptosis (Figure 7.5B), and counts of pyknotic and TUNEL-positive nuclei are remarkably similar.

The magnitude of cell death in the proliferative zone appears to be similar to that observed in most postmitotic populations. About 70% of the subventricular zone cells in newborn rats can be double-labeled with the TUNEL technique and BrdU, a marker of cell proliferation. This suggests that cells die during the early G1 phase of the mitotic cycle (Thomaidou et al., 1997). Little is known about the signals that lead proliferating cells to enter the apoptotic pathway. However, evidence from cultured quail neural crest cells indicate that cell–cell contact is involved. When neural crest clusters were grown on a nonadhesive substrate to prevent then from dispersing, the cells displayed a marked increase in TUNEL-labeling as compared to dissociated crest cells that were permitted to disperse (Maynard et al., 2000).

The presence of apoptosis in the proliferative zone suggests that the total number of neurons in the brain is regulated, in part, by the elimination of stem cells. It also raises the interesting possibility that these two stages of development—birth and death—share certain molecular pathways, a concept that is discussed below (see Intracellular Signaling).

HOW MANY DIFFERENTIATED NEURONS DIE?

It might seem a straightforward matter to determine how many neurons are being added or removed from a population: Simply count the neurons in a young animal, and subtract this number from an identical

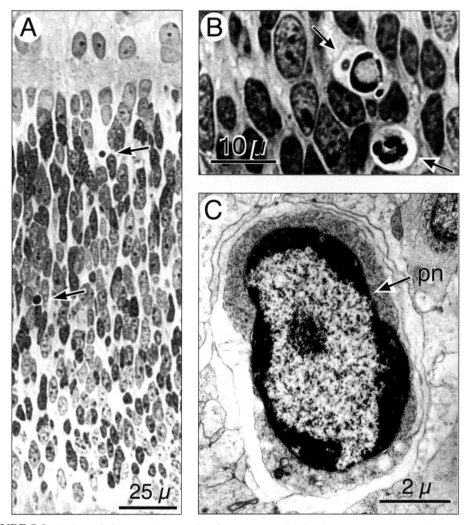

FIGURE 7.2 Light and electron micrographs of apoptosis in the developing cat retina. A. A low magnification photomicrograph of the retina at embryonic day 57 contains two cells in the ventricular layer (arrows). The retinal ganglion cell layer is at the top. B. A high-power photomicrograph shows two neurons with condensed chromatin in their nuclei (arrows). C. An electron micrograph shows a degenerating retinal ganglion cell with a clearly pyknotic nucleus (pn). (Reprinted from Wong and Hughes, 1987).

count obtained in an adult. If the number is positive, then neurons must have been born. If the number is negative, then neurons must have died. However, obtaining an accurate neuron count is trickier than one might suppose. For example, if neurogenesis and cell death overlap in time, then cell counts can remain relatively stable and this will conceal the existence of cell addition or elimination. A second difficulty revolves around the counting strategy. Since it is often too laborious to count each cell in a neuronal structure, estimates are made from tissue sections, and changes in the size or packing density of cell bodies can each influence the final counts. Finally, neurons are not the only type of cell to die during development. For example, about 50% of oligodendrocytes in the rat

optic nerve die during development, and their survival depends upon the presence of retinal axons (Barres et al., 1992). Therefore, those who tally up cell bodies must be careful to discriminate neurons from glia.

After it became clear that cell death was a general feature of the developing nervous system, its magnitude became an issue, and a number of rigorous cell counting studies appeared. One of the most convincing ways to demonstrate that neurons are dying is to count both the healthy cells and the pyknotic cells in the same tissue section (Hughes, 1961). Counts of spinal motor neuron (MN) cell bodies in the chick and frog demonstrate that the decrease in number of healthy-looking MNs is perfectly correlated with the appearance of

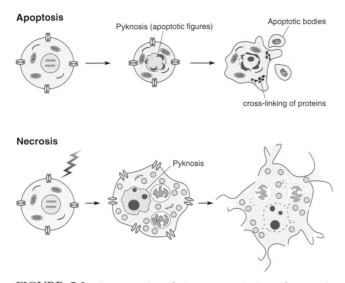

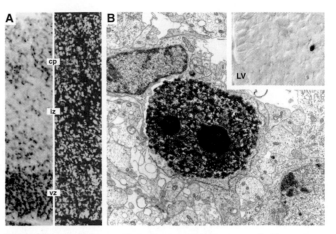

FIGURE 7.5 Dying cells in the proliferative zone. A. Two micrographs of the embryonic day 18 mouse cortex. To the left, apoptotic cells are labeled with a technique similar to TUNEL, called ISEL. To the right, all cells are revealed with fluorescent labeling of nuclei. Note the prominent ISEL labeling in the ventricular zone where proliferation occurs (cp, cortical plate; iz, intermediate zone; vz, ventricular zone) (Blaschke et al., 1996). B. An electron micrograph from the ventricular zone of an E16 rat cerebral cortex shows a cell with the histological features of apoptosis. The inset shows a TUNEL-positive cell in the subventricular zone of a newborn rat (LV, lateral ventricle) (Thomaidou et al., 1997).

FIGURE 7.3 A comparison between apoptosis and necrosis. Naturally occurring cell death is usually accomplished through a process called *apoptosis*. During apoptosis, the neuron begins to shrink, and the nuclear matter becomes condensed, forming crescent-shaped figures. As proteins become cross linked at the membrane, small apoptotic bodies break off and are phagocytized. In contrast, injured neurons tend to die through a process of *necrosis*. During necrosis, the mitochondria stop functioning, and neurons cannot maintain an osmotic balance. The cell swells up, undergoes autolysis, and finally bursts open.

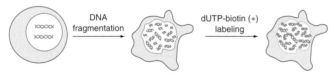

FIGURE 7.4 During apoptosis, DNA is broken down by endonuclease activity to produce double-stranded, low-molecular-weight fragments. These DNA fragments can be detected by a labeling technique called TUNEL. A modified nucleotide, such as dUTP-biotin, is catalytically attached to the free 3'-hydroxyl end of each DNA fragment by the enzyme, terminal deoxynucleotidyl transferase. Therefore, the nuclei of dying neurons can be detected before the cells break up and are phagocytosed.

within two weeks of birth, and a 50% loss is observed in embryonic chick ciliary ganglion after the neurons have projected to the iris and choroid (Potts et al., 1982; Landmesser and Pilar, 1974). While it is impractical to count the total number of neurons in any one area of cerebral cortex, it is estimated that 20 to 50% of postmigratory neurons are eliminated. However, this may depend on the type of cell, the region of cortex, and the time of birth (Miller, 1995; Finlay and Slattery, 1983).

SURVIVAL DEPENDS ON THE SYNAPTIC TARGET

The overproliferation of neurons in most areas of the CNS suggests that it is a valuable mechanism. It has been suggested that cell death ensures that the number of afferents is well matched to the size of the target population. This theory makes the rational assumption, known to every woodworker, that it is easier to trim off the excess than to paste on a little bit extra later. Thus, we would expect a far greater number of motor neurons to innervate the bulky leg muscles of an elephant than would innervate the short, skinny legs of a mouse. There are many interesting examples of this principle, such as the limbless lizard, *Anguis fragilis*, that does produce a set of motor

pyknotic cells (Figure 7.6). At first glance, it may not be clear why so few pyknotic neurons are observed during the period of maximal neuron elimination. This is due to the rapid removal of cell debris, which has been variously estimated to occur in as little as 3 hours.

The magnitude of cell death is impressive. The data in Figure 7.6 show that the motor neuron population innervating frog hind limbs declines from an initial size of about 4000 to a final size of 1200. Thus, well over half of the differentiated motor neurons are eliminated from the nervous system. A similar amount of cell death has been detected at all levels of the nervous system. About 50% of rat retinal ganglion cells die

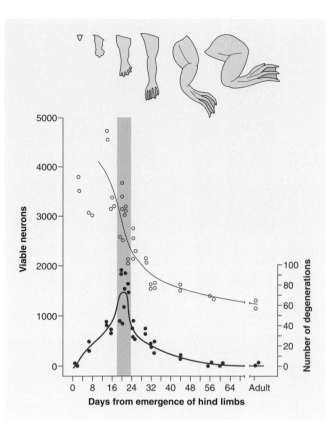

FIGURE 7.6 The period of cell death in frog motor neurons. The graph shows the total number of healthy (black) and pyknotic (red) motor neurons that innervate the frog hind limb during each of several developmental stages. At the top is a picture of leg size during this period. The number of pyknotic (degenerating) neurons reaches a peak at precisely the time when the loss of healthy (viable) neurons is most rapid, as indicated by the bar. (Adopted from Hughes, 1961)

neurons in the limb region of its spinal cord which then proceed to die during development (Raynaud et al., 1977). Therefore, neurogenesis provides a primary point of regulation for setting neuron number (see Chapter 3), and cell death provides a second. This concept is raised again when we discuss the overproliferation and elimination of synaptic contacts (see Chapter 9). In fact, cell death has been suggested as one way to get rid of extra synapses during development.

If apoptosis is a developmental mechanism for matching the size of a presynaptic population to its target, then it should occur at a discrete time and place. This was explored in dorsal root ganglia (DRG), where about 30% of neurons die during normal development (Figure 7.7). The first observation is that neurons die during a specific time interval. In the chick, DRG neurons begin to degenerate at embryonic day 4.5, and this continues for 2½ days. The second observation is that natural cell death occurs predominantly in those

DRGs that innervate axial musculature. The DRGs that innervate wings and legs have more tissue to innervate, and the amount of cell death is much reduced in these ganglia. The central and peripheral projections of DRG neurons make contact with their target before cell death occurs. Therefore, naturally occurring cell death does appear to be an important mechanism for thinning out neuron populations with less target to innervate (Ernst, 1926; Hamburger and Levi-Montalcini, 1949).

There is only one system in which cell death has been quantified in both the pre- and postsynaptic neuronal population (Figure 7.8). This provides a natural situation in which to ask whether a decrease in the target population is correlated with cell loss in the innervating neurons. The presynaptic population of cells, called *nucleus magnocellularis* (NM), is the first region of the chick brain to receive input from the ear, and these cells project to a second-order auditory nucleus, called *nucleus laminaris* (NL). Both groups of cells undergo their final mitosis, migrate to their final positions, and begin a period of normal cell death that occurs primarily from embryonic day 11 to 13. The percentage of cells that die in the two nuclei is quite similar. About 18% of the presynaptic NM neurons die during this interval, while 19% of the postsynaptic NL neurons are wiped out (Rubel et al., 1976; Solum et al., 1997). However, cell death occurs more rapidly in NM than in NL, as highlighted by the bar at 11–13 days (Figure 7.8). Thus, although the magnitude of cell death is well-correlated in these two interconnected cell groups, the relative timing suggests that target size is not the sole determinant of neuron survival.

Even before normal cell death was a well-accepted event, it was known that developing nerve cells died when their target was removed. A common manipulation was to remove a limb bud prior to the time of innervation (i.e., no direct damage to the axons) and then examine the motor neurons or DRG cells that would have made synapses there. These studies were performed on amphibian or chick embryos because it was relatively easy to carry out the surgeries. The removal of an appendage was usually devastating to the pool of presynaptic neurons (Figure 7.9). In the salamander, *Amblystoma*, the sensory ganglia that normally innervate a limb are much smaller when the limb is excised. In contrast, sensory ganglia that normally innervate axial musculature are much larger than normal if provided with a transplanted limb (Detwiler, 1936).

The relationship between motor neuron number and target size is remarkably linear in the chick. As greater and greater amounts of muscle are removed from developing chick embryos, the ventral horn of

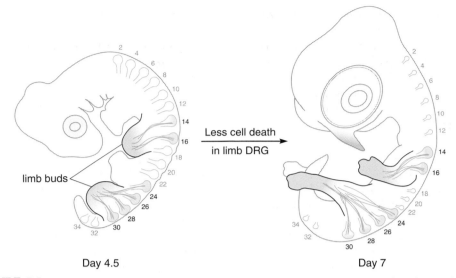

FIGURE 7.7 The pattern of naturally occurring cell death in chick dorsal root ganglia. There is less cell death among neurons in the DRGs that innervate limb buds (blue filled), compared to those that innervate the axial musculature (unfilled). The numbers refer to the somitic segment of the spinal cord. Thus, the size of DRGs at segments 14–16 decrease only slightly during development, whereas DRGs at segments 18–22 become much smaller. This result shows that neuron survival is correlated with the amount of target tissue. (Adapted from Hamburger and Levi-Montalcini, 1949)

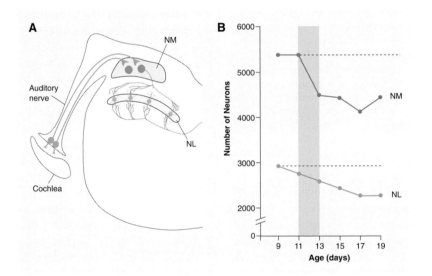

FIGURE 7.8 The period of cell death in pre- and postsynaptic nuclei. A. A transverse hemisection through the chick auditory brainstem. The nucleus magnocellularis (NM) is a central auditory nucleus that is innervated by auditory nerve terminals. It projects to a second-order nucleus, called nucleus laminaris (NL). B. The graph shows the total number of cell bodies in NM and NL during the latter period of embryonic development. In both nuclei, about 20% of the neurons are lost between embryonic day 9 and 17. (Adapted from Rubel et al., 1976 and Solum et al., 1997)

the spinal cord becomes smaller (Figure 7.9). When an extra limb bud is transplanted next to the original one, providing a larger than normal target region, developing processes grow into the added target, and the population of motor and DRG neurons is found to be much larger than normal. The addition of an extra limb

bud saves up to 25% of the motor neurons that would otherwise die. Experimental alteration in the size of the periphery led to the hypothesis that the target provides a survival factor. When targets are eliminated, there are fewer neurons to be found in the adult, and when targets are enlarged neurons are found in greater

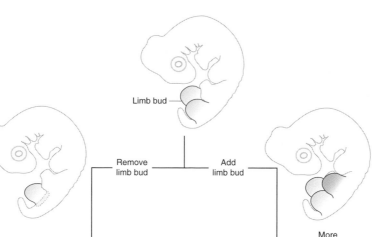

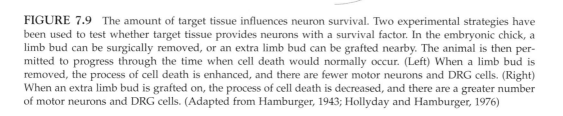

FIGURE 7.9 The amount of target tissue influences neuron survival. Two experimental strategies have been used to test whether target tissue provides neurons with a survival factor. In the embryonic chick, a limb bud can be surgically removed, or an extra limb bud can be grafted nearby. The animal is then permitted to progress through the time when cell death would normally occur. (Left) When a limb bud is removed, the process of cell death is enhanced, and there are fewer motor neurons and DRG cells. (Right) When an extra limb bud is grafted on, the process of cell death is decreased, and there are a greater number of motor neurons and DRG cells. (Adapted from Hamburger, 1943; Hollyday and Hamburger, 1976)

than normal numbers (Hamburger, 1943; Hollyday and Hamburger, 1976).

A complementary experiment can be performed by reducing the number of neurons projecting to a target, and determining whether the remaining cells died off anyway. In one such experiment, about two-thirds of the ciliary ganglion neurons innervating one eye are killed off by cutting their axons. After the normal period of cell death has ended, the number of ganglion cells remaining is almost 40% greater than expected. Furthermore, several of the surviving axons sprout into the peripheral territory that is vacated by the death of axotomized ciliary neurons. Therefore, particular neurons are not preordained to die but appear to do so through some sort of competition for a feature of the target (Pilar et al., 1980).

A basic question that arises from these studies is whether the target influences neuron proliferation or cell death. By carefully studying the pattern of degeneration of DRG neurons following wing bud removal in the chick, Viktor Hamburger and Rita Levi-Montalcini (1949) demonstrated that target removal leads to an increase in the number of dying neurons. One wing

bud was removed at about 3 days of incubation, and the dorsal root ganglia were examined ipsilateral and contralateral to the ablation. Within 2–3 days, the ganglia ipsilateral to the extirpated wing buds were much smaller than normal and contained a large number of darkly stained pyknotic cells (Figure 7.10). Although they also reported the number of mitotic cells to vary with target size, subsequent studies show that target removal has little or no affect on the amount of tritiated-thymidine incorporated into DRG neurons (Carr and Simpson, 1978).

The concept of target-dependence extends to the earliest period of maturation, well before growth cones have reached their synaptic target. During embryonic development, commissural neurons in the dorsal spinal cord are attracted to the floorplate where they eventually cross the midline (see Chapter 5). The floorplate region also appears to provide a survival signal. When E13 commissural neurons are are grown in vitro, they all die within 2 days, but they can survive for several days when grown in the presence of floorplate-conditioned medium (Wang and Tessier-Lavigne, 1999). In the zebrafish, the death of vental primary

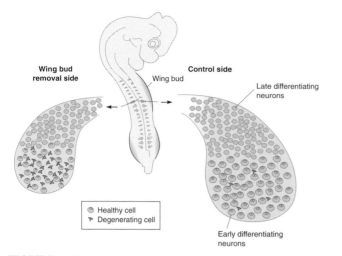

FIGURE 7.10 The target influences neuron survival. Following unilateral limb bud removal in a 3-day chick embryo (left side), there is an increase in the number of degenerating cells in the DRG ipsilateral to the ablated limb, as compared to the control side (right). The number of viable DRG neurons is reduced dramatically. (Adapted from Hamburger and Levi-Montalcini, 1949)

(VaP) motor neurons depends on contact with a transient population of muscle cells. If these muscle cells are ablated a few hours before being contacted by VaP, then the VaP motor neurons live and innervate the ventral musculature (Eisen and Melancon, 2001).

NGF: A TARGET-DERIVED SURVIVAL FACTOR

Neuron survival clearly depends on the presence of target tissue, but what is being procured? One simple hypothesis is that the target cells secrete a chemical that presynaptic neurons require for their survival. In fact, an extraordinary series of experiments, coupled with a few strokes of serendipity, led to the first endogenous neurotrophic substance to be discovered, the nerve growth factor (NGF). NGF has since been shown to largely control the survival of sympathetic neurons and contribute to the survival of sensory DRG neurons during development. Although NGF turns out to be the tip of an enormous iceberg of growth and survival factors (below), we are going to examine its discovery in some detail because it remains the best understood system.

How did scientists arrive at the neurotrophic theory of cell survival? Viktor Hamburger (1934) first suggested that the target produces a factor that is retrogradely transported by the innervating neurons and influences their development. Initially, it was not clear whether this hypothetical substance upregulated neurogenesis, or recruited cells to differentiate as neurons, or prevented differentiated neurons from dying. As described above, two sets of careful observations strongly suggested that the hypothetical substance worked by maintaining the survival of differentiating neurons (Levi-Montalcini and Levi, 1942; Hamburger and Levi-Montalcini, 1949).

By modern standards, the next step would be to harvest the target tissue (e.g., muscle) and try to isolate a soluble substance that enhances survival. However, most of the necessary biochemical tools did not yet exist in the 1950s. The isolation of a neurotrophic factor took a few decades to achieve, and it began with a surprising set of observations. In an effort to provide neurons with an "unlimited" amount of target tissue, various mouse tumors were implanted into the chick hind limb (Bueker, 1948; Levi-Montalcini and Hamburger, 1951; Levi-Montalcini and Hamburger, 1953). One tumor, a connective tissue cell line called sarcoma, grew rapidly and was invaded by nerve fibers. Within five days of the transplant, there was a dramatic increase in the survival of sensory and sympathetic neurons, while motor neurons were unaffected (Figure 7.11). When the tissue was examined in a little more detail, a key observation was made: ganglia with no apparent physical connection to the tumor were also greatly enlarged. This provided the first indication that cell survival was mediated by a diffusible chemical.

A more direct demonstration came from experiments in which tumor cells were placed on a vascularized respiratory membrane in the chick egg called the chorioallantois. In this case, the tumor was not in contact with sympathetic and sensory ganglia, but it did share the same blood supply. Even though the tumor was physically isolated from the nervous system, it was able to elicit a strong growth-promoting effect (Figure 7.11). Thus, sarcoma tumor cells must have released a soluble factor that could be transported to the neurons through the circulatory system.

As a first step toward isolating the putative survival factor found in mouse sarcoma, an in vitro assay system was developed. Sympathetic ganglia were obtained from chick embryos and placed in a tissue culture dish, either by themselves or next to mouse sarcoma tumor cells. When grown next to tumor, the neurons survived and grew a dense halo of axons within hours, providing a simple and convenient assay system. Although biochemical isolation was a slow process, it was possible to obtain a tumor cell fraction that had only proteins and nucleic acids. In order to determine whether either of these components contained the growth factor, a biochemical trick was

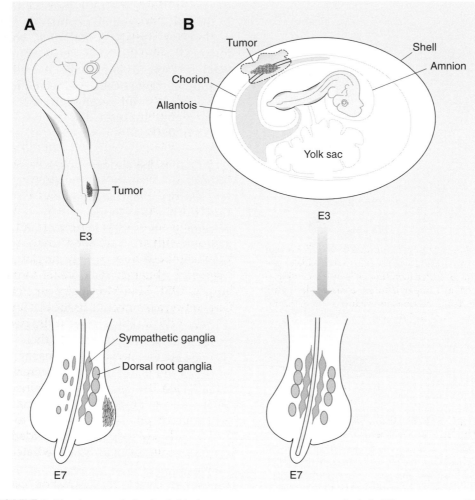

FIGURE 7.11 A target-derived soluble factor can support neuron survival. A. When a tumor cell line is placed in the chick embryo at E3, the size of sympathetic ganglia and DRGs is much larger ipsilateral to the tumor by E7. (B) When the same tumor cell line is placed on the chorioallantoic membrane at E3, such that nerve fibers have no direct access, all of the symapthetic and dorsal root ganglia are much increased in size by E7. Thus, the tumor must have secreted a soluble factor that enhanced neuron survival. (Adapted from Bueker, 1948; Levi-Montalcini and Hamburger, 1951; Levi-Montalcini and Hamburger, 1953)

employed. Snake venom was known to contain high levels of an enzyme that breaks down nucleic acids (phosphodiesterase). Therefore, it was added to the extract to determine whether this class of molecules mediated the trophic effect (Cohen and Levi-Montalcini, 1956; Levi-Montalcini and Cohen, 1956). If the biological activity was lost, then one could conclude that growth factor contained nucleic acids. Surprisingly, the tumor fraction containing the snake venom was even more potent than the origin protein-nucleic acid fraction. Even more curious, the snake venom itself was found to support nerve growth (Figure 7.12).

As it turned out, the discovery of a growth-promoting effect in snake venom was extremely fortunate. It suggested that growth-promoting activity would also be found in a mammalian analog, the salivary

gland. In fact, the mouse submaxillary gland proved to be a wonderful source for the Nerve Growth Factor, and this eventually led to its complete isolation and sequencing.

Once the NGF was purified, it was possible to perform two critical experiments in vivo to determine whether this protein is both necessary and sufficient to keep sensory and sympathetic neurons alive during development. First, the NGF protein that was purified from snake venom was injected directly into neonatal rodents, and it did produce a dramatic increase in the size of sensory and sympathetic ganglia (Levi-Montalcini and Cohen, 1956). In addition to keeping neurons alive, it is also clear that NGF promotes process outgrowth. For example, when NGF is injected into neonatal rodents, sympathetic nerve fibers are no

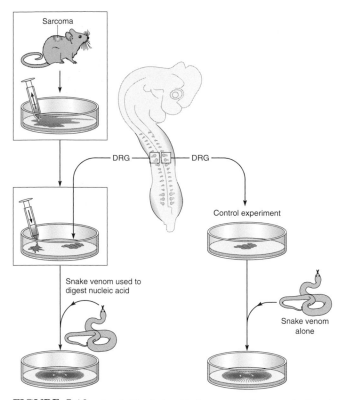

FIGURE 7.12 A soluble factor that supports the survival and growth of DRG neurons is discovered in a mouse sarcoma and, later, in snake venom. (Left) DRG neurons obtained from chick embryos were placed in a tissue culture dish, and conditioned medium from a mouse sarcoma was added. The venom of a snake was added to the culture to determine whether nucleic acids mediate the trophic effects. The DRG neurons survived and grew processes under these conditions. (Right) When the control experiment was performed, in which only snake venom was added to the DRG neuron cultures, a surprising discovery was made. The DRG neurons survived and grew, indicating that the snake venom must also have contained a soluble survival factor. (Adapted from Cohen and Levi-Montalcini, 1956; Levi-Montalcini and Cohen, 1956)

longer restricted to their normal synaptic target, but grow widely in the peripheral field and can even invade blood vessels or the central nervous system (see Chapter 5).

In a second experiment to determine whether endogenous NGF is necessary for survival, an antibody directed against the NGF protein was injected into neonatal rodents. This leads to the loss of almost all sympathetic neurons (Levi-Montalcini and Booker, 1960). It was later found that DRG cells are no longer dependent on NGF at the age when antibody was administered, but they can be destroyed by prenatal exposure to NGF antibody (Johnson et al., 1978). In fact, not all sensory neurons are dependent on NGF for survival. Those sensory neurons that derive from sensory placodes (e.g., nodose ganglion), rather than

the neural crest, are unresponsive to NGF treatment. In the DRG, only small peptidergic neurons that carry nociceptive signals to the spinal cord are killed following loss of the NGF signal. More recently, it has been possible to reproduce the effects of antibody treatment in genetically engineered mice. When a deletion is made in the coding sequence of the NGF gene, homozygous animals display profound cell loss in both sympathetic and sensory ganglia (Crowley et al., 1994).

If NGF is the endogenous survival factor, then it is important to verify its presence at the sympathetic and sensory ganglion target regions at an appropriate time during development. Although NGF levels are extremely low (except in the fortuitous case of the male mouse salivary gland, from which it was purified), it has been possible to localize the protein with immunohistochemical staining and the NGF mRNA with in situ hybridization. For example, trigeminal axons arrive at their cutaneous target just before the NGF mRNA and protein are manufactured, and the initial outgrowth of trigeminal axons is NGF-independent (Davies et al., 1987), suggesting that the maintenance of trigeminal neurons depends on NGF derived from their target. The success with NGF was achieved by 1960, and the expectation was that many other neurotrophic substances would quickly be found in the central nervous system. While several growth and survival factors were discovered in nonneuronal systems, the search for another bona fide neurotrophic substance was, at first, somewhat frustrating.

THE NEUROTROPHIN FAMILY

The full amino acid sequence of NGF was obtained by 1971, yet a decade elapsed before a second neurotrophic factor was identified. The search began with the simple observation that, in contrast to its effect on the retinae from lower vertebrates, NGF does not stimulate neurite outgrowth from cultured rat retina. Working under the assumption that there must be a growth factor or factors for mammalian retina, a soluble extract was prepared from the entire pig brain. This extract did, in fact, stimulate retinal process outgrowth in a dose-dependent manner. When the active substance, named brain derived growth factor (BDNF), was purified and its amino acid sequence determined, its structure displayed a striking similarity to that of NGF (Turner et al., 1982; Leibrock et al., 1989).

Several members of the neurotrophin family have now been isolated, and they are found in both the peripheral and central nervous systems. The more

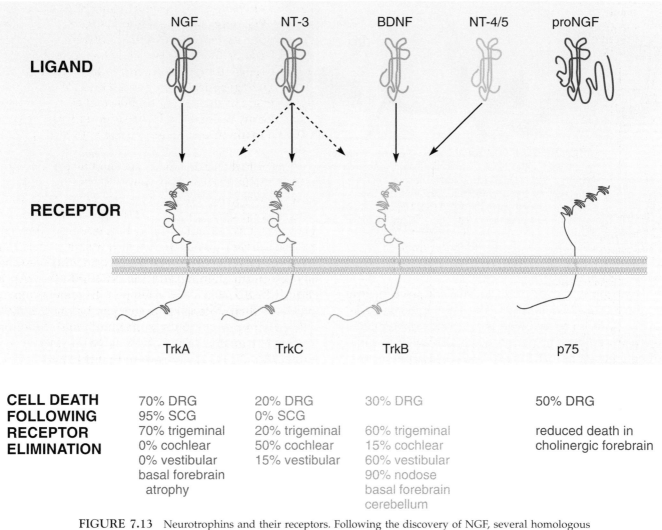

LIGAND	NGF	NT-3	BDNF	NT-4/5	proNGF

RECEPTOR	TrkA	TrkC	TrkB		p75

CELL DEATH FOLLOWING RECEPTOR ELIMINATION	70% DRG 95% SCG 70% trigeminal 0% cochlear 0% vestibular basal forebrain atrophy	20% DRG 0% SCG 20% trigeminal 50% cochlear 15% vestibular	30% DRG 60% trigeminal 15% cochlear 60% vestibular 90% nodose basal forebrain cerebellum		50% DRG reduced death in cholinergic forebrain

FIGURE 7.13 Neurotrophins and their receptors. Following the discovery of NGF, several homologous proteins were found, including NT-3, BDNF, and NT-4. Each of these proteins binds selectively to a member of the Trk receptor tyrosine kinase family, as illustrated. In addition, there is a low-affinity receptor, called p75[NTR]. The effect of eliminating the neurotrophin or its receptor in mice is shown beneath each pair.

recent additions to the family have been given the less colorful names: neurotrophin-3 (NT-3), NT-4/5, NT-6, and NT-7; the latter two are found only in fish (Figure 7.13). In each case, a precursor protein of about 250 amino acids, called the *proform*, is processed post-translationally to produce active peptides of about 120 amino acids. These peptides form homodimers and become biologically active. The family members share about 50% sequence homology with one another, particularly within six hydrophobic regions that are responsible for linking the two protomers together. Each neurotrophin also contains a unique amino acid sequence, and it is this variable region that is responsible for binding to a specific receptor (Ibanez, 1994).

Each of the neurotrophins has been shown to play a role in the survival of specific peripheral neuron pop-ulations. As with NGF, two general classes of experiment have been performed: One can provide excess neurotrophin (in vivo or in vitro), or one can decrease the amount of endogenous neurotrophin, commonly by single-gene knockout experiments (Chapter 2). These experiments indicate that BDNF is a necessary endogenous signal for the survival of vestibular ganglia, while NT-3 is an endogenous survival signal for the cochlear ganglion. A comparison between the two types of experiments also show that positive results must be treated cautiously. While BDNF is able to save chick motor neurons when administered during the period of naturally occurring cell death, there is no effect on motor neuron survival in BDNF knockout mice (Oppenheim et al., 1992; Ernfors et al., 1994, 1995). Both BDNF and NT-3 also contribute to the

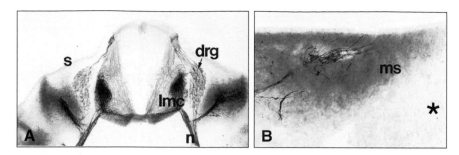

FIGURE 7.14 The pattern of NT-3 expression is revealed by a *lacZ* reporter gene, and nerves are counterstained with a neurofilament antibody. A. A transverse section through the thoracic region of an E11 mouse embryo shows DRGs (drg), and peripheral nerves (n). NT-3 expression (blue) is prominently around the DRG and sensory-motor projection. B. Peripheral axons within the forelimb growing through mesenchyme (ms) are surrounded by NT-3 expression. There is almost no NT-3 expression at the end of the limb (asterisk), which is not yet innervated. Lmc = lateral motor column, s = skin. (Reprinted from Farinas et al., 1996).

survival of neurons in the sensory, trigeminal, and nodose ganglia.

Neurotrophins may promote survival even before an axon has reached its target. The development of NT-3 expression can be followed with a *lacZ* reporter gene during the embryonic period when DRG neurons first extend their axons. NT-3 is expressed heavily along the path of growth (Figure 7.14), and this may explain why loss of NT-3 expression can affect neuron survival before target innervation has occurred (Farinas et al., 1996).

THE TRK FAMILY OF NEUROTROPHIN RECEPTORS

Even before a receptor for NGF was discovered, it was known that NGF binds to a site on the axon terminal with very high affinity. The β subunit of NGF can be labeled with ^{125}I, and used to perform binding studies on freshly dissociated chick sensory neurons. These experiments reveal two types of binding sites (Sutter et al., 1979). The first displays a lower affinity for NGF (e.g., nanomolar concentrations saturate the binding sites), while the second displays a higher affinity for NGF (e.g., picomolar concentrations saturate the binding sites). In fact, there are two different types of NGF receptor that are associated with these binding kinetics, and each one has now been isolated.

The high-affinity receptor was discovered through a series of interesting observations (Figure 7.15). Initially, it was found that NGF exposure induces rapid phosphorylation of proteins on their tyrosine residues, suggesting that the receptor might be a kinase (Maher, 1988). Soon after, an oncogene was discovered in human colon carcinoma cells, and this turned out to contain a putative transmembrane protein containing a tyrosine kinase on its cytoplasmic tail (Martin-Zanca et al., 1986). The oncogene apparently results from a genetic rearrangement that fuses a tyrosine kinase with part of a nonmuscle tropomyosin sequence, leading to its name: tropomyosin related kinase or Trk (later called TrkA). The *trk* proto-oncogene was cloned, and the distribution of its mRNA was examined in vivo. The highest levels of expression are confined to the cranial sensory, dorsal root, and sympathetic ganglia (Martin-Zanca et al., 1990). Most importantly, the TrkA protein (also called p140) is a high-affinity binding site for NGF, and the binding event induces tyrosine kinase activity (Kaplan et al., 1991a).

Three high-affinity neurotrophin Trk receptors have now been isolated: TrkA, TrkB, and TrkC (Figure 7.13). The latter two were discovered by taking the *trk* sequence and performing low-stringency binding screens with cDNA libraries. In this manner, two sequences were isolated that encoded for 145 kD receptor tyrosine kinases, named TrkB and TrkC (Barbacid, 1994). Each Trk receptor has two immunoglobulin-like repeats in the extracellular domain and a tyrosine kinase with autophosphorylation sites in the cytoplasmic domain. The extracellular domains are about 50% homologous, but each Trk displays a specific affinity for one or two of the neurotrophins: TrkA is specifically activated by NGF, Trk B is specifically activated by BDNF or NT-4, and TrkC is specifically activated by NT-3. Trk receptors may also be activated in the absence of neurotrophins from within the neurons, a process called *transactivation*. For example, the small neurotransmitter, adenosine, can produce Trk phosphorylation through its G protein-coupled receptor. The actived Trk is then able to promote survival of PC12 cells and hippocampal neurons in vitro (Lee and Chao, 2001).

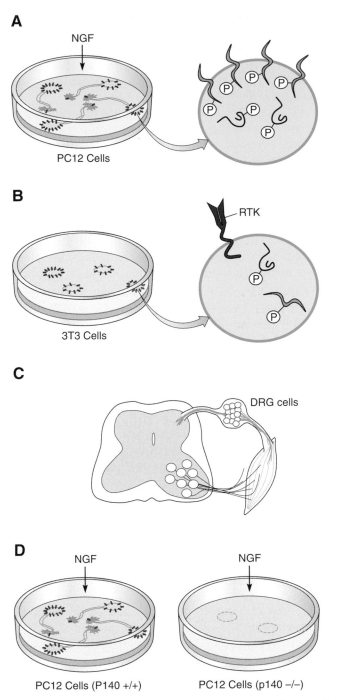

More than three receptor proteins are expressed in vivo because each Trk receptor gene is differentially spliced. In fact, the *trkC* genes may encode for up to eight different TrkC receptor proteins. There are a number of truncated Trk receptors (those missing the tyrosine kinase domain) that are able to bind to their cognate ligand, and these are generally expressed only by glial cells during development. Differential splicing also leads to differences in the extracellular domains, and these can influence ligand binding.

Several lines of evidence indicate that Trk receptors do mediate a survival signal when bound to neurotrophin. A mutant line of PC12 cells that lack TrkA protein are unresponsive to NGF, but they can be rescued if they are transfected with expression vectors encoding the full-length rat *trk* cDNA (Figure 7.15). Perhaps the most compelling evidence is that transgenic mice lacking Trk receptors display extensive death in specific neuron populations (Figure 7.13). TrkA$^{-/-}$ mice exhibit large-scale cell death in sympathetic and dorsal root ganglia, in agreement with earlier experiments that eliminated NGF with function-blocking antibodies (Smeyne et al., 1994). Targeted disruption of the *trkB* gene in mice seems to be particularly devastating in that all mice die within two days of birth. Several peripheral populations are affected, such as the trigeminal, nodose, and vestibular ganglia. In contrast, TrkC receptor deletion leads to a 50% loss of cochlear ganglion neurons but spares most nodose and trigeminal neurons. The developmental effects of eliminating a neurotrophin or its cognate Trk receptor are not necessarily identical (Klein et al., 1993; Smeyne et al., 1994). For example, the cell death that results from TrkB disruption could be quantitatively greater than BDNF disruption because TrkB also serves as a receptor for NT-4/5.

Expression of TrkB and TrkC is widely distributed in the CNS, and levels remain quite high into adulthood (Barbicid, 1994). However, there is little evidence that central neurons depend on neurotrophin signaling for survival during development. Cholinergic cells of the basal forebrain are NGF-sensitive, and exogenous NGF is able to keep them alive when their axons are cut (Gage et al., 1988). In addition, *trkA*$^{-/-}$ mice have fewer axonal projections from cholinergic basal forebrain neurons to the hippocampus and cortex, suggesting a role in process outgrowth (see Chapter 4). In *trkB*$^{-/-}$ or *trkB/trkC*$^{-/-}$ mice, increased apoptosis has been reported in the developing hippocampus, cerebellum, cortex, striatum, and thalamus. However, this is not necessarily associated with the period of naturally occurring cell death (Minichiello and Klein, 1996; Alcantara et al., 1997).

FIGURE 7.15 The high-affinity NGF receptor was discovered through a series of disparate observations. A. NGF was found to elicit protein phosphorylation in PC12 cells. B. The oncogene in a cancer cell line was found to be a transmembrane receptor tyrosine kinase (RTK). C. The messenger RNA for this RTK was found in extremely high levels in DRG neurons. D. When this RTK, called p140, was eliminated from PC12 cells, they became unresponsive to NGF. (Adapted from Maher, 1988; Martin-Zanca et al., 1986; Martin-Zanca et al., 1990; Loeb et al., 1991)

HOW DOES THE NEUROTROPHIN SIGNAL REACH THE SOMA?

While the interaction of each neurotrophin with its receptor at the cell membrane is critical for neuron survival, it is still not clear how the survival signal is conveyed to the cell body. Exposure of the growing tips of axons to NGF is sufficient to prevent cell death. This was demonstrated in an elegant tissue culture study where sympathetic neuron cell bodies were placed in a central chamber that was physically isolated from the growth media that bathed the neuritic processes (Figure 7.16). When NGF is provided only to neurites that grow out and reach one of the isolated side chambers, 95% of the neurons survive; few survive without NGF (Campenot, 1977, 1982). This result suggests that the signal for survival is somehow relayed back to the cell body.

Retrograde transport of some signal is clearly important to neuron survival. Sympathetic neurons die when vinblastine is used to disrupt their retrograde transport, and the cells can be saved with NGF treatment (Johnson, 1978). The original concept was that NGF, itself, carried the signal to the cell body. Both NGF and activated TrkA receptor are internalized, and retrogradely transported to neuron cell bodies in vivo. At least some of the neurotrophin and receptor are retrogradely transported together, perhaps in the bound state (Hendry et al., 1974; Johnson et al., 1978; Bhattacharyya et al., 1997; Ehlers et al., 1995; Tsui-Pierchala and Ginty, 1999; Watson et al., 1999).

A sustained role for NGF within the neuron remains controversial. To test whether retrograde transport of NGF is necessary for survival, rat sympathetic neurons were grown in the presence of NGF that was covalently linked to 1 μm beads to prevent internalization (Figure 7.17). Once again, only the neuronal processes had access to the NGF. The bead-linked NGF was almost as effective as free NGF: 81% of the neurons survived. This experiment emphasizes the importance of local signaling at the distal axon membrane, but it does not exclude an important role for internalized NGF towards achieving cell survival. For example, preventing NGF internalization leads to a clear loss of phosphorylated transcription factor back in the cell body. To test whether retrogradely transported NGF influences survival, a function blocking antibody was introduced in the somata of sympathetic neurons (Figure 7.17). In this case, only 60% of the neurons survive. There is clearly some discrepancy between the outcome of preventing internalization with beads and blocking somatic NGF with antibody. At present, the most parsimonious explanation is that NGF participates in both distal and somatic signaling, but the relative impact of this signal may vary with the culture or in vivo conditions (MacInnis and Campenot, 2002; Riccio et al., 1997; Ye et al., 2003).

THE P75 NEUROTROPHIN RECEPTOR

A second neurotrophin receptor was originally described as a low-affinity binding site for NGF. Now known as $p75^{NTR}$ (75 kD neurotrophin receptor), this glycoprotein is a member of the tumor necrosis factor (TNFR) family of receptors (Johnson et al., 1986; Locksley et al., 2001). It binds to each of the neurotrophins, as well as to other proteins, at nanomolar levels. For example, the beta amyloid peptide also binds to the $p75^{NTR}$, and this may contribute to degeneration of cholinergic brainstem neurons in Alzheimer's disease (Yaar et al., 1997). It now appears that the uncleaved proform of NGF is biologically active. The proforms of NGF and BDNF are released and cleaved in the extracellular space, and proNGF is a high-affinity ligand for the $p75^{NTR}$ receptor (Lee et al., 2001).

The $p75^{NTR}$ receptor does not have a uniform role in naturally occurring cell death. It can both improve and

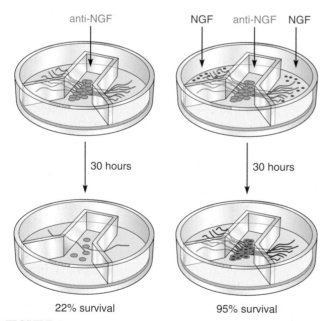

FIGURE 7.16 The NGF signal can be transduced at the tips of growing neuronal processes. Sympathetic neurons were placed in a special tissue culture system that permitted the cell bodies and neurites to be bathed in different media. (Left) Most neurons died when grown in the absence of NGF for 30 hours. (Right) Neurons could be kept alive by adding NGF only to the compartments with growing neurites. In both cases, an antibody against NGF was added to the central compartment to prevent activation of TrkA. (Adapted from Campenot, 1977, 1982; MacInnis and Campenot, 2002)

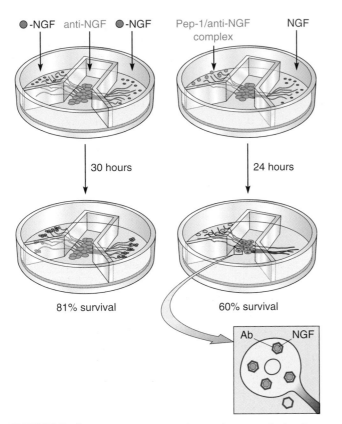

FIGURE 7.17 Experiments were designed to test whether internalized NGF contributes to sympathetic neuron survival. (Left) NGF is covalently bound to beads, preventing internalization but permitting local activation of the TrkA receptor. In this case, 81% of the neurons survive for 30 hours. (Right) Sympathetic neuron cell bodies are exposed to a protein delivery system (Pep-1-anti-NGF complex) that permits delivery of anti-NGF antibody to enter the cells. With NGF neutralized within the soma, 40% of the sympathetic neurons died. (Adapted from MacInnis and Campenot, 2002; Ye et al., 2003)

worsen the chance for neuron survival. There is evidence that p75^{NTR} collaborates with Trk receptors to enhance ligand binding and phosphorylation (Chao, 1994). Cutaneous sensory trigeminal neurons cultured from p75^{NTR} knockout mice require a fourfold greater concentration of NGF in order to survive (Davies et al., 1993). Consistent with this survival-promoting role, the complete loss of p75^{NTR} function in mice leads to a 50% reduction in lumbar DRG neuron number (von Schack et al., 2001).

Unlike Trk receptors, the p75^{NTR} does not have an intracellular catalytic domain (Figure 7.13), but it does activate several intracellular signaling pathways. One intriguing clue about the function of p75^{NTR} comes from its homology to other members of the TNF receptor family. Each of these proteins has a "death domain" on the cytoplasmic tail which is similar to the *reaper* gene product in *Drosophila*. The deletion of *reaper* blocks most cell death in the embryonic fly nervous

system, and its overexpression in the retina leads to the complete loss of cells (White et al., 1996).

As suggested by its homology to the *reaper* gene product, p75^{NTR} activation has been shown to promote apoptosis in several developing neuronal populations. In p75^{NTR} knockout mice there is less cell death in the retina, the cholinergic brainstem, and the spinal cord mantle zone (Frade and Barde, 1999; Naumann et al., 2002). The death-promoting activity of the p75^{NTR} may be evoked by its high-affinity ligand, the uncleaved neurotrophin. When SCG neurons are grown in the presence of mature NGF, there is virtually no cell death, but the neurons display a significant increase in TUNEL labeling when grown in the presence of pro-NGF (Figure 7.18).

Most experiments have examined the effect of NGF only (because it was not yet known that pro-NGF was a good ligand). In these experiments, the relative amount of TrkA and p75^{NTR} activity determines whether life or death occurs. Rat brain oligodendrocytes grown in culture express the p75^{NTR} receptor, but not TrkA, and NGF treatment kills the majority of cells (Casaccia Bonnefil et al., 1996). In the developing retina, the depletion of endogenous NGF with antibody results in better survival of retinal neurons. The ability of NGF to kill retinal neurons is apparently mediated by p75^{NTR} because antibodies against this receptor prevents the cell death (Frade et al., 1996). In cultured sympathetic neurons that express both TrkA and p75^{NTR}, BDNF is able to kill these cells because it selectively activates the p75^{NTR} receptor and not TrkA (Bamji et al., 1998). Thus, the relative amount of each ligand (proneurotrophin and neurotrophin), and each receptor (Trks and p75^{NTR}) makes it tricky to predict the outcome for any neuron population.

Additional members of the TNFR family contribute to neuronal cell death during development. In the cortex, some neuroblasts express the TNFR family member called Fas, and neighboring cells express its ligand, FasL. Activation of this receptor produces cell death in the primary cultures of cortex neuroblasts. Spinal motor neurons also express Fas and its ligand during the embryonic period of normal cell death. Primary cultures of motor neurons can be kept alive when grown in the presence of a compound that blocks Fas activation (Cheema et al., 1999; Raoul et al., 1999).

Members of the TNFR family initiate the cell death process by recruiting a broad range of intracellular messengers (Figure 7.18). The Fas receptor binds to an adapter protein (FADD) that leads to the activation of caspase-8, an enzyme that is directly involved in promoting cell death (see Caspases: Agents of Death). The survival-promoting influence of p75^{NTR} may employ a transcription activator called nuclear factor κB (NF-κB) which is recruited to enter the nucleus (Fig 7.18).

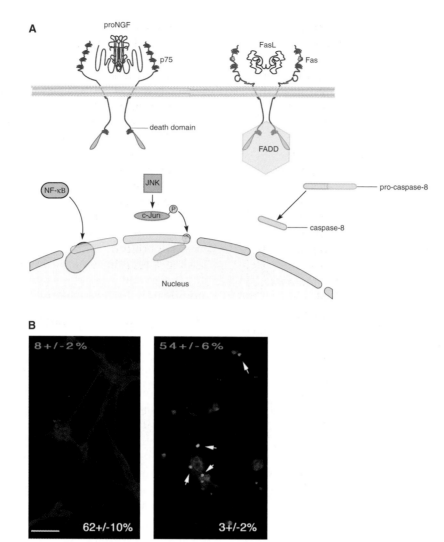

FIGURE 7.18   Members of the TNF receptor family influence neuron survival. A. A schematic shows two members of the TNFR familly, p75[NTR] and Fas. The p75[NTR] is a high-affinity receptor for the proneurotrophin, proNGF, and it may mediate either survival or apoptosis. Survival may involve the recruitment of a transcription activator, nuclear factor κB (NF-κB), while apoptosis may employ Jun kinase (JNK), which phosphorylates the transcription factor, c-Jun. The Fas receptor is a high-affinity site for the Fas ligand (FasL), and it may facilitate cell death through recruitment of the adaptor protein, Fas-associated death domain (FADD), and the subsequent activation of caspase-8 from its proform. B. The images show two sympathetic neuron cultures grown either in the presence of NGF (left) or proNGF (right). The NGF promotes survival and axon outgrowth, whereas the proNGF induces more apoptosis, as assessed with TUNEL labeling (green dots) through activation of the p75[NTR]. (Adapted from Lee et al., 2001)

The death-promoting influence of p75[NTR] can employ the c-Jun N-terminal kinase (JNK) and its target (the transcription factor, c-Jun) to kill cells. This pathway can even lead to increased production of FasL (Le-Niculescu et al., 1999). In cultures of NGF-dependent sympathetic neurons, expression of c-Jun is induced immediately after withdrawal of the neurotrophin. Experimental reduction of c-Jun is sufficient to rescue NGF-deprived neurons, and constitutive overexpression of c-Jun is sufficient to kill neurons in the presence of NGF (Ham et al., 1995). Similarly, rat spinal motor neurons can be kept alive in vitro by blocking JNK activity (Maroney et al., 1998).

THE EXPANDING WORLD OF SURVIVAL FACTORS

The old adage about taking care not to wish for something lest it come true has some validity in the world of survival factors. As research on the NGF and its receptors accelerated, it became difficult to reconcile why other survival factors had not been found for the many other cells that die during development. Whereas scientists were vigorously searching for even a single endogenous neuron survival factor in 1980, there is now evidence that many families of factors and receptors influence survival. Unfortunately, the trophic influence of most factors has been tested in relatively few brain regions. Of the neurons that have been investigated, most are influenced by more than one trophic factors, and the array of factors (or receptors) can vary during the course of development.

In recent years, several cytokines have been found to keep neurons alive in dissociated primary culture. Cytokines are a diverse family of secreted proteins that were originally discovered as growth factors in lymphocyte cultures, and many of these have turned out to have a primary role in neuron survival as well. The names of individual cytokines derive from the first biological activity that they were discovered to have, such as killing tumors (tumor necrosis factor) or promoting mitosis of hematopoietic stem cells (colony stimulating factor). One of the most thoroughly studied cytokines, Ciliary neurotrophic factor (CNTF), supports the in vitro survival of autonomic, DRG, hippocampal, and motor neurons. CNTF binds to an intrinsic membrane protein, called CNTFR α, and this binding event recruits two other transmembrane proteins (gp130 and LIFRβ) that form the β subunit of the receptor complex (Figure 7.19). The α subunit provides specificity to the trimeric receptor, while the β subunits are responsible for signal transduction (Sleeman et al., 2000). When the receptor complex forms, a tyrosine kinase (member of the Jak family) that is associated with the cytoplasmic tail of each β subunit becomes activated, and phosphorylates a DNA-binding protein (p91) that translocates to the nucleus and activates transcription (Bonni et al., 1993).

In contrast to CNTF, the TGF-β family of cytokines and their receptors seem to be involved in promoting the death of sympathetic, sensory and motor neurons in chicks. When all three isoforms of TGF-β were neutralized with antibody treatment in vivo, virtually all of the normally occurring cell death was prevented (Krieglstein et al., 2000).

The influence of newly discovered factors on motor neuron survival is of particular interest because normally occurring cell death in this population has been well characterized and closely linked to the target (Figures 7.6 and 7.9). When chick embryos are treated with human recombinant CNTF, half of the naturally occurring motor neuron death is prevented (Oppenheim et al., 1991). Surprisingly, parasympathetic, sympathetic, and sensory neuron cell death is unaffected.

Although CNTF knockout mice display little effect on cell survival during development, including motor neurons, the functional loss of CNTF receptors does increase cell death. Loss of CNTFRα increases motor neuron death by about a third, and similar observations have been made on LIFRβ and gp130 knockout mice (DeChiara et al., 1995; Li et al., 1995; Nakashima et al., 1999). This result implies that there is at least one target-derived cytokine that supports motor neuron survival through activation of the CNTFRα. In fact, cardiotrophin-1 (CP-1) a cytokine that is expressed in embryonic skeletal muscle, contributes to the survival of embryonic motor neurons. CP-1 deficient mice display 20-40% greater motor neuron loss during the normal period of cell death. However, CP-1 does not bind to CNTFRα, and so the hunt continues for the

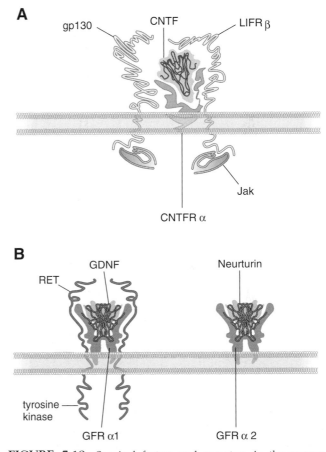

FIGURE 7.19 Survival factors and receptors in the nervous system. A. Cytokine signaling: CNTF binds first to an intrinsic membrane protein called CNTFRα. This event causes two other transmembrane proteins, gp130 and LIFR, to form the β subunit of the receptor complex. The activated receptor complex signals via a tyrosine kinase (Jak) that is associated with the cytoplasmic tail of each β subunit. B. GDNF signaling: GDNF and Neurturin bind first to their cognate GFRα receptor. This complex then recruits the receptor tyrosine kinase, RET, to homodimerize and to become autophosphorylated, leading to intracellular signaling.

array of factors and receptors that keep motor neurons alive during development (Oppenheim et al., 2001).

A second substance, glial cell line-derived neurotrophic factor (GDNF), has been identified as preventing naturally occurring motor neuron death in vivo. GDNF was initially characterized by its ability to keep midbrain dopaminergic cells alive in vitro. This assay was chosen because Parkinson's disease involves the death of these dopaminergic neurons, and a survival factor may have important therapeutic value (Lin et al., 1993). Four members of the GDNF ligand family have now been isolated, and they all belong to the TGF-β superfamily. Each ligand binds to a specific ligand recognition α subunit (GFRα1 through 4). The GFRα subunits are attached to the membrane by a glycosyl phosphatidylinositol anchor (Figure 7.19). The ligand-receptor complex becomes associated with a transmembrane tyrosine kinase, called RET, and leads to its activation (Airaksinen and Saarma, 2002).

There is now strong evidence that GDNF is an endogenous, target-derived survival factor for a subpopulation of motor neurons. GDNF mRNA is found in the limb, and GFRα and RET are expressed by subsets of chick motor neurons during the period of normal cell death (Homma et al., 2003). GDNF treatment prevents naturally occurring motor neuron death in chick embryos, and overexpression of GDNF in the musculature of mice increased survival in most motor neuron populations. Conversely, motor neuron death was increased in GDNF-deficient mice (Oppenheim et al., 1995; Oppenheim et al., 2000). The GDNF signaling pathway influences more than just motor neuron survival. Mice lacking either GDNF, Neurturin (a second ligand), GFRα1, GFRα2, or RET also exhibit specific loss of parasympathetic and enteric neurons (Huang and Reichardt, 2001).

It is clear that motor neurons are a diverse population, particularly in their dependence on trophic substances. Yet a third motor neuron trophic factors has been reported, a cytokine called hepatocyte growth factor/scatter factor (HGF/SF). As with GDNF, HGF/SF influences the survival of only a subset of motor neurons. In the chick, only lumber motor neurons are dependent on HGF/SF for their survival (Ebens et al, 1996; Yamamoto et al., 1997; Novak et al., 2000).

Despite the expansion of candidate growth factors, few of them seem to have an effect on the survival of CNS neurons when they are eliminated from the developing organism. One hypothesis is that central neurons, unlike peripheral ganglion cells, have multiple targets and afferents, perhaps giving them access to many different growth factors during development.

The prediction from this hypothesis is that one must eliminate two or more growth factors or receptors in order to disrupt survival. It is also likely that many survival factors have yet to be identified. For example, the survival of embryonic retinal ganglion cells is enhanced by tectal cell-conditioned media in a manner that cannot be duplicated by CNTF or the neurotrophins (Meyer-Franke et al., 1995).

A final consideration is that many trophic factors also promote certain aspects of differentiation (see Chapters 2-4), and progress along these pathways may be entangled with the decision to live or die. An interesting example of this occurs in the developing fly eye. The epidermal growth factor receptor (EGFR) mediates a survival signal from a nearby cluster of postmitotic cells in the fly retina. In its absence, the number of omatidial cells declines significantly. However, the EGFR is also playing an important role in progression through the cell cycle during this same period of time (Baker and Yu, 2001).

ENDOCRINE CONTROL OF CELL SURVIVAL

Hormonal signaling controls many aspects of development, including cell survival. There are now several examples of brain structures that are quantitatively different in males and females of the same species, often referred to as a sexual dimorphism. These sexual dimorphisms are thought to arise from regional differences in the amount of steroid hormones or their receptors. Steroid hormones (e.g., estrogens and androgens) are lipid soluble molecules that bind to cytoplasmic receptors, and these receptors can translocate to the nucleus where they regulate gene transcription. For example, normal cell death among developing superior cervical ganglion (SCG) neurons is greater in female rats than in males. Furthermore, castration of neonatal male rats significantly increases the number of dying neurons, suggesting that a gonadal hormone may be responsible for better neuron survival in the male SCG. In fact, treatment of neonatal animals with a sex hormone (estradiol or testosterone) improves SCG neuron survival, even in female animals (Wright and Smolen, 1987). Although such discoveries have sparked much interest and theorizing about the neural substrates of male- and female-specific behavior, there remain few solid examples that correlate structure to function. These are considered in more detail below (see Chapter 10).

The survival of some motor neurons is also dependent on the presence of specific sex hormones. In the lumbar spinal cord of male rats, there are two motor nuclei that innervate striated muscle of the penis: the spinal nucleus of the bulbocavernosus (SNB) and the dorsolateral nucleus. As one might expect, these nuclei and the muscles that they innervate are present in males but are nearly absent in female rats. This sexual dimorphism arises from the selective loss of motor neurons in female rats. During the first 10 postnatal days, the total number of SNB neurons decrease by nearly 70% in females, although the number only decreases by about 30% in males (Figure 7.20).

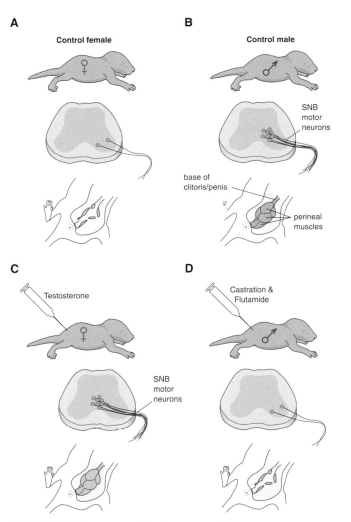

FIGURE 7.20 Hormonal influence on SNB motor neuron survival. A. Areas of the lumbar spinal cord that innervate perineal muscles in male rats are nearly absent in female rats, owing to motor neuron death during development. B. Male rats exhibit a greater number of SNB motor neurons. C. When neontal female rats are treated with the androgen steroid hormone, testosterone, SNB motor neuron death is decreased. D. In contrast, when neontaal males are castrated and reared with an androgen antagonist, flutamide, SNB motor neurons display increased cell death. The target muscle size is affected by each treatment and appears to explain the effect of hormone. (Adapted from Nordeen et al., 1985; Breedlove and Arnold, 1983)

However, when females are treated with the androgen steroid hormone, testosterone, the amount of cell death is decreased and resembles the pattern seen in males (Nordeen et al., 1985). When males are castrated and reared with an androgen antagonist, flutamide, their SNB neurons are as scarce as in female rats (Breedlove and Arnold, 1983). The endogenous androgen signal probably keeps more male SNB neurons alive by keeping the target muscles alive and preserving target-derived survival factors. In fact, when females are treated with androgen, but also receive a treatment that blocks signaling at either TrkB, TrkC, or the CNTFR, there is very little sparing of SNB motor neuron number (Freeman et al., 1997, Xu et al., 2001).

Endocrine signals can lead to extensive remodeling throughout the nervous system. In the moth, a great many neurons are necessary in larvae, or are required for the process of metamorphosis from caterpillar to pupa to moth. A surge of the steroid hormone, 20-hydroxecdysone (20E), triggers each molt, during which the larva sheds its cuticle and grows. The 20E also acts as a trigger for normal cell death. The small pulse of 20E that triggers the formation of a pupa, also acts directly on a specific set of abdominal motor neurons to bring about their death. These motor neurons are not dependent on their target for survival, in contrast to those in vertebrates. Instead, these motor neurons can be placed in culture, and killed by exposure to 20E at a specific time in their development (Hoffman and Weeks, 1998). The final prolonged elevation of 20E causes the adult moth to emerge from its pupa. As the level of 20E falls, there is a second wave of cell death during which nearly 40% of abdominal neurons are lost (Truman and Schwartz, 1984; Zee and Weeks, 2001). In fruit flies, there are about 300 neurons in the ventral cord that express much higher levels of the 20-HE receptor (EcR), and these are precisely the cells that die at metamorphosis (Robinow et al., 1993). Apparently, the cells that are to be eliminated become extremely dependent on 20E. This period of cell death can be delayed by treatment with 20E in both the moth and fruit fly. The molecular signals that mediate ecdysone-induced cell death are quite similar to those found in other systems (Hoffman and Weeks, 2001; Buszczak and Segraves, 2000).

Endocrine signaling is also responsible for neuron death during vertebrate metamorphosis (Decker, 1976; 1977). The surge of thyroxine that initiates metamorphosis from tadpole to frog causes lysosomal activity to increase in motor neurons. The motor neurons that innervate regressing tail musculature of tadpoles are eliminated directly by a thyroid hormone signal. Thyroxine exposure has also been shown to elicit cell death in a region of the adult zebra finch forebrain that is involved in song production (Tekumalla et al., 2002). Interestingly, the EcR is almost identical to the thyroid receptor.

Endocrine signals have also been implicated in cell survival among the sexually dimorphic telencephalic nuclei of songbirds: those species where males learn to produce mating calls, while females vocalize little, if at all. In canaries and zebra finches, at least three areas of the brain that support song production are much larger in males than females (Nottebohm and Arnold, 1976). In some nuclei such as the *robustus* nucleus of the *archistriatum* (RA), which shares some features with motor cortex, the differences in neuron number arise from a selective loss of cells in the female (Nordeen and Nordeen, 1988; Kirn and DeVoogd, 1989). What is the evidence that steroid hormones influence cell survival in males? If developing females are treated with testosterone or its active metabolite, estrogen, then the number of neurons in RA becomes masculinized, and the birds acquire male-like vocalizations (Gurney, 1981). This effect is mediated by androgen receptor since it can be prevented with a specific antagonist, Flutamide (Grisham et al., 2002). More recently, the idea that steroid hormones can account for sexual dimorphism of songbird vocal nuclei has been challenged. For example, when genetic females are "engineered" to grow testicular tissue that secretes androgens, their vocal nuclei do not become masculinized (Wade and Arnold, 1996). Therefore, there are strong reasons to think that steroid hormones play a role in control of cell number in male and female songbirds, but the precise mechanisms remain elusive (see Chapter 10).

CELL DEATH REQUIRES PROTEIN SYNTHESIS

One might suppose that when a neuron is deprived of a trophic factor, it fails to maintain normal levels of synthesis and metabolism, and simply "passes away." In fact, neurons collaborate in their own death by activating genes and synthesizing protein that injure the cell. That is, they "commit suicide." The first indication that an active process could account for cell death came from studies of nonneuronal cells. For example, cultured tadpole tail cells die when exposed to thyroxine, but this can be prevented by blockers of RNA and protein synthesis (Tata, 1966). A major breakthrough came from genetic studies of cell death in the nematode, *Caenorhabditis elegans* (Driscoll and Chalfie, 1992). About 10% of cells die during development in

C. elegans, most of them being neurons, but inactivation of two specific genes (called *ced-3* and *ced-4*) rescues all of these cells, including neurons (discussed in more detail below). Is it possible that neuron cell death can actually be prevented by blocking protein or RNA synthesis?

This hypothesis was tested by asking whether neurons are rescued by protein synthesis inhibitors. As described above, embryonic sympathetic neurons are able to survive in vitro when grown in the presence of NGF. When NGF is removed from the culture media, few neurons remain after two days. Therefore, the first experiment determined whether inhibitors of RNA or protein synthesis could save NGF-deprived neurons (Figure 7.21). Actinomycin-D blocks transcription by binding to DNA and preventing the movement of RNA polymerase, while cycloheximide prevents translation by blocking the peptidyl transferase reaction on ribosomes. Each of these treatments completely rescued sympathetic neurons following NGF deprivation, demonstrating that new RNA and proteins must be manufactured to bring about cell death. To determine when the harmful phase of translation occurs, cycloheximide was delivered at several times after NGF deprivation, and it was found that the cell death-promoting proteins are produced at about 18 hours (Martin et al., 1988). If all the molecular machinery for cell death was present in the cytoplasm, one would expect that neurons would be committed to die within a few hours.

To determine whether mRNA and protein synthesis are general features of cell death in vivo, animals were treated with synthesis inhibitors at the age when neurons are normally lost (Figure 7.21). When chick embryos are treated with either cycloheximide or actinomycin D on embryonic day 8, the time of maximum motor neuron and DRG cell death, they exhibit a striking reduction in the number of dying neurons (Oppenheim et al., 1990). Similarly, the cell death that occurs in response to declining levels of 20-HE in moths can be reduced by RNA or protein synthesis inhibitors (Fahrbach et al., 1994). These studies suggest that trophic signals may stimulate the production of proteins that protect the neuron from death, and in the absence of a trophic signal harmful proteins may be synthesized. The search for such proteins is discussed below.

Much of the cell death machinery seems to be present at all times in a neuron's cytoplasm. However, there is good evidence from several neuronal systems that transcription or translation is necessary to activate this existing machinery. In fact, recent studies have shown that, under certain conditions, cells can undergo apoptosis even if their nucleus is removed, or in the presence of macromolecular synthesis. It will, therefore, be important to study naturally occurring cell death in many areas of the nervous system in order to determine whether protein synthesis is an obligate part of the death pathway.

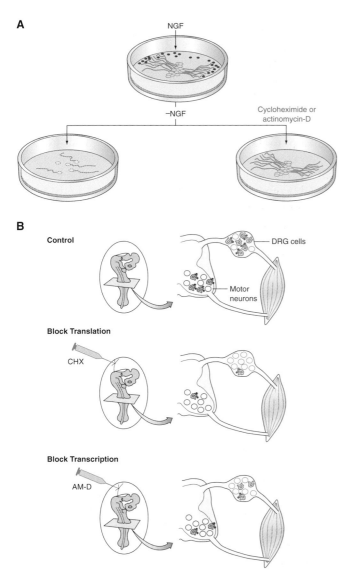

FIGURE 7.21 Neuron cell death can be delayed by blocking protein synthesis. A. Sympathetic neurons die within two days, when NGF is removed from the culture media (left). When either a translation blocker (cyclohexamide, CHX) or a transcription blocker (actinomycin-D, AM-D) was added to NGF-deprived cultures, sympathetic neurons were rescued (right). B. The synthesis of mRNA and protein is also required for naturally occurring cell death in vivo. In control chick embryos, pyknotic motor neurons and DRG neurons are counted during normal development (top). When chick embryos are treated with CHX during the time when motor neuron and DRG cell death is at its greatest, the number of pyknotic neurons is decreased (middle). Similarly, when chick embryos are treated with AM-D, there is a reduction in the number of pyknotic neurons. (Adapted from Martin et al., 1988; Oppenheim et al., 1990).

INTRACELLULAR SIGNALING

A full description of the molecular pathways that support apoptosis can quickly overpower the reader with a list of acronyms. In simplified form, the developing neuron contains two types of proteins in its cytoplasm: those that maintain survival and those that mediate death. When a survival factor binds to its receptor, anti-apoptotic proteins are activated by phosphorylation. Moreover, the expression of anti-apoptotic proteins increases. When survival factors are withdrawn, pro-apoptotic proteins are activated by phosphorylation, and their expression increases. Therefore, the vital purpose of a survival factor is to upregulate the function or expression of pro-apoptotic proteins and suppress the anti-apoptotic proteins.

Although growth factors influence developmental events besides cell death (e.g., mitosis, differentiation, and axon outgrowth), many of the intracellular signals have been identified during studies of cancer cell growth. To make this discussion somewhat manageable, we will only cover the cytoplasmic signals that are most relevant to the survival of neurons during normal development.

Neurotrophin signal transduction, the best characterized pathway in neural tissue, involves the sequential recruitment and activation of several kinase pathways. Activation ultimately leads to modification of existing cytoplasmic or membrane proteins, and regulation of gene transcription. The binding of NGF to the TrkA protein induces receptor dimerization, followed by the rapid phosphorylation of 5 tyrosine residues on the cytoplasmic tail by neighboring Trk receptors (Kaplan et al., 1991b). Many components of this intracellular pathway are identical to those that are recruited by the *sevenless* receptor tyrosine kinase and participate in fly eye differentiation (see Chapter 4). The phosphorylation sites on the Trk receptor serve as docking sites for the cytoplasmic molecules that will propagate the signal toward cytoplasmic kinases and the nucleus (Figure 7.22). Shc is one of the first adaptor proteins that binds to the Trk receptor, and is phosphorylated by it. In fact, the Shc binding site appears to be critical to the survival of NT-4-dependent sensory neurons in vivo (Minichiello et al., 1998).

As shown in Figure 7.23, the first major kinase pathway is activated by the recruitment of a second adaptor protein (Grb2, growth factor receptor-bound protein-2) and a docking protein (Gab1, Grb2-associated Binder-1). In this pathway, a phosphotidylinositol 3-kinase (PI3K) activates a serine/threonine kinase, called Akt. The PI3K-Akt pathway provides a crucial intracellular signal for keeping NGF-dependent

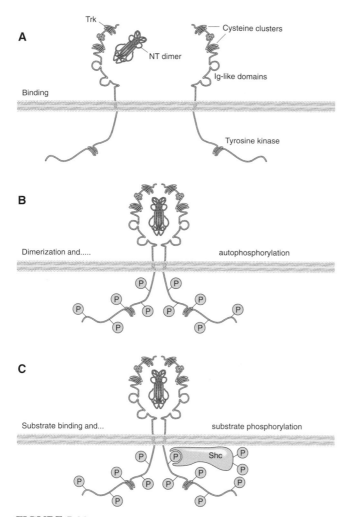

FIGURE 7.22 Neurotrophin-Trk receptor interaction. A. The biologically active forms of neurotrophins are dimers of identical 13 kDa peptide chains. The neurotrophin dimer binds to the Trk protein. B. The binding induces 2 Trk receptors to dimerize. The ligand-receptor complex leads to transphosphorylation of tyrosine residues on the cytoplasmic tail of neighboring receptors. (C) Cytoplasmic adaptor proteins (Shc) bind to specific phosphorylation sites on the cytoplasmic tail, and these substrates are then phosphorylated.

neurons alive (Figure 7.24). Specific blockade of either PI3K or Akt activity kills the primary cultures of sympathetic neurons even in the presence of NGF, while constitutive activation of either kinase keeps the neurons alive even in the absence of NGF (Crowder and Freeman, 1998).

There is a second major pathway that also leads to phosphorylation of cyoplasmic components and that participates in cell survival. In this case, the adaptor and docking proteins recruit a guanine nucleotide exchange factor (SOS) to activate a membrane-associated G-protein called Ras (Figure 7.23). Ras is activated when it exchanges a GDP for a GTP, whereupon it

binds to and activates a serine/threonine kinase, called Raf. Raf activation leads to activation of the mitogen-activated protein kinase cascade (MAPK), eventually resulting in translocation into the nucleus (Figure 7.23).

The potential importance of Ras-Raf signaling to neuron survival is demonstrated in experiments where the Ras protein is injected directly into cultured chick DRG neurons. Although DRG neurons depend on NGF for their survival, the Ras protein is sufficient to prevent cell death (Borasio et al., 1989). Furthermore, the same treatment promotes the survival of BDNF-dependent nodose ganglion neurons and CNTF-dependent ciliary ganglion neurons.

Which MAPKs are responsible for the positive effects of NGF binding, and which ones produce the negative effects following its withdrawal? In PC12 cells, there is evidence for both types of signal (Xia et al., 1995). NGF promotes the survival of PC12 cells, and this is accompanied by activation of several MAPKs. Within six hours of NGF withdrawal, PC12 cells begin to die in great numbers, and there is a prominent decrease in ERK activity. To test whether ERK is responsible for the positive effect of NGF, PC12 cells were engineered to produce constitutively high levels of ERK activity (Figure 7.24). These cells survived much better following NGF withdrawal. There is also a MAPK signal, called JNK, that appears to turn on following NGF withdrawal. In fact, PC12 cell death will occur in the presence of NGF if JNK is constitutively activated (Figure 7.24). Thus, NGF may upregulate proteins needed for cell survival through ERK, while proteins needed for suicide are upregulated by JNK.

How do activation of the two major kinase pathways, PI3K-Akt and MAPK, prevent cell death? Akt can promote cell survival by inactivating pro-apoptotic proteins in the cytoplasm and increasing transcription of anti-apoptotic proteins. For example, RSK phosphorylates, and inactivates, one of the pro-apoptotic regulatory proteins in the Bcl-2 family. These regulatory proteins are discussed below (see Regulating Death Proteins). The Ras-MAPK pathway promotes survival in the fly using a similar strategy. When active Ras is ectopically expressed in the developing eye, it decreases the expression of pro-apoptotic proteins and prevents normal cell death (Kurada and White, 1998).

As discussed above, NGF withdrawal and the p75[NTR] lead to activation of the JNK-c-Jun pathway, resulting in neuron death. Therefore, it seems important for the survival pathway to suppress this system as well. In fact, the active Ras-Raf pathway promotes survival by blocking the JNK pathway (Figure 7.23),

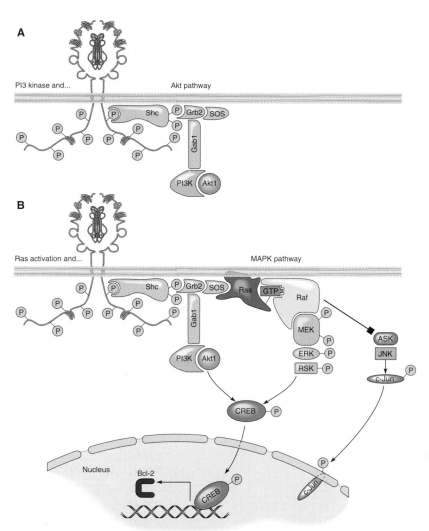

FIGURE 7.23 Trk intracellular signaling. A. The first pathway leads to activation of the Akt Kinase. Two proteins (Grb-2, Gab-1) are recruited to the receptor complex, resulting in the activation of a phosphatidyli-nositol-3 kinase (PI3K). PI3K generates phosphoinositide phophases that activate the serine/threoninee kinase, Akt. B. The second pathway leads to activation of the mitogen-activated protein kinase (MAPK) cascade. In this pathway, a membrane-associated G-protein, Ras, first becomes activated through GDP-GTP exchange, mediated by a guanine nucleotide exchange factor (Sos). The activated Ras phosphorylates several substrates, including a serine/threonine kinase, Raf. Raf can inhibit a proapoptotic pathway consisting of two kinases (ASK, JNK) and a trascription factor (c-Jun). Raf also initiates the MAPK cascade, concluding with the activation of ribosomal S6 protein kinase (RSK). Both Akt and RSK can phosphorylate the transcription factor, cyclic AMP response element binding protein (CREB). CREB enters the nucleus and increases the expression of antiapoptoptic proteins, such as Bcl-2.

possibly through phosphorylation of a JNK-activator. Thus, when sympathetic neurons are transfected with constitutively active Ras, the level of JNK and c-Jun declines, and the neurons survive in the absence of NGF. This pathway appears to play role in the survival mechanism during in vivo maturation. The genetic elimination of p53, which is known to be a downstream effector of c-Jun, results in increased survival of SCG neurons (Mazzoni et al., 1999; Aloyz et al., 1998; Kanamoto et al., 2000).

The PI3K-Akt and MAPK pathways also promote survival at the level of gene transcription. The transcription factor, cyclic AMP response element binding protein (CREB), is phosphorylated by RSK2. The activated CREB enters the nucleus and binds to DNA and increases the expression of a pro-apoptotic regulatory protein (Bcl-2) that is discussed below (see Regulating Death Proteins). The mechanism appears to support the in vitro survival of both sympathetic neurons through NGF activation of TrkA, and cerbellar granule

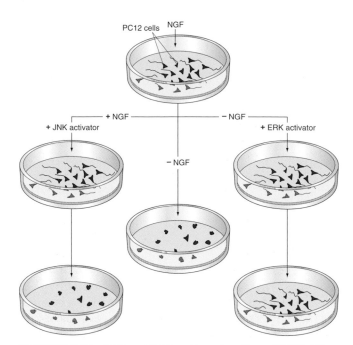

FIGURE 7.24 ERK and JNK mediate the effects of NGF. When NGF is withdrawn from the culture media, PC12 cells die (center). NGF is not able to support PC12 cell survival when the MAP kinase, JNK, is constitutively activated (left). In contrast, when ERK is constitutively activated in PC12 cells, they are able to survive even in the absence of NGF (right). (Adapted from Xia et al., 1995)

cells through BDNF activation of TrkB (Xing et al., 1996; Riccio et al., 1999; Bonni et al., 1999).

The extent to which CREB-dependent transcription contributes to neuron survival in vivo remains to be determined (Lonze and Ginty, 2002). In any event, other transcription pathways are likely to play a significant role. For example, Akt phosphorylates a transcription factor, called forkhead, preventing it from translocating to the nucleus and upregulating Fas ligand (Brunet et al., 1999).

Although we have tried to present a unified picture of cell survival signaling, it is likely that different neurons will employ specific cytoplasmic mechanisms. For example, Ras appears to be a principal cytoplasmic signal for NGF signaling in DRG neurons. However, it is also known that NGF-dependent sympathetic neurons from caudal regions of nervous system are *not* saved by Ras injection, whereas sympathetic neurons from the rostral SCG are saved (Markus et al., 1997).

The many molecular pathways between receptor and nucleus interact with one another, producing another level of complexity. For example, activation of ERK requires binding and activation of both Shc and phospholipase C-γ1 (PLC-γ1) (Stephens et al., 1994). However, the PLC-γ1 pathway seems to be more involved in neurite outgrowth (see Chapter 5). Therefore, it will be crucial to

understand how receptors are activated in vivo if we are to evaluate the contribution of each of the cytoplasmic kinases.

A complete description of the intracellular pathways mediating survival may still be lacking, but many of the candidates that pass the signal from membrane to nucleus have now been identified. The importance of each candidate molecule is likely to range broadly from one neuron type to another. Fortunately, these remaining mysteries need not be solved in order for us to consider the cytoplasmic weapons that neurons use to kill themselves.

CASPASES: AGENTS OF DEATH

The discovery of specific genes that are directly involved in apoptosis was made in experiments on the nematode, *Caenorhabditis elegans*. In 2002, Robert Horvitz shared the Nobel Prize for Physiology or Medicine for uncovering the discrete genetic steps that support apoptosis, which came to be known as programmed cell death (PCD). Two genes, *ced-3* and *ced-4*, *must* be expressed by each *C. elegans* cell if it is to die during development (*ced* stands for cell death abnormal) (Yuan and Horvitz, 1990). When either of these genes is mutated, almost all of the PCD is prevented (Figure 7.25). Analysis of mosaic animals (i.e., animals in which the *ced* gene is expressed in only a few identified cells) indicates that the gene product acts within the cell that produces it, showing that PCD proceeds by suicide rather than a violent neighbor. What are these gene products, and how do they control the life or death decision?

Proteases appear to be a common weapon of choice for cell death. Although it has been known for some time that one can rescue injured cells by blocking proteolytic enzymes, the evidence for protease involvement in normally occurring cell death came only recently. In fact, one of the gene products that kills *C. elegans* neurons, CED-3, turns out to be a cysteine protease, an enzyme that specifically cuts up proteins after an aspartate residue. In mammals, there is a CED-3-like protease called interleukin-1β-converting enzyme (ICE). While ICE plays a life-affirming role in processing pro-interleukin-1β for the purposes of blood cell production, it also serves as an angel of death in the developing nervous system. New cysteine proteases have been identified in many species, and they are now generally referred to as **C**ysteine requiring **ASP**artate prote**ASE**s, or caspases (Figure 7.25). Thus, ICE is now called caspase-1.

In *C. elegans*, the *ced-4* gene product is also a necessary constituent of the death pathway. CED-4 binds to the

A

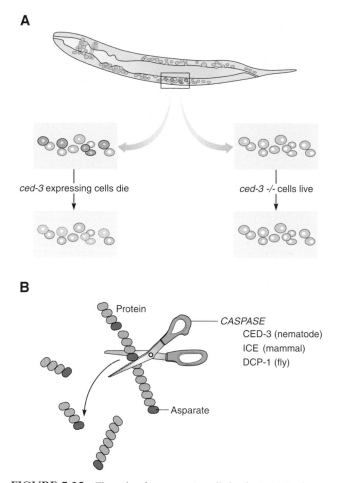

ced-3 expressing cells die   *ced-3 -/-* cells live

B

Protein

CASPASE
CED-3 (nematode)
ICE (mammal)
DCP-1 (fly)

Asparate

FIGURE 7.25 The role of caspases in cell death. A. In *C. elegans*, CED-3-expressing cells (yellow cytoplasm) die during development, but almost all cell death is prevented when the *ced-3* gene is mutated. B. CED-3 is a member of the caspase family, enzymes that specifically cleave proteins after an aspartate residue. There are homologs of CED-3 in mammals (ICE) and fruit flies (DCP-1). (Panel A adapted from Yuan and Horvitz, 1990)

inactive form of CED-3 and leads to its activation (Figure 7.26). In mammals and flies, the pathway involves *ced-3* and *ced-4* homologs, but the mechanism leading to caspase activation begins at the mitochondrion. Various stimuli, such as the withdrawal of growth factor, lead to increased permeability of the outer mitochondrial membrane. When this occurs, a member of the electron transport chain, called cytochrome *c* (cyt *c*), leaks out of the mitochondrion. As it enters the cytoplasm, cyt *c* binds tightly to an adaptor protein called apoptosis protease activating factors-1 (Apaf-1). This is the mammalian homolog of *ced-4*. Procaspase-9 is then recruited to this complex, which is called an *apoptosome*. This is the site where caspase-9 becomes activated autoproteolytically. The activated caspase-9, in turn, cleaves pro-caspase-3, resulting in its activation. Mutations of this Apaf lead to

a decrease in cell death in vivo (Li et al., 1997), similar to the effects of *ced-4* inactivation.

A caspase begins with the release of a mitochondrial flavoprotein, called apoptosis-inducing factor (AIF). In this case, the AIF enters the nucleus and initiates DNA cleavage (Figure 7.27).

Several caspases have now been implicated in neuron cell death. There are 14 members in mammalian genome, 6 in flies, and only 3 in worms. There are many lines of evidence that caspase activity is required for neuronal apoptosis. When a caspase inhibitor (a cytokine response modifier, crmA) is microinjected into chick DRG neurons in vitro, they can survive the withdrawal of NGF (Gagliardini et al., 1994). Members of the caspase family may also mediate the death-promoting effect of "death domain" containing proteins, such as p75[NTR], and the *Drosophila* protein caller Reaper (discussed above). For example, Reaper overexpression in the *Drosophila* eye causes all the cells to die, but a caspase inhibitor is able to block this effect (White et al., 1996; Vernooy et al., 2000).

To determine whether caspases are involved in the normal period of cell death, chick embryos were treated with a synthetic peptide inhibitor of caspase on embryonic day 8, the peak of motor neuron death. After 24 hours, the number of pyknotic cell bodies was cut in half compared to animals treated with a less selective protease inhibitor (Milligan et al., 1995). However, it remains possible that the synthetic peptide inhibitor blocked more than one member of the caspase family. To examine the effect of a single protease, transgenic mice were produced lacking caspase-3, the closest mammalian homolog of CED-3. The brains of these animals are disorganized, and there are few signs of the pyknotic cell clusters that accompany nervous system morphogenesis, suggesting a decrease in normal cell death. This is most apparent in the proliferative zone or immature populations in the forebrain (Kuida et al., 1996, Pompeiano et al., 2000). In contrast, cell death of motor, sensory, and sympathetic neurons proceeds unchecked when caspase-3 activity is genetically eliminated. Interestingly, electron micrographs of the dying neurons suggest that they die in a somewhat different manner than those in control mice; there is little sign of chromatin condensation or fragmentation into apoptotic bodies. Together, these results suggest that other caspases, or caspase-independent pathways, mediate cell death in many developing neuronal populations. They also suggest that TUNEL labeling alone does not characterize cell death. It is possible that cell death is occurring, but without DNA fragmentation, in cells that are not stained with TUNEL (Oppenheim et al., 2001). It is not too surprising to learn that nerve cells commit suicide by destroying their own proteins. This

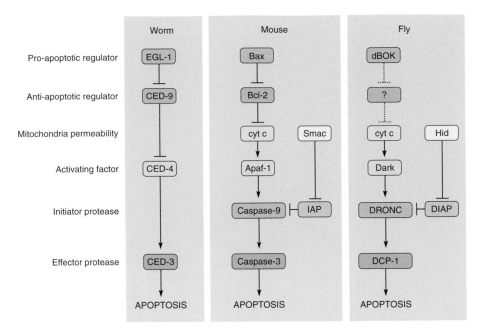

FIGURE 7.26 Regulation of cell death machinery in three species. (Left) In *C. elegans*, the effector protease that leads to apoptosis (CED-3) is activated by CED-4. An anti-apoptotic regulator (CED-9) can complex with CED-4 and interfere with CED-3 activation. A pro-apoptotic regulator (EGL-1) can bind to CED-9, and facilitate the processing of CED-3. (Middle) Many of the same components are present in mammals, including pro- and anti-apoptotic regulators (Bax and Bcl-2). However, Caspase-3 is activated by an initiator protease (Caspase-9) that is processed in molecular complex with cytochrome c and Apaf-1. An additional regulatory component consists of IAP, which prevents Caspase-9 activation, and Smac, which can inhibit IAP and permit the apoptotic pathway to progress. (Right) In the fruit fly, the pathway is relatively similar to that described for mammals.

raises a question about the specificity of caspases: is their activity imprecise, or do they break down specific substrates? The evidence suggests great specificity, particularly for proteins that are involved in genome regulation, such as DNA repair, DNA replication, and RNA splicing enzymes (Lazebnik et al., 1994; Loetscher et al., 1997; Nicholson and Thornberry, 1997). Structural proteins of the nucleus and cytoskeleton, such as actin and fodrin, are also targets for cleavage. One example of caspase target is the DNA repair enzyme called poly (ADP-ribose) polymerase (PARP), suggesting that cell death is achieved by compromising the neuron's transcription machinery. A second set of targets, of some interest to those studying Alzheimer's disease, are the transmembrane proteins called presenilins that are apparently involved in the Notch signaling pathway (see Chapter 2: Induction). Although NGF-deprived PC12 cells normally die, they can be rescued by transfection with presenilin 2 antisense mRNA.

While no final arrests have been made, caspases appear to be a primary felon in the developing nervous system. However, conspiracy theorists can take comfort in the many other death mechanisms that underlie normal cell death. We have learned that the release of a mitochondrial flavoprotein, AIF, can act directly to fragment DNA without the intermediate involvement of caspases. A second caspase-independent pathway involves the regulation of superoxide ($O_2 \cdot^-$), which accumulates as a result of oxygen usage in the mitochondrial respiratory chain. Free radicals such as $O_2 \cdot^-$ have unpaired electrons, making them an extremely reactive species. Excess $O_2 \cdot^-$ can disrupt membrane integrity, inhibit pumps, and fragment DNA. Superoxide dismutase (SOD) is the endogenous enzyme that eliminates $O_2 \cdot^-$ by catalyzing a reaction to O_2 and H_2O_2. Interestingly, sympathetic neurons can survive for a longer period of time after NGF deprivation if injected with SOD. Although a caspase mediates the cell death process initiated by trophic factor deprivation, it is *not* responsible for the death initiated by free radicals (Troy et al., 1997).

Thus, for each developing population of neurons, the naturally occurring period of cell death may invoke a distinctive set of molecular mechanisms. In fact, one of the most varied features of cell death involves the regulatory proteins that determine whether or not caspases cross the threshold to their active state.

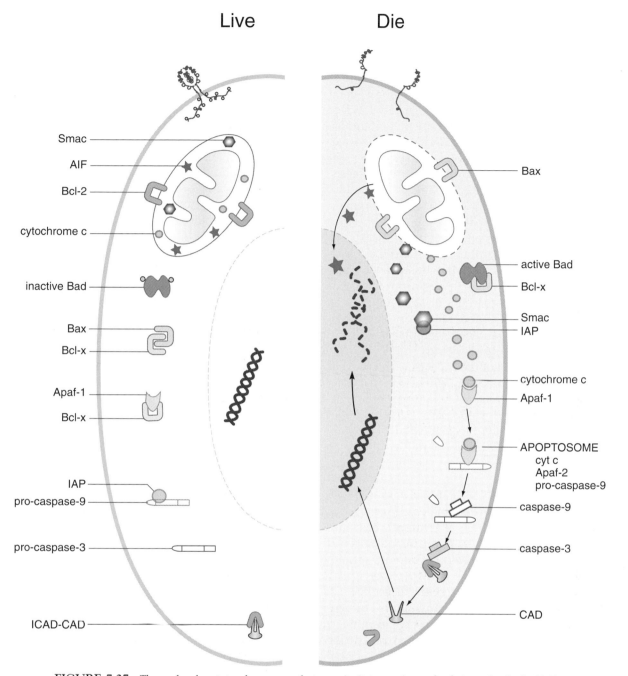

FIGURE 7.27 The molecular state of a neuron that permits it to survive or leads it to death. (Left) The living cell contains mitochondria that are preserved in a nonpermeable state due to the presence of an anti-apoptotic regulator, Bcl-2. A second anti-apoptotic regulator, Bcl-x, complexes with Apaf-1, preventing the activation of caspase-9. Bcl-x also binds to a pro-apoptotic regulator, Bax, and prevents it from influencing the mitochondrion. A different pro-apoptotic regulator, Bad, is inactive, having been phosphoryated by neurotrophin-elicited kinase activity. IAP binds to pro-caspase-9, and this also serves to prevent activation. A nuclease that is responsible for DNA fragmentation, caspase-activated deoxyribonuclease (CAD), is bound by its inhibitor, ICAD. (Right) In dying neurons, Bad becomes active when it is dephosphorylated, and it binds to Bcl-x. This permits Bax to associate with the mitochondrion, leading to the release of cytochrome c, AIF, and Smac. It also permits Apaf-1 to form an apoptosome with cyt c, process caspase-9, and activate caspase-3. Smac binds to IAP, which also permits the processing of pro-caspase-9. One target of caspase-3 is ICAD, leading to the release of CAD, and the fragmentation of DNA. In a caspase-independent pathway, AIF also enters the nucleus and fragments DNA.

BCL-2 PROTEINS: REGULATORS OF APOPTOSIS

The continuous presence of death-promoting molecules in the cytoplasm is rather like keeping several loaded guns scattered about the house; safety locks of some sort are essential. In fact, there are several important checks and balances to ensure that only the correct neurons die. Once again, the molecular acronyms will fly, so there is some appeal in grasping the basic mechanism first. There are two key points of regulation. First, the mitochondrion outer membrane must remain relatively impermeable so that cytochrome c does not leak into the cytoplasm. Second, inactive pro-caspases must be constrained so that their threshold for activation remains relatively high. Mitochondrion permeability is controlled by a large family of proteins that interact with the outer mitochondrial membrane as well as with one another. The activation of pro-caspases is regulated by two interacting molecules, one of which is released from mitochondria.

In *C. elegans*, activation of the *ced-9* gene can prevent apoptosis in all cells. Conversely, mutations that inactive *ced-9* lead to death among cells that would normally survive through development (Hengartner et al., 1992). CED-9 apparently blocks death by complexing with CED-4 and interfering with its ability to activate the protease, CED-3 (Figure 7.26). The mammalian homolog of CED-9 is a membrane-associated protein called Bcl-2 (named for its discovery in B Cell Lymphoma cells) that was originally discovered in studies of tumor formation. The Bcl-2 family has since been found to include members that promote survival (anti-apoptotic), as well as those that promote apoptosis (pro-apoptotic). To date, 12 pro-apoptotic and 7 anti-apoptotic members have been described in mammals, although few have been evaluated as participants in naturally occurring neuron cell death.

In healthy neurons, anti-apoptotic members of the Bcl-2 family (Bcl-2 and Bcl-x) are closely associated with mitochondria (Figure 7.27). Bcl-x apparently prevents the release of cy$t$ c by interacting with a voltage-dependent anion channel in the mitochondrial outer wall, causing it to remain closed (Shimizu et al., 1999). Bcl-x also associates with two other proteins to keep them inactive: a pro-apoptotic member of the Bcl-2 family (Bax) and the caspase activator (Apaf-1).

One of the more impressive displays of Bcl-2 influence on neuron survival comes from transgenic mice that overexpress this protein. These mice have much bigger brains than controls, and cell counts in the facial nucleus and retina reveal a 40% increase in neuron number (Martinou et al., 1994). However, genetic studies suggest that there is some redundancy among survival-promoting members of the Bcl-2 family: Targeted disruption of the *bcl-2* gene does *not* have an effect on neuron survival, despite its high expression in the developing nervous system (Veis et al., 1993). In contrast, disruption of a different anti-apoptotic gene, *bcl-x*, does produce a clear increase of apoptosis in the nervous system (Motoyama et al., 1995).

In dying neurons, the pro-apoptotic members of the Bcl-2 family bind to members that maintain survival (i.e., Bcl-2 and Bcl-x), and then mount an assault on the mitochondria (Figure 7.27). One subset of pro-apoptotic proteins (those containing only a single Bcl-2 homology domain, such as Bad, Bid, and Bim) recruit a second subset into play (Bax subfamilymembers such as Bax and Bak). This latter set of pro-apoptotic proteins oligomerizes and becomes associated with the mitochondrial membrane. At this point, Bax and Bak apparently interact with a voltage-dependent anion channel within the mitochondrial wall and facilitate its opening (Shimizu et al., 1999). This increases mitochondrion permeability and permits cyt c to be released. At the same time that the mitochondria are under attack, pro-apoptotic proteins are dimerizing with anti-apoptotic members and inactivating them. Bad binds to Bcl-x, and Bax binds to Bcl-2. As the caspases become activated, anti-apoptotic members of the Bcl-2 family are themselves a substrate. When Bcl-2 is cleaved by a caspase, its protective influence is obliterated (Cheng et al., 1997).

The pro-apoptotic members have a clear influence on neuron cell death in vivo. Apoptosis in ganglia and motor neurons is virtually eliminated in *bax* knockout mice, and it is significantly reduced in many areas of the CNS (White et al., 1998). The functional interaction between pro- and anti-apoptotic members is illustrated in a mouse with deletions of one or both members (Figure 7.28). Mice deficient for a *bcl-x* exhibit increased apoptosis, suggesting that pro-apoptotic proteins are no longer being suppressed. To test this idea, a double knockout mouse deficient for both *bcl-x* and the pro-apoptotic member, *bax*, was examined. When apoptosis was examined in the spinal cord of bcl-$x^{-/-}/bax^{-/-}$ mice, the level had returned to normal (Shindler et al., 1997).

One recent experiment has tied together elements of the neurotrophin withdrawal response, including elevation of pro-apoptotic Bcl-2 family members. When sympathetic neurons were transfected with virus that expresses a dominant negative form for c-Jun, the cells could survive NGF withdrawal. Without active c-Jun, there were two basic changes to the cells physiology when NGf was withdrawn. First, the expression of a pro-apoptotic Bcl-2 family member, Bim, did not increase as it normally does. Second, cyt c was not

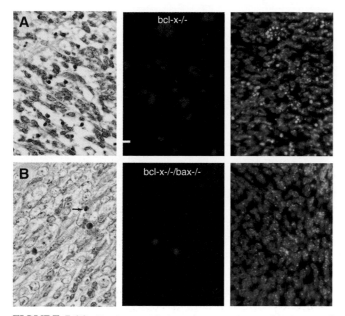

FIGURE 7.28 In vivo regulation of neuron survival by pro- and anti-apoptotic regulatory proteins. A. A spinal cord tissue section from an E12 bcl-x-/- mouse shows many pyknotic nuclei, indicative of massive cell death (left). TUNEL-positive cells (red) are common in the E12 bcl-x-/- spinal cord (middle). A nuclear stain (bright blue) indicates condensed chromatin (right). B. A spinal cord tissue section from an E12 bcl-x-/-/bax-/- double mutant mouse shows few pyknotic nuclei, indicating that cell death is curtailed (left). In contrast to the bcl-x-/- cord, there are few TUNEL positive cells in bcl-x-/-/bax-/- mice (middle). In contrast to the bcl-x-/- cord, the nuclear stain reveals few cases of condensed chromatin in bcl-x-/-/bax-/- mice (right). (Adopted from Shindler et al., 1997, by permission of Soc of Neuroscience)

released from the mitochondria. This result suggests that neurotrophins, at least, prevent cell death by suppressing the pathway that leads to pro-apoptitic protein expression (Whitfield et al., 2001).

The second major point of regulation is located at the caspases themselves. One family of regulators, called inhibitors of apoptosis proteins (IAPs), act directly on the caspases (Figure 7.26). IAPs have been identified that bind directly to caspase-3 or caspase-9, and block their proteinase activity. One striking indication that IAPs keep neurons alive comes from a clinical observation. A member of the IAP family is deleted in humans with a disorder called spinal muscular atrophy, a condition in which spinal cord motor neurons gradually die (Roy et al., 1995). There are also antagonists to the IAPs. A mitochondrial protein called Smac/DIABLO, which is also released when the outer membrane becomes permeable, can bind to the apoptosome and inhibit IAP. This promotes caspase activation and apoptosis.

Therefore, there are two safety latches on caspase activation. The first is control of cyt *c* release from mitochondria, and the second is IAP (Figure 7.27). In fact, both safety latches must be removed to kill sympathetic neurons grown in the presence of NGF. Injection of cyt *c* alone is not able to activate caspases and produce apoptosis, but if Smac is co-injected, then caspases are activated and the neurons die (Du et al., 2000; Verhagen et al., 2000; Deshmukh et al., 2002). In *Drosophila*, loss of IAP function alone leads to unrestrained activation of caspases and death. Conversely, the loss of gene products that inhibit IAP (*grim*, *reaper*, and *hid*) lead to almost no apoptosis (White et al., 1994; Hay et al., 1995).

SYNAPTIC TRANSMISSION AT THE TARGET

Some of the earliest experimental manipulations of target size suggested that functional synaptic contacts were correlated with survival. Removal of the nasal placodes in salamander embryos did not, at first, decrease the size of the innervating forebrain region. However, after the system became functional, loss of the target did produce a hypoplasia (Burr, 1916). Observations from the NMJ also suggest that functional synapses are involved in motor neuron survival. There is a very tight correlation between the onset of neuromuscular activity in chicks and the onset of normal motor neuron cell death.

If synapse activity at the target is necessary for survival, then one would predict that more neurons would die in its absence. To test this hypothesis, chick embryos were treated with an acetylcholine receptor (AChR) antagonist (curare or bungarotoxin) during a four-day period that overlapped the normal period of motor neuron death. Curare was quite effective at blocking neuromuscular transmission, as spontaneous movements were virtually eliminated for much of the treatment period. Rather than increasing cell death, the surprising result was that synapse blockade saved motor neurons (Figure 7.29). Over 90% of the motor neurons that would have died were still alive after the period of normal cell death ended and the curare had been removed. This effect is due to the interaction at the neuromuscular junction because agents that block AChRs in the CNS do not prevent cell death (Pittman and Oppenheim, 1979; Oppenheim et al., 2000). Furthermore, curare produced a threefold increase in the number of motor axon branches and synapses during the period when normal cell death occurs (Dahm and Landmesser, 1991). A similar decrease in normal cell death is found in the isthmo-optic nucleus when activity is blocked in its target, the eye, by injecting TTX during development (Péquignot and Clarke, 1992).

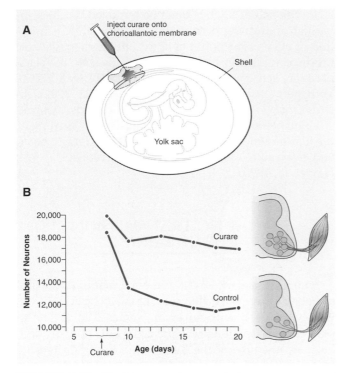

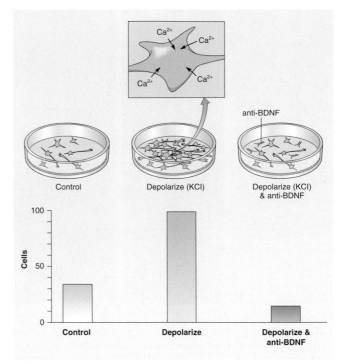

FIGURE 7.29 Blocking synaptic transmission prevents normal motor neuron cell death. A. Neuromuscular transmission can be blocked by applying curare onto the chorioallantoic membrane of chick embryos. B. In control animals, over 30% of motor neurons die after embryonic day 5. When animals are treated with curare from E6-9, the magnitude of normal cell death is greatly diminished. (Panel B adopted from Pittman and Oppenheim, 1979)

FIGURE 7.30 Electrical activity enhances the survival of embryonic cortex neurons by way of a neurotrophic signal. When the cultures are depolarized by adding KCl to the culture media, calcium enters the neurons, and the level of BDNF expression increases, leading to greater neuron survival compared to control media. The trophic influence of depolarization is eliminated by adding a function-blocking anti-BDNF antibody to the growth medium. (Adopted from Ghosh et al., 1994)

Thus, neuron survival may depend on proper access to the target-derived survival factor rather than on the total amount of factor produced by the target (Oppenheim, 1989). Access could result from a greater number of synapses. In fact, several observations suggest that synapse activity and survival factor expression are entwined with one another. Embryonic hippocampal neurons grown in dissociated culture for seven days make numerous synaptic contacts with each other, and the expression of NGF and BDNF mRNA gradually increases over this period. This neurotrophin expression is influenced by synaptic activity: NMDA receptor blockade decreases neurotrophin expression, whereas GABA$_A$ receptor blockade increases it (Zafra et al., 1991). Muscle cell expression of the neurotrophin, NT-4, is also regulated by synaptic activity, and the morphology and function of neuromuscular synapses depends on this signal (Funakoshi et al., 1995; Belluardo et al., 2001; Gonzalez et al., 1999). It is possible that activity-dependent regulation of this neurotrophin permits motor neuron synapses to obtain access to a second factor that mediates survival.

The activity-dependent expression of BDNF has been shown to support embryonic cortex neuron survival in a culture dish (Figure 7.30). BDNF expression and survival depend on the entry of calcium into the neurons when they are depolarized in a medium containing a high potassium ion concentration. When function-blocking antibodies directed against BDNF are added to these cultures, the trophic effect of depolarization is eliminated (Ghosh et al., 1994). Therefore, neurons may have some influence over the survival factors that they seek from a target: increased branching may provide better access, and synaptic transmission can regulate the amount of factor produced.

AFFERENT REGULATION OF CELL SURVIVAL

Neuron growth and differentiation occur without synaptic contacts at first, but maturation and survival quickly become dependent on neurotransmission and electrical activity. If the amount of synaptic transmission is too low during development, then postsynaptic neurons can cease protein synthesis, become atrophic, and may even die. Paradoxically, too much excitatory activation has been shown to kill neurons by loading their cytoplasm with calcium (Nicholls and Ward, 2000; Duchen, 2000). In this section, we discuss the relationship between innervation, synaptic activity, and neuron survival, and ask what trophic signal is being provided by afferent terminals.

Many of the original studies involved removing centrally projecting axons to see whether the central target developed properly in their absence. For example, Larsell (1931) removed on eye in tree frog larvae and found that its target, the contralateral optic tectum, had many fewer cells than expected. However, these studies were not able to discriminate between effects on neurogenesis or migration versus effects on neuron survival.

One of the best studied cases of afferent regulated survival is the *nucleus magnocellularis* (NM) in the chick central auditory system (Figure 7.31). Just before taking up her studies of NGF (above), Rita Levi-Montalcini had been studying the effect of cochlear nerve fibers on the survival of NM neurons and other brain stem nuclei. These studies have fascinated students of biology because they were performed with very little equipment in the countryside of Italy while World War II raged around her. In spite of these privations, Levi-Montalcini (1949) was able to show that the period of normal cell death is elevated when the cochlea is removed. Although there was little sign of degeneration at E11, the age at which auditory nerve fibers first activate NM neurons, there was a dramatic loss of cells by E21. Subsequent studies showed that about 30% of NM neurons are lost following cochlear ablation, and the effect of denervation is much reduced in adult animals (Parks, 1979; Born and Rubel, 1985). In fact, neuron survival can change from being dependent upon afferent innervation to being completely independent over the course of a few days (Figure 7.32). When the cochlea is removed in P7 gerbils, about 50% of the postsynaptic cochlear nucleus neurons are lost. However, when the cochlea is removed just two days later, at P9, there is no neuronal cell loss (Tierney et al., 1997).

Survival in other peripheral and central neurons also depends, in part, on afferent connections during development (Linden, 1994). However, surgical removal of the afferent population does not really tell us much about the trophic signal. Does the synapse provide a survival factor such as NGF? Does the neurotransmitter itself enhance survival? To address this question, intact afferent pathways were treated with

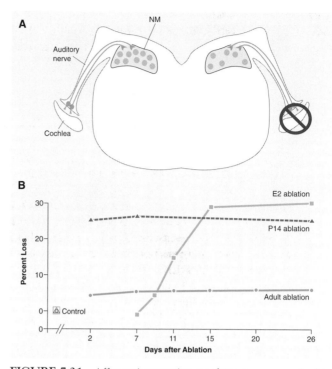

FIGURE 7.31   Afferent innervation regulates neuron survival in a chick central auditory nucleus. A. Auditory neurons from the cochlea innervate the nucleus magnocellularis (NM) in the chick auditory brain stem. The removal of a cochlea (right) completely denervates NM neurons on the ipsilateral side. B. When a cochlea is removed at embryonic day 2 (E2), about 30% of NM neurons are lost during the ensuing two weeks, although cell death does not begin until E10. When the cochlea is removed at posthatch day 14, about 25% of neurons die within two days. In adults, cochlear ablation results in the loss of only about 5% of NM neurons. (Adapted from Parks, 1979; Born and Rubel, 1985)

agents that block neuronal activity (Maderdrut et al., 1988; Born and Rubel, 1988). In the chick ciliary ganglion, cell death is increased when transmission is blocked, although neurogenesis and migration proceed normally. Similarly, action potential blockade in the cochlea for 48 hours is sufficient to increase normal cell death in the chick NM. Therefore, synaptic activity seems to play a critical role in postsynaptic neuron survival.

What is it about synaptic activity that promotes neuron survival? One possibility is that synaptic transmission provides a positive survival signal. Alternatively, synaptic transmission may evoke action potentials, and the associated voltage-gated currents may affect survival. To distinguish between these two possibilities, brain slices containing the chick NM and its auditory nerve afferents were placed in vitro and provided with two different stimulation protocols (Figure 7.33). Although the experimental period was too brief to observe dying neurons, denervation of NM

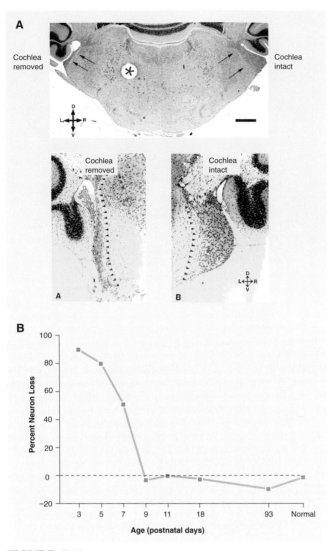

FIGURE 7.32 The survival of neurons in the gerbil cochlear nucleus depends on afferent innervation from the cochlea only through the first postnatal week. A. Cross section through the brainstem showing the ventral cochlear nucleus that was deafferented by cochlea removal (left) and the cochlear nucleus that received normal innervation (right). The cochlea was removed at P3. Shown below are higher magnification images of a deafferented (left) and control (right) cochlear nucleus from an animal in which the cochlea was removed at P5. There are very few neurons remaining on the cochlea removal side. B. The plot shows the percentage of nerve cells lost on the cochlea removal side, as compared to the control side, for different ages of surgical removal. The number of neurons lost is 50% or greater when the cochlea is removed by postnatal day 7. However, there is no cell death when the cochlea is removed after P9. (Adapted from Tierney et al., 1997).

neurons leads to a rapid decrease in protein synthesis that is thought to be a condition preceding cell death in this and many other neural systems. When the auditory nerve is stimulated, NM neurons receive synaptically evoked activity, and protein synthesis is maintained. In contrast, when NM axons are stimu-

lated to produce antidromic action potentials in their cell body, protein synthesis is not maintained (Hyson and Rubel, 1989). Thus, the preservation of postsynaptic neuron metabolism, and presumably its survival, depend on the release of something from the synaptic terminal.

Although we do not yet know what the trophic substance might be, there is some indication that neurotrophins may be released by the afferent terminal (Wang et al., 2002). The neurotransmitter itself may also play a primary role in cell survival. Auditory nerve fibers release glutamate at their synapse on NM neurons, and this transmitter produces large excitatory postsynaptic potentials. However, glutamate also acts to limit the amount of calcium that enters NM neurons by activating a metabotropic glutamate receptor (mGluR). When auditory nerve fibers are stimulated in the brain slice preparation, NM cytoplasmic calcium levels remain low, but the addition of an mGluR antagonist during continued orthodromic stimulation results in a rapid increase in calcium (Zirpel and Rubel, 1996). It is interesting that blockade of ionotropic glutamate receptors in vivo can decrease normal cell death in NM and block it entirely in nucleus laminaris (Solum et al., 1997). Thus, excitatory afferents may jeopardize survival through ionotropic glutamate receptor activation but promote survival through activation of mGluRs.

A precipitous rise in cytoplasmic calcium may dispose a cell toward death, but we have also learned that it can elicit increased BDNF production and increased survival, at least under culture conditions. How can these two findings fit with one another? One possibility is that the location of calcium entry is crucial to gauging its effect (Sattler et al., 1998). In cultured hippocampal neurons, activation of synaptic glutamate receptors, called NMDA receptors, leads to calcium entry and permits the cells to survive an insult. However, when nonsynaptic NMDA receptors are activated, again leading to calcium entry, the neurons die. Therefore, the location at which calcium enters the neuron appears to determine its effect on neuron survival (Hardingham et al., 2002).

Although it seems as if the synaptic mechanisms supporting neuron survival are quite different in the chick cochlear nucleus and in hippocampal or cortical cultures, there is one important similarity. In both types of neurons, survival depends on CREB activation. In the hippocampal cultures, synaptic NMDA receptor activity increases CREB phosphorylation, leading to transcription of BDNF. Given the primary role of CREB in mediating neurotrophin-dependent survival (Figure 7.27), it would not be surprising if synaptically mediated rises in phosphorylated CREB

Anterograde stimulation (synaptic)

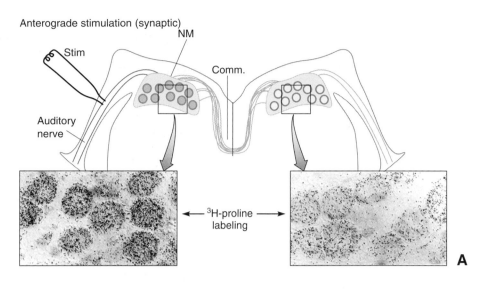

Retrograde stimulation (action potentials only)

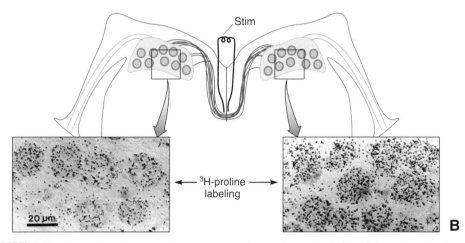

FIGURE 7.33 Synaptic activity regulates postsynaptic protein synthesis in chick auditory brainstem. A. Brain slices containing the chick nucleus *magnocellularis* (NM) and its auditory nerve afferents were incubated in an oxygenated solution containing ^{3}H-proline. Synaptic transmission in one NM was elicited by electrically stimulating the auditory nerve. When the tissue was processed for autoradiography, it was found that synaptically stimulated NM neurons incorporated far more proline (black dots) into newly synthesized proteins compared to the control side. B. When the axons of NM neurons were stimulated at the commissure, evoking a retrograde action potential in the cell body, protein synthesis was not maintained. (Adapted from Hyson and Rubel, 1989)

work through similar cytoplasmic effectors. Interestingly, the activity of nonsynaptic NMDA receptors suppresses CREB phosphorylation and increases apoptosis. In the chick cochlear nucleus, deafferentation initially results in calcium elevation, and it is this calcium that appears to be responsible for killing 30% of the neurons. However, the 70% of neurons that survive may well depend on CREB, which is phosphorylated within one hour of deafferentation. The source of calcium is from AMPA-type glutamate receptors. The survival pathway may be quite similar to that discussed for neurotrophins: Within 6 hours of deaf-

ferentation, the expression of an anti-apoptotic gene, *bcl-2*—known to be regulated by CREB—has increased (Zirpel et al., 2000; Wilkinson et al., 2002).

Although it is not yet clear how a localized calcium entry biases a neuron toward life or death, it may have something to do with the cytoplasmic localization of calcium sensors. For example, when a calcium-dependent kinase (CaMKII) becomes autophosphorylated, it becomes associated with a specific NMDA receptor subunit (Strack and Colbran, 1998).

It is possible that afferents release trophic factors, along with neurotransmitters, during development.

Neurotrophins, such as NT-3, are produced in the developing retina, and they are transported anterogradely down retinal ganglion cell axons to the optic tectum (von Bartheld et al., 1996). Since the survival of optic tectum neurons depends on both axonal transport and electrical activity by retinal axons, it seems possible that NT-3 mediates this afferent regulation (Catsicas et al., 1992). Of course, if neurotrophins provided an anterograde signal, then the target neurons would be expected to have Trk receptors at the synapse. In fact, an electron microscopic study of TrkB and TrkC receptors shows that they are located at postsynaptic profiles in the developing (and adult) central nervous system (Hafidi et al., 1996).

It is also interesting to consider that membrane depolarization, which is often found to enhance the survival of cultured neurons, actually promotes the expression of survival factor receptors. In a sympathetic neuron cell line, membrane depolarization causes the cells to produce TrkA receptor, and this allows NGF to become an effective survival factor (Birren et al., 1992). In fact, some neurons depend on sufficient levels of cytoplasmic calcium even before they become dependent on neurotrophins (Larmet et al., 1992). Thus, synapses may be employing more than one mechanism in keeping neurons alive, and these may include activation of ionotropic and metabotropic receptors, postsynaptic depolarization (and calcium homeostasis), and activation of neurotrophin receptors.

SUMMARY

Naturally occurring cell death claims up to 80% of the neuroblasts and differentiating neurons and glia in the developing brain. Depending on the particular group of neurons, survival may rely on target-derived trophic factors, synaptic activity, hormonal signals, and other cues. The diversity in survival factors is mirrored by a diversity of cytoplasmic mechanisms for dying. However, all forms of normal cell death require either the production or activation of proteins that can do damage to the neuron, such as caspases. Given the danger of keeping death machinery in place, neurons also express a broad range of regulatory proteins which ensure that apoptosis occurs only under the appropriate conditions.

Despite the wealth of candidate mechanisms that have been shown to mediate cell death in a few model systems, the process remains poorly understood in the CNS. Perhaps there are CNS trophic factors that have yet to be discovered. Alternatively, the survival of CNS neurons may be distributed among the many afferents and targets to which they connect, each one contributing a survival signal. This would make it difficult to detect the involvement of any single factor through a loss-of-function experiment.

As new trophic factors and intracellular mechanisms are put forward, we must be cautious of evidence supporting a role for each individual factor in supporting neuron survival. There are a number of interesting methods for sustaining neurons that have been removed from the body and placed in culture. However, some of these methods do not duplicate the strategy used by developing neurons to survive their natural environment, just as parachutes and life preservers are relevant to our survival only in specific situations. Conversely, the failure to observe an effect in genetically altered animals should not remove a factor from the list of candidates because CNS survival may depend on several, functionally overlapping signals.

The role of electrical activity in cell survival points out the tremendous plasticity of the developing nervous system. Small perturbations of synaptic activity can have a profound impact on the number of cells and the amount of postsynaptic membrane on which the synapses form (see Chapter 9). It is not too difficult to imagine that these mechanisms are necessary to optimize the diverse kinds of neural circuitry found within each animal.

8

Synapse Formation and Function

The formation of functional connections between nerve cells distinguishes neural development from that of all other tissues. This process begins with axonal growth cones following a set of extracellular cues to reach a precise location within a distant target (see Chapters 5 and 6). When the growth cone finally comes in contact with an appropriate postsynaptic cell, a decision is made to stop growing and to differentiate into a presynaptic terminal. Almost simultaneously, the target neuron begins to create a minute specialization that will serve as the postsynaptic site. In fact, both the growth cone and postsynaptic neuron generate many of the components needed for neurotransmission well before innervation occurs, and the formation of a functional contact can be remarkably swift.

One general problem in studying synapses at any age is that they are extremely small, often having a contact length of less than 1 μm. This makes them nearly impossible to see with a light microscope, and one might wonder how they were discovered in the first place. In fact, at the turn of the twentieth century, one group of biologists believed that neuronal processes fused with one another to produce long fibers with a continuous protoplasm, called a *syncytium* (Figure 8.1). Another group of scientists felt that neurons remained separate, as had been shown for other cell types, and that they must be in contact with one another at the tips of their processes (Ramón y Cajal, 1905). The great interest that was then focused on the tips of neuronal processes led both to the discovery of growth cones in very young tissue (see Chapter 5) and to the first descriptions of presynaptic terminals in older animals (Held, 1897). Charles Sherrington, winner of the 1932 Nobel Prize in Medicine, realized that a separation between nerve cells would allow for a new form of intercellular communication (cf. chemical transmission), and he popularized the term *synapse* (Sherrington, 1906).

The average mammalian neuron receives synapses along its soma and dendrites, Some of these synapses release glutamate, which excites the postsynaptic neuron; others release GABA, which acts to inhibit the neuron; and still others release various modulatory transmitters. At a single glutamatergic synapse, there may be several types of receptors; some will gate open ion channels while others can activate a second messenger system. The change in membrane potential brought about by this synaptic transmission quickly recruits nearby ion channels (see Box: Maturation of Electrical Properties). In the case of voltage-gated sodium channels, this leads to an action potential.

This sanitized description of synaptic organization highlights many of the challenges to a developing nervous system. For example, the specific receptors for GABA must be placed in the correct patch of postsynaptic membrane. At the same time, each growth cone must identify an appropriate patch of membrane on which to differentiate. In the cortex, most glutamatergic synapses are located on postsynaptic specializations, called *spines*; GABAergic synapses tend to form on the cell body and proximal dendrite. A tight little cluster of $GABA_A$ receptors on a dendritic spine head would be of little use to the glutamate-releasing terminals that are located there. There must also be a mechanism to control the total number of synapses that can form on any one neuron. That value can vary tremendously, from over 10,000 for a cortical pyramidal neuron to only a single excitatory synapse on neurons of the medial nucleus of the trapezoid body.

To make some sense of this complexity, we will consider separately how presynaptic terminals and postsynaptic specializations arise. Three general observations

MATURATION OF ELECTRICAL PROPERTIES

Information processing by the central nervous is based on electrical signals. The developmental regulation of each neuron's resting potential and voltage-gated ion channels is essential for the emergence of adult function. The resting membrane potential becomes more negative during development (Kullberg et al., 1977; Burgard and Hablitz, 1993; Tepper and Trent, 1993; Sanes, 1993; Ramoa and McCormick, 1994; Warren and Jones, 1997). This is due to regulation of extracellular K$^+$ by glial cells which are proliferating and differentiating throughout the brain (Connors et al., 1982; Skoff et al., 1976; Syková et al., 1992). For example, extracellular K$^+$ drops from about 35 mM in the cortex of newborn rabbits to 3 mM in adults (Mutani et al., 1974). This difference translates into a shift of almost 35 mV in membrane potential.

A few simple properties determine the size and speed of electrical events (Figure 8.A). The first, membrane input resistance, determines how much the membrane potential will change for a given current pulse. The second, membrane time constant, determines how rapidly the membrane will reach a new potential when current is injected. Both of these properties tend to decrease with age, probably reflecting an increase in cell size (i.e., total membrane). Thus, input resistance decreases because the number of resistors (i.e., channels) increase as membrane is added to a cell. For example, the potassium channel-blocker, cesium, has the greatest effect on input resistance and time constant just when these values are decreasing in developing motor neurons (Cameron et al., 2000). Thus, the mature neuron is able to process information rapidly and accurately because the synaptic currents elicit brief changes in membrane potential.

The Action Potential: Sodium and Potasssium Channels

When a neuron becomes slightly depolarized, perhaps owing to a synaptic potential, the opening of voltage-gated sodium channels permits a large depolarizing current due to the relatively high extracellular sodium concentration. As the neuron depolarizes, a second set of voltage-gated channels are activated that permit potassium to leave the cell, thus returning the membrane potential to rest. In many cases, the initial depolarization recruits a third type of voltage-gated channel that permits calcium to enter the neuron. When do these channels appear during development?

In many developing systems, the action potential is first carried by calcium ions. Since the calcium channels tend to remain open for a longer time, the action potentials can be very slow. Thus, *Xenopus* neurons begin life with 60–90 ms action potentials, although they quickly decrease to about 1 msec in duration (Figure 8.B). Two basic changes explain this decrease. First, sodium channels become the primary conduit for inward current (Spitzer and Lamborghini, 1976; Baccaglini and Spitzer, 1977). Second, there is a 3.5-fold increase in a potassium channel current, called the *delayed rectifier*, that is activated during membrane depolarization (Barish, 1986). The maturation of this large outward current brings the

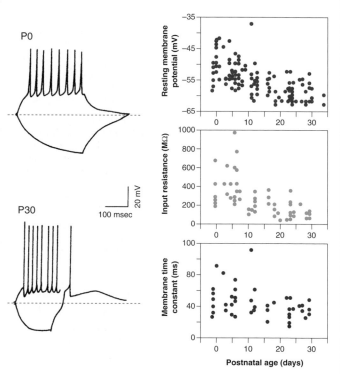

FIGURE 8.A   Development of passive membrane properties. (Left) The intracellularly recorded voltage response to positive and negative current pulses in a P0 and a P30 neuron from ferret lateral geniculate nucleus (LGN) brain slices. The P0 neuron displayed a longer time constant and larger voltage deflection. (Right) Plots of membrane potential, input resistance, and time constant from LGN neurons during postnatal development. Membrane potential becomes about 10 mV more negative, input resistance decreases by about 200 MΩ, and the time constant decreases by about 10 ms. (From Ramoa and McCormick, 1994)

BOX (cont'd)

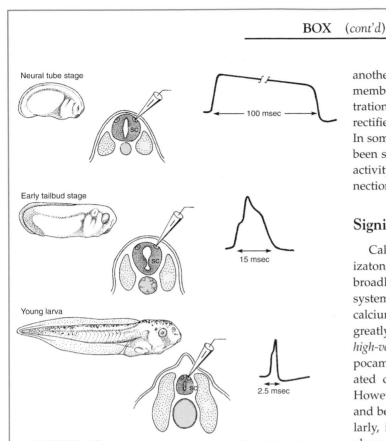

Neural tube stage

sc

100 msec

Early tailbud stage

sc

15 msec

Young larva

sc

2.5 msec

FIGURE 8.B Action potentials are initally calcium-dependent. (Top) When intracellular recordings were made from spinal cord (sc) Rohon-Beard neurons in neural tube stage *Xenopus* embryos, depolarizing current injection produces long-lasting calcium action potentials. (Middle) In early tailbud embryos, current injection evokes a mixed sodium/calcium response. (Bottom) In the young larva, current evokes a brief sodium-dependent action potential. (Adopted from Spitzer, 1981)

cell back to rest and limits the amount of calcium that enters during an action potential.

The sodium and potassium currents, as measured in dissociated *Xenopus* spinal neurons, increase dramatically within about 24 hours of their terminal mitosis (O'Dowd et al., 1988). Similar observations have been made in explants of chick cortex (Mori-Okamoto et al., 1983). In acutely dissociated rat cortical neurons, the sodium current density increases six-fold during the first two postnatal weeks (Huguenard et al., 1988). However, there is no uniform order of channel appearance in the nervous system. Chick motor neurons have significant sodium and delayed rectifying potassium currents from the outset, and there is a relatively late appearance of at least one type of potassium channel, and two types of calcium channel (McCobb et al., 1989; McCobb et al., 1990). Yet

another potassium channel which depends upon both membrane potential *and* the intracellular calcium concentration to open are generally expressed after the delayed rectifier (O'Dowd et al., 1988; Dourado and Dryer, 1992). In some systems, the appearance of specific channels has been suggested to play a role in generating spontaneous activity that is essential for maturation of synaptic connections (Vasilyev and Barish, 2002; Picken et al., 2003).

Significance of Calcium Channel Expression

Calcium currents that are activated by small depolarizatons, called *low-voltage activated (LVA)* or *T currents*, are broadly expressed in developing tissue. As the nervous system matures, there is an increasing prominence of calcium channels that activate only when the cell is greatly depolarized (Figure 8.C). These are referred to as *high-voltage activated (HVA)* or *N and L currents*. When hippocampal neurons from E19 rats are placed in a dissociated culture, only LVA currents are recorded at first. However, HVA currents appear over the next few days and become a major contributor (Yaari et al., 1987). Similarly, it is the LVA calcium currents that are primarily observed when neurons from chick dorsal root ganglia, ciliary ganglia, or ventral horn are first recorded from (Gottmann et al., 1988; McCobb et al., 1989). These are overtaken by HVA currents within about 24–48 hours.

The initial appearance of LVA calcium channels can contribute greatly to a neuron differentiation. For example, spontaneous calcium transients in developing *Xenopus* spinal neurons, largely carried by LVA calcium channels, have been implicated in the acquisition of GABAergic phenotype and process outgrowth (Spitzer, 1994). In fact, these calcium transients regulate the maturation of electrical properties, including a switch in potassium channel isoforms. The rate of activation for single potassium channels also increases by two to three times as *Xenopus* spinal neurons mature in vitro. This transition in channel kinetics is dependent upon calcium influx and can be induced by activation of a protein kinase C (Desarmenien and Spitzer, 1991).

Regulation of Ionic Channel Expression

The addition of new channels to the membrane is necessary for most increases in current density. For example, when *Xenopus* neurons were grown in the presence of RNA or protein synthesis inhibitors, the transition from

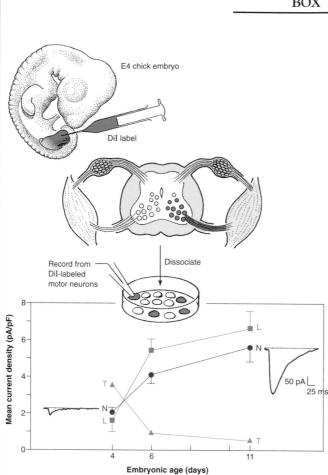

FIGURE 8.C Two calcium currents increase while a third declines in developing spinal motor neurons. A. To obtain identified motorneurons, a dye (DiI) was injected into the leg bud, and this dye was retrogradely transported by motor neurons. Thus, when the tissue was dissociated, it was possible to identify motor neurons because they carried the DiI label. B. When calcium currents were recorded from the dissociated motor neurons, it was found that T-type calcium channels declined with age, while N- and L-type channels increased. (Adopted from McCobb et al., 1989)

calcium- to sodium-dependent action potentials is prevented (Blair, 1983; O'Dowd, 1983). In like manner, transcription blockers prevent the normal increase in potassium current density (Ribera and Spitzer, 1989). The signal to increase production of potassium channels is present for only a brief period of time. A 9-hour exposure to an RNA synthesis inhibitor prevents the normal increase in potassium current density, even though RNA synthesis resumes upon withdrawal of the inhibitor. However, the appearance of A currents are not permanently blocked by transcription inhibitors.

If transsynaptic signals regulate ion channel maturation, they remain largely unknown. However, the glycoprotein that stimulates AChR synthesis in muscle cells, neuregulin, can induce a twofold increase in sodium channels (Corfas and Fischbach, 1993). Other well-described growth factors, such as FGF, can upregulate the density of calcium channels in dissociated cultures of hippocampal neurons, and the effect requires protein synthesis (Shitaka et al., 1996). Electrical activity itself may affect the expression level of certain channels. Action potential blockade delays or prevents the normal increase in sodium and potassium current density in *Xenopus* myocytes in vitro (Linsdell and Moody, 1995).

The extrinsic signals that regulate ion channel expression are beginning to be understood in parasympathetic neurons of the chick ciliary ganglion. The expression of an A-type (I_A) and a calcium-activated potassium current ($I_{K[Ca]}$) is reduced when ciliary neurons are grown in dissociated culture, in the absence of their target or preganglionic afferents (Duorodo and Dryer, 1992). To determine whether synaptic connectivity influences potassium channel expression, in vivo manipulations were performed in which either the optic vesicle containing the target tissue was removed or a portion of the midbrain primordium containing the preganglionic nucleus was removed (Duorodo et al., 1994). The neurons were then acutely dissociated so that whole-cell voltage-clamp recordings could be obtained easily. The density of I_A was unaffected by either manipulation, although the channels did appear to open and close more rapidly than normal. In contrast, $I_{K[Ca]}$ was reduced by 90 to 100% following either target removal or deafferentation. Blocking the spontaneous activity of chick lumbar motor neurons does lead to a dramatic reduction in I_A expression, both in ovo and in vitro (Casavant et al., 2004).

A factor has now been isolated from a target of the ciliary ganglion, the iris, that is able to upregulate the density of $I_{K[Ca]}$ (Subramony et al, 1996). When neurons are cultured in the presence of iris extract, the density of $I_{K[Ca]}$ reaches normal levels within 7 hours (Figure 8.D). This factor turns out to be a TGFβ₁. When an antibody directed against the TGFβ family is added to iris extract or injected into the eye, the expression of $I_{K[Ca]}$ is inhibited (Cameron et al., 1998). Interestingly, transcripts of the calcium-activated potassium channel are present in cultured ganglia before the current can be recorded, and the effect of iris extract does not require protein synthesis. However, brief exposure to TGFβ₁ also elicits a persistent increase

BOX *(cont'd)*

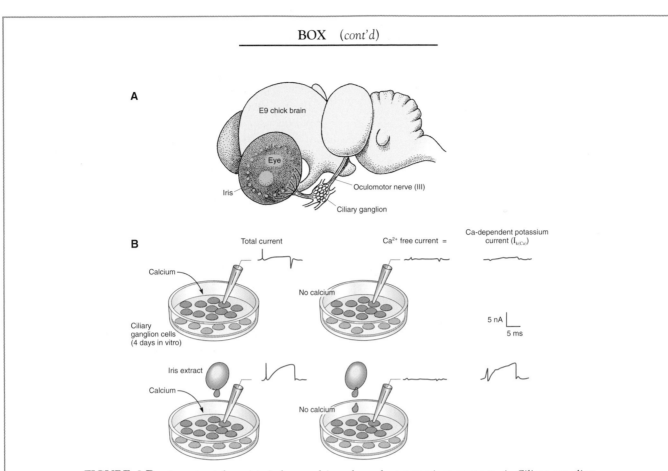

FIGURE 8.D An extract from iris induces calcium-dependent potassium currents. A. Ciliary ganglion neurons, which innervate the iris, were isolated from E9 chicks and placed in dissociated culture. B. When the cultures are grown in control culture medium, one can record very little calcium-dependent potassium current ($I_{K(Ca)}$). This is shown by recording total current and subtracting the current in Ca^{++}-free media. When an extract from the iris was added to the cultured neurons, a much larger $I_{K(Ca)}$ is subsequently recorded. (Adapted from Subramony et al., 1996)

in $I_{K[Ca]}$ that depends on transcription (Lhuillier and Dryer, 2000). These results suggest that retrograde signals can affect the translation, insertion or modification of potassium channels with a very short latency. The regulation of $I_{K[Ca]}$, and many other channels, is likely to depend on several signals. For example, there is a second isoform of TGFβ that inhibits the functional expression of $I_{K[Ca]}$, and an afferent-derived signal (neuregulin-1) participates in upregulating this channel (Cameron et al., 1999, 2001).

emerge from this section. First, neurons manufacture many of the synaptic building blocks even before making contact with one another. Second, intercellular signaling induces the differentiation of newly formed synapses from the moment of contact. Signals from glia, extracellular matrix, and neighboring neurons all participate in synaptogenesis. Third, synapses do not function in a mature manner for quite some time after they are fabricated. We will discuss how their functional properties change with development and how this might explain some of the behavioral limitations that young animals display (see Chapter 10).

WHAT DO NEWLY FORMED SYNAPSES LOOK LIKE?

Studies of the synapse began in earnest during the early 1950s with the arrival of two new techniques.

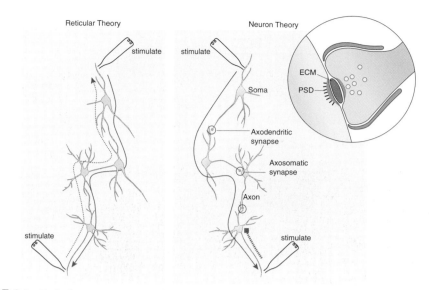

FIGURE 8.1   Reticular versus neuron theory. Over a century ago, the nervous sytem was thought to be a syncytium (left) of cells that were joined together by their processes. This arrangement would permit electrical activity to travel through the syncytium (arrows) in either direction upon stimulation. As evidence mounted that neurons were separate cells (right), it was recognized that a chemical synapse (inset) would permit electrical activity to travel in only one direction. ECM, extracellular matrix; PSD, postsynaptic density.

Electron microscopy permitted neuroscientists to see the complex structure of synaptic contacts for the first time. Intracellular recordings allowed one to observe their electrical behavior (Palade and Palay, 1954; Fatt and Katz, 1951). Together, these techniques established the benchmarks by which we determine whether two nerve cells are, in fact, connected to one another.

For the sake of simplicity, we begin with an anatomical description of synaptogenesis; the molecular and physiological transformations that accompany them are considered below. At the ultrastructural level, there are three clear signs of a synaptic specialization: small vesicles accumulate at the presynaptic membrane, a narrow cleft filled with extracellular matrix is found between pre- and postsynaptic membranes, and the postsynaptic membrane appears thickened owing to the accumulation of membrane associated proteins and cytoskeletal elements, called the postsynaptic density (PSD). In contrast, newly formed synapses have few, if any, vesicles in the presynaptic terminal profile (Figure 8.2). In the rodent cortex, the average number of vesicles found in a synaptic profile increases almost threefold during the first postnatal month (Dyson and Jones, 1980). A second characteristic of newly formed synapses is the close apposition of pre- and postsynaptic membranes, referred to as a tight junction (Figures 8.2 and 8.3). Finally, the postsynaptic membrane does not yet display a PSD. Therefore, newly formed synapses do not display any of the adult anatomical features, making them almost unrecognizable.

Even the highest power electron microscope cannot detect the onset of synaptogenesis between a growth cone and postsynaptic cell (Vaughn, 1989). There is simply not much to be seen. More importantly, the morphology cannot tell us how the synapse is working. To get around these problems, many scientists have turned to the tissue culture technique where it is possible to watch cells come into contact with one another, and monitor synapse morphology and function from moment to moment. The earliest tissue culture studies demonstrated that mature synapses could form in isolated pieces of neural tissue, but the temporal resolution was comparable to the best in vivo studies. When it became possible to observe the growth cone approaching a postsynaptic target neuron, then observations were first made close to the onset of synaptogenesis.

One of the first in vitro systems consisted of a piece of fetal rat spinal cord plated next to dissociated neurons from the superior cervical ganglion, a target of autonomic motor neurons (Rees et al., 1976). Within the first several hours of contact, there are only subtle changes in morphology to indicate that synapse formation is underway (Figure 8.3). Of course, this is precisely why in vivo observations could not detect the very onset of synapse formation. At first, the growth cone loses its filopodia and forms a punctate contact that is unusually close to the postsynaptic cell membrane (about 7 nm, less than the diameter of a hemoglobin molecule). This suggests that an adhesive interaction may be involved

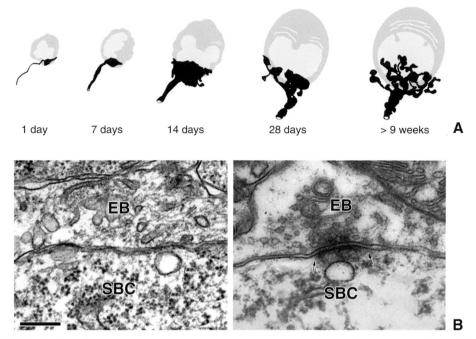

FIGURE 8.2 Development of synapse morphology. A. The contact of an auditory nerve endbulb (EB) onto its postsynaptic target, the spherical bushy cell (SBC), is shown at several ages. The end bulb initially forms a small contact on the spherical bushy cell in newborn mice. In adults, the endbulb forms an extremely large ending on the spherical busy cell. B. When contacts from newborn animals are examined at the ultrastructural level, there is little evidence of synapse formation. (Right) In adults, these contacts display many signs of mature synapses, including presynaptic vesicles and a postsynaptic membrane density. (Limb and Ryugo, 2000)

in the initial stages of synaptogenesis, as discussed below. There are also many examples of presynaptic protrusions being engulfed by muscle membrane, termed a *coated pits*. These observations show an intense interaction at the initial site of contact. The first sign of differentiation is found below the postsynaptic membrane, where the Golgi apparatus accumulates and coated vesicles proliferate, both of which probably contribute to the construction of the postsynaptic density. As for the presynaptic terminal, it is only after about 24 hours of contact that vesicles begin to accumulate at the site of contact. Thus, the *structure* of a synapse appears to mature over a relatively long period in these in vitro studies, and the postsynaptic cell is the first to display any sign of differentiation.

The location of synapse formation is extremely important to the operation of the nervous system. Even at the nerve–muscle junction, motor synapses form at distinct central locations on the myofibers. On the typical central neuron there are many more options. Synapses that form near the soma are thought to have a greater voice in deciding whether the neuron will fire an action potential. Thus, inhibitory synapses are often found nestled up around the cell body so that they can halt activity efficiently. In contrast, many synapses form on dendritic spines where their activity

provides tiny potentials that must be summed together to produce a significant change in the neuron's membrane potential. When many different types of afferents synapse on a postsynaptic neuron, each with a distinct functional role, then the problem becomes quite difficult indeed. Does synaptogenesis proceed in any particular sequence, and how does each synapse know where to form on the postsynaptic cell?

Early observations from Golgi stained spinal cord material showed axonal growth cones and dendritic growth cones seemingly reaching for one another, suggesting that axodendritic synapses result from an early trophic interaction. The electron microscopist has been able to show where synapses are added because both the presynaptic terminal and the postsynaptic location (e.g., soma, dendrite, spine) are identifiable at high magnification. In many systems, including the NMJ, the spinal cord, the hippocampus, and the cortex, contacts seem to form initially on postsynaptic processes. For example, muscle cells extend tiny processes, called *myopodia*, just before motor terminals arrive, and these postsynaptic processes are the prefered site of contact (Ritzenthaler et al., 2000). In the embryonic mouse spinal cord, nearly 75% of axodendritic synapses are found on dendritic growth cones (Vaughn et al., 1974). Even in the cortex and hippocampus, axodendritic

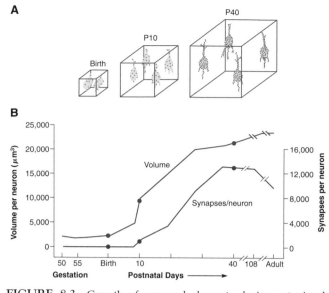

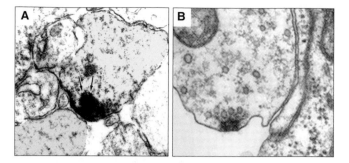

FIGURE 8.4 Pre- and postsynaptic differentiation without a partner. A. An electron micrograph showing clustering of α2 adrenergic receptor (arrows) in a postnatal day 4 rat visual cortex neuron. Both of the red-tinted structures are dendrites. B. An electron micrograph showing an apparent presynaptic terminal adjacent to hemolymph. This terminal is made by a *Drosophila* motoneuron in a mutant strain, twisted (*twi*), that does not generate postsynaptic muscle cells. (Panel A from C. Aoki, unpublished observations; Panel B adapted from Prokop et al., 1996)

FIGURE 8.3 Growth of neuronal elements during cat visual cortex development. A. From birth until postnatal day 40 (P40), the density of neuronal cell bodies decreases as gliogenesis and angiogenesis occurs. During this same period, dendritic arbors are expanding, and synaptic terminals (black dots) are accumulating on the postsynaptic membrane. B. The total volume of visual cortex occupied by each neuron increases by almost 10-fold during the first postnatal month. When neuron packing density is taken into consideration, the accumulation of synapses can be expressed as synapses per neuron. As shown in the graph, there is a dramatic increase in synapses from P10 to P30, and significant decline after P108 (adapted from Cragg, 1975).

synapses are present in newborn tissue, while few axosomatic synapses are found until two to three weeks later (Pappas and Purpura, 1964; Schwartz et al, 1968). In fact, when a postsynaptic marker protein (PSD-95) is visualized in dendrites as they grow within the zebrafish midbrain, the new postsynaptic sites appear first on dendritic filopodia (Niell et al., 2004). Therefore, it is the most recently generated postsynaptic structures that are first contacted by axonal growth cones (Fiala et al., 1998).

It is also possible that some membrane compartments are not available for contact. For example, excitatory connections to an auditory brainstem nucleus called the *medial superior olive* (MSO) are at first restricted to the dendritic regions of the cell. At this stage, MSO cell bodies are completely surrounded by glial membrane. As the glial membrane regresses, synapses are formed on the MSO cell bodies (Brunso-Bechtold et al., 1992). In this regard, it is interesting that elimination of a putative cell adhesion molecule at the neuromuscular junction (cf, s-laminin) permits glial processes to invade the synaptic region and impeed synapse maturation (Noakes et al., 1995).

Therefore, glial cells may serve as gatekeepers, determining when and where synapses can be formed.

During the time when synapses are forming between nerve cells, it is quite common to see pre- and postsynaptic structures all by themselves: essentially, synapses to nowhere. From the amphibian spinal cord to the rodent olfactory cortex, presynaptic-like structures with an accumulation of vesicles apparently develop in the absence of a postsynaptic cell. Similarly, postsynaptic densities that are not in contact with a presynaptic terminal have been found in the olfactory bulb and cortex (Figure 8.4A). These lonesome structures indicate that growth cones and dendrites are poised on the brink of differentiating into synaptic specializations (Hayes and Roberts, 1973; Newman-Gage et al., 1987; Westrum, 1975; Hinds and Hinds, 1976).

Presynaptic terminal differentiation has even been found to occur on nonneuronal cells. Transient presynaptic-like terminals have been identified on glial cells during axon ingrowth, particularly in areas without dendritic processes (Hendrikson and Vaughn, 1974). In a *Drosophila* mutant that has no mesoderm, and therefore no muscle, motor axons continue to form presynaptic-like profiles on glia and other cells (Figure 8.4B). However, some cells appear to be crucial for synapse differentiation. Ablation of selected embryonic muscle precursors in *Caenorhabditis* results in gaps in a set of dorsal muscles and prevents presynaptic varicosity formation (Plunkett et al., 1996). Thus, growing neuronal processes can display a synaptic morphology with little encouragement from its appropriate partner.

Ultrastructural studies also provide the best description of the time period when synapses are added in the peripheral and central nervous system.

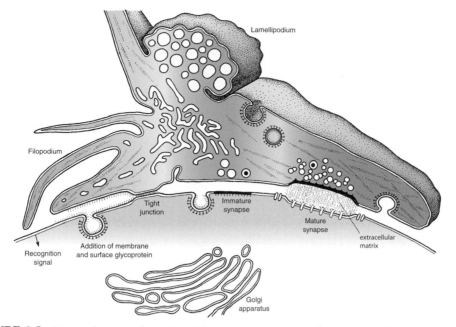

FIGURE 8.5 Stages of synapse formation. When a presynaptic growth cone comes into contact with the postsynaptic membrane, its filipodia retract (left), and the membranes become tightly apposed to one another. Vesicles are found in both pre- and postsynaptic elements, possibly to add membrane and extracellular glycoproteins. The immature synapse may display a few vesicles presynaptically and a small postsynaptic density (center). The mature synapse (right) exhibits an accumulation of presynaptic vesicles, a dense extracellular matrix in the cleft, and a postsynaptic density. (Adapted from Rees, 1978)

Bursts of synapse formation are found throughout the nervous system, but the timing and duration varies greatly (Vaughn, 1989). In the mouse olfactory bulb, synaptic profiles can first be recognized in electron micrographs at embryonic (E) day 14. The total number of synaptic profiles increases exponentially through the first postnatal week and then continues to increase at a lower rate over the next several weeks (Figure 8.5). Therefore, new synaptic contacts continue to be manufactured over a long time period after axons invade their target. One reason for this extended period of synaptogenesis is that dendrites are still growing, and the addition of postsynaptic membrane may attract new contacts. It is also likely that certain afferent projections may arborize at different times. In the rat visual cortex, where the synaptic profiles of excitatory and inhibitory synapses can be recognized, their increase in number occurs at different times. Other areas display a steady increase in synapse number, such as the the rat superior cervical ganglion, where the process occurs gradually from innervation at E14 to over one month after birth (Smolen, 1981).

An important caution is that neither anatomy nor physiology alone is sufficient to identify a developing synapse. Purely anatomical measures of synapse formation can be misleading because synaptic physiology can develop rapidly (see below), with little evidence of specialized morphology. On the other hand, an exclusively functional assay of synapse formation may create problems because there is evidence that "silent" synapses exist in the CNS, which nonetheless display normal structure. Therefore, a precise chronology of synapse addition is still missing for most regions of the CNS.

THE FIRST SIGNS OF SYNAPSE FUNCTION

At the moment a growth cone comes in contact with its postsynaptic target, it begins a metamorphosis, transforming from a spindly pathfinding organelle to a bulbous presynaptic terminal. One surprise is that the growth cone comes equipped with a rudimentary transmitter-releasing mechanism (Young and Poo, 1983; Hume et al., 1983). This was first demonstrated in a primary culture of *Xenopus* spinal neurons and myocytes. During the first two days of culture, spinal neurons produce growth cones, extend neurites, and form functional cholinergic synapses with neighboring muscle cells. To detect the release of ACh, a special type of recording electrode is manufactured (Figure 8.6). The electrode has an excised piece of muscle cell membrane at its tip, and this membrane contains ACh

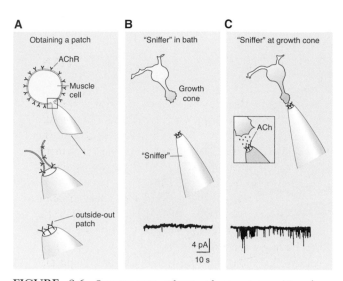

FIGURE 8.6 Spontaneous release of neurotransmitter from growth cones. A. A biological sensor for ACh (a "sniffer") was created by excising a patch of membrane from a muscle cell with a recording pipet. The membrane contained AChRs that were facing outward. B. Recording of ACh-evoked currents (downward deflections) when the "sniffer" pipet was distant from the growth cone, and C. when it was within a few microns of the growth cone. The increased activity indicates that the growth cone was releasing ACh. (From Young and Poo, 1983)

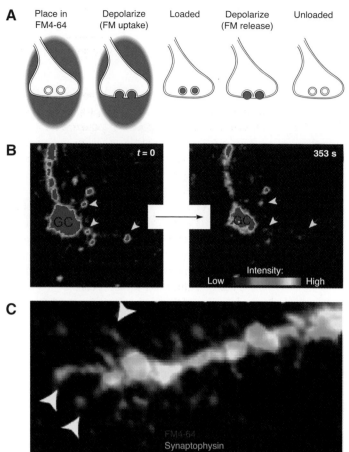

FIGURE 8.7 A presynaptic vesicular release mechanism is present in growth cone filopodia. A. A schematic of the FM dye technique. The dye is taken up into vesicles as they fuse with the membrane. When the terminal is depolarized again, the vesicles fuse with the membrane again and the terminal is unloaded of dye. B. Release of FM4–64 from growth cone filopodia in response to depolarization (yellow arrows). C. Co-localization of FM4–64 (red) and the synaptic vesicle protein, synaptophysin (green), in growth cone filopodia. (Panels B and C from Sabo and McAllister, 2003)

receptors (see BOX: Biophysics: Nuts and Bolts of Functional Maturation). These electrodes are brought within a few microns of a growth cone before it has contacted a myocyte. If the growth releases ACh, then the ACh receptor-coupled channels open, and current flows across the electrode.

The release of transmitter is probably a general property of all growth cones. For example, a different neurotransmitter (GABA) is released from growth cones of mammalian CNS neurons (Gao and van den Pol, 2000). Growth cones are also able to release transmitter in response to electrical stimulation of their cell bodies. Therefore, some of the presynaptic neurotransmission machinery is present even before synaptogenesis occurs, albeit in an immature form. It is not yet known whether growth cones release transmitter via the fusion of synaptic vesicles, as do mature synapses. To demonstrate that vesicular release can occur from growth cone filopodia (see Chapter 5), an optical technique was employed. As shown in Figure 8.7A, neurons are incubated in the presence of a fluorescent dye (FM4-64) that does not cross the cell membrane. However, when the neuron is depolarized, the dye enters vesicles that have fused transiently with the membrane. Vesicles that are loaded with fluorescent dye in this manner will then release the dye when they next fuse with the membrane; that is, when the mem-

brane is next depolarized. In fact, the growth cone filopodia of cortex neurons can incorporate and release dye in response to depolarization, suggesting that a vesicular mechanism is present (Figure 8.7B). Several synaptic vesicle proteins are co-localized within growth cone filopodia (Figure 8.7C), raising the possibility that neurotransmitter release occurs at the moment of contact (Sabo and McAllister, 2003).

In some systems, particularly in the peripheral nervous system or the NMJ, the timing of axon ingrowth and synaptic transmission has been followed with great precision in vivo. In the rat superior cervical ganglion, axons first enter the target between embryonic (E) days 12 and 13, and afferent-evoked postsynaptic potential are recorded by E13. Similarly,

BOX

BIOPHYSICS: NUTS AND BOLTS OF FUNCTIONAL MATURATION

The study of neural tissue development is unique because the cells possess diverse electrical properties. These properties result from two essential components. First, the neuron must produce batteries by selectively pumping ions from one side of the membrane to the other. Second, the neuron must produce switches, commonly referred to as voltage- and ligand-gated channels, that allow the batteries to discharge (i.e., ionic currents flow due to an electrochemical gradient) across the membrane. To determine how pumps and channels operate, one must be able to record from a single neuron (or a portion of it), and to control the neurons environment. The most important parameters that must be controlled include ionic composition, voltage across the membrane, and the presence of ligands. The technical challenges presented by these requirements have largely been overcome in the past two decades, providing some fundamental discoveries about developing neurons.

To study the voltage-gated channels, one must be able to move the membrane potential to different holding voltages (voltage-clamp), and then observe whether current flows across the membrane. Thus, if one depolarizes an axon, voltage-gated sodium channels will open at some criterion voltage, termed *threshold*, and Na^+ will enter the cell (i.e., inward current). A novel set of recording techniques, called *patch-clamping*, was introduced to fully characterize different types of channels. Patch-clamp electrodes can form high-resistance seals ("giga-seals") with small areas of membrane, and these patches of membrane can then be excised from the cell (Hamill et al., 1981). This approach has several advantages. Small patches of membrane often contain single channels, they are relatively easy to voltage-clamp, and either side of the membrane may be exposed to the defined media. These techniques allow one to determine a channel's characteristic properties: the voltage at which activation and inactivation occur, the mean channel open time, the mean current amplitude, the relative permeability to different ions, and the pharmacological profile. Finally, when the excised patches of tissue contain a known class of neurotransmitter receptors, then the recording pipet may be used to detect the release of neurotransmitter ("sniffer pipets"). This approach has led to the discovery that growth cones release transmitter (Young and Poo, 1983; Hume et al., 1983).

It is also possible to form a giga-seal with the neuron of interest, and then rupture the membrane, forming a whole-cell recording configuration. Although this technique is qualitatively similar to a standard sharp electrode intracellular recording, there are added benefits. The tip of the recording electrode is much larger than that of the sharp electrode, both improving the signal-to-noise ratio and allowing for relatively large current injections. The large tip diameter translates into a large hole in the membrane through which the patch pipet solution travels quite easily, allowing the intracellular composition to be controlled within a matter of minutes. In a more elegant form of this technique, a perfusion system is added to the recording pipet so that intracellular composition can be altered during a recording session (Chen et al., 1990).

Although the patch-clamp techniques offer rigid biophysical measures, they seldom allow one to evaluate the movement of a single type of ion. One common strategy requires the use of several antagonists to block the contribution of contaminating ions (e.g., magnesium ions block the flow of calcium). A second approach makes use of a novel group of electrodes, each of which is responsive to changes in the concentration of a specific ion, such as potassium (Syková, 1992). The tips of these electrodes are filled with a liquid membrane that is selectively permeable to one species of ion, so that local changes in concentration result in the net movement of that ion across the membrane, resulting in a detectable potential difference. When employed in the central nervous system, these electrodes reveal substantial developmental changes in the regulation of extracellular potassium and pH (Connors et al., 1982; Davis et al., 1987; Jendelova and Syková, 1991).

The fields of electrophysiology and image processing have found a productive relationship in the area of membrane channels. The introduction of ion-sensitive fluorescent dyes has provided a noninvasive means of assessing functional properties, while providing a high degree of spatial resolution. Each of these dyes emits light at a specific wavelength when activated with a beam of light at a different exciting wavelength. The amount of emitted light is proportional to the free concentration of a specific species of ion. That is because a dye's absorption or emission properties is altered when it binds to the ion. Selective indicator dyes now exist for a wide range of ions including Na^+, Ca^{2+}, Cl^-, and H^+. The indicator dye, fluo-3, has been used to demonstrate an elevation of Ca^{2+} immediately following contact between growth cone and target cell (Dai and Peng, 1993). A novel variation of this technology makes use of compounds that exist in a "caged" configuration and that only become activated when exposed to light of a specific wavelength. In this manner, one may elevate the concentration of a specific substance with great temporal and spatial resolution.

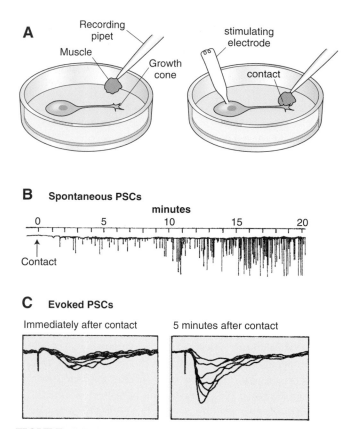

FIGURE 8.8 Muscle cell contact enhances spontaneous and evoked transmission. A. Cultures of *Xenopus* spinal neurons were grown in culture, and whole cell pipets were used to record from round muscle cells and to manipulate them into contact with the neuron. B. A continuous recording from a muscle cell shows spontaneous transmission (downward deflections) during the first 20 minutes after contact. C. Nerve evoked postsynaptic currents (downward deflections) increased in amplitude from the moment of contact to 5 minutes later. (Adapted from Evers et al., 1989; Xie and Poo, 1986)

motor axons grow out of the *Xenopus* spinal cord and form functional synapses on the developing myotubes over a period of hours (Kullberg et al., 1977). In the fruit fly, it takes only eight hours for neuromuscular transmission to reach a mature level of function (Broadie and Bate, 1993a). However, it is nearly impossible to record from a cell at the exact moment that it is first contacted by a growth cone in vivo.

Fortunately, the appearance of synaptic transmission can be explored with great accuracy in dissociated cultures. When intracellular recordings were obtained from isolated *Xenopus* muscle cells, and the formation of a neurite contact was visually monitored on a microscope, it was found that synaptic potentials could be elicited within minutes of lamellopodial contact (Kidokoro and Yeh, 1982). To provide even better temporal resolution, muscle cells were manipulated into contact with growing neurites while they were being

recorded from (Figure 8.8A). The tight seal between the large tip of a whole-cell recording electrode and the muscle membrane (see Box: Biophysics: The Nuts and Bolts of Functional Maturation) permits the recordings to continue while a small round muscle cell, called a *myoball*, is detached from the substrate and repositioned in the culture dish. By using this technique, it is possible to observe spontaneous synaptic events within seconds of contact (Figure 8.8B), and they continue to increase in both rate and amplitude over the first 10 to 20 minutes (Xie and Poo, 1986). Nerve-evoked synaptic transmission that is great enough to elicit an action potential can be found within 15s of nerve–muscle contact. However, in most cases, evoked synaptic responses continue to increase during the first 15 minutes of contact (Figure 8.8C; Sun and Poo, 1987; Evers et al., 1989). Certain adult-like characteristics of synaptic transmission, such as depression and facilitation, are also present immediately after contact in the neuromuscular system. Clearly, functional maturation proceeds briskly at the NMJ in vitro. However, most analyses of the mammalian CNS, both in vitro and in vivo, indicate that synaptic properties take days or weeks to reach maturity (see below).

In comparing the development of synaptic structure and function, it is interesting that the maturation of transmission seems to evolve far more rapidly. In the chick ciliary ganglion, synaptic potentials can be recorded before synapses are detected with an electron microscope (Landmesser and Pilar, 1972). Similarly, when a muscle is manipulated into contact with a growth cone in a *Xenopus* culture, the recorded synaptic currents can be quite large at contacts that show no appreciable differentiation (Buchanan et al., 1989). Therefore, a rapid phase of functional maturation occurs over minutes, and is due primarily to developmental events that have preceded contact: the expression of a neurotransmitter release mechanism by the growth cone and neurotransmitter receptors by the postsynaptic cell.

THE DECISION TO FORM A SYNAPSE

Growth cones usually slow down when they enter their target, and this may involve a signal to halt growth cone motility and encourage synapse formation (see Chapter 6). Evidence for a target stop signal was found in a system where newborn mouse basilar pontine nuclei were co-cultured with their target neurons, the granule cells of the cerebellum (Baird et al., 1992). Pontine neurites grow rapidly on cerebellar glial cells (greater than 100 μm/hr), indicating that glia do not provide the stop signal. However, when pontine nuclei were cultured on a bed of glia along with dissociated granule cells, the outgrowth of neurites was depressed. By closely examining individual neurites it was found that the decreased growth was due to contact with granule cells (Figure 8.13A). Neurites that did not come upon a granule cell during their outgrowth continued to grow for a normal distance. Moreover, when granule cells were suspended above the pontine explants, the neurites grew at a normal rate. Thus, growth cones can be terminated at the appropriate target by a contact-dependent mechanism.

The dialogue between pre- and postsynaptic cells begins as soon as the growth cone filopodia makes a contact. Calcium levels suddenly increase in the growth cone (Dai and Peng, 1993; Zoran et al., 1993). This was determined for both frog and snail motor neurons that were grown in dissociated tissue culture and filled with a Ca^{2+}-sensitive indicator dye. When a muscle cell is manipulated into contact with a growth cone, the Ca^{2+} increases locally within seconds (Figure 8.9). This response exhibits some target-specificity. The Ca^{2+} rise only occurs when appropriate postsynaptic cells are manipulated into contact with the motor neuron. This mechanism is similar to that observed when growth cones collapse as they contact specific pathfinding cues, which is often accompanied by an elevation of intracellular Ca^{2+} (see Chapter 5). It is not yet clear how calcium levels increase, but one possibility is that calcium is released from internal stores. In the rat central nervous system, IP_3 receptors, which transduce calcium release from endoplasmic reticulum, are upregulated during the period of intense synaptogenesis (Dent et al., 1996).

What is the evidence that a contact-evoked rise in Ca^{2+} provides a signal for growth cone differentiation? Intracellular Ca^{2+} can be manipulated in growth cones by exposing them to an ionophore such as A23187, a molecule that spontaneously inserts into a neuron membrane, allowing Ca^{2+} to pass freely into the cell (Mattson and Kater, 1987). As calcium rises, growth cones are often found to slow down and to assume a rounded appearance. The effect of increased calcium can even be detected on growth cones that have been isolated from their cell body, indicating that calcium acts locally. When the Ca^{2+} concentration within the growth cone is adjusted to differerent levels by setting extracellular Ca^{2+} concentration, cultured chick DRG growth cones became stationary in all but a limited range of Ca^{2+} concentrations, from 200 to 300nM (Lankford and Letourneau, 1991).

The formation of new synapses may be regulated by the presence of astrocytes. When retinal ganglian cells are isolated and grown in a defined medium, they

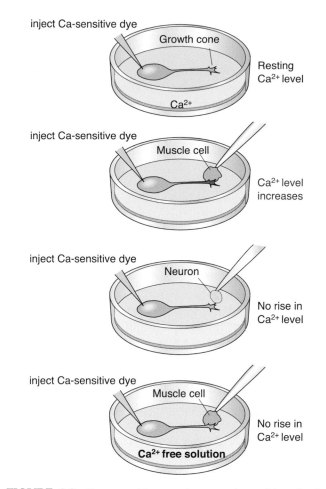

FIGURE 8.9 Contact with target increases free calcium in the growth cone. Dissociated neurons were filled with a Ca^{2+}-sensitive dye (top), and the growth cone was imaged while either a muscle cell or a neuron was brought into contact (middle). Intracellular free calcium increased only during contact with the muscle (red). The muscle-evoked rise in Ca^{2+} did not occur when the cells were bathed in a Ca^{2+}-free medium (bottom), indicating the involvement of calcium channels. (Adapted from Dai and Peng, 1993)

display little synaptic activity. However, the addition of astrocytes from their target region leads to a dramatic increase in the number and strength of synaptic contacts (Ullian et al., 2001).The glial cells need not be in contact with retinal neurons to elicit this response, suggesting that they release a soluble factor. A search for the synapse-inducing activity led to the discovery that an essential membrane constituent, cholesterol, plays an important role in synapse-formation. Apparently, developing neurons manufacture only enough cholesterol to survive and grow dendrites, but depend on the delivery of additional cholesterol from astrocytes to produce synapses (Mauch et al., 2001). There are likely to be many other glial signals; a large protein, thrombospondin, can partly explain the ability of astrocytes to induce synapse formation.

THE STICKY SYNAPSE

Adhesion between growth cones and target cells increases rapidly upon contact (Evers et al., 1989). To demonstrate this sort of adhesion in vitro, round muscle cells, known as myoballs, were lifted off the culture dish with an electrode and placed in contact with the growing tip of a *Xenopus* spinal neuron (Figure 8.10). At either 1.5 or 15 minutes after a contact was initiated, the intercellular adhesion was categorized by observing how much the neurite was deformed by the retracted myoball. A low level of adhesion is evident after 1.5 minutes, and the percentage of tightly adherent contacts more than doubles during the first 15 minutes of contact.

What kinds of adhesion molecules are involved in the formation of early contacts? At the nerve–muscle junction, NCAM is gradually lost during innervation but reappears at the endplate following denervation. This suggests that NCAM facilitates synapse formation, but nerve-muscle contacts appear to develop normally in NCAM knockout mice (Covault and Sanes, 1985; Moscoso et al., 1998). The initial development of motor synapses in fruit flies is also normal in the absence of FasII, a homolog of NCAM (Schuster et al., 1996a). However, the synapses retract from the muscle in slightly older animals, suggesting that this cell adhesion molecule is required for the stabilization of connections. A second set of adhesion molecules that could facilitate synapse differentiation are found in the extracellular matrix (see Chapter 5). Muscle cells synthesize and deposit the ECM molecule, s-laminin, in the synaptic cleft, which can inhibit neurite outgrowth. This growth-inhibiting activity may promote the transformation of a growth cone into a presynaptic terminal (Porter et al., 1995). In s-laminin knockout mice, there is dramatic decrease of vesicle-associated phosphoproteins, called *synapsins*, in presynaptic motor terminals (Noakes et al., 1995).

Several groups of cell adhesion molecules are localized at developing synapses, and many of these contain immunoglubin-domains. For example, synCAM is a brain-specific adhesion molecule that was discovered by searching for a mammalian homolog of FasII, the *Drosophila* cell adhesion molecule that contributes to nerve–muscle synaptogenesis. SynCAM expression gradually increases in rat brain during the first three postnatal weeks and is highly enriched at both pre- and postsynaptic plasma membrane. When nonneuronal cells are transfected with SynCAM, they can induce presynaptic differentiation in cultured hippocampal neurons, and these terminals are able to release glutamate (Biederer et al., 2002).

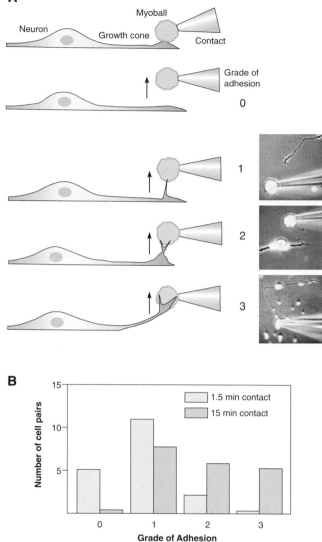

FIGURE 8.10 Rapid adhesion between growth cone and postsynatpic muscle cell. A. A muscle cell was manipulated into contact with a growth cone in a dissociated cultures of *Xenopus* spinal cord. After 1.5 or 15 minutes, the muscle cell was withdrawn, and the degree of adhesion was graded: (0) no attachment, (1) filamentous attachment, (2) deformation of growth cone, and (3) detachment of growth cone from substrate. B. After 1.5 minutes of contact, most pairs exhibited only grade 0–1 adhesion. However, after 15 minutes of contact, the level of adhesion shifted to grade 1–3. (Adapted from Evers et al., 1989)

A family of calcium-dependent cell adhesion molecules, called cadherins, are located at many different central synapses, including the cerebellum and hippocampus. Cadherins interact with actin filaments via catenins. In the hippocampus, two members of the family, N-cadherin and E-cadherin, are restricted to separate synapses along the dendrite, suggesting a role in specific innervation patterns (Fannon and Colman, 1996).

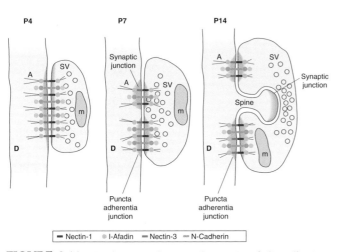

P4   P7   P14

- Nectin-1   I-Afadin   - Nectin-3   - N-Cadherin

FIGURE 8.11 A schematic showing the nectin–afadin adhesion system during the formation of a synapse in developing hippocampal pyramidal neurons. The nectin–afadin system organizes adherens junctions cooperatively with the cadherin–catenin system. Nectin is an immunoglobulin-like adhesion molecule, and afadin is an actin filament–binding protein that connects nectin to the actin cytoskeleton. During development, nectin-1 and -3 localize at both the puncta adherentia junctions (i.e., mechanical anchoring sites) and at synaptic junctions. This changes during development such that nectin–afadin comes to be localized around the synaptic active zone. The cadherin–catenin system is likely to co-localize with the nectin–afadin system at each stage. D, dendritic trunks of pyramidal cells; SV, synaptic vesicles; A, actin filaments. (Adapted from Mizoguchi et al., 2002)

When cadherins were blocked in a hippocampal culture, neither the postsynaptic spines nor the presynaptic terminals were as differentiated as in control cultures (Togashi et al., 2002). The nectins are another family of Ig-like CAMs that are connected to the cytoskeleton by the actin filament-binding protein, l-afadin. The nectin-afadin system is co-localized with the cadherin-catenin system during synapse formation in the hippocampus. Initially, these proteins are found at the site of contact between growth cone and postsynaptic neuron, a close apposition of membrane called an *adherens junction* (see EM of neonatal synapse in Figure 8.2). As the synapse matures, the adhesion molecule gradually become localized adjacent to the synapse (Figure 8.11). As with the cadherin system, blockade of nectin-1 function has been shown to affect the size and number of synapses (Mizoguchi et al., 2002).

CONVERTING GROWTH CONES TO PRESYNAPTIC TERMINALS

Although the growth cone has some ability to release neurotransmitter, there is a dramatic improve-

ment of neurotransmission soon after the initial contact is made (Figure 8.8). In fact, the presynaptic release site (called the active zone) is preassembled and shipped down that axon (Ahmari et al., 2000; Shapira et al., 2003). Dense core vesicles are found to contain most, if not all, of the proteins that are necessary for synaptic vesicle release. These include presynaptic cytoskeleton-associated proteins (Piccolo and Bassoon) and a regulator of vesicle fusion (Rim). It is estimated that a new active zone can be established by the insertion of only two to three of these vesicles. To determine how quickly these proteins accumulate at a presynaptic contact, new active zones were labeled with the FM4–64 method shown in Figure 8.7 and then counterstained for pre- and postsynaptic marker proteins (Friedman et al., 2000). Within 45 minutes of detecting a new release site, about 90% of them have already accumulated Bassoon. In contrast, less than 30% of the postsynaptic sites are labeled for the postsynaptic density protein, PSD-95 (Figure 8.12). Thus, presynaptic sites display rapid molecular development, which seems to precede postsynaptic maturation.

Cell adhesion molecules (above) may explain some aspects of development, but it seems necessary to invoke an asymmetric signal—that is, a signal that provides different instructions to the growth cone and the postsynaptic membrane. Two signaling systems have now been identified that exhibit this sort of asymmetry, and each one can recruit synaptic vesicles as well as the associated transmitter-release machinery (Figure 8.13).

When pontine neurons are grown in vitro, their axons come to a halt and accumulate synaptic vesicles when they contact their postsynaptic target neurons, cerebellar granule cells (Figure 8.13A). The growth cones of these axons express the cell surface protein, β-neurexin, which serves as a receptor for neuroligin, a ligand that is expressed in the developing cerebellum. This signaling system is sufficient to induce presynaptic differentiation. When pontine axons contact nonneuronal cells that are expressing neuroligin-1 or -2, they stop growing and accumulate presynaptic protein (e.g., synapsin) and synaptic vesicles (Figure 8.13B). This effect can be blocked with the addition of soluble neurexin to the culture media (Scheiffele et al., 2000; Dean et al., 2003). Granule cells also release the soluble factor that can induce presynaptic differentiation, *Wnt-7a*. Pontine axons that express the *Wnt* receptor, *Frizzled*, accumulate synapsin when grown in the presence of *Wnt-7a*-transfected cells. There is now evidence that *Wnt* supports synapse differentiation in other systems (Hall et al., 2000; Krylova et al., 2002; Packard et al., 2002).

Many of the trophic factors that support target invasion and neuron survival (Chapters 6 and 7) also play an important role during synapse formation; these

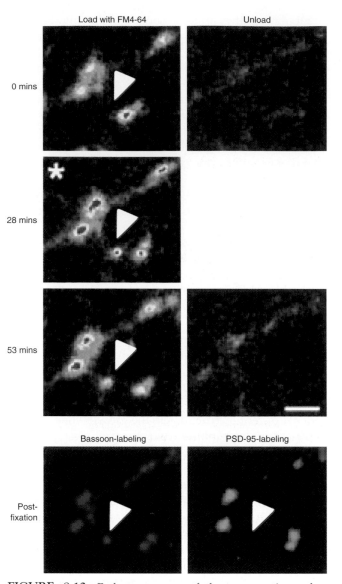

FIGURE 8.12 Early appearance of the presynaptic marker, Bassoon, at a new release site. FM 4–64 is used to label sites of vesicular release, and it can be unloaded by stimulating the synapses in the absence of the dye. A new release site (i.e., presynaptic bouton) appears between 0 and 28 minutes (arrowhead), and this site is retained at 53 minutes. The neurons are then fixed, and immunohistochemistry is performed on the same region. The new bouton is associated with an aggregate of Bassoon, but not with the postsynaptic density protein, PSD-95. Scale bar is 3 μm (Friedman et al., 2000)

nerve–muscle cultures are exposed to these proteins, there is an increase in the rate of spontaneous synaptic events and an increase in the amplitude of evoked synaptic currents (Figure 8.14). BDNF exerts its effect at the synaptic terminal through a calcium-dependent process, whereas CNTF seems to act at the soma. It takes approximately 10 minutes for BDNF or CNTF to effect transmission, and neurotransmission remains altered for hours after these compounds are removed. These effects may depend on local protein synthesis at the synapse (Stoop and Poo, 1996; Zhang and Poo, 2002).

Several intracellular signals are also involved in the transition from growth cone to terminal. By observing the giant growth cones produced by cultured *Aplysia* bag cells, one finds that microtubules extend toward the site of contact with a target, and filamentous actin begins to accumulate (Forscher et al., 1987). A similar transformation can be produced by raising cAMP levels within the growth cone. The cytoskeleton reorganizes and neurosecretory granules invade the growth cone's lamellapodia, resulting in a presynaptic-like morphology. Activation of a second intracellular signaling pathway, protein kinase C (PKC), results in the rapid appearance of new calcium channels at the edge of the grow cone (Knox et al., 1992).

Calcium, PKC, and cAMP may work in tandem to support the accumulation of secretory vesicles and ion channels at the site of contact. In fact, the contact-evoked increase in calcium (Figure 8.9) may actually be a result of cAMP signaling. When an identified snail motor neuron is manipulated into contact with its normal target in vitro, it exhibits an increase in calcium (Funte and Haydon, 1993). This rise is mimicked by a membrane permeable analog of cAMP and is prevented by injecting an inhibitor of cAMP-dependent protein kinase (PKA) into the motor neuron. How are these intracellular signals activated during growth cone differentiation? Certain neurotransmitter receptors can produce a Ca^{2+} influx, and some of these have been shown to inhibit growth cone motility (Mattson and Kater, 1989). Cell adhesion molecules are also capable of transducing cell surface signals to produce an elevation of internal Ca^{2+} (Doherty et al., 1991).

RECEPTOR CLUSTERING SIGNIFIES POSTSYNAPTIC DIFFERENTIATION AT NMJ

The aggregation of neurotransmitter receptors beneath the presynaptic terminal is the most obvious hallmark of synaptogenesis. Is receptor clustering a

range from the induction of new synapses to the upregulation of neurotransmitter release. For example, the addition of BDNF can increase the number of synapses, both in cultures and in vivo (Vicario-Abejon et al, 1998; Marty et al. 2000; Alsina et al., 2001). Members of the neurotrophin family of growth factors (BDNF and NT-3) and CNTF are also able to potentiate the release of ACh by presynaptic terminals. When

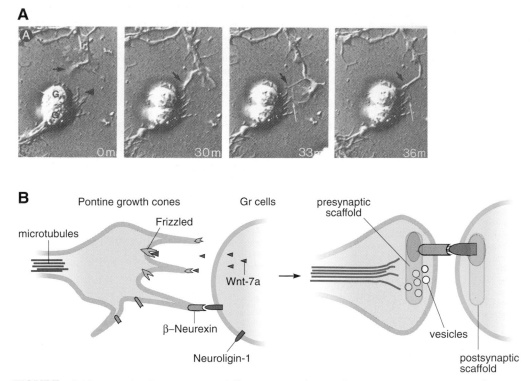

FIGURE 8.13 Signals that promote differentiation of growth cones into presynaptic boutons. A. Pontine growth cones (arrows) stop when they contact granule cells (arrowhead) in a dissociated culture system. B. Pontine growth cones express two receptors, frizzled and β-neurexin, on their surface. Granule cells express their respective ligands: *Wnt-7* and Neuroligin-1. These signaling pathways recruit synaptic vesicles and presynaptic proteins to the site of contact. (Panel A from Baird et al., 1992)

cell-autonomous process, or is it induced by the presynaptic terminal? At first inspection, the postsynaptic site appears to be produced in an autonomous fashion. Structures that resemble postsynaptic densities, but that do not appear to contact a presynaptic element, have also been found in the developing olfactory bulb and visual cortex during early development (Hinds and Hinds, 1976; Bahr and Wolff, 1985). In fact, most neurotransmitter receptors are expressed before innervation occurs; they can even be found in clusters, similar in appearance to a postsynaptic site (Figure 8.4A). Acetylcholine receptors (AChRs) also form small clusters on the muscle cell membrane even before the motor axon terminals arrive (Fischbach and Cohen, 1973).

At the time of innervation, the muscle cell membrane still displays an immature distribution of AChRs. This was originally demonstrated by recording the response from rat muscle cells in vivo as ACh was applied at different places along the myofiber surface (Diamond and Miledi, 1962). Early in development, ACh application at each site evokes a similar shift in membrane potential. As the muscle became innervated, the ACh-evoked response becomes much larger at the site of innervation, and the response at extrasynaptic regions declines.

Soon after, it became possible to visualize the distribution of AChRs by labeling them with radioactive α-bungarotoxin (α-Btx), a high-affinity peptide from the venom of the Taiwanese cobra (Bevan and Steinbach, 1977; Burden, 1977a). Consistent with the electrophysiological measures, α-Btx labeling is broadly distributed at first and then became highly localized to the synapse. The process of clustering leads to a dramatic disparity in receptor concentration: there are >10,000 AChRs/μm^2 at the synaptic region but <10/μm^2 in extrasynaptic regions (Fertuck and Salpeter, 1976; Burden, 1977a; Salpeter and Harris, 1983).

While these observations suggested that the motor axons induce receptor clustering, higher AChR concentrations occur at the center of skeletal muscle fibers in the absence of motor innervation. In *HB9* mutant mice, the phrenic nerve fails to develop, and the diaphragm muscle remains uninnervated during embryonic development (Yang et al., 2001). Despite this, AChRs are concentrated in the central region of the muscle by E18.5 (Figure 8.15). However, there are signs that a motor terminal does organize postsynaptic structure from the earliest stage. *Drosophila* muscle fibers produce tiny processes, called *myopodia*, that interact

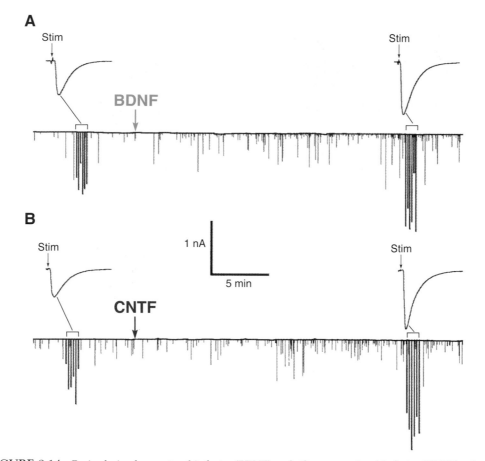

FIGURE 8.14 Brain-derived neurotrophic factor (BDNF) and ciliary neurotrophic factor (CNTF) enhance evoked synaptic responses at developing *Xenopus* neuromuscular synapses. The continuous traces show membrane currents recorded from innervated myocytes under voltage-clamp. Downward events are inward synaptic currents. Spontaneous events occur throughout the recording period, and evoked responses are indicated by the brackets (single evoked currents are shown on an expanded time scale above each bracket). The arrow marks the time of addition of either BDNF (A) or CNTF (B) to the culture medium. In each case, there is a significant increase in the size of evoked synaptic currents, and this effect lasts several hours. (From Poo and Stoop, 1996)

with motor nerve terminals. These myopodia gradually become restricted to the site of innervation. When the motor axons are delayed, as in the *prospero* mutant, the myopodia no longer become clustered (Ritzenthaler et al., 2000).

PRESYNAPTIC TERMINALS INDUCE RECEPTOR AGGREGATION

If one carefully watches growth cones in tissue culture, they are seen to induce new postsynaptic sites at their initial sites of contact. When dissociated cultures are made of cholinergic spinal cord neurons and myocytes, it is possible to follow the clustering of AChRs during the period of innervation by labeling the muscles with a fluorescent α-Btx (Anderson and

Cohen, 1977; Cohen et al., 1979). Although small AChR clusters are seen prior to innervation, they do not serve as preferential sites of innervation. Rather, the growing neurites induce the rapid accumulation of AChRs as they extend across the muscle (Figure 8.16A). When the temporal relationship between innervation and AChR cluster formation is estimated in the *Xenopus* embryo, nerve terminals are found to precede the appearance of AChR clusters by about three hours (Chow and Cohen, 1983). Indeed, the clustering of GABA$_A$ and AMPA receptors is also correlated with the presence of presynaptic contacts in hippocampal cultures, although precise measures have not yet been made (Killisch et al., 1991; Craig et al., 1993).

The ability to induce AChR clusters is specific to certain types of presynaptic neurons (Figure 8.16B). In *Xenopus* dissociated culture, dorsal root and sympathetic ganglion neurites can contact muscle cells, but

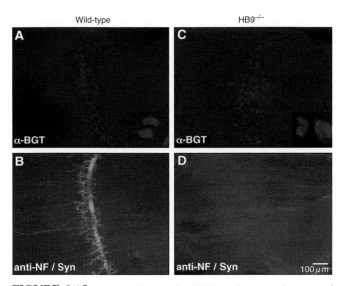

FIGURE 8.15 Accumulation of AChRs in the central region of uninnervated muscle. Diaphragm muscle from wild-type (A and B) or *HB9* mutant (C and D) embryos were stained with Texas red-α-bungarotoxin (αBGT) and antibodies to neurofilament and synaptophysin (anti-NF/Syn). At embryonic day 18.5, there is dense innervation at the center of the wild-type muscle but no motor innervation in the *HB9* mutants. (A few sensory and/or autonomic axons can be seen at the edge of *HB9* muscle in panel D.) The insets show individual AChR clusters at high magnification in wild-type (A) and mutant (C) embryos. The bar is 100 μm for the low magnification images. (Adapted from Yang et al., 2001)

they are associated with few AChR clusters. In contrast, neurites from spinal cord neurons, which presumably include motor neurons, are associated with AChR clusters at 70% of contact sites (Cohen and Weldon, 1980; Kidokoro et al., 1980). When spinal motor neurons are selectively prelabeled in vivo and then dissociated in the presence of myocytes, their ability to induce AChR clusters can be compared to the unlabeled spinal interneurons (Role et al., 1985). Motor neurons and ciliary ganglion cells, both of which secrete ACh, are able to induce AChR clusters on all contacted myocytes, whereas the interneurons do not induce significant AChR clustering.

The studies on NMJ suggest that appropriate nerve terminals (i.e., cholinergic) support the induction of AChR clusters. Does the neurotransmitter receptor have to be activated in order to cluster? When myocytes and spinal cord are cultured in the presence of an AChR antagonist, D-tubocurarine, they develop an identical number of functional contacts compared to control cultures (Cohen, 1972). To directly assess AChR clustering during the blockade of neurotransmission, fluorescently labeled α-Btx was first applied to visualize the AChRs that were present on muscle cell membrane, and excess unlabeled α-Btx was then added to prevent visualization of newly inserted AChRs (Anderson and

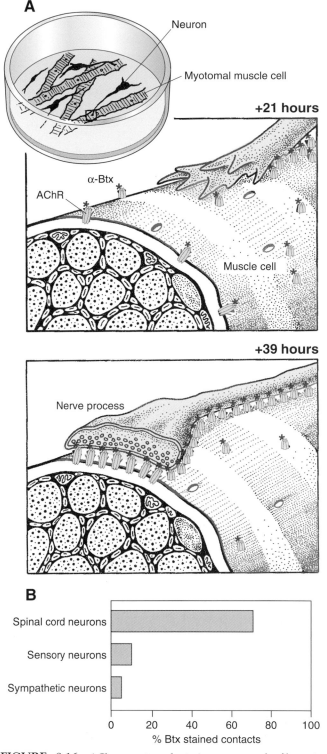

FIGURE 8.16 ACh receptor clustering on muscle fibers is induced by contact with spinal neurites. A. A culture of spinal neurons and muscle cells was labeled with a fluorescent α-Btx (red) at 21 and 39 hours after plating. Soon after the spinal neurite grew across the muscle surface, fluorescent α-Btx appeared at the contact site, indicating that AChR aggregation is induced. B. The ability to induce AChR clusters is neuron-specific. When muscle cells were cultured with spinal neurons, DRG neurons, and sympathetic neurons, only the spinal neurons induced significant α-Btx labeling. (Adapted from Anderson and Cohen, 1977; Cohen and Weldon, 1980)

Cohen, 1977). In these experiments, the clustering of AChRs occurs normally at the site of neurite contact in the absence of cholinergic transmission.

AGRIN, A TRANSYNAPTIC CLUSTERING SIGNAL

The clustering of AChRs is a result of both migration within the muscle membrane and, after several hours, the insertion of newly synthesized protein. The studies discussed above strongly suggest that nerve terminals produce a signal that initiates receptor clustering at the postsynaptic cell. Interestingly, AChR clustering can also be produced by the basal lamina, an extracellular matrix that ensheathes each muscle cell. When muscle cells are damaged in adult frogs, they degenerate, leaving behind the basal lamina (Figure 8.17). New myofibers then form beneath this basal lamina, and AChR clusters form at the original synaptic sites along the basal lamina, even if motor nerve terminals are absent (Burden et al., 1979). These results motivated a search for a "clustering" signal among the extracellular matrix molecules. A proteoglycan that is able to mimic the clustering ability of nerve terminals or the basal lamina, named Agrin, was subsequently isolated from the electric organ of the marine ray *Torpedo californica*, a site rich in cholinergic synapses (Godfrey et al., 1984; Nitkin et al., 1987). Monoclonal antibodies directed against Agrin have been used to localize this protein to motor neuron cell bodies, the synaptic basal lamina, and muscle cells (Reist et al., 1987; Magill-Solc and McMahon, 1988; Fallon and Gelfman, 1989).

To test whether release of neuronal Agrin (called z-agrin in mammals) is responsible for AChR cluster formation, polyclonal antibodies were used to block its function in vitro (Reist et al., 1992). The polyclonal antibodies are raised against *Torpedo* Agrin, and they bind selectively to chick Agrin, blocking cluster formation in chick nerve–muscle cultures. However, muscle cells also produce Agrin, and they could also induce clustering. Since the polyclonal antiserum does not block rat Agrin, a co-culture experiment was designed that made use of tissue from chicks and rats (Figure 8.18). The antiserum does prevent clustering on rat muscle cells that are innervated by chick neurons (Figure 8.18). However, it does not block AChR cluster formation on chick muscle cells that are innervated by *rat* neurons. Thus, rat neurons must be releasing the Agrin that elicits cluster formation. A neuron-specific isoform of Agrin, generated by alternative splicing of the mRNA, has since been shown to

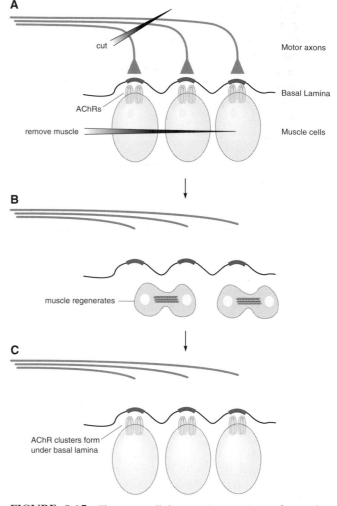

FIGURE 8.17 The extracellular matrix contains a factor that induces AChR clustering. A. In the adult frog cutaneous pectoris muscle, the motor axons were cut and the muscle cells were destroyed, leaving only the basal lamina which contains extracellular matrix molecules. B. New myofibers are generated as a result of cell division. C. AChR clusters (green) form on the regenerated muscle fibers, directly beneath the synaptic portion (red) of the basal lamina. (Adapted from Burden et al., 1979)

have greater cluster-inducing activity than that found in muscles (Ruegg et al., 1992; Ferns et al., 1993).

The results suggest that neuron-derived Agrin is necessary for AChR cluster formation in vivo, and its contribution to neuromuscular formation in situ was confirmed with homologous recombination. Agrin knockout mice demonstrate that this signal is not required for the embryonic aggregation of AChRs to the central region of the muscle (Figure 8.15) but is necessary for their alignment with the motor nerve terminal. Furthermore, postsynaptic sites appear to be less mature than normal, suggesting that this signaling

Controls

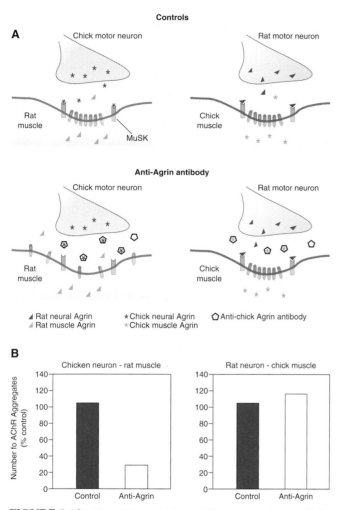

FIGURE 8.18 Neural Agrin induces ACh receptor clusters. A. To test whether neural Agrin or muscle Agrin induced AChR clusters, mixed chick–rat cultures were produced. This strategy was used because a function-blocking antibody (black pentagon) exists for *chick* Agrin. Normal AChR aggregation was observed in control heterospecific cultures (Controls, top). The antibody blocked AChR clustering when the motor neuron was from chick (Anti-Agrin antibody, left) but not when the muscle cells were from chick (right). MuSK is the Agrin receptor. B. These results were quantified for each type of culture. The anti-Agrin antibody decreased the number of AChR aggregates to 25% in chick motor neuron cultures. (Adapted from Reist et al., 1992)

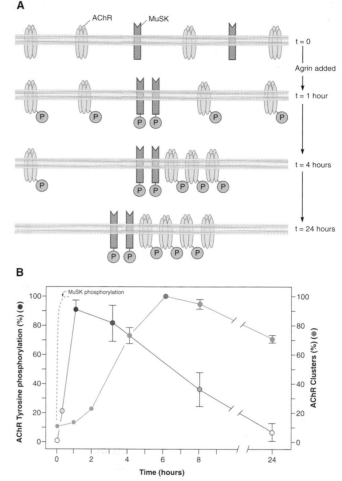

FIGURE 8.19 Agrin induces ACh receptor phosphorylation prior to clustering. A. When cultures are exposed to Agrin, the Agrin receptor (MuSK) is phosphorylated within minutes (blue line), and the AChRs are maximally phosphorylated within an hour (red line). Receptor aggregation occurs over the next few hours (green line), but the level of phosphorylation begins to decline. B. The plot shows a gradual decline in phosphorylation over 24 hours. In contrast, AChR clustering is maximal by six hours after Agrin exposure and declines slightly by 24 hours. (Adapted from Ferns et al., 1996; Glass et al., 1996)

pathway regulates more than just receptor clustering (Gautam et al., 1996).

POSTSYNAPTIC RESPONSE TO AGRIN

Although there is strong evidence that neural Agrin initiates AChR clustering at the developing NMJ, the signal transduction mechanism is not fully resolved. It is known that Agrin induces postsynaptic tyrosine phosphorylation, and the AChR β- and δ-subunits are sites of action in chick and mouse muscle cells (Wallace et al., 1991; Qu and Huganir, 1994). In fact, an inhibitor of tyrosine kinase can prevent AChR clustering in response to Agrin (Wallace, 1994; Ferns et al., 1996). The temporal relationship between β-subunit phosphorylation and receptor clustering is quite close. Receptor phosphorylation reaches a peak within one hour, and receptor clustering then occurs over the next 6 hours. By the time that clustering has reached a maximum, phosphorylation is in steep decline (Figure 8.19). The amount or location of receptor subunit phosphorylation may also interfere with cluster formation. For example, a phosphatase inhibitor that leads to increased β-subunit phosphorylation can actually prevent Agrin-induced receptor clustering (Wallace, 1995). Although a number of cytoplasmic enzymes, such as Src or PKC, have been implicated in phosphorylating AChR subunits, the key events are poorly understood (Huganir and Greengard, 1983; Huganir et al., 1984). At the onset of synaptogenesis, the AChR clusters are labile, and will disperse if the muscle is denervated. When α-Btx labeled muscles from embryonic mice are denervated and placed in calcium-depleted culture media, the AChR clusters are lost (Bloch and Steinbach, 1981). By birth, the labeled AChR clusters have become resistant to this treatment.

During a widespread search for the postsynaptic Agrin receptor, one candidate has emerged as the most important transducer. A muscle-specific kinase (MuSK), identified during a search for novel receptor tyrosine kinases in denervated muscle, drew attention because it is localized to the synaptic junction (Figure 8.20A) (Valenzuela et al., 1995). MuSK is expressed very early in development, beginning at around E13 in rats, when motor axons are first growing into the muscle. Agrin induces the tyrosine phosphorylation of MuSK within minutes (Figure 8.19), and this leads rapidly to MuSK clustering (Moore et al., 2001). The importance of MuSK was verified in a targeted disruption of the gene in mice (Figure 8.20B). These animals display a dramatic loss of postsynaptic maturation, including the loss of AChR clustering (DeChiara et al., 1996). Although Agrin and MuSK display strong binding kinetics, this is only apparent when MuSK is expressed in muscle cells, suggesting that the protein must form a complex with one or more accessory proteins (Glass et al., 1996). The MuSK protein is also necessary for the early prepatterning of AChRs that is observed in the absence of motor axons (Figure 8.15). The lasting importance of MuSK activity is clearly seen in a group of patients suffering from the autoimmune disease, myasthenia gravis. In most of these patients, auto-antibodies against the AChR are made, leading to a decrease in synaptic transmission and muscle weakness. However, some patients make autoantibodies against the MuSK protein, and this also leads to severe problems with neuromuscular transmission (Hoch et al., 2001).

The transduction process from MuSK activation to receptor clustering is not worked out, but activation of a Src-like kinase and GTP-binding proteins are probably required (Wallace, 1994; Ferns et al., 1996; Weston et al., 2000; Mittaud et al., 2001). One crucial peripheral membrane protein, called Rapsyn, is clearly required to mediate AChR clustering (Figure 8.20A) (Sealock et al., 1984). Messenger RNA for Rapsyn is present in muscle cells prior to AChR cluster formation, and the protein co-localizes with newly formed clusters in vivo (Noakes et al., 1993). Rapsyn contains one domain that mediates self-association, a second that associates with the main intracellular loop of AChRs, and a third that interacts with α-dystroglycan (Ramaroa et al., 2001). When AChR subunits were introduced into cells that do not ordinarily express this molecule, no clusters formed. However, the co-expression of Rapsyn is sufficient to promote AChR clustering (Frohner et al., 1990; Phillips et al., 1991). The phenotype of Rapsyn-deficient mice is fully consistent with a primary role in cluster formation (Gautam et al., 1995). AChR mRNA and protein are restricted to the central region of muscle fibers but do not aggregate at the site of neural contact. Rapsyn probably acts as more than just an intermediate signal for MuSK. The presence of Rapsyn is able to induce MuSK clusters in a fibroblast expression system, and it is also able to activate tyrosine kinase activity (Gillespie et al., 1996). Several other constituents accumulate at the synaptic cleft, including acetylcholinesterase and s-laminin. Therefore, many signaling pathways regulate postsynaptic maturation.

The dystrophin-associated glycoprotein, α-dystroglycan, is a second postsynaptic Agrin-binding protein (Figure 8.20A) (Gee et al., 1994; Campanelli et al., 1994; Bowe et al., 1994; Sugiyama et al., 1994). However, Agrin mutants that have poor affinity for α-dystroglycan are nonetheless able to induce AChR clusters (Meier et al., 1996; Hopf and Hoch, 1996). Therefore, α-dystroglycan is probably not directly involved in this part of synaptogenesis. The absence of dystrophin in Duchenne's muscular dystrophy leads to reduced expression of associated proteins in the sarcolemma, resulting in damage during contraction, poor calcium homeostasis, and eventual necrosis (Davies et al., 1995). Agrin also binds to laminin, a major component of the basal lamina (Figure 8.20A). Laminin can also induce AChR clustering in cultured muscle cells in a MuSK-independent manner (Sugiyama et al., 1997).

It seems likely that intercellular signaling influences AChR function. Calcitonin gene-related peptide (CGRP) is released from motor terminals and is able to rapidly increase the mean open time of AChR channels in *Xenopus* nerve-muscle cultures (Lu et al., 1993). This effect appears to be mediated through an elevation of cAMP in the muscle cell and is blocked by inhibitors of cAMP-dependent protein kinase (PKA). A second protein kinase, PKC, is also able to modulate the kinetics of low-conductance AChRs (Fu and Lin, 1993). Therefore, phosphorylation of "immature" AChRs may prolong their open state, thereby increasing the size of transmitter-evoked postsynaptic potentials at the time of innervation.

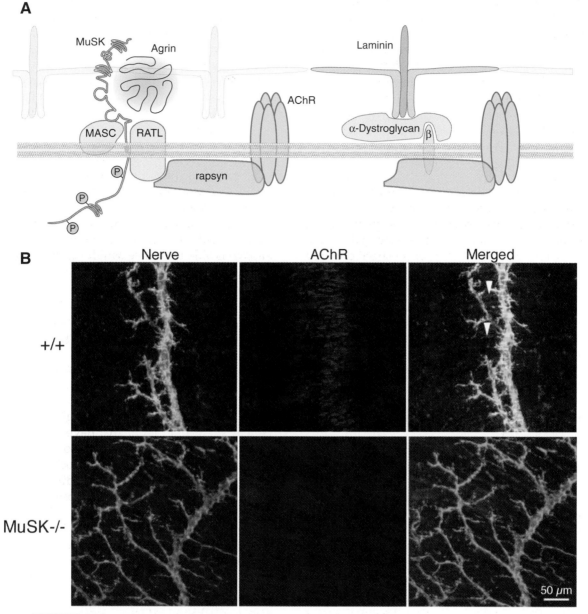

FIGURE 8.20 Agrin binds to a receptor complex, and MuSK is required for clustering. A. The schematic shows that Agrin activates a receptor complex that includes muscle-specific kinase (MuSK) and an unidentified accessory protein (MASC). The intracellular peripheral membrane protein, Rapsyn, is required for the Agrin-mediated MuSK activation to produce AChR phosphorylation, and it participates in receptor clustering. There are other Agrin binding sites, including laminin and α-dystroglycan. Laminin has been shown to induce AChR clusters independent of MuSK signaling. B. In wild-type mice (+/+), AChRs aggregate at the nerve–muscle junction (merged), but in MuSK-deficient mice (MuSK–/–) there is no clustering and the postsynaptic site does not differentiate. (Adapted from Lin et al., 2001, Nature 410: 1057–1064)

RECEPTOR CLUSTERING SIGNALS IN THE CNS

Since receptor clustering is the most identifiable postsynaptic feature of central synapses, it is reasonable to ask whether Agrin-like molecules play a role. Agrin mRNA has been detected in the brains of embryonic rats and chicks, and four splice variants of Agrin are distributed throughout the adult rat brain (Rupp et al., 1991; Tsim et al., 1992; Ferns et al., 1992). However, differing results have been reported for an influence of Agrin on synaptogenesis in the CNS (Serpinskaya et al., 1999; Bose et al., 2000). Therefore, the search for receptor clustering signal in the CNS has focused on three different pathways. First, a secreted pentraxin (i.e, a family of proteins sharing a discoid arrangement of five noncovalently linked subunits), Narp, is concentrated at excitatory synapses in cultured spinal and hippocampal neurons, and ultrastructural studies from hippocampal neurons in vivo show that Narp is present on both the presynaptic surface and the dendritic shaft. To determine whether endogenous Narp participates in glutamate receptor clustering, the C terminus region that supports axon transport was disrupted such that axonal transport and secretion at the synapse could be prevented. When cultured spinal neurons were transfected with this mutant Narp, there was a dramatic reduction in the clustering of AMPA-type glutamate receptors at axon–dendrite contacts (O'Brien et al., 1999; 2002).

A second system that is able to cluster glutamate receptors is the EphrinB-EphB signaling pathway which plays such an important role in axon pathfinding and target selection (Chapters 5 and 6). When cultured cortical neurons are exposed to ephrinB1, they rapidly display clusters of the ephrin receptor, EphB2, followed by the appearance of NMDA-type glutamate receptor clusters (Figure 8.21A) (Dalva et al., 2000). Furthermore, when hippocampal neurons are cultured from mice that lack 3 EphB receptors, there is a dramatic loss of normal spine morphology and glutamate receptor clustering (Figure 8.21B). However, this signaling pathway does not appear to disrupt inhibitory GABAergic synapse formation, presumably because these contacts are not made on dendritic spines (Henkemeyer et al., 2003).

A third system that participates in receptor clustering is the neurotrophin-Trk signaling pathway, which also plays a fundamental role in target selection and neuron survival (Chapters 6 and 7). Once again, the assay system is a dissociated culture of embryonic hippocampal neurons which express the BDNF/NT-4 receptor, TrkB, diffusely along their dendrites and soma. When these neurons are exposed to BDNF, the number of NMDA receptor clusters doubles, and they are far more likely to be located adjacent to a presynaptic terminal (Elmariah et al., 2004). Conversely, decreasing endogenous BDNF in the cultures produces a loss of NMDA receptor clusters. Unlike the ephrinB-EphB system, the TrkB pathway fails to increase AMPA receptor clustering but does support GABA$_A$ receptor cluster formation. To summarize, several candidate signals may be released by excitatory and inhibitory nerve terminals and may recruit the aggregation of postsynaptic receptors, similar to the role of Agrin at the nerve–muscle junction.

Several studies suggest that the neurotransmitter itself can promote receptor accumulation. For example, the largest glutamate receptor clusters occur opposite the presynaptic terminals that release the most glutamate at the fly neuromuscular junction (Marrus and DiAntonio, 2004). Furthermore, when spontaneous glutamate release is blocked during development, glutamate receptor clusters do not form (Saitoe et al., 2001). An effect of activity has also been observed for glycine receptors and glutamate receptors in tissue culture preparations (Kirsch and Betz, 1998; Shi et al., 1999; Liao et al., 2001; Lu et al., 2001). It is not yet clear whether activity-dependent receptor clustering is a developmental mechanism, or is used largely to adjust synaptic strength throughout the animal's life (Aoki et al., 2003).

It must be emphasized that an equal number of studies report little influence of synaptic transmission on receptor clustering, and this area of research remains unsettled. In *C. elegans*, GABA is not required to obtain synaptically clustered GABA$_A$ receptors (Gally and Bessereau, 2003). A careful in vitro study showed that glutamate receptors are evenly distributed along the dendrite at a low density ($\approx 3/\mu m^2$) prior to innervation but form high-density ($\approx 10,000/\mu m^2$) aggregates at the site of presynaptic contacts (Figure 8.22). These glutamate receptor clusters appear even when synaptic or electrical activity is blocked (Cottrell et al., 2000).

INTERNAL MEMBRANE PROTEINS AND RECEPTOR AGGREGATION IN THE CNS

Whatever the clustering signal may be at each central synapse, there must be a molecular mechanism to hold the receptors together, similar to the way that Rapsyn restricts AChR mobility. In fact, a tremendous array of proteins are located at the cytoplasmic surface and bind to both membrane receptors and cytoskeletal elements. Perhaps the strongest case can be made for a protein called Gephyrin that was discovered during the purification of glycine receptor subunits (Kirsch et al., 1993a,b). Gephyrin displays a high affinity for polymerized tubulin, and a C-terminal domain can bind with high affinity to the cytoplasmic loops of 2 glycine receptor β subunits (Schrader et al., 2004). Spinal neurons grown in dissociated culture normally display glycine receptor clusters. However, when the cells are grown in the presence of a gepherin antisense nucleotide, which presumably prevents the translation of gepherin mRNA, then clusters do not form at the membrane (Figure 8.23). Similarly, glycine receptors do not aggre-

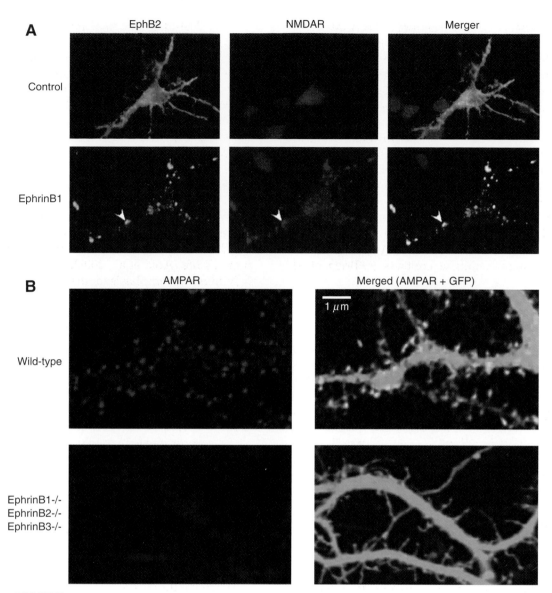

FIGURE 8.21 EphB signaling influences glutamate receptor aggregation in the central nervous system. A. Cortical neurons were transfected with labeled EphB2 (green, left images) constructs, and exposed to EphrinB1 or a control solution for one hour. The cultures were then fixed, and NMDA receptors (NMDAR) were stained with antibody (red, center images). The merged images (right) show that EphrinB1 induced NMDAR clusters on cortical neuron dendrites. B. Hippocampal neurons were cultured from wild-type or triple EphB-deficient mice (EphB1–/–,EphB2–/–, EphB3–/–). The cells were transfected with green fluorescent protein (GFP) so that the dendrite could be observed, and AMPA receptors (AMPAR) were stained with antibody (red, left images) after 21 days in vitro. Wild-type neurons exhibited punctate labeling of AMPAR clusters all along the dendrites (merged images, right). In contrast, AMPAR clustering was not observed in EphB-deficient neurons. (Adapted from Dalva et al., 2000; Henkemeyer et al., 2003)

gate within the membrane in gephyrin-deficient mice (Feng et al., 1998). Gephyrin is also required for the clustering of some (Kneussel et al., 1999), but not all (Levi et al., 2004), GABA receptors.

A family of molecules with similar functional properties, called membrane-associated guanylate kinases (MAGUKs), have been implicated in synapse forma-

tion throughout the nervous system (Cho et al., 1992; Kistner et al., 1993). Synaptic channels and receptors bind to amino acid sequences called the PDZ domains, which are found on the N-terminal side. Similar to the NMJ, cluster formation is regulated by protein phosphorylation. There is a specific 4 amino acid sequence on the C-terminal tail of receptors and channels, called

A

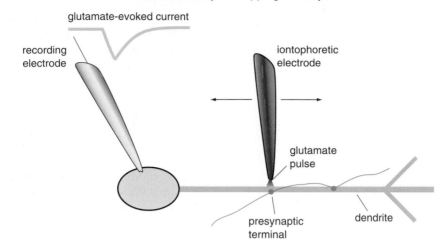

B

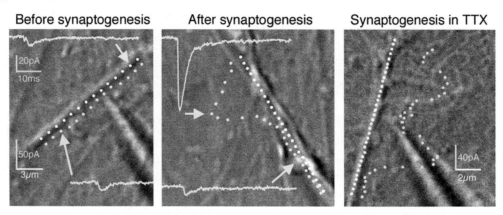

FIGURE 8.22 Mapping glutamate receptor location during synaptogenesis. A. To map the location of glutamate receptors, an iontophoretic pipette (red) ejects glutamate focally, and the evoked response is recorded at the soma. B. Glutamate-evoked currents become restricted to the site of synaptic contacts (green) after the dendrites become innervated. White dots show the positions of glutamate application; the relative distance of the yellow dot indicates the evoked current magnitude for each position. Representative glutamate-evoked currents are shown (cyan). When neurons are cultured in the presence of TTX to block all action potentials, the synaptic localization of glutamate receptors occurs nonetheless. (Adapted from Cottrell et al., 2000)

ET/SXV, that serves as an important phosphorylation site (Niethammer et al., 1996; Cohen et al., 1996).

One of the first MAGUKs to be isolated was postsynaptic density protein-95 (PSD-95; aka, SAP-90). PSD-95 is located in the postsynaptic densities of hippocampal neurons, and some evidence suggests that it participates in clustering both NMDA receptors and potassium channels (Kim et al., 1995; Kornau; et al., 1995; Kim and Sheng, 1996; Niethammer et al., 1996). However, hippocampal synapses look normal and cluster NMDA receptors in PSD-95 knocked mice (Migaud et al., 1998);

studies suggest that its role may be to decrease internalization of NMDA receptors (Roche et al., 2001). In contrast, deletion of a second MAGUK, called PSD-93, leads to decreased expression of NMDA receptor subunits at the membrane surface and smaller NMDA receptor-mediated synaptic potentials (Tao et al., 2003). This result implies that PSD-93 is necessary for the insertion of new NMDA receptors; to the extent that PSD-93 itself is localized to the postsynaptic site, it will promote receptor clustering. A third member of the family, SAP-97, is driven into the spine heads when CaMKII is acti-

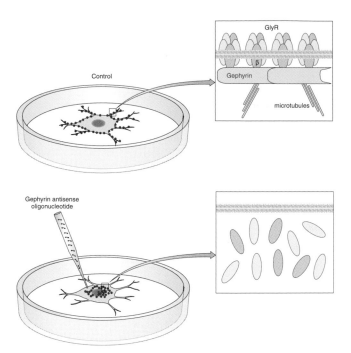

FIGURE 8.23 Gephyrin is required for glycine receptor clustering. When spinal neurons are grown in culture, the peripheral membrane protein, Gephyrin, co-localizes with glycine receptor clusters (top). When translation of the Gephyrin protein is blocked with an antisense oligonucleotide (bottom), the glycine receptors do not form clusters in the neuronal membrane. (Adapted from Kirsch et al., 1993)

vated, and this is directly correlated with an increase in AMPA-type glutamate receptors at these spines (Mauceri et al., 2004). This may provide a mechanism by which activity does lead to receptor clustering.

There is evidence that GABA$_A$ receptor clustering involves a novel internal membrane protein, called GABARAP, which was isolated on the basis of its interaction with the $\gamma 2$ subunit of GABA$_A$ receptors. An N-terminal subdomain of GABARAP binds microtubules, suggesting that it may stabilize receptors at the membrane (Wang et al., 1999; Wang and Olsen, 2000). Although GABARAP has been shown to aggregate GABA$_A$ receptors and enhance their conductance in tissue culture (Chen et al., 2000; Everitt et al., 2004), its role in normal development has yet to be established.

Clustering proteins are conserved across species. A MAGUK family member, called discs-large (DLG), co-localizes with glutamate receptors at the *Drosophila* nerve–muscle junction. When the *dlg* gene is inactivated, synaptic structure and function are profoundly altered (Woods and Bryant, 1991; Lahey et al., 1994; Budnick et al., 1996). The synaptic localization of DLG is regulated by CaMKII activity (Koh et al., 1999). DLG is necessary for the localization of FASII, a cell adhesion molecule that regulates synapse formation (see

Chapter 9). As with the mammalian family members, the DLG protein has a PDZ binding domain, and the membrane proteins that it anchors have the conserved C-terminal motif. The developmental importance of MAGUK family proteins is underlined by studies showing moderate to severe mental retardation when one of the genes is truncated (Tarpey et al., 2004).

THE EXPRESSION AND INSERTION OF NEW RECEPTORS

Even while receptor clustering is underway, the synthesis of new synaptic proteins increases dramatically. The majority of AChRs within a cluster are newly inserted a short time after innervation (Salpeter and Harris, 1983; Ziskind-Conhaim et al., 1984; Role et al., 1985; Dubinsky et al., 1989). The contribution of existing and newly inserted AChRs was examined at newly formed synapses in chick nerve–muscle cultures by labeling the receptors before and after innervation (Figure 8.24). Before innervation, all of the AChRs present on the muscle membrane surface were labeled with α-Btx ("old" AChRs). Following the addition of neurons, a monoclonal antibody directed against an extracellular AChR epitope was applied to label all AChRs ("old" and "new" AChRs). In this way, it was possible to determine the contribution of both "old" receptors (i.e., α-Btx labeled) and "new" receptors (i.e., antibody-labeled minus α-Btx labeled). Within eight hours of neuron addition, more than 60% of the AChRs are newly inserted into the muscle membrane, indicating that synthesis is rapidly upregulated (Role et al., 1985). These results suggest that synthesis is regulated by the presynaptic terminal. Similarly, when innervation of the *Drosophila* neuromuscular junction is delayed or prevented in *prospero* mutants, the normal increase in functional glutamate receptors fails to occur (Broadie and Bate, 1993b).

Further evidence that motor neuron terminals selectively regulate AChR synthesis comes from the localized expression of the specific mRNAs. Muscle cells are polynuclear, and the nuclei that lay directly below the synaptic cleft are distinctive from those found extrasynaptically in that they preferentially transcribe AChR mRNA; this is controlled by the presence of motoneuron terminals (Klarsfeld et al., 1991; Sanes et al., 1991; Simon et al., 1992; Merlie and Sanes, 1985).

As new receptors are added, they also become more stable. This has been shown by measuring how long the receptors remain in the membrane before being replaced. In the chick, the rate of AChR turnover gradually increases from from a half-life of $\approx$30 hours at the

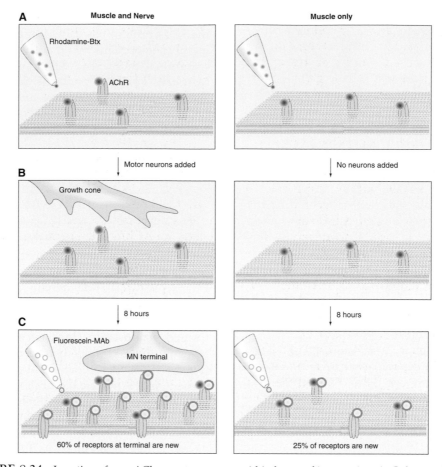

FIGURE 8.24 Insertion of new ACh receptors occurs within hours of innervation. A. Cultures of muscle cells were prelabeled with rhodamine-conjugated bungarotoxin (red). B. In one set of cultures, motor neurons were added (left), while a second set of cultures remained without neurons (right). C. After eight hours, both cultures were labeled with a fluorescein-conjucated antibody against AChRs (yellow). The cultures with motor neurons contained many AChRs that were labeled with only antibody (yellow), indicating that they had been newly inserted after the addition of motor neurons. The muscle cell cultures had AChRs that were primarily labeled by both Rhod-Btx (red) **and** Fluor-MAb (yellow). (Adapted from Role et al., 1985)

time of synaptogenesis to ≥50 days at 3 weeks posthatch (Burden, 1977a, 1977b). In the rat diaphragm muscle, the AChRs that appear during synaptogenesis at E15 also have a half-life of ≈30 hours, and this increases to a mature half-life of 6–11 days by E21 (Reiness and Weinberg, 1981). The signal that leads to increased receptor stability is not yet known, but it may involve common second messenger systems. Receptor half-life is prolonged either by the influx of postsynaptic calcium or a rise in cAMP (Rotzler et al., 1991; Shyng et al., 1991).

Innervation can also regulate the expression of receptors at neuron–neuron contacts. In co-cultures of chick spinal cord and dissociated sympathetic neurons, the ACh-evoked response recorded in sympathetic neurons increases almost 10-fold after innervation (Role, 1985). This effect can be produced with spinal cord-conditioned

media, suggesting that the signal is a soluble factor (Gardette et al., 1991). The influence of innervation on receptor synthesis can be quite specific, as revealed by culturing chick motoneurons in the presence or absence of spinal cord interneurons. Motoneurons were first selectively labeled with fluorescent dye in vivo, such that they could be identified in a dissociated cell culture (O'Brien and Fischbach, 1986a, 1986b). When cultured in the absence of spinal interneurons, dissociated chick motoneurons exhibit much smaller glutamate-evoked currents, but their sensitivity to GABA and glycine is unaffected. The presence of interneurons serves to localize glutamate sensitivity to the motoneuron processes, whereas the somata were maximally sensitive to glutamate in sorted cultures.

Are receptors synthesized only within the cell body and then transported to distant synapses? Protein synthesis occurs in neuronal dendrites, often near synapses, and polyribosomal aggregates appear in dendritic spines during development (Miyashiro et al., 1994; Steward and Shuman, 2001). However, it has been difficult to determine whether the synthesis of new glutamate receptor subunits occurs subsynaptically in the dendrites of central neurons (Steward, 1994; Craig et al., 1993). Injection of mRNA directly into isolated dendrites showed that local synthesis was possible (Kaharmina et al., 2000).

An imaginative set of techniques was used to prelabel existing AMPA receptors in cultured hippocampal neurons, and then counterstain with a second label to identify the location of newly synthesized subunits. In concept, this approach is nearly identical to that employed on AChRs by Role et al. (see Figure 8.24). Hippocampal neurons were first transfected with a glutamate receptor subunit (GluR1 or GluR2) that was modified to contain a tetracysteine motif on its intracellular C-terminal (Figure 8.25). This permitted the use of two different dyes that fluoresced (red or green) only when bound to tetracysteine. When exposed to the red dye, the majority of GluRs expressed over a 24-hour period were labeled in the dendrites. The red dye was then removed, and the neurons were exposed to the second green dye eight hours later. Newly expressed

GluRs were labeled primarily in the soma, the presumed site of synthesis, and a few were found in the dendrite, presumably due to rapid transport. To show that some receptors are synthesized locally, the dendrite was transected from the soma after prelabeling with red dye. Thus, any GluR aggregates that were subsequently labeled with only the green dye must have been synthesized within the dendrite (Figure 8.25). This turned out to be the case (Ju et al., 2004). Finally, it was possible to show that the GluRs synthesized in the dendrite were inserted into the membrane. Since the mRNA for GABA receptors is located within the dendrites of cortical pyramidal neurons (Costa et al., 2002), it is likely that receptor expression can be regulated by a local cue, as occurs for AChRs at the neuromuscular junction.

NEURONAL ACTIVITY REGULATES RECEPTOR EXPRESSION

The increase in receptor synthesis that accompanies synapse formation suggests that the presynaptic terminal initiates this process. One simple possibility is that the transmitter itself can regulate expression of synaptic proteins. At the neuromuscular junction,

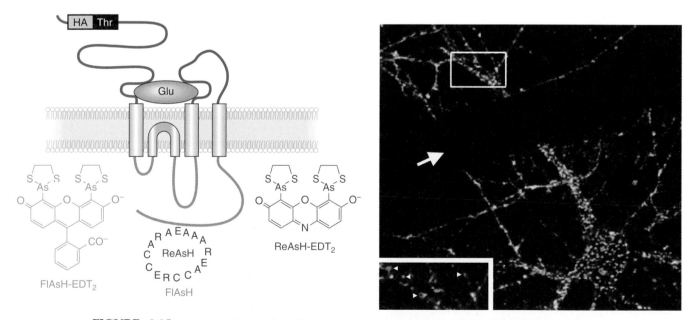

FIGURE 8.25 Local synthesis of AMPA receptors in the dendrite. (Left) Schematic showing the membrane topology of GluR1/2 and the location of the intracellular tetracysteine and extracellular HA/thrombin tags. The tetracysteine binds to FlAsH-EDT$_2$ or ReAsH-EDT$_2$, yielding a green or red fluorescent signal, respectively. (Right) An example of a hippocampal neuron in which all GluR1 protein was prelabeled with red dye, followed by transection of the dendrite (white arrow) and labeling with green dye. Note that some GluR aggregates are labeled only with the green dye (inset, white arrowheads), indicating that they must have been synthesized after the dendrite was detached from the soma. Preexisting aggregates that are labeled with both red and green dye are orange. (Ju et al., 2004)

synaptic activity actually inhibits AChR synthesis in the extrasynaptic region. For example, when adult cat muscle is denervated, the muscle cells become highly responsive to ACh applied at any position along the surface. This is referred to as denervation supersensitivity, and it requires new receptor synthesis (Axelsson and Thesleff, 1959; Merlie et al., 1984). Denervation supersensitivity can also be produced by decreasing the transmission of an intact terminal (Figure 8.26). When presynaptic action potentials are blocked or cholinergic transmission is eliminated, there is a dramatic increase in AChRs (Lømo and Rosenthal, 1972; Berg and Hall, 1975). The oppositive manipulation, direct electrical stimulation of muscle cells in vitro, produces a decrease in AChR synthesis (Shainberg and Burstein, 1976). At least in muscle cells, synaptic activity limits receptor synthesis through increasing postsynaptic calcium (Figure 8.26). The effect may depend in part on the CaMKII activation and phosphorylation of the muscle transcription factor, myogenin (Klarsfeld et al., 1989; Laufer et al., 1991; Huang et al., 1992; Tang et al., 2004).

A similar sort of regulation probably occurs in neurons. Unlike the NMJ, supersensitivity cannot be observed as directly in the central nervous system because neurons are embedded in a web of glia, ECM, and blood vessels. However, many areas of the nervous system express high levels of the NMDA receptor during development, and this expression seems to be regulated by innervation. For example, the functional expression of NMDA receptors decreases with age in the visual cortex of normal kittens, but when animals are reared in complete darkness to decrease visually driven activity, NMDAR-mediated transmission remains at an unusually high level (Fox et al., 1992). Similar sorts of observations have been made for AMPA receptors.

NEUREGULIN, A REGULATOR OF POSTSYNAPTIC TRANSCRIPTION

If synaptic activity represses receptor synthesis, how does the presence of motor nerve terminals cause an increase AChR synthesis by the postsynaptic muscle cells? Neonatal AChRs are initially composed of four subunits, $\alpha_2\beta\gamma\delta$, but during the first two postnatal weeks in rat, the many nuclei beneath each synapse stop expressing the γ-subunit (Figure 8.27A). There is a gradual increased expression of ε-subunit transcripts, resulting in a new heteromeric receptor, $\alpha_2\beta\varepsilon\delta$ (Gu and Hall, 1988). The basal lamina has previously been shown to also contain a signal that

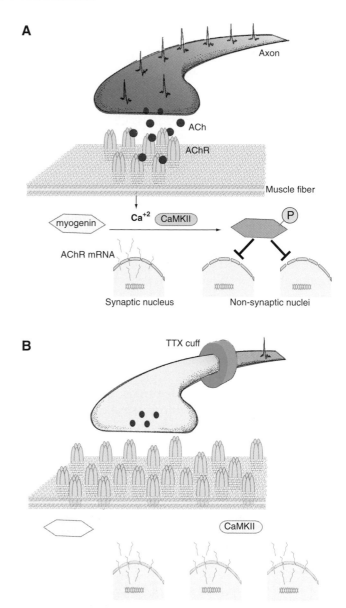

FIGURE 8.26 Extrasynaptic ACh receptors accumulate when the nerve is inactive. A. At the control nerve–muscle junction, the electrically active terminal releases ACh and the receptors are clustered at the postsynaptic membrane. The activity-dependent signal that suppresses extrajunctional receptors involves calcium influx and activation of the calcium calmodulin-dependent protein kinase II (CamKII). A transcription factor found in muscle (myogenin) is phosphorylated and blocks transcription in extrasynatpic nuclei. B. When motor axon activity is blocked with the sodium channel blocker, tetrodotoxin (TTX), extrajunctional ACh receptors are distributed over the entire muscle surface. (Adapted from Lømo and Rosenthal, 1972)

activates AChR transcription in the absence of motor nerve terminals (Goldman et al., 1991).

An initial screen of soluble factors present in the chick brain revealed a substance that could stimulate AChR synthesis in isolated myotubes (Jessell et al.,

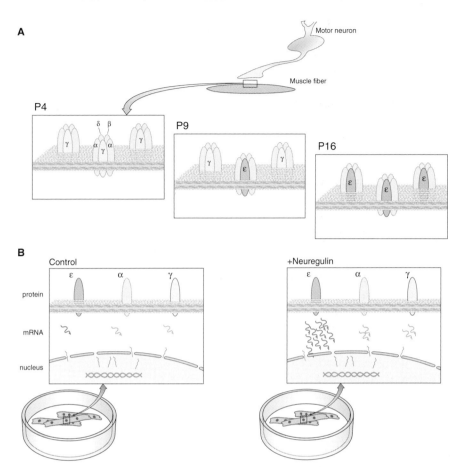

FIGURE 8.27 Substitution of ACh receptors subunit during development. A. In rat muscle, AChRs are composed of α, β, δ, and γ subunits at postnatal (P) day 4. By P9, there are a mix of receptors: some have the initial complement of subunits, and other have substituted the ε subunit in place of the γ subunit. At P16, all receptors contain the ε subunit. B.Neuregulin selectively upregulates the transcription of ε subunit mRNA. In control muscle cell cultures, approximately equal amounts mRNA for the ε, γ, and α AChR subunits are produced. The addition of neuregulin increases mRNA for all subunits, but the ε subunit is selectively enhanced. (Adapted from Gu and Hall, 1988 and Martinou et al., 1991)

1979). A 42 kD glycoprotein was isolated from the chick central nervous system, and subsequently found to be a member of the Neu protooncogene ligand family (Usdin and Fischbach, 1986; Falls et al., 1993; Jo et al., 1995). In the nervous system, members of this protein family are now referred to as the neuregulins (NRGs), and they play a broad role in neural development, including migration and glial differentiation (Chapters 3 and 4). More than 20 NRG1 isoforms are produced from the single *NRG1* gene. The receptors for neuregulins are actually members of another large family: the EGF receptor tyrosine kinase family (erbBs). At the neuromuscular junction, NRG-1 co-localizes with at least three of these erbB receptors. Neuregulin signaling may involve a common kinase pathway. In one cell line, neuregulin stimulates tyrosine phosphorylation of erbBs and mitogen-activated protein kinase (MAPK). Furthermore, a specific inhibitor of MAPK abolishes neuregulin-induced AChR subunit expression (Si et al., 1996).

NRG1 message is localized to motor neuron cell bodies, and the protein is transported to the synaptic junction. Electron microscopic images show that NRG1 becomes concentrated on the presynaptic side of the basal lamina after it is released at the neuromuscular junction (Goodearl et al., 1995). NRG1 remains in the synaptic basal lamina even after denervation (Falls et al., 1993; Sandrock et al., 1995; Jo et al., 1995). The NRG-erbB signaling pathway is recruited, in part, by Agrin. Even in the absence of nerves, Agrin is able to cluster both NRG and erbB in muscles (Rimer et al., 1998). Furthermore, one of the NRG receptors, erbB3, does not accumulate at synapses in rapsyn -/- mice (Moscoso et al., 1995).

The increase in AChR synthesis results largely from an accumulation of mRNAs for specific AChR subunits, particularly the ε-subunit (Figure 8.27B) (Harris et al., 1988; Martinou et al., 1991). Mice deficient for NRG1 or ErbB2 die prior to muscle formation and the influence on synaptogenesis cannot be determined. However, the heterzygotes of a NRG1 knockout mouse express less neuregulin than controls, and display a significant reduction of AChRs (Sandrock et al., 1997). This was studied by measuring the size of individual synaptic events, called *quanta*, which are probably due to the release of ACh from a single synaptic vesicle. These quantal events are smaller in heterozygote mice, presumably because there is less AChR to transduce the signal. Similarly, when erbB2 is inactivated selectively in muscle, there is a modest decline in miniature endplate currents and in AChR number at the synapse (Leu et al., 2003).

Unfortunately, the NRG-erbB signaling pathway is also essential to Schwann cell survival, and Schwann cell-derived cues are not available to the motor axons. Furthermore, muscle cells contain and release NRG1, which can induce AChR gene expression (Meier et al., 1998; Yang et al., 2001). Finally, a second neuregulin (NRG2) is expressed by motor neurons and Schwann cells and is found at the neuromuscular junction (Rimer et al., 2004). Despite these complications, most evidence suggests that NRG regulates postsynaptic transcription.

Several members of the NRG and erbB families are found in the chick and mammalian CNS during development, suggesting a role in neuron–neuron synapse formation. In developing chick sympathetic ganglia, a specific NRG1 isoform (type III) selectively upregulates the transcription of the α3 AChR subunit (Yang et al., 1998). In the developing cerebellum, NRG1 has been shown to increase the expression of an NMDA receptor subunit and a $GABA_A$ recetor subunit (Ozaki et al., 1997; Rieff et al., 1999). Intrestingly, NRG1 has been shown to decrease $GABA_A$ receptor expression in the hippocampus without affecting glutamate receptors (Okada and Corfas, 2004).

MATURATION OF TRANSMISSION AND RECEPTOR ISOFORM TRANSITIONS

Synapse formation is rapid, but adult functional properties emerge only gradually during development. One of the most common observations is that the duration of excitatory or inhibitory synaptic potentials declines over the course of days (Figure 8.28). For example, in the rat neocortex, the duration of excitatory postsynaptic potentials (EPSPs) decreases from approximately 400 to 100 ms during the first two post-

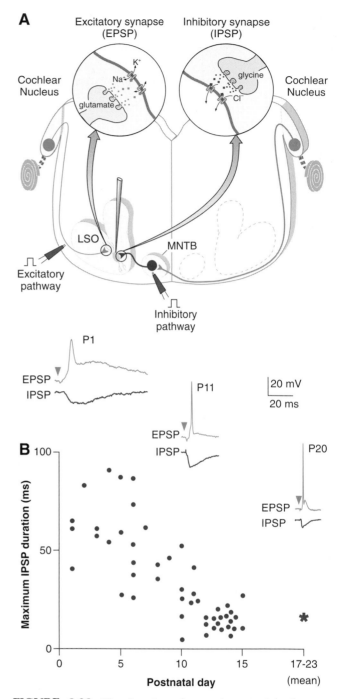

FIGURE 8.28 The duration of synaptic potentials decreases during development. A. A schematic of a central auditory nucleus, the lateral superior olive (LSO), that receives excitatory synapses from the ipsilateral cochlear nucleus and inhibitory synapses from the medial nucleus of the trapezoid body (MNTB). The inset at left shows that excitatory terminals release glutamate and open receptors that are permeable to Na^+ and K^+. The inset at right shows that inhibitory terminals release glycine and open receptors that are permeable to Cl^-. B. When intracellular recordings are made from LSO neurons during the first three postnatal weeks, the afferent-evoked EPSP and IPSP durations decline by about 10-fold. Examples for postnatal day 1, 11, and 20 are shown at the top, and a summary of all IPSPs is plotted in the graph. (Adapted from Sanes, 1993)

natal weeks of development (Burgard and Hablitz, 1993), and synaptic potentials in the rat hippocampus display a similar schedule of maturation. Even synapses in the brainstem display marked alterations during postnatal development. In the lateral superior olive, the maximum duration of both glutamatergic EPSPs and glycinergic inhibitory postsynaptic potentials (IPSPs) declines approximately 10-fold during the first three postnatal weeks (Sanes, 1993). The reduction in IPSP and EPSP duration has a similar rate of development, suggesting that some of the underlying mechanisms are the same. These long-lasting synaptic potentials probably limit the behavior capabilities of young animals (see Chapter 10).

Why are synaptic potentials of such long duration in developing neurons? One common difference is that young synapses usually express a unique form of the neurotransmitter receptor, called a *neonatal isoform*. These transiently expressed receptors often have different functional properties than the receptor that is expressed by adult neurons. In particular, the receptor-coupled ion channels in young cells tend to remain open for a longer period of time, compared to those in mature cells. In mammalian muscle cells, recordings were made from single channels with the patch-clamp recording technique (see BOX: Biophysics: Nuts and Bolts of Functional Maturation), and the mean channel open time was found to decline from about 6 to 1 ms during development (Siegelbaum et al., 1984; Vicini and Schuetze, 1985). This functional change is due, in part, to the molecular composition of AChRs in neonates (γ-subunit) versus that of adults (ε-subunit) (Figure 8.27A).

Even though nerve cells limp along on one nucleus in their soma, it appears that they are able to respond to innervation by altering the receptor isoform expression. When chick sympathetic ganglion neurons are innervated, their sensitivity to ACh is enhanced, and this is correlated with increased expression of 5 AChR transcripts (Moss and Role, 1993; Corriveau and Berg, 1993). At E11, only 30% of neurons have significant AChR activity, and each individual patch of membrane contains a mixture of AChRs. At E17, the great majority of patches have a single functional type receptor. Similar changes in specific AChR subunits have been observed in rat brainstem, spinal cord, and dorsal route ganglia during prenatal development.

A developmental switch in receptor subunits has now been demonstrated in nearly every transmitter system in the central nervous system. For example, the adult form of the glycine receptor heteromer involves the substitution of a 48 kD ligand-binding subunit for a neonatal isoform (Becker et al., 1988). Recordings from rat dorsal spinal cord neurons during develop-

ment showed that there is a complementary change in function (Takahashi et al, 1992). The glycine-gated channels from young animals (<P5) open for a much longer period of time and pass a greater amount of current, compared to older postnatal animals (Figure 8.29). By examining the properties of two different glycine receptor subunits in a *Xenopus* expression system, it was determined that a transition from the $\alpha 2$ to the $\alpha 1$ subunit could explain the functional change.

When a receptor family has many subunits, the type of receptor that is produced becomes a combinatorial problem. The temporal and regional expression of 13 different GABA$_A$ receptor subunits in the developing rat brain provides an interesting example (Laurie et al., 1992). The expression patterns are determined by in situ hybridization, a technique in which radiolabeled antisense oligonucleotides are used as probes for each species of mRNA (Figure 8.30). The onset of expression and the adult level of expression can vary greatly for a single subunit, depending on location. Moreover, there are a large number of subunits that are transiently expressed within a given structure. As one of many examples, the onset of $\gamma 2$ subunit expression occurs throughout the brain at embryonic day 17.

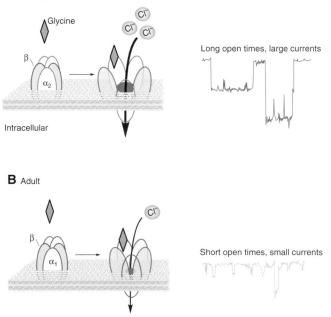

FIGURE 8.29 Neonatal glycine receptors have immature functional properties. A. In neonatal mammalian neurons, the glycine receptor is composed of β and α_2 subunits. When bound by glycine, the receptors remain open for a relatively long time and pass a relatively large current. B. In adult neurons, the glycine receptor contains a β subunit, but the neonatal isoform, α_2, is replaced by the α_1 subunit. These receptors open briefly and pass less current than the neonatal form. (Adapted from Takahashi et al., 1992)

P6

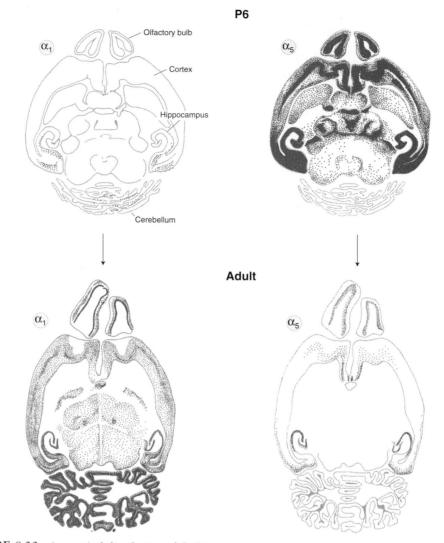

Adult

FIGURE 8.30   Anatomical distribution of GABA$_A$ receptor subunits changes dramatically during development. Each panel displays the staining pattern of an antibody directed against either the α_1 or the α_5 GABA$_A$R subunits. (Top) At P6, there is little α_1 in the brain, whereas α_5 is heavily expressed in the hippocampus and cortex. (Bottom) In adult, α_1 is heavily expressed, and α_5 is nearly absent. (Adapted from Laurie et al., 1992)

Whereas the level of $\gamma 2$ expression gradually increases in the hippocampus and cerebellum, it ceases to be expressed in the cortex and thalamus.

The long duration of excitatory synaptic events in many regions of the CNS is at least partly due to a neonatal form of the NMDA-gated glutamate receptor (Figure 8.31). The duration of afferent-evoked excitatory postsynaptic currents (EPSCs) in the rat superior colliculus that are mediated by NMDA receptors declines several fold during the first three postnatal weeks (Hestrin, 1992). Similar observations have been made in the ferret lateral geniculate nucleus and rat cortex. NMDAR subunit composition probably affects

other functional properties of the receptor. For example, NMDARs are sensitive to the presence of both ligand (glutamate) and membrane depolarization in adults, but voltage sensitivity may be absent in the neonatal hippocampus (Ben-Ari et al., 1988). Apparently, the neonatal receptors are less sensitive to Mg^{2+}, the ion that must be expelled from the channel pore during depolarization, thus permitting Na^+ and Ca^{2+} to pass through (Bowe and Nadler, 1990; Kirson et al., 1999). Since resting membrane potential is generally more depolarized early in development, excitatory transmission through the NMDAR may contribute a larger fraction of the the synaptic current.

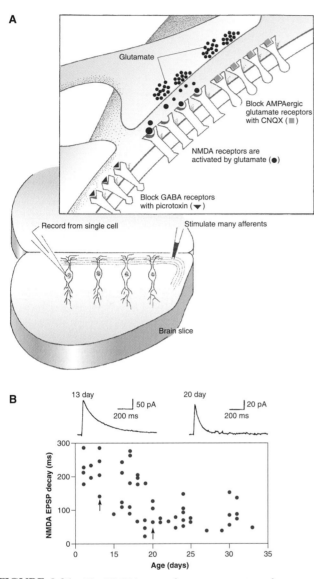

FIGURE 8.31 The NMDA-type glutamate receptors close more rapidly with age. A. Intracellular recordings were obtained from rat hippocampal neurons in a brain slice preparation, and AMPA-type glutamate receptors and GABA receptors were blocked. Thus, stimulation of afferents evoked glutamate release, and only postsynaptic NMDA-type receptors were activated. B. The afferent-evoked EPSPs were longer lasting in neurons from young neurons due to the slow decay time. (Adapted from Hestrin, 1992)

The signals that are responsible for receptor transitions may come from neurotransmission itself. For example, the waning of NMDAR-mediated responses in many areas of the brain (see above) is often accompanied by increased transmission through a second class of glutamate receptors, called *AMPA receptors*. Functional AMPARs can be rapidly recruited by NMDAR activity. In the neonatal rat, glutamatergic

synaptic transmission appears to be absent because most synapses have only functional NMDARs, and NMDA receptors tend to remain closed at the resting membrane potential. These are sometimes referred to as "silent synapses." When NMDARs are permitted to be active by stimulating the synapse during depolarizing current pulses, the synapses are soon found to have functional AMPARs (Durand et al., 1996). A similar pattern of maturation occurs in the optic tectum of *Xenopus* tadpoles. When calcium-calmodulin-dependent protein kinase II (CaMKII) is constitutively expressed in the tectal neurons, the appearance of AMPAergic transmission can be facilitated. This suggests that calcium entry through NMDARs may activate CaMKII, which mediates the recruitment of functional AMPARs (Wu et al., 1996). Activation of "silent" synapses has also been observed at slightly later periods of development and may, in fact, underlie certain forms of learning or memory (see Chapter 9).

MATURATION OF TRANSMITTER REUPTAKE

The time that the neurotransmitter remains in the synaptic cleft will also affect the duration of synaptic potentials, and the development of transmitter uptake systems is critical for the emergence of mature function. Neurotransmitter transporter protein development has been studied by expressing polyadenylated brain RNA (polyadenylation, or the addition of about 200 adenylate residues, is a common modification to transcripts in eukaryotic cells) in *Xenopus* oocytes (Blakely et al., 1991). Messenger RNA was obtained from animals of different ages and placed in a *Xenopus* oocyte expression system. The amount of transport was quantified by incubating the oocyte in a radiolabeled amino acid neurotransmitter, such as ^{3}H-glycine, and the amount of ^{3}H was quantified with a liquid scintillation counter. By using this assay, it was found that glutamate and GABA transporters first appear in the cortex at postnatal day 3 and increase to adult levels over the next two weeks. In the brainstem, the expression of a glycine transporter gradually increases to adult levels over the first three postnatal weeks.

A number of amino acid transporters have now been identified at the molecular level, and a few studies have traced their developmental appearance using in situ hybridization. The excitatory amino acid transporters, mEAAT1 and mEAAT2, are first found in the proliferative zone of mouse forebrain and midbrain during gliogenesis (E15-E19). However, mEAAT2 mRNA continues to increase in many areas of the CNS during the first two to three postnatal weeks (Sutherland et al., 1996). Transcripts for the Na$^+$/Cl$^-$-dependent glycine transporter (GlyT1), found almost exclusively in glial cells, achieve maximal levels in E13 mice, much earlier in neural development (Adams et al., 1995). Although the presence of transporter mRNA suggests that neurotransmitter could be efficiently cleared at the onset of synaptogenesis, studies of amino acid transporter function show that their physiology remains immature for some time (Blakely et al., 1991). Therefore, the maturation of transporter proteins probably limits the kinetics of synaptic transmission.

SHORT-TERM PLASTICITY

To this point we have considered only the most basic response of a synapse: the release of transmitter to a single action potential and the postsynaptic current that it produces. Of course, neurons will fire many times per second under realistic conditions, and the synaptic response may become facilitated or depressed over time. These changes in synaptic response are called short-term plasticity, and their maturation depends on the development of presynaptic release properties and the complement of postsynaptic receptors and ion channels.

A simple approach to examine short-term plasticity involves taking relatively thick (300–500 μm) slices of brain tissue at increasing postnatal ages and recording the synaptic response that is elicited when trains of stimuli are delivered to the afferent pathway. To examine short-term plasticity, synaptic currents were recorded from MNTB neurons in response to stimulation of excitatory afferents from the cochlear nucleus (see schematic in Figure 8.28A). MNTB neurons are innervated by only a single glutamatergic afferent that makes a large synapse on the cell body, called the *endbulb of Held*. When these synapses are stimulated at 200 Hz in young MNTB neurons (P5), they display a rapid depression of the postsynaptic response, and there are complete failures where the transmitter is apparently not released by the endbulb of Held (Figure 8.32A). However, when the same stimulus is delivered to MNTB neuron afferents at P14, the response does not display as much depression, and there are no failures of transmitter release (Joshi and Wang, 2002). Several mechanisms may account for this maturation. In young animals, the endbulb of Held produces an action potential that lasts a relatively long time, and this prevents it from responding to each stimulus. It is

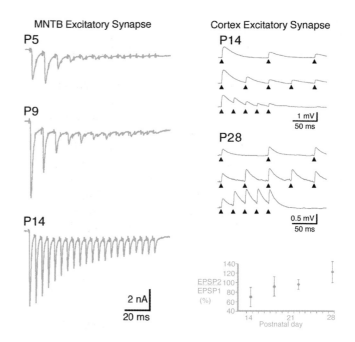

FIGURE 8.32 Development of short-term synaptic plasticity. A. AMPA receptor-mediate EPSCs are recorded in MNTB neurons in response to a 200 Hz stimulus train. Examples are shown from neurons at P5, P9, and P14. There is a significant reduction in the extent of synaptic depression and failures. B. EPSPs are recorded in Layer 5 pyramidal neurons in response to stimulation of a second Layer 5 cell. In P14 cortex, stimulation of the presynaptic neuron at an increasing rate (responses to 10, 20, and 40 Hz are shown) evoked EPSPs that declined in amplitude. In P28 cortex, stimulation of a presynaptic Layer 5 neuron evoked EPSPs in a postsynaptic Layer 5 cell that facilitated at that same stimulus rates. The graph shows summary data from P14 (*n* = 52), P18 (*n* = 9), P22 (*n* = 6), and P28 (*n* = 10) rats. (From Reyes and Sakmann, 1999; Joshi et al., 2002)

also likely that the pool of vesicles available for release during rapid stimulation increases with development. Postsynaptic mechanisms could also contribute to this depression, including desensitization of glutamate receptors.

The developing cortex displays an even greater transformation in short-term plasticity (Figure 8.32B). When two interconnected neurons are recorded in P14 sensorimotor cortex, it is found that excitatory synapses display a depression when stimulated between 10 and 40 Hz. When a similar pair of neurons is recorded at P28, the excitatory connections display facilitation (i.e., the second postsynaptic response is larger than the first) (Reyes and Sakmann, 1999). Once again, the developmental switch from depression to facilitation may be caused by several factors including regulation of presynaptic Ca^{+2} concentration and glutamate receptor desensitization.

APPEARANCE OF SYNAPTIC INHIBITION

Up to this point, our discussion has focused on excitatory synapses; these connections have provided the great majority of information on synaptogenesis, and most of that from the cholinergic NMJ. Initially, it was thought that inhibitory synapses, those releasing GABA or glycine, matured after excitatory synapses. This is because IPSPs are often not observed in neontal animals. For example, intracellular recordings from the kitten visual cortex demonstrate that afferent-evoked IPSPs are absent from over half the neurons during the first postnatal week, whereas all neurons display IPSPs by adulthood (Komatsu and Iwakiri, 1991). Similar observations have been made on the developing rat neocortex (Luhman and Prince, 1991).

However, synaptic inhibition appears with a similar time course as synaptic excitation in diverse areas such as the spinal cord, cerebellar nuclei, olfactory bulb, lateral superior olive, and somatosensory cortex (Oppenheim and Reitzel, 1975; Sanes, 1993). Inhibitory events are probably more difficult to detect in young animals, both because they are concealed by excitatory events (Agmon et al., 1996) and their equilibrium potential is close to the resting membrane potential (Zhang et al., 1991). Therefore, it is likely that inhibitory synapses are present from the outset, but their functional properties are immature.

IS INHIBITION REALLY INHIBITORY DURING DEVELOPMENT?

In adult animals, inhibitory synaptic potentials are generally hyperpolarizing. This is because the receptor is coupled to a Cl^- channel and the Cl^- equilibrium potential is more negative than the cells resting potential. However, inhibitory synaptic transmission usually produces *depolarizing* potentials during the initial phase of development (Obata et al., 1978; Bixby and Spitzer, 1982; Mueller et al., 1983, 1984; Ben-Ari et al., 1989). For example, during the first postnatal week, rat hippocampal neurons display large spontaneous and evoked depolarizations that are blocked by the $GABA_A$ receptor antagonist, bicuculline (Figure 8.33A).

These depolarizing IPSPs are apparently large enough to open voltage-gated calcium channels. In dissociated cultures obtained from embryonic rat hypothalamus, intracellular free calcium is decreased by bicuculline during the first 10 days in vitro (Obri-

etan and van den Pol, 1995). As the cultures mature, bicuculline increases calcium, presumably by allowing excitatory synaptic acitivity to have a greater depolarizing influence (Figure 8.33B). Therefore, inhibitory synapses may provide a qualitatively different input to postsynaptic neurons during development.

Inhibitory postsynaptic potentials gradually become hyperpolarizing during development, as has been demonstrated in the spinal cord, brainstem, hippocampus, and cortex (Kandler and Friauf, 1995; Agmon et al., 1996; Zhang et al., 1991). The depolarizing inhibitory potentials seen in young animals are probably due to the outward flow of Cl^- through $GABA_A$ or glycine receptor-coupled channels (Reichling et al., 1994; Owens et al., 1996). Therefore, intracellular chloride must be elevated in young neurons, and it is important to understand how chloride is distributed across the membrane.

Intracellular chloride $[Cl^-]_i$ is regulated primarily by two cation-chloride cotransporter family members: a Na-K-2Cl cotransporter (NKCC1) leads to cytoplasmic accumulation of chloride, and a K-Cl cotransporter (KCC2) extrudes chloride (Payne et al., 2003). During early development. $[Cl^-]_i$ is relatively high due to NKCC1 activity (Clayton et al., 1998; Kanaka et al., 2001). In LSO neurons, NKCC1 transports Cl^- into the cell, particularly in immature neurons, and this contributes to the depolarizing IPSPs (Kakazu et al., 1999). Similarly, Cl^- is transported into Rohon-Beard cells in the developing *Xenopus* spinal cord, leading to GABA-evoked depolarizations (Rohrbough and Spitzer, 1996).

As KCC2 expression increases, $[Cl^-]_i$ drops below the electrochemical equilibrium (Lu et al., 1999; DeFazio et al., 2000; Hübner et al., 2001). This event plays the greatest role in the transition from inhibitory synapse-evoked depolarizations to hyperpolarizations (Owens et al., 1996; Ehrlich et al., 1999; Kakazu et al., 1999; Rivera et al., 1999). The presynaptic terminal can influence chloride transporter expression or function. In hippocampal cultures, $GABA_A$ receptor activation facilitates KCC2 expression and the appearance of $GABA_A$-mediated hyperpolarizations (Ganguly et al., 2001).

Neuron-specific KCC2 has a tyrosine phosphorylation consensus site (Payne, 1997), and its function may be modulated during development. For example, KCC2 is expressed at high levels in LSO neurons during the time when they display depolarizing IPSPs (Balakrishnan et al., 2003), suggesting that a post-translational modification must be involved. It has been found that cultured hippocampal neurons initially expressed an inactive KCC2 protein, which becomes activated during maturation (Kelsch et al., 2001). Activation of KCC2 in immature neurons can be

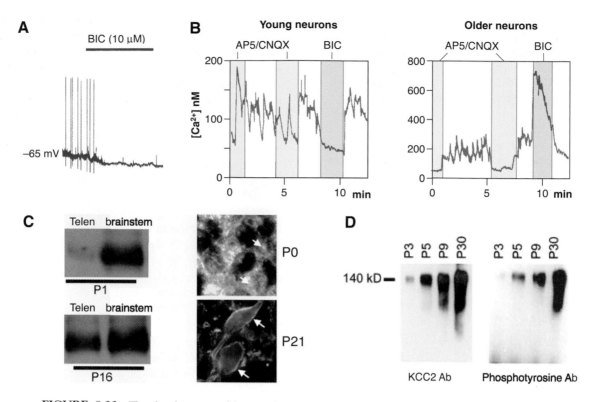

FIGURE 8.33 The development of hyperpolarizing inhibition. GABAergic inhibition initially evokes membrane depolization and calcium entry. A. Intracellular recording from a neonatal rat hippocampal neuron shows that the GABA_A receptor antagonist, bicuculline (BIC), blocks the spontaneous action potential and causes the cell to hyperpolarize. B. Intracellular free calcium was monitored in hypothalamic cultures during exposure either to glutamate receptor antagonists (AP5/CNQX) or to a GABA_A receptor antagonist (BIC). After 8 days in vitro (young neurons), only BIC produced a decrease in calcium. At 33 days in vitro (older neurons), AP5/CNQX produced a decrease in calcium, and BIC increased calcium. C. Western blot shows that KCC2 protein is developmentally upregulated in the telencephalon but not in the brainstem, where expression is high from the outset (left). Immunohistochemical staining of rat LSO neurons (right) shows that the KCC2 protein labeling is primarily intracellular at P0 (arrows), but the signal is at the plasma membrane surrounding the somata and proximal dendrites at P21 (arrows). D. KCC2 was immunoprecipitated with a KCC2 antiserum in cortex tissue from P3, P5, P9, and P30 mice. A Western analysis was performed on the immunoprecipitate using either a KCC2 antibody (left) or a phosphotyrosine-antibody (right). (Adopted from Ben-Ari et al., 1989; Obrietan and van den Pol, 1995; Balakrishnan et al., 2003; Stein et al., 2004)

induced by IGF-1 or a Src kinase, whereas membrane-permeable protein tyrosine kinase inhibitors deactivate KCC2. Therefore, endogenous protein tyrosine kinases may mediate the developmental switch of inhibitory responses by modifying KCC2. In fact, there is an increase in both KCC2 protein expression and tyrosine phosphorylation during normal development of the mouse cortex (Stein et al., 2004). Finally, the neurotrophin signaling system has been implicated in regulating chloride transport. In transgenic embryonic mice that overexpress BDNF under the control of the *nestin* promoter, KCC2 expression increases dramatically (Aguado et al., 2003). Interestingly, it appears that BDNF may exert the opposite influence in older animals (Rivera et al., 2002).

SUMMARY

The generic cortical neuron with which we began the chapter somehow manages to express just the right complement of receptors and channels, and place them at the correct part of the cell. As this chapter makes clear, the differentiation of synapses and electrical properties depends upon an ongoing discussion between neuronal connections. Fortunately, we now have a basic understanding of synapse formation, including a few of these transynaptic signalling pathways. As extraordinary as these accomplishments are, it is important to recognize that we have ignored most of the modulatory afferents, many of the neurotransmitter receptors, and several of the cytoplasmic signalling pathways. We have begun to understand how the most basic attributes of transmission and electrogenesis develop, yet we have little understanding of how these functional building blocks shape the computational properties of a developing neuron. After all, that is the goal of neural development. Why is 10,000 synapses the correct number for a cortical pyramidal neuron, while an MNTB neuron receives only a single synapse? Would each of these neurons perform adequately with a different number of inputs? Similarly, how does the number of glutamatergic synapses influence the number of GABAergic or serotonergic synapses? How would a neuron operate if inhibitory synapses formed on dendritic spines, instead of excitatory synapses? As we start to understand how individual synapses are constructed, it becomes critical to explore the activity-dependent mechanisms that regulate their placement and strength (Chapter 9).

Refinement of Synaptic Connections

The process of development would appear to be complete once all neurons are born, differentiate, and connect with one another. Yet even as the number of synapses increases in the target region (Chapter 8), a separate event is set in motion that leads to the *elimination* of some existing synapses. For example, thalamic projections arborize in the kitten visual cortex before birth, and a postnatal burst of synaptogenesis leads to an increase in the number of synapses per neuron from a few hundred to about 12,000. As new synapses are added, individual afferent projections from the thalamus begin to retract some of their branches from neighboring regions of the cortex (Cragg, 1975; LeVay et al., 1978). This eventually leads to a "striped" pattern of innervation that has been studied intensively. Some afferent connections are eliminated entirely during development. Turning again to the kitten visual cortex, it has been shown that commissural afferents from the opposite side of the brain are lost in great numbers during the first three postnatal months.

Why are synapses being assembled and disbanded at the same time, particularly when the pathfinding and mapping mechanisms produce such accurate results? The central nervous system is a tissue designed for continuous modification, from birth into adulthood (cf. learning and memory). The addition and loss of synapses may reflect a major goal of development: that is, to optimize behavioral performance in a particular environment. One might argue that there is no better time for learning and optimizing performance than during development (see Chapter 10). Therefore, both synapse addition and synapse elimination can improve the specificity of neural connections. The correct complement and strength of synaptic connections should optimize the computational properties of each neuron.

What distinguishes developmental plasticity from learning and memory? The clearest difference is that the developing nervous system is altered permanently by some manipulations that have little effect on the adult. As discussed below, there is often a critical period of development during which synaptic connections or function can be altered by manipulations to the sensory environment. Experimental manipulations of this sort probably alter the normal amount or the pattern of synaptic transmission and action potentials, and this altered activity state somehow influences the growth and differentiation of synaptic connections. Even though the immature nervous system is particularly sensitive to these manipulations, we will see that developmental plasticity and adult learning share several molecular mechanisms.

THE EARLY PATTERN OF CONNECTIONS

Three general patterns of innervation distinguish the developing nervous system from that of the adult. First, individual axons that arborize in the correct topographic position (see Chapter 6) may spread out further than they do in the adult, perhaps a few tens of microns past their proper boundary (Figure 9.1A). While this may seem to be a trivial distance, if billions of neurons make projections of this sort, then neural computations could be adversely affected. A second way in which innervation may be immature occurs when a postsynaptic neuron receives synapses from more afferents than in the adult (Figure 9.1B). The ratio of innervating afferent axons per postsynaptic neuron, called *convergence*, varies greatly in the nervous system. At the mammalian nerve–muscle junction there is one

A   Refinement of topography

B   Refinement of convergence

C   Refinement of postsynaptic compartment

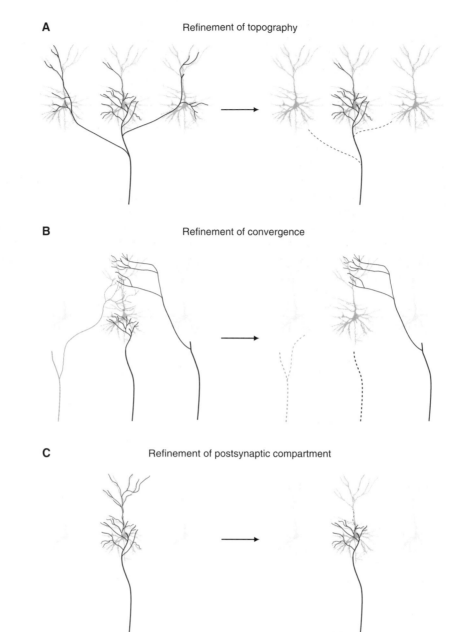

FIGURE 9.1   Three kinds of immature afferent projections during development. A. The projection of three afferents to the cortex is shown, and each one centers its arborization at the topographically correct position in the target. However, one of the arbors initially extends too far, and three local branches are eliminated during development. B. A single neuron is shown to receive input from three afferents initially, and two of these inputs are eliminated during development. C. A projection is shown to innervate the soma and dendrite of a postsynaptic neuron initially, but the dendritic innervation is eliminated during development.

motor axon synapse per muscle cell. In the cerebellum, each Purkinje cell receives innervation from a single climbing fiber axon (cf. climbing fiber convergence is 1), but is contacted by thousands of parallel fiber synapses on its dendritic tree (cf. parallel fiber convergence is ≈200,000). As we shall see, many postsynaptic neurons attain the adult number of afferents only after a fraction of the functional contacts are eliminated. A third way that innervation may become more specific

during development is through the elimination of terminals from one region of the postsynaptic neuron (Figure 9.1C). In one auditory brainstem nucleus, inhibitory terminals are eliminated from the dendrite and gradually become restricted to the cell body.

The addition or elimination of synapses during development is the most extreme way to modify a neural circuit. However, the postsynaptic response magnitude that is produced by an individual synapse (cf. synaptic strength) can also be regulated. In the adult nervous system, the strength of synaptic transmission changes dramatically with use, and these alterations support the storage of memories (see BOX: Remaining Flexible). The first studies to draw a strong causal relationship between environmental stimulation and the development of connections were performed in the cat visual system (Wiesel and Hubel, 1963, 1965; Hubel and Wiesel, 1965). In control animals, extracellular recordings from cortex show that most neurons fire action potentials in response to stimulation of both eyes. However, when visual stimulation to one eye is decreased during development, there is a dramatic loss in the ability of that eye to activate cortical neurons. The result suggests that synapses driven by the closed eye were either eliminated or weakened. Even though the initial connectivity of the visual pathway is quite accurate (Chapter 6), it is apparently not stable. Synaptic connections can be altered permanently by a developmental mechanism that makes use of electrical activity.

What is the evidence that synapses are eliminated in the developing nervous system? How widespread is this mechanism? Two experimental approaches have been taken to determine whether a loss of synapses occurs during development. First, intracellular recordings show changes in the number of functional afferents per postsynaptic neuron. Second, anatomical studies reveal that single axonal arbors become spatially restricted within the target population. But how do these detailed synaptic decision impact on nervous system performance? To answer this question, we will shift our attention from the molecular level to experiments that explore nervous system function and behavioral performance. One way to examine whether all of the synapses are working together correctly is to study the response of single neurons to sensory stimuli, such as light or sound. Many auditory neurons respond with great accuracy to the location of a sound source in space, and this reflects both the number and strength of its synaptic inputs. If there are immature patterns of connectivity, then one might expect that auditory neurons will respond to an unusually broad range of spatial stimuli. Therefore, a neuron's computational abilities are a sensitive assay of synaptic refinement.

FUNCTIONAL SYNAPSES ARE ELIMINATED

The developmental loss of synaptic contacts has been observed throughout the nervous system, from the nerve–muscle junction of invertebrates to the cerebral cortex of primates. These changes are not obvious by merely looking at neonatal and adult tissue sections under the microscope. Instead, quantitative comparisons must be made using measurements from many neurons. How is it possible to count the number of afferents per postsynaptic neuron? An imaginative approach to this problem, first employed in the early 1970s, used intracellular recordings and electrical stimulation of the afferent pathway. The basic assumptions are that each axon will evoke a postsynaptic potential (PSP) when stimulated, and that the PSPs will summate linearly, in discrete steps, as each additional fibers is recruited by the electric stimulus (Figure 9.2). Therefore, the increments in PSP size provide an estimate of the number of axons making a functional contact on a single muscle fiber or neuron.

When this experiment was performed at the mature neuromuscular junction, a single large PSP was recorded, indicating that the muscle fiber was innervated by a single motor nerve terminal. However, when the same experiment was performed in neonatal animals, the PSP size first doubled and then tripled in amplitude as the stimulus activated two, then three, motor axons (Figure 9.3). Similar observations have been made in developing chick, rat, and kitten muscle (Redfern, 1970; Bagust et al., 1973; Bennett and Pettigrew, 1974). The elimination of convergent motor axons at the rat soleus muscle results in a decrease

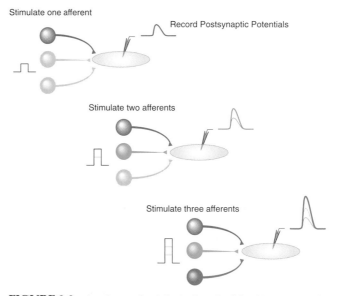

FIGURE 9.2 An electrophysiological method for determining the number of inputs converging onto a neuron. A stimulating electrode is placed on the afferent population while an intracellular recording is obtained from the postsynaptic cell. As the stimulation current is increased, the afferent inputs are recruited to become active. When a single afferent is active (top), the postsynaptic potential (PSP) is small. When two (middle), and then three afferents are activated (bottom), the PSP become quantally larger. One can estimate the number of inputs by counting the number of quantal increases in PSP amplitude, in this case three.

from three axons per muscle fiber to only one during the second postnatal week.

The precise time course over which synapse elimination occurs, and the number of afferents lost, vary greatly between areas of the nervous system, even within a single species. In the rat cerebellum, the elimination of climbing fiber synapses onto Purkinje cells occur during the second postnatal week (Mariani and Changeux, 1981). In contrast, the elimination of preganglionic synapses onto neurons of the rat submandibular ganglion occurs over at least five postnatal weeks (Lichtman, 1977), far longer than is required for elimination at the neuromuscular junction (Figure 9.3). The number of cochlear nerve synapses on neurons of the chick cochlear nucleus declines rapidly, from about four to two afferents (Figure 9.3), and reaches a mature state even before hatching (Jackson and Parks, 1982).

As the number of afferents increases, it becomes difficult to resolve small differences in PSP size. However, functional estimates of synaptic convergence suggest that elimination occurs even in systems that remain multiply innervated as adults (Lichtman and Purves, 1980; Sanes, 1993). This physiological method is not sensitive to certain forms of synapse elimination. For example, climbing fiber axons innervate the soma and dendrite of cerebellar Purkinje cells in neonates, but the somatic synapses are eliminated during develop-

ment (Altman, 1972). A novel method for studying the functional elimination of synapses makes use of a technique in which glutamate is focally elevated in the projecting population while an intracellular recording is made from a postsynaptic neuron. This technique was used to demonstrate a fourfold decrease in the number of MNTB neurons innervating a single LSO neuron during the first postnatal week in rats (Kim and Kandler, 2003). Therefore, synapse elimination appears to be a widespread phenomenon, although there are no general rules about the percentage of afferents that are lost or the duration of time required.

AXONAL ARBORS ARE REFINED OR ELIMINATED

How is it possible to know whether axon or synapse elimination occurs at central neurons that receive contacts from hundreds or thousands of afferents? The size of individual PSPs is too small and their amplitude is too variable, leaving one with a cloud of synaptic potentials that do not increase in crisp steps. One approach is to count all the synaptic contacts on a neuron at several postnatal ages. For example, measures of synapse number and length taken from motor neurons provide an estimate that 50% of synaptic contacts are lost during development. Similar estimates have been made for human cortex (Conradi and Ronnevi, 1975; Huttenlocher and de Courten, 1987).

Impressive as these numbers are, there are two potential problems. First, they do not tell us whether the actual number of axons making synapses onto postsynaptic neurons change during development. For example, a single afferent could initially make 100 weak synapses that gradually transform into 50 strong ones. Second, *total* synapse number may increase in some systems during the time when a small fraction of synapses are being eliminated, and this would go unnoticed.

If we assume for a moment that axons make synapses wherever they arborize, then it should be possible to study synapse elimination indirectly by looking at the amount of territory occupied by the axonal arborization. The most obvious case occurs when a projection is eliminated entirely due to the death of nerve cell bodies. In chicks, there is a projection from a brainstem nucleus, the isthmo-optic nucleus, to the retina in which nearly 60% of the projecting cells die, particularly those projecting to the wrong place in the retina (Casticas et al., 1987). The elimination of errant projections from the olfactory epithelium to the olfactory bulb may also result from the selective loss of sensory neurons (Zou et al., 2004).

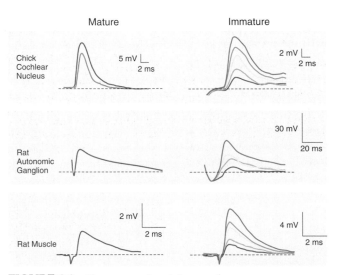

FIGURE 9.3 Three examples of decreased convergence as measured electrophysiologically (see Figure 9.2). On the right are shown the increases in afferent-evoked PSP recorded in immature neurons of the chick cochlear nucleus, rat autonomic ganglion, and rat NMJ. There are 3–5 quantal increases in PSP amplitude. On the left are shown the increases in afferent-evoked PSP amplitude in mature neurons. There are 1–2 quantal increases in PSP amplitude, indicating the functional elimination of inputs. (Adapted from Jackson and Parks, 1982; Lichtman, 1977; O'Brien et al., 1978)

Furthermore, these errant projections remain into adulthood when one nare is closed during development, suggesting that sensory experience is required for the formation of adult maps (Chapter 6).

Elimination of commissural axons, projecting from one side of the cerebral cortex to the other, occurs in the absence of cell death (Innocenti et al., 1977). This was demonstrated in rats by labeling commissural neurons twice, once early in development and then once again after certain axons retract (O'Leary et al., 1981). Neurons that projected to the other hemisphere were retrogradely labeled at birth by injecting a dye on one side of the brain and allowing commissural axons to transport it back to the cell body (Figure 9.4). Two weeks later, a second dye was injected in the same spot, and the remaining commissural axons retrogradely transported it to their cell bodies. When the tissue was examined, many cells were stained with only the first dye (Figure 9.4, green), but only a fraction of these labeled cells also contained the second dye (Figure 9.4, red). The green cells must have sent axons through the commissure at birth, but some of the cells had apparently retracted their axons before the red dye was injected. Therefore, many cortical neurons generate transient projections through the cerebral commissure (corpus callosum), and some of these axons are eliminated during development.

The wholesale withdrawal of axons provides a wonderful example of developmental refinement, but unfortunately for the anatomist this is not the norm. Changes in terminal arbor morphology are usually quite subtle. An axon tends to innervate the correct target region (Chapter 6), extend a bit beyond the correct topographic position, and then pull back to the adult boundary. Perhaps the best characterized examples of axon terminal elimination come from the developing visual pathway. In cats and primates, retinal ganglion cells from each eye project to separate layers in the lateral geniculate nucleus (LGN). The LGN neurons then project to Layer IV of the visual cortex, forming segregated eye-specific termination zones, called ocular dominance columns or "stripes" (Figure 9.5A).

It is possible to visualize the projection pattern of an entire eye by injecting ^{3}H-proline into the eye cup. This label is taken up by retinal ganglion cells and transported down their axons to the LGN where it crosses the synapse, enters the postsynaptic LGN neuron, and is carried by the axons to their terminals in the cortex. Autoradiographic images were obtained from the cortex, which showed the termination pattern of LGN axons originating in one eye-specific layer (Figure 9.5B). The LGN afferents from one eye were widespread in cortical Layer IV at 7 days post-

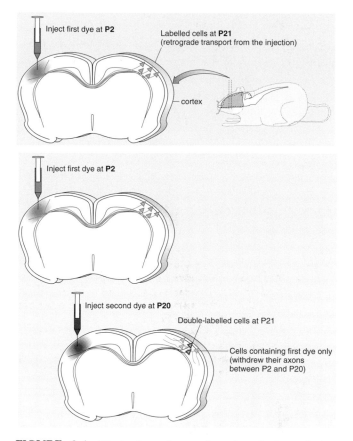

FIGURE 9.4 Elimination of commissural projections during development. At postnatal day 2 (P2), a dye (green) was injected into the cortex, and it was retrogradely transported by commissural neurons. At P21, the animal was sacrificed, and the tissue was processed to reveal the labeled neurons. In a second experiment, the first (green) dye was injected at P2, and a second (red) dye was injected at P20. When the tissue was processed, some commissural neurons were double labeled (green/red), whereas others contained only the green dye. Thus, the green-labeled cells projected to the contralateral cortex at P2, but retracted their axons during postnatal development and maintained local connections. (Adapted from O'Leary et al., 1981)

natal. Over the next several weeks, this diffuse label breaks up into discrete patches that represent eye-specific termination zones (LeVay et al., 1978; Crair et al., 2001). The light-evoked responses of Layer IV visual cortex neurons are consistent with the anatomy. By monitoring an intrinsic optical signal that is produced by electrically active brain tissue (see BOX: Watching Neurons Think), it is possible to show that visual cortex responds uniformly to stimulation of one eye at postnatal day 8. However, the same stimulus to one eye begins to activate patches of cortex by postnatal day 14 (Crair et al., 2001). Therefore, individual LGN arbors from one eye must retract a portion of their terminals during development or there is a selective elimination of entire geniculate axons. The

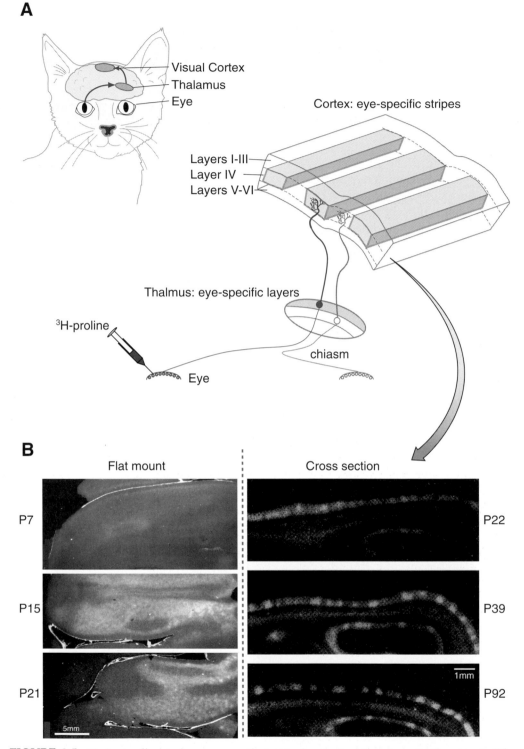

FIGURE 9.5 Thalamic afferents form eye-specific terminal zones in cat primary visual cortex. A. When ^{3}H-proline is injected into one eye, it is transported transynaptically to the cortex. A schematic shows the visual pathway from eye to cortex. Retinal ganglion cells project to an eye-specific layer in the thalamus, and thalamic neurons project to eye-specific stripes in cortex Layer IV. B. Autoradiograms of ^{3}H-proline in postnatal cat visual cortex show that the terminal field from one eye initially spreads across the entirety of Layer IV (7 days. left) but gradually becomes restricted to stripes. The left-hand column of images was obtained by flat mounting the entire cortex and looking down on the surface of Layer IV. This technique provides greater sensitivity as compared to the individual coronal sections (right-hand column) through the cortex. (Adapted from LeVay et al., 1978; Crair et al., 2001)

BOX

WATCHING NEURONS THINK: FUNCTIONAL PROPERTIES OF NEURON ENSEMBLES

A major goal of neurophysiology is to demonstrate a causal relationship between CNS function and animal behavior. While our success has been limited, there have been a number of exciting strategies that should bring us closer to the goal. For almost a century, neurophysiologists have been recording electrical activity from the nervous system, at first from large populations of cells with scalp electrodes (cf. electroengephalograms) and eventually with small extracellular electrodes that monitor the action potentials from a single neuron. For those interested in sensory coding and perception, the extracellular electrodes are usually lowered into the brain of an anesthetized animal, and a neuron's activity is recorded while stimuli are delivered to the ears, eyes, or other receptor populations. In this way, we learn how environmental stimuli are converted into a neuronal code. For example, if an electrode is placed in the visual cortex and stimuli are delivered to each eye, we find that some neurons respond to bars of light that are vertically oriented, whereas others are driven best by horizontal bars. The neurophysiologist would call this "orientation selectivity," and such response properties usually find their way into theories on the neural basis of visual perception. Therefore, single neuron recordings provide an extremely sensitive measure of whether the building blocks have been assembled correctly during development.

Of course, it would be most compelling to record neural activity while the animal is actually processing a stimulus or moving a limb. In an early approach, animals were injected with a tritium-labeled sugar molecule, ^{3}H-2-deoxyglucose, that was taken up by nerve cells that were very active and required energy (Kennedy et al., 1975). It is now possible to measure neural activity and behavior simultaneously using several different techniques. Electrodes can be permanently mounted in the nervous system during an initial surgery, and these electrodes are then used to monitor neural activity when the animal recovers. Arrays of such electrodes are now used to record from the hippocampus of freely moving rats as they explore their environment and learn new tasks. It is also possible to stimulate or inactivate a region of the brain in awake-behaving animals, including humans during the course of neurosurgical treatment, and to monitor the effects on motor function or sensory perception (Penfield and Rasmussen, 1950; Riquimaroux et al., 1991).

There are several ways to monitor brain activity in awake animals, including humans, that can be performed without exposing the brain. Although these techniques have not been applied widely to developing animals, they will probably play an important role in our future understanding of plasticity. One technique that offers <1 mm spatial resolution, called *function magnetic resonance imaging* (fMRI), uses a very strong magnetic field (15,000 times the earth's magnetic field) to detect oxygen content. Since deoxyhemoglobin is paramagnetic relative to oxyhemoglobin and surrounding brain tissue, brain activity commonly produces a local increase in oxygen delivery. For example, it has recently been possible to visualize activity in a single barrel field in rat somatosensory cortex (Yang et al., 1996). Another technique, magnetoencephalography (MEG), uses superconducting detectors to monitor the magnetic fields produced by a population of active neurons. This technique provides information about the timing, location, and magnitude of neural activity. For example, word-specific responses in the inferior temporo-occipital cortex are slow or absent in dyslexic individuals compared to control subjects, suggesting a specific neural impairment in this developmental disorder (Salmelin et al., 1996). Finally, there are changes in light absorbance that are well correlated with neuronal activity, and it is possible to illuminate the surface of the brain and measure the reflected light while the system is processing information, referred to as differential optical imaging. By using these signals, many features of visual cortex development have been observed, including ocular dominance orientation selectivity (Blasdel et al., 1995).

Obviously, the challenge to find causality between brain function and behavior is magnified during development when nonspecific behavioral factors (cf. level of arousal) and the fragility of nerve cell function (cf. rapid fatigue) introduce great restraints. However, the study of immature brains, and the behaviors that they manufacture, should prove useful because it is the only means of correlating changes in the two without imposing surgery, drugs, or bad genes.

morphology of single LGN axon terminals have been examined in the visual cortex of cats and primates, and individual LGN terminals display a significant refinement during postnatal development, consistent with the autoradiographic studies (Florence and Casagrande, 1990; Antonini and Stryker, 1993).

The development of retinal arbors in the LGN illustrates how targeting errors are prevalent, yet subtle in appearance. In mammals, retinal axons grow to the correct area of the LGN from the outset, but they also make two kinds of inappropriate targeting projections. First, nearly all of the axons produce a few small collaterals, about 10 to 20 µm in length, in a part of the LGN that will eventually be innervated solely by axons from the other eye. Interestingly, the formation of an eye-specific innervation pattern in the LGN begins prior to visual experience. In the cat, these side branches are selectively eliminated in utero (Figure 9.6A and B).

A second phase of refinement occurs postnatally when retinal terminals become more focused within a portion of the correct eye-specific layer (Figure 9.6C). Retinal ganglion cells that have small visual receptive fields (cf. X-cells) decrease the width of LGN tissue that they innervate by a small (60 µm) but significant amount (Sur et al., 1984; Sretevan and Shatz, 1986). This would be the equivalent of pulling into your neighbor's driveway after completing a trip from hundreds of miles away. Thus, immature projections are numerous but modest in size.

Are synapses actually being formed by these transient projections? The answer can be found by first filling entire axons with a tracer, such as horseradish peroxidase, and then examining the terminals with an electron microscope. In fact, labeled presynaptic terminals have been found in parts of the target where they never remain in the adult, indicating that these structurally mature synapses will eventually have to be broken (Reh and Constantine-Paton, 1984; Campbell and Shatz, 1992).

The natural elimination of neuromuscular synapses can be observed over several days in living animals (Figure 9.7). Two strains of mice are genetically engineered to express a unique fluorescent protein in their motor neurons. When the two strains are mated, muscle cell fibers can be identified that are co-innervated by motor terminals that contain one or both fluorescent proteins (Walsh and Lichtman, 2003). By imaging the muscle fiber in vivo over successive days, one can observe the withdrawal of one synapse and the enlargement of the other to occupy the entire endplate (termed *takeover*).

Just how commonplace is axonal refinement and the loss of synaptic connections? It is found in a great

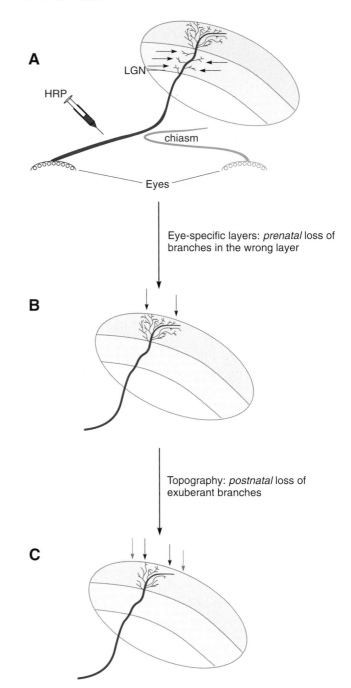

FIGURE 9.6 Development of retinal ganglion cell terminals in the cat lateral geniculate nucleus (LGN). A. When individual retinal fibers are labeled with horseradish peroxidase (HRP) at embryonic days 43–55, they have many side branches in the inappropriate layer of LGN (arrows). B. By birth, most of the side branches have been eliminated, and terminals arborizations have been restricted to the correct layer. However, the terminal zone remained wider in the eye-specific lamina at 3–4 weeks postnatal (arrows). C. When fibers of retinal X-cells were filled in adult cats, they were found to have retracted (black arrows). (Adapted from Sur et al., 1984; Sretevan and Shatz, 1986)

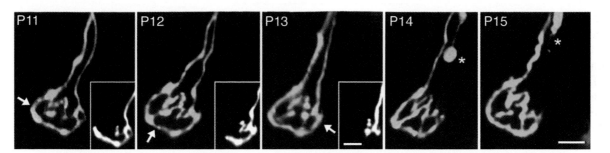

FIGURE 9.7 In vivo imaging of the same multiply innervated neuromuscular junctions in a neonatal mouse. Two transgenic mouse lines, each expressing fluorescent proteins in their motor neurons, were mated, and the progeny examined. Images were obtained from the same sternomastoid muscle fiber over the course of several days. In this example, one of the motor terminals (blue and insets) occupies a larger percentage (70%) of the postsynaptic territory at P11. It gradually withdraws from the junction (arrows) over the next four days. The withdrawing axon is marked by an asterisk at P14 and 15. Scale bars are 10 μm. (From Walsh and Lichtman, 2003)

variety of species and neural structures, although there are exceptions. There is significant remodeling of retinal axons during topographic map formation in the visual midbrain (Reh and Constantine-Paton, 1984; Simon and O'Leary, 1992). In frogs, single retinal afferents innervate successively posterior locations within the tectum as development proceeds. In mammals, retinal axons innervate topographically incorrect positions at birth, and these terminals are gradually withdrawn over two weeks. Synapse elimination is also found in invertebrates. In the cricket, single sensory neurons are functionally connected with two different interneurons during an early stage of development. Gradually, the strength of one connection doubles while the other connection is completely eliminated (Chiba et al., 1988). As we will see below, excitatory synaptic transmission has been implicated in the rearrangement of connections. However, inhibitory afferents can also become more refined during development. Inhibitory projections from the MNTB to the LSO (see Figure 8.28) undergo a striking refinement during the first postnatal week, followed by a decrease in the arbor size of single afferents by about 25% along the tonotopic axis (Sanes and Siverls, 1991; Kim and Kandler, 2003).

While the functional implications of terminal expansion are not clear, some neural circuits probably require broad connectivity with the sensory world. Many systems may employ a mixed strategy, with some axons expanding in territory while others retract. In the chick auditory system, two sets of axons converge on the *nucleus laminaris*, one innervating the dorsal dendrites and the other the ventral dendrites. Both sets of afferents spread out along the tonotopic axis during the embryonic period analyzed, suggesting that synapses are being added (Young and Rubel, 1986). However, the ventral terminal arbors form directly above their parent axon at first, but nearly half of them come to lay at a significant distance later in development. Since it is unlikely that the entire axon shifts its position, individual synaptic contacts are probably eliminated from one region of the terminal while new synapses are added at a different position. The connections from sensory axons to motor neurons appear to be maintained from the outset. Inappropriate monosynaptic connections from sensory axons to spinal motor neurons are not found with intracellular recordings, although there may be significant immaturities within the polysynaptic pathways (Seebach and Ziskind-Conhaim, 1994; Mears and Frank, 1997).

SOME TERMINALS EXPAND OR REMAIN STABLE

There are a few afferent pathways where *synapse elimination is either scarce or absent*. In contrast to the axonal arbors described above (see Figures 9.5 and 9.6), there are retinal cells with large visual receptive fields (Y-cells) whose arbors expand during development (Sur et al., 1984; Florence and Casagrande, 1990).

NEURAL ACTIVITY REGULATES SYNAPTIC CONNECTIONS

We encounter use-dependent changes of our nervous system on a regular basis. Sensory information is processed by our nervous system, and stored as memories (see BOX: Remaining Flexible). Our hands gradually become more adept at working with a newly

REMAINING FLEXIBLE: ADULT MECHANISMS OF LEARNING AND MEMORY

The cellular mechanisms responsible for changes of synaptic strength have been explored most thoroughly in adult animals because of their importance in learning and memory. Synaptic plasticity has been studied in a wide variety of neuronal systems, from molluscan ganglia to mammalian cerebral cortex, yet the catalog of cellular, molecular, and genetic mechanisms has grown steadily. This is probably good news for those interested in developmental plasticity because many of the ideas and techniques have been imported successfully.

The very first inquiries into synaptic mechanisms of plasticity demonstrated that synaptic transmission could be enhanced for about one minute following a period of intense stimulation. Recordings from muscle cells revealed that post-tetanic facilitation occurred because more neurotransmitter was released from the presynaptic terminal (Larrabee and Bronk, 1947; Lloyd, 1949). For the most part, contemporary studies continue to rely on intracellular recordings, usually in conjunction with a drove of "magic bullets" that are designed to block the function of a specific molecule. A relatively new approach makes use of genetic manipulations in the fruit fly and the mouse to provide an important experimental link between gene products, synaptic function, and behavior.

The modern era of cellular research began in the 1960s when the classical conditioning paradigm of paired stimuli was applied directly to a molluscan nervous system. In an intact sea slug, *Aplysia*, it is possible to enhance a touch-evoked withdrawal of the siphon when the tactile stimulus is paired with an electric shock during a training period. To study the neural basis of this sensitization, afferent-evoked EPSPs were recorded intracellularly from an identified neuron in the abdominal ganglion, and stimuli were delivered to both afferent pathways simultaneously. Following paired stimulation, one of the synapses produced much larger EPSPs, and this effect lasted for up to 40 minutes (Kandel and Tauc, 1965b). The increase was termed *heterosynaptic facilitation* because synaptic transmission at one set of synapses modified the functional status of a second, independent set.

One of the most compelling examples of synaptic plasticity, called *long-term potentiation* (LTP), was first identified in the early 1970s. By recording extracellularly from the hippocampus of anesthetized rabbits, it was found that a brief, high-frequency stimulus to the afferent pathway resulted in an enhancement of the evoked potential that lasted for hours to days (Bliss and Lømo, 1973). Over the next few years, intracellular recordings from mammalian brain slice preparations demonstrated that the size of EPSPs also increased following tetanic afferent stimulation. LTP is now thought to be one mechanism by which synapses store information because humans with hippocampal lesions display memory deficits, and a drug that blocks LTP in vitro is also able to impair spatial learning in rodents.

The discovery of a cellular analog of learning has raised many questions about the cellular mechanisms and the molecular pathways involved. Recent studies have indicated that two types of changes can occur at a potentiated synapse: increased transmitter release and enhanced postsynaptic response. One likely scenario for LTP in the hippocampus has glutamatergic transmission and postsynaptic depolarization combining to activate NMDARs, allowing calcium to flood the postsynaptic cell. NMDA receptor-dependent learning has also been demonstrated both in *Aplysia* (Murphy and Glanzman, 1999) and *Drosophila* (Xia et al., 2005). Thus, this molecular mechanism may be an evolutionarily conserved form of synaptic plasticity.

The influx of calcium activates one or more kinases which, in turn, phosphorylate proteins at the synapse. Although it is still not clear how many proteins are modified, there is evidence that functional glutamate receptors are added to the membrane, thus enhancing the postsynaptic response. A very simple form of learning in *Aplysia*, long-term facilitation of transmitter release, illustrates another important molecular pathway. An increase in presynaptic cAMP leads to the activation of a cAMP-dependent protein kinase (PKA). Once activated, the PKA subunit travels to the nucleus where it phosphorylates a transcription factor. The facilitated transmitter release involves new gene expression and protein synthesis (Kaang et al., 1993).

The cAMP signaling pathway seems to be a primary bridge to the formation of long-term memories in fruit flies and mice. There are two cAMP-dependent transcription factors (CREB), one that activates gene expression and a second that represses it. Thus, when transgenic flies are bred to express the activator, they remember an odor with much less training. However, flies that express the repressor are unable to store long-term olfactory memories (Yin et al., 1994, 1995). The many experimental studies on CREB (in *Aplysia*, fly, mouse and rat) established that the nucleus was involved in long-term memory formation. This fact presented an interesting question: How are these nuclear signals—common to all synapses of a given neuron—give rise to synapse-specific structural and functional modifications? The emerging answer seems to involve the "pumilio/staufen" pathway (Dubnau et al., 2003), which is involved in subcellular transport of mRNA and the local control of protein translation. The genetic approach to learning and memory clearly holds the promise of uniting cellular and behavioral findings, and it is likely that studies of developmental plasticity will profit as well.

purchased tool or joystick. When the level of noise or illumination changes abruptly, our eyes, ears, and nervous system adjust to maintain the fidelity of the signal. Similarly, we generally accept the notion that our rearing environment influences much of our adult behavior (see Chapter 10). To take an obvious example, we produce the language to which we were exposed as infants, whether it was Hindi, American Sign Language, or Spanish. But how much is the developing nervous system actually altered by the environment? Are synapse formation and elimination influenced directly by a use-dependent process?

An early approach to this problem, and one that is still regularly employed, involved the elimination of sensory structures. Denervation studies demonstrate how neuron growth and survival depend on intact connections during development (Chapter 7). The next experimental step was to change nervous system activity to find out whether improperly used synapses became weak or lost entirely. With this goal in mind, Wiesel and Hubel (1963a, 1965) began to explore the effects of monocular and binocular deprivation on the development of visual coding properties in the CNS. Their results clearly showed that synaptic activity influenced the maintenance or elimination of neural connections during development.

Before we can understand the functional changes brought about by visual deprivation, it is necessary to review some basic properties of the visual cortex. As described above, each neuron in Layer IV of the visual cortex receives projections that are largely driven by one eye or the other (Figure 9.5). When visual stimuli

are delivered to the appropriate eye, Layer IV neurons respond with a burst of action potentials. Cortical neurons that respond to only one eye are referred to as monocular. The majority of cortical neurons, particularly those laying outside of Layer IV, are activated by both eyes, and are referred to as binocular. Thus, when an extracellular electrode passes through the visual cortex, recording from many neurons in succession, most cells are found to be binocular (i.e., most neurons are recorded outside of Layer IV). Hubel and Wiesel (1962) divided the cortex neurons into seven groups, based on the relative ability of each eye to evoke a response. For example, if a neuron was driven solely by the contralateral eye, then it was assigned to group one. If it was driven equivalently by each eye, then it was assigned to group four, and so forth. This data can be conveniently represented as a histogram of ocular dominance (Figure 9.8). Judged by its continued use during the past 35 years, ocular dominance histograms provide a sensitive measure of the innervation pattern in the visual cortex.

A series of experiments were performed in which light-evoked activity was decreased by keeping the eyelid closed, referred to as visual deprivation (Wiesel and Hubel, 1963a, 1963b, 1965). This manipulation does not damage the retina, and LGN neurons remain responsive to visual stimulation after the eyelid is reopened. At first, a single eyelid was kept closed for a few months, and recordings were made from the visual cortex after the eye was reopened. The effect was unmistakable: Most cortical neurons no longer responded to stimulation of the deprived eye (Figure

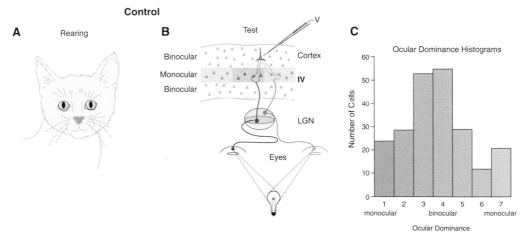

FIGURE 9.8 Response properties of visual cortex neurons in normally reared cats. A. The visual system received normal stimulation until the time of recording. B. Single neuron recordings were made with an extracellular electrode that passed tangentially through the cortex. Neurons respond to only one eye in Layer IV (monocular). In Layers I-III and V-VI, neurons respond to both eyes (binocular) due to convergent connections. In normal cats, the terminal stripes from each eye-specific layer of the LGN occupy a similar amount of space. C. Each neuron was characterized as responding to a single eye (monocular), or responding to both eyes (binocular). In normal animals, most visual cortical neurons are binocular. (Adapted from Hubel and Wiesel, 1962)

9.9), although they continued to respond to the unmanipulated eye.

Reasoning that disuse must have weakened the synapses from the deprived eye, Wiesel and Hubel recorded from binocularly deprived kittens, expecting to see a total absence of visually evoked activity. It came as a great surprise, then, that most cortical neurons remained responsive to stimuli through both eyes. That is, the ocular dominance histogram obtained from binocularly deprived animals resembled that of normal animals (Figure 9.10). "It was as if the expected ill effects from closing one eye had been averted by closing the other" (Wiesel and Hubel, 1965). However, many neurons displayed abnormal responses, and a large fraction of neurons were completely unresponsive to light, as originally predicted (Sherman and Spear, 1982).

The total amount of evoked activity does not necessarily predict whether a synapse will be strong or weak. Rather, differences in the amount of synaptic activity seem to determine the strength of a connection. This idea came to be known as the *competition hypothesis*. Under this proposal, retinal synapses in the LGN should not be affected by deprivation because LGN neurons are monocular and receive afferents that are either uniformly active or uniformly deprived of light. Similarly, binocular deprivation evens the playing field in the cortex because all afferents should have a similar low level of activity. Monocular deprivation creates a situation in which cortical neurons receive a set of active afferents from the open eye and a group of afferents with lowered activity from the closed eye, placing the latter at a disadvantage.

In a pivotal test of the competition hypothesis, kittens were raised with an artificial strabismus (cf. misalignment of the eyes), produced by surgically manipulating one of the extraocular muscles (Hubel and Wiesel, 1965). This manipulation mimics a clinical condition in humans, called *amblyopia*, which commonly results in the suppression of vision through one of the eyes, presumably to avoid double vision. In strabismic kittens, visual stimuli activate different positions on the two retinas, and cortical neurons are rarely activated by both eyes at the same time. Following several months of strabismus, recordings were once again made from the visual cortex. This time, both eyes effectively activated neurons in the cortex, but most cortical neurons respond to stimulation of one eye or the other (Figure 9.11). Few binocular neurons were observed. Therefore, an equivalent *amount* of activity in the two pathways is not sufficient to explain the results. Instead, it seems that the *timing* of synaptic activity must somehow be involved in allowing inputs to remain active on cortical neurons. Synapses from each eye must be active at nearly the same instant if both are to keep strong, functional contacts with the same postsynaptic neuron. Small disparities in the timing of synaptic activity may determine the strength of a synapse, an idea that has since been tested in tissue culture (below).

The ocular dominance columns (stripes) formed by geniculate terminals have served as an important anatomical model of synapse elimination. In normal animals, when ^{3}H-proline is injected into one eye, there is a periodic variation in silver grain density in Layer IV of the cortex. Labeled and nonlabeled regions are

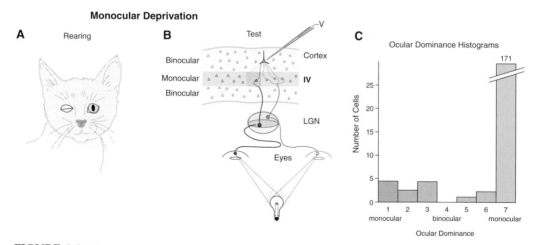

FIGURE 9.9 Response properties of visual cortex neurons in monocularly deprived cats. A. The visual system received normal stimulation through one eye, and the other eye was kept closed until time of recording. B. The terminal stripes from the deprived eye became much narrower, and the visual response of neurons outside of Layer IV was more responsive to the open eye. C. In monocularly deprived cats, the vast majority of cortical neurons responded to the open eye only. (Adapted from Wiesel and Hubel, 1963a)

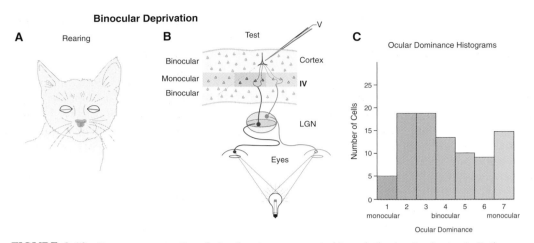

FIGURE 9.10 Response properties of visual cortex neurons in binocularly deprived cats. A. Both eyes were kept closed until time of recording. B. The terminal stripes from each eye occupied a similar amount of space. C. The majority of visually responsive neurons were driven by both eyes. (Adapted from Wiesel and Wiesel, 1965)

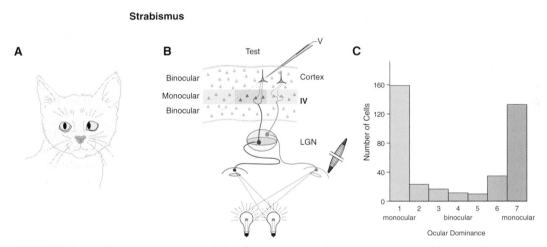

FIGURE 9.11 Response properties of visual cortex neurons in cats reared with artificial strabismus. A. One eye was surgically deflected, such that a visual stimulus activated different topographic position on each eye. Thus, cortical neurons were not activated by both eyes at the same time. B. The terminal stripes from each eye occupied a similar amount of space. However, neurons outside of Layer IV did not receive convergent input from both eyes. C. In strabismic cats, the vast majority of cortical neurons responded to either one eye or the other. (Adapted from Hubel and Wiesel, 1965)

about 500 μm wide (Figure 9.5). Following monocular deprivation, the LGN afferents from the nondeprived eye come to occupy the majority of Layer IV, while LGN afferents from the deprived eye occupy narrower regions (Figure 9.9, middle panel). This suggests that synapses from the nondeprived eye fail to undergo their normal process of elimination. In contrast, a greater than normal number of synapses from the deprived eye must be lost.

The emergence of stripes does occur in cat cortex during binocular deprivation over the first three postnatal weeks. When the deprivation is extended into the fourth postnatal week, the striping pattern begins to deteriorate (Crair et al., 1998). Pattern vision is unnecessary for the segregation of thalamic afferents into stripes, but maintenance of this striping pattern does require normal visual experience. Surprisingly, thalamic afferents can segregate into a striping pattern even when one or both eyes are removed (Crowley and Katz, 1999, 2000). This was demonstrated in ferrets by injecting a tracer directly into an eye-specific layer of the LGN. When animals are enucleated, an afferent striping pattern forms in the cortex with the same dimensions found in control animals. While these results suggest that synapse elimination in Layer IV proceeds in the absence of visually evoked activity, the reason for it is not clear. One possibility is that spontaneous activity within corticothalamic circuits that is independent of visual input has an impact on synapse development, as discussed below.

A very direct measure of synaptic pruning was achieved by imaging individual retinal ganglion cell (RGC) arbors as they compete for postsynaptic space within the frog optic tectum (Ruthazer et al., 2003). When RGC axons from both eyes are induced to innervate the same tectal lobe, the axons compete for postsynaptic space and gradually segregate from one another (Figure 9.12A). A single ipsilateral RGC was labeled with a fluorescent dye during this period of segregation, and imaged over an eight-hour period. The addition and retraction of each axonal branch was followed during this interval (Figure 9.12B). The brain was then fixed, and RGC projections from each eye were bulk-labeled with two different dyes. The procedure permitted one to characterize the innervation density of each eye, a measure that is analogous to ocular dominance. New branches that persist for the entire recording period, called *stabilized branches*, are more numerous when formed in a tectal region dominated by the same eye (Figure 9.12C). Conversely, RGC axons preferentially retract branches from territory that is already dominated by the other eye. This mechanism was eliminated in the presence of an antagonist to a class of glutamate receptors called NMDA receptors (see below). This result shows that branch addition and retraction (and presumably synapse formation and elimination) can occur simultaneously and depend on synaptic transmission.

The majority of binocular neurons in the cortex are created by local projections from one cortical neuron to its neighbors, and these projections also become refined during development. Since neurons in all layers of the cortex are monocular following strabismus, it would be interesting to know what happens to these intracortical projections. Do they now become more segregated than normal, extending the striped pattern throughout the entire cortical depth? This seems to be precisely what happens. Small injections of dye were made in the cortex, retrogradely labeling neurons that form local projections to this area (Figure 9.13). The animals were also injected with 2-deoxyglucose (see BOX: Watching Neurons Think) and stimulated through one eye to label all areas of cortex that were driven by that eye. With this double-labeling technique, one can learn whether local projection neurons are found exclusively above one ocular dominance column or both. In normal animals, the local projections come from both columns, and the cells that

they innervate are binocularly driven. In strabismic cats, the local projections originate exclusively above one column (Löwel and Singer, 1992). Furthermore, this alteration of horizontal projections occurs very rapidly; after only two days of strabismus, one can detect the loss of horizontal projections (Trachtenberg and Stryker, 2001). Thus, activity influences the development of synaptic connections not only in ascending sensory projections, but also in many of the intracortical projections.

The central concept to emerge from these studies is that coactive synapses are stabilized, while inactive synapses, particularly those that are inactive while others are firing, become weakened and in many cases are eliminated. As one might expect, the development of a complicated structure such as the cortex is unlikely to be explained by one tidy hypothesis. We have just learned that activity-dependent changes in connectivity are occurring simultaneously at several locations. Functional and structural changes are also found in the LGN following lid suture or strabismus, and the extent to which these changes influence cortical development is not clear. Proprioceptive feedback from the eye muscles also contributes, and its blockade somehow prevents monocular deprivation from altering synaptic connections in the cortex. This is not to lose sight of the forest, but rather to say that there are some large trees that must be carefully examined.

Given the great complexity of cortex circuitry, it would be nice to have a simpler model system for synaptic plasticity. As we learned in the last chapter, the neuromuscular junction (NMJ) is the most accessible and well studied of all synapses, and it has served as the mascot of synaptic plasticity for decades. In mammals, there is only a single type of synapse on fast muscle fibers, and only a single fiber ends up innervating each muscle cell in adults (above). If that were not good enough, it is also extremely easy to record intracellularly from muscle cells, to manipulate the nerve or muscle cells, and to place the entire system in tissue culture. Of course, these advantages also serve as the limitations. For example, there are no inhibitory synapses, and the postsynaptic cell does not have dendrites or spines, as found in many areas of the central nervous system.

Despite these differences, it is uncanny how much NMJ developmental plasticity resembles that observed in cortex. As we learned earlier, mammalian muscle cells are innervated by more than one axon at birth, but after synapse elimination only a single axon remains. When action potentials are blocked with TTX during the normal period of synapse elimination, muscle cells remain polyneuronally innervated (Figure 9.14) (Thompson et al., 1979). The same kind of results are

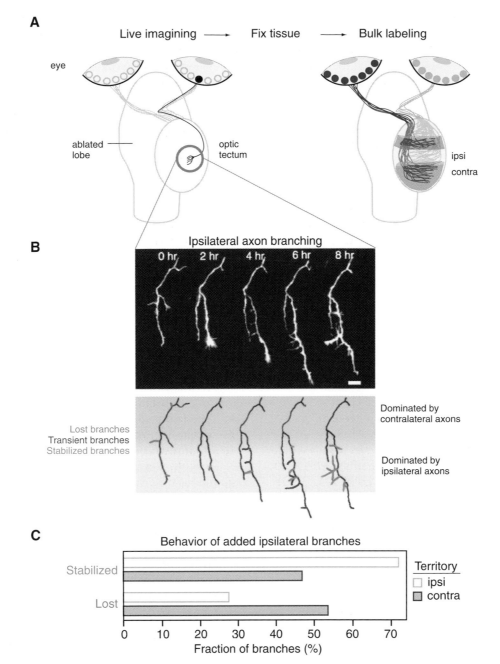

FIGURE 9.12 Selective addition and retraction of axon branches during competition for postsynaptic space. A. The retinotectal projection in *Xenopus* tadpoles is normally to the contralateral tectum, but ablation of one tectal lobe causes RGC axons from the ipsilateral eye to innervate the spared tectum. A single RGC axon is labeled with a fluorescent dye and imaged over an eight-hour period. Following this, the brain is fixed, and the projection from each eye is completely labeled with two different dyes (purple and yellow) such that the innervation pattern in the tectum is evident. B. Time-lapse images (top) and drawings (bottom) of a single ipsilateral eye axon in a living tadpole are shown. Branches are color coded for their behavior during the recording period: lost branches (green), transient branches (red), and stabilized branches (blue). In this case, the largest number of stabilized branches was formed in the region dominated by ipsilateral axons. C. New ipsilateral axon branches are more likely to remain for the entire recording period (i.e., stabilized) if they are formed in territory dominated by ipsilateral axons. In contrast, they are preferentially eliminated from the contralateral territory. (From Ruthazer et al., 2003)

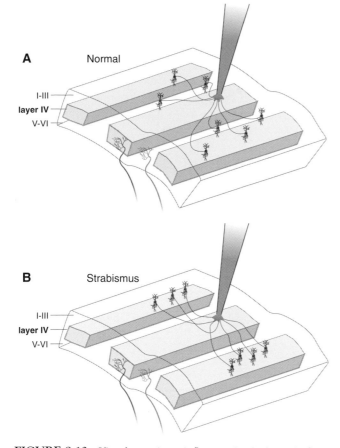

FIGURE 9.13 Visual experience influences intrinsic cortical projections. A. When dye injections were made into superficial layers of the visual cortex of normal cats, the label was retrogradely transported by neurons in both ocular dominance columns. B. In cats reared with artificial strabismus, the label was retrogradely transported only by neurons that shared the same ocular dominance. (Adapted from Löwel and Singer, 1992)

obtained in the developing cat visual pathway, where TTX blocks the segregation of retinal afferent in eye-specific layers of the LGN, and also the segregation of LGN afferents in the cortex. To test whether the temporal pattern of activity is important (as suggested by the strabismus results), two sets of motor axons innervating the same muscle were stimulated stimulated in synchrony (Busetto et al., 2000). This manipulation preserved polyneuronal innervation of muscle fibers.

If too many synapses remain when activity is blocked or synchronized, then one might predict that unsynchronized postsynaptic activity could speed up the process of synapse elimination. In fact, direct electrical stimulation of the muscle induces the early loss of motor synapses (Figure 9.15). This result is particularly intriguing because it suggests that *postsynaptic* electrical activity can determine whether presynaptic terminals survive (O'Brien et al., 1978). Together, these

experiments show that synaptic transmission can influence the process of synapse elimination at the NMJ. In fact, plasticity at the NMJ is a lifelong matter. Adult motor terminals will sprout to innervate adjacent muscle cells when neuromuscular transmission is blocked, and this polyneuronal innervation can be reduced when the muscle cells are stimulated directly (Jansen et al., 1973; Holland and Brown, 1980).

SENSORY CODING PROPERTIES REFLECT SYNAPSE REARRANGEMENT

If synapse strength and elimination are influenced by neural activity, then we would expect sensory coding properties (see BOX: Watching Neurons Think) to be altered when the pattern of environmental stimulation is manipulated. In fact, there are many examples of animals being reared in an altered sensory environment, and an influence on neuronal function is often found. Even when the sensory environment is well defined, it is practically impossible to figure out how a stimulus will affect the neural activity pattern produced throughout the nervous system. Nevertheless, this issue has been addressed successfully by two experiments in the central auditory system, one using sound stimulation and the other using electrical stimulation of the cochlea. Central auditory neurons usually respond to a limited range of frequencies because the auditory nerve fibers from the cochlea project topographically in the central nervous system (cf. tonotopy). A range of sound frequencies plotted against the sound intensities at which each frequency evokes a threshold response from a neuron, called a *frequency tuning curve*, provides a good measure of afferent innervation. When many areas of the cochlea project to a central neuron, then its frequency tuning curve is broad. When only a small region of the cochlea projects to a neuron, then its frequency tuning curve is narrow.

To test whether the timing of neural activity influences the development of frequency tuning, mice were reared in a sound environment consisting of repetitive clicks for a few weeks (Sanes and Constantine-Paton, 1985b). This type of sound evokes synchronous activity in a large population of cochlear nerve axons (Figure 9.16A and B). When frequency tuning curves were obtained from the inferior colliculus of normal mice and compared to those reared in repetitive clicks, the latter group had significantly broader curves (Figure 9.16C). A similar experiment has been performed in rats using pulses of noise, and the normal

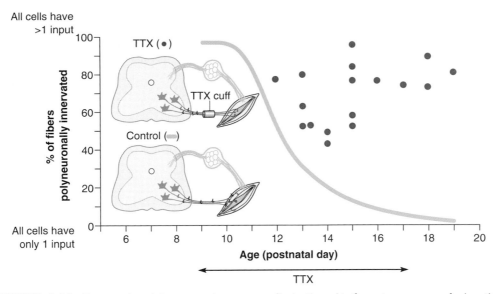

FIGURE 9.14 Decreased activity prevents synapse elimination. At the rat nerve–muscle junction, polyneuronal innervation declines between 10 and 15 days postnatal (green line). When a TTX cuff was placed around the motor nerve root from postnatal day 9–19 and action potentials were eliminated, polyneuronal innervation remained high (red circles). (Adapted from Thompson et al., 1979)

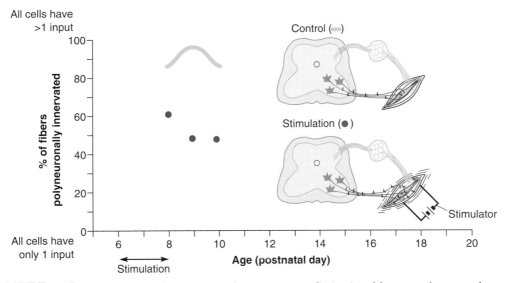

FIGURE 9.15 Increased muscle activity accelerates synapse elimination. Most rat soleus muscles are polyneuronally innervated between postnatal days 8–10 (green line). When a stimulating electrode was implanted in the leg to activate the sciatic nerve and muscle from postnatal days 6–8, there was a severe decline in the number of polyneuronally innervated muscle cells (red circles). (Adapted from O'Brien et al., 1978)

sharpening of tuning curves was prevented (Figure 9.16D). Noise pulse-rearing had no effect on tuning curves from mice older than 30 days, suggesting a specific developmental effect (Zhang et al., 2002). Therefore, when afferents all have the identical pattern of activity, they are apparently unable to segregate properly along the frequency axis.

A similar finding is obtained in cats that were deafened by damaging the inner hair cells, and then stimulated electrically within the cochlea. This is a highly relevant experimental model because children with profound hearing loss are now routinely implanted with cochlear prostheses, and they hear the acoustic world via electrical stimulation of their cochlea

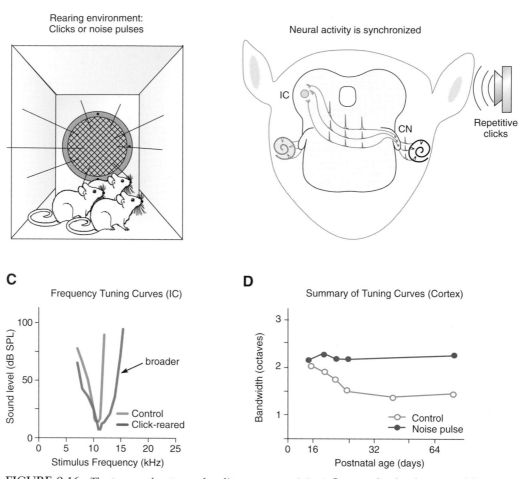

FIGURE 9.16 The temporal pattern of auditory nerve activity influences the development of frequency tuning. A. Mice or rats were reared with repetitive broadband stimuli (clicks or noise pulses). B. Stimuli of this sort activate many hair cells in the cochlear and synchronize the discharge pattern of auditory nerve fibers. C. Single neurons were recorded in the mouse inferior colliculus (IC), and those from click-reared animals responded to a broader range of frequencies compared to those from controls. D. Recordings were also made in the rat auditory cortex, and broader frequency tuning properties were maintained during development in noise pulse-reared animals. A larger bandwidth, plotted in octaves, indicates that the neurons were responding to a greater range of frequencies. (Adapted from Sanes and Constantine-Paton, 1985; Zhang et al., 2002)

(Snyder et al., 1990). In the experimental situation, the dendritic endings of the auditory sensory neurons can be stimulated directly with an electrode that is implanted within the cochlea. When animals were reared with repetitive electrical pulses to their cochlea, a synchronous activity pattern was produced. Although it was not possible to record frequency tuning curves because the animals were deafened, a single position along the cochlea was stimulated electrically while an electrode was lowered through the inferior colliculus. In normal animals, a single point on the cochlea evokes a response within a limited region of tissue, referred to as a *spatial tuning curve*. In stimulated animals, the spatial tuning curves were much broader, as if a single position in the cochlea now pro-

jected to a much wider area of the tonotopic map in the inferior colliculus. Consistent with findings from the visual pathway, these experiments support the idea that coactive synapses establish strong connections during development.

In the visual system, many cortical neurons respond to stimuli of a specific orientation, or to stimuli that are moving in a specific direction. Many investigators have asked whether a specific visual experience can influence the maturation of complex response properties, although the activity pattern produced by each environment is seldom known (Blakemore and Cooper, 1970; Cynader et al., 1973). When kittens are reared in a visual environment consisting entirely of vertical or horizontal stripes, and neurons are later

A

Goggles:
Kittens see vertical orientiation

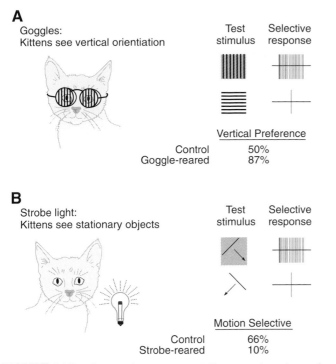

	Test stimulus	Selective response
	▓▓▓▓▓	╫╫╫╫╫
	≡≡≡	+

Vertical Preference
Control 50%
Goggle-reared 87%

B

Strobe light:
Kittens see stationary objects

	Test stimulus	Selective response
	◤	╫╫╫╫╫
	◣	+

Motion Selective
Control 66%
Strobe-reared 10%

FIGURE 9.17 The visual environment influences orientation and motion processing. A. Kittens were reared with goggles that permitted visual experience with only vertically oriented stimuli. The response to oriented stimuli was then obtained from single visual cortex neurons. The cell shown responds best to vertical stimuli. In goggle-reared cats, 87% of neurons were selective for vertical stimuli, compared to only 50% in controls. B. Kittens were reared in stroboscopic light that permitted visual experience with only stationary objects. The response to moving stimuli was then obtained from single visual cortex neurons. The cell shown responds best to stimuli moving down and to the right. In strobe-reared cats, only 10% of neurons were motion selective, compared to 66% in controls. (Adapted from Stryker et al., 1978; Pasternak et al., 1985)

recorded in the cortex, the majority of them responded selectively to the orientation that was present in the rearing environment (Figure 9.17A). To determine whether moving visual stimuli also influences the development of motion selectivity, kittens were raised in stroboscopic light. As nightclub enthusiasts know, this stimulus permits one to see a full range of shapes and colors, but it eliminates smooth motion. When cortical neurons were recorded after a period of strobe-rearing, the majority could no longer respond selectively to the direction of movement (Figure 9.17B).

As with ocular dominance columns, sensory experience appears to alter normal development, rather than shape the adult pattern. For example, orientation columns form normally in the kitten visual cortex in the absence of experience (Crair et al., 1998).

These results, and many others like them, show that environmental stimuli influence a broad range of functional properties. Although this discussion has focused on the cortex, sensory experience may influence

sensory coding properties at the most peripheral level of the nervous system. For example, mouse retinal ganglion cells (RGC) initially respond to both an increase and decrease in luminescence, called an *ON-OFF response*. During postnatal development, the RGC dendritic arbor is refined, and most neurons end up producing either ON or OFF responses to visual stimulation, but not both. When mice are reared in the dark, the RGC dendrites are not refined, and the cells continue to produce ON-OFF responses (Tian and Copenhagen, 2003). Therefore, environmental rearing studies may influence maturation at many levels, and the cumulative effect is assessed in the cortex.

ACTIVITY CONTRIBUTES TO THE ALIGNMENT OF SENSORY MAPS

When peripheral sensory axons reach the central nervous system, they usually innervate the target in an orderly manner, forming topographic maps that are quite accurate due to molecular gradients that direct axons to an approximate location within the target (Chapter 6). However, synaptic activity also influences the development of topography. This is most evident when two maps must become aligned with one another in the same structure. For example, binocular neurons in the frog optic tectum respond to the same visual position in space when activated through each eye. We have already studied the direct contralateral projection from retina to tectum in Chapter 6. The ipsilateral eye activates the tectum via an indirect projection. Tectal neurons project to a structure called *nucleus isthmus*, and isthmal neurons project to the contralateral tectal lobe (Figure 9.18A, left panel). Therefore, there are two perfectly aligned maps of visual space in the tectum, one from the contralateral eye and one from the ipsilateral eye.

In order to test whether visual activity plays a role in this precise alignment, frogs were reared in the dark. Direct retinal projections from the contralateral eye continue to form a precise map, but the indirect projection via the nucleus isthmus is poorly organized (Keating and Feldman, 1975). That is, the formation of one map depends largely on molecular cues, while the other map depends on activity. The effect is so powerful that one can actually cause isthmal axons to move to a new location in the tectum when the contralateral map is disrupted (Figure 9.18A, center panel). If the contralateral eye is rotated by 180°, a single point in space will activate different positions in the tectum via each eye. When this manipulation is performed while the animals are still tadpoles, then the isthmal projection to the tectum will shift so that the ipsilateral retinal map comes into register with the direct

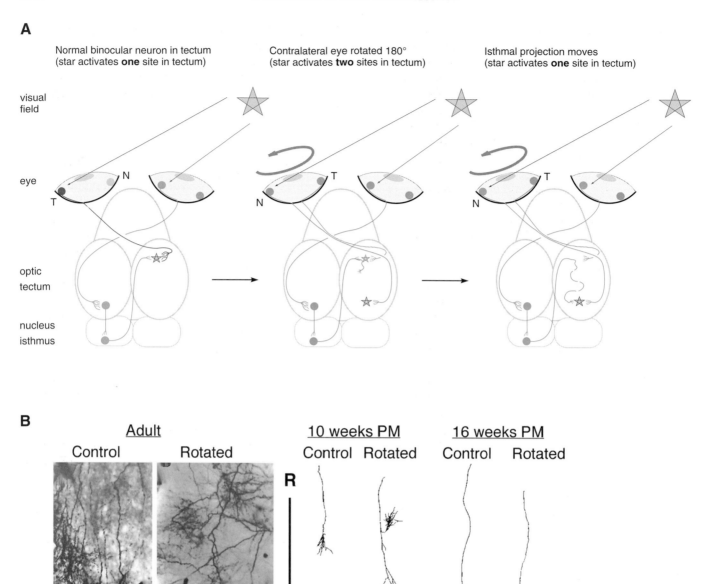

FIGURE 9.18 Remapping in the frog visual pathway. A. In adult frogs (left), the optic tectum receives two retinotopic projections: a direct projection from the contralateral eye and an indirect projection from the ipsilateral eye (via the nucleus isthmus). These two projections are aligned such that a visual stimulus activates the same position in the tectum through either eye. In this example, the temporal eye activates the anterior tectum. When the contralateral eye is rotated 180°, a visual stimulus activates two different positions in the tectum, one via the eye and the other via the projection from nucleus isthmus (middle). This is because the retinal ganglion cells in nasal eye project to the posterior tectum. Over time, the projection from nucleus isthmus to the tectum adjusts its position so that a visual stimulus once again activates the same tectal position through either eye (right). B. Photographs of axons from nucleus isthmus as they course through the tectum of an adult control, and an eye-rotated animal (left). Note that control axons grow parallel to one another from rostral to caudal. Axons from eye-rotated animals display less ordered growth. Drawings of individual axons from nucleus isthmus in control and eye-rotated animals are shown at two postmetamorphic (PM) ages (right). Note that the axons of eye-rotated animals can make two terminal zones. (Adapted from Udin and Keating, 1981; Guo and Udin, 2000)

contralateral projection (Figure 9.18A, right panel). Thus, isthmotectal axons can be induced to innervate a part of the tectum that they would not ordinarily contact (Udin and Keating, 1981). When nucleus isthmus axons are labeled in normal animals they are oriented very accurately along the rostro-caudal axis of the tectum (Guo and Udin, 2000). In eye-rotated animals, these isthmal axons appear disordered, often terminating in two locations or meandering within the tectum (Figure 9.18B).

In some cases, maps from two different sensory systems are found within the same structure, and they must come into alignment during development. In the optic tectum of barn owls, there are maps of both the visual and auditory world. The maps are aligned such that neurons respond to acoustic and visual stimuli from the same position in space. For example, neurons that respond to visual stimuli directly in front of the animal (0°) will also respond to a sound stimulus that arrives at each ear simultaneously (i.e., 0 μs interaural time difference). Neurons that respond to visual stimuli at 20° to the right will also respond to a sound stimulus that arrives first at the right ear and then at the left (60 μs interaural time difference). Does the alignment of these maps also depend on neural activity? To test this idea, owls were reared with prismatic glasses that displace visual stimuli by 23° (Figure 9.19A). As in the retinotectal system, the physical connections between eye and tectum are unaltered in prism-reared animals. If no compensation were to occur, then visual stimuli and auditory stimuli from the same position in space would activate different loci in the optic tectum. That is, maps of auditory space and visual space would be out of alignment. In fact, the auditory map adjusts to remain in register with the visual map. When the prisms are removed, tectal neurons that responded to visual stimuli directly in front of the animal (0°) are now found to respond to an auditory stimulus to one side. Therefore, auditory connections must have changed in response to visual activity (Brainard and Knudsen, 1993).

To determine how this happened, a tracer was injected in the central nucleus of the inferior colliculus, the structure that responds to interaural time differences (Figure 9.19B) (DeBello et al., 2001). In control animals, tracer injected at a specific position labels fibers that project to a unique location in a second auditory nucleus, called ICX (Figure 9.19C, left). It is the ICX neurons that will project to visually responsive neurons in the optic tectum. In prism-reared animals, the labeled ICC axons project to a different position within ICX, permitting optic tectum neurons to integrate a new auditory spatial position (Figure 9.19C, right). Thus, the elimination and addition of synapses

depends on their activity pattern for some, but not all, afferent pathways. Even in the visual cortex, geniculate afferents from the contralateral eye appear to establish their innervation pattern first, and those from the ipsilateral eye may be more dependent on activity (Crair et al., 1998).

SPONTANEOUS ACTIVITY AND AFFERENT SEGREGATION

Orientation and motion selectivity mature rapidly from the onset of sight, and it is possible that normal neuronal activity simply maintains a precise but unstable set of connections (Blakemore and Van Sluyters, 1975). Alternatively, it is possible that the innervation patterns that underlay these coding properties may not attain their mature pattern without the proper stimulation. In simpler terms, when does activity begin to influence neural development? In humans, visual- and auditory-evoked brain activity can be demonstrated in utero using the noninvasive brain imaging techniques, MEG and fMRI (see BOX: Watching Neurons Think) (Schneider et al., 2001; Eswaran et al., 2002; Fulford et al., 2003, 2004). Furthermore, electrical activity is present in the nervous system even in the absence of visual or auditory stimulation. This "spontaneous" activity could have a profound influence on synapse formation and elimination.

We previously learned that stripes in Layer IV formed in binocularly deprived cats (Figure 9.10). To test whether spontaneous retinal activity might be responsible for the segregation of thalamic afferents in the cortex, retinal activity was completely eliminated by injecting both eyes with TTX from about 2–6 weeks postnatal (Figure 9.20). In TTX-reared cats, the LGN afferents fail to segregate into stripes in Layer IV of the cortex (Stryker and Harris, 1986). Similarly, suppression of spontaneous retinal activity prevents the elimination of small retinal afferent sidebranches in the inappropriate layer of LGN. However, the TTX injections were made after the thalamic afferents had begun to segregate in the cortex, and it is possible that this manipulation leads normally refined afferent projections to sprout, once again, into an inappropriate target region (Chapman 2000).

Since spontaneous activity may influence synaptic development, we should be able to record the amount and the pattern. Retinal ganglion cells fire action potentials before the system is activated by light. In the embryonic rat, single retinal neurons discharge about once every second, but occasionally fire short bursts of almost 100/sec (Galli and Maffei, 1988). However, the

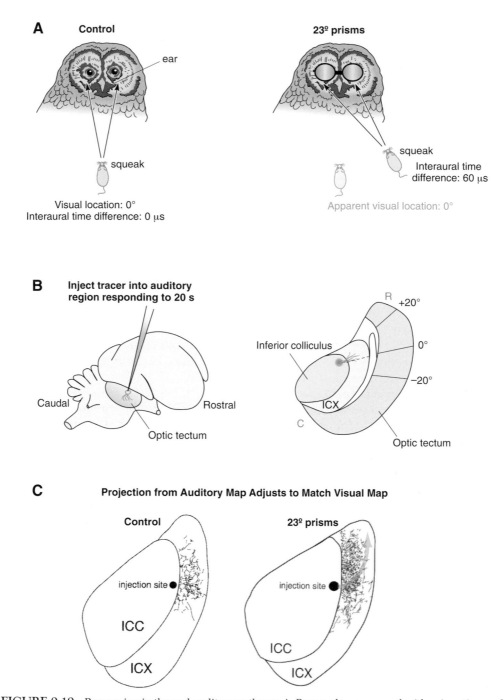

FIGURE 9.19 Remapping in the owl auditory pathway. A. Barn owls were reared with prismatic goggles that shifted the visual field by 23°. In control animals, a squeaking mouse in front of the owl would appear at 0°, and the sound would reach both ears simultaneously (0 μsec interaural time difference). With prisms in place, a squeaking mouse at 23° to the left would *appear* at 0°, and the squeak would reach the left ear first (60 μsec interaural time difference). B. A tracer was injected into the auditory nucleus (ICC, central nucleus of the inferior colliculus) that encodes the ITD of the sound stimulus (left). The IC relays this information to the ICX where the map of auditory space is assembled. ICX then projects to the optic tectum where visual and auditory information is first integrated (right). C. In control animals, the projection from a region in ICC that responds to a 20 μs interaural time difference projects to a narrow region in the ICX (and ICX then projects to an optic tectal location that responds to visual cues about 8 degrees from the midline). In prism-reared animals, ICC projects to a more rostral region of ICX. Thus, a new projection is formed within the auditory space map which compensates during prism rearing such that sound and sight can once again be integrated. (Adapted from DeBello et al., 2001)

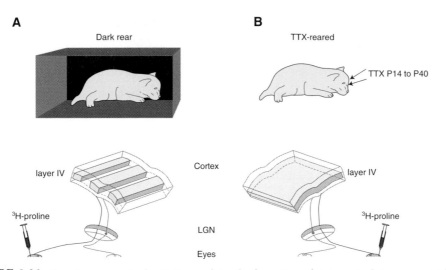

FIGURE 9.20 Spontaneous retinal activity regulates the formation of stripes. A. Cats were reared in the dark, and ³H-proline was injected into one eye to visualize ocular dominance columns. Although the columns were slightly degraded, they did form. B. Bilateral intraocular TTX injections were performed beginning at postnatal day 14. When ³H-proline was injected into one eye to visualize ocular dominance columns, it was found that segregation of geniculate afferents into stripes failed to occur. Thus, spontaneous retinal activity is sufficient to influence stripe formation. (Adapted from Stryker and Harris, 1986)

competition hypothesis suggests that synapse modification depends on the pattern of activity. Is it possible that spontaneous activity has a temporal or spatial pattern? To address this question, the entire retina was isolated from an embryo and placed in a perfused recording chamber, where it is possible to record from many retinal ganglion cells at the same time (Figure 9.21A). Synchronous bursts of action potentials were recorded from many neurons about every minute or two, and regions of maximal activity moved slowly across the retina (Meister et al., 1991) (Figure 9.21B). Recordings from both the ferret and the mouse each show that activity waves sweep across the entire retina during the first few postnatal weeks (Wong et al., 1993). This patterned activity gradually breaks down at about the time of eye opening, but visual experience does not seem to terminate the waves. They are lost at about the same time even in dark-reared animals (Demas et al., 2003).

Spontaneous action potentials have also been observed throughout the developing nervous system and often displays interesting temporal or spatial patterns (Lippe, 1994; O'Donovan et al., 1994; Kotak and Sanes, 1995). Recordings from the nerve–muscle junction show that spontaneous activity is very low at birth and gradually rises during the first two postnatal weeks (Personius and Balice-Gordon, 2001). Interestingly, motor axon firing is highly correlated during the first postnatal week (that is, motor synapses are firing in synchrony), perhaps preserving polyneuronal innervation at the muscle. The period of synapse elimination occurs when motor activity becomes tempo-

rally decorrelated. In the cortex, large-scale waves of activity have been found to move slowly, from caudal to rostral, during the first postnatal week (Figure 9.21C). This activity pattern involves the vast majority of neurons and seems to end when inhibitory synapses become hyperpolarizing (Garaschuk et al., 2000).

Are the temporal or spatial patterns of spontaneous retinal activity required for the formation of specific connections in the central nervous system? It turns out that the activity waves within the retina depend on cholinergic transmission. When cholinergic transmission is disrupted in mice by knocking out the β2 AChR subunit, the retinal waves are lost. Retinal ganglion cells continue to burst, but the rate of bursting differs across the retina, and neighboring retinal neurons no longer fire at the same time. Under these conditions, projections from posterior retinal ganglion cells do not become restricted to the anterior superior colliculus, as they would in wild-type animals (McLaughlin et al., 2003). The projection from retina to LGN is also less precise as revealed by fine-grained electrophysiological mapping (Grubb et al., 2003). The absolute levels of spontaneous activity from each eye are crucial to establishing the mature lamination pattern in the LGN. When the spontaneous activity of one eye is experimentally increased by elevating cAMP levels, its projection within the LGN is expanded (Stellwagen and Shatz, 2002). However, the segregation of thalamic afferents into eye-specific layers may not depend on the precise pattern of retinal activity in the postnatal period (Huberman et al., 2003).

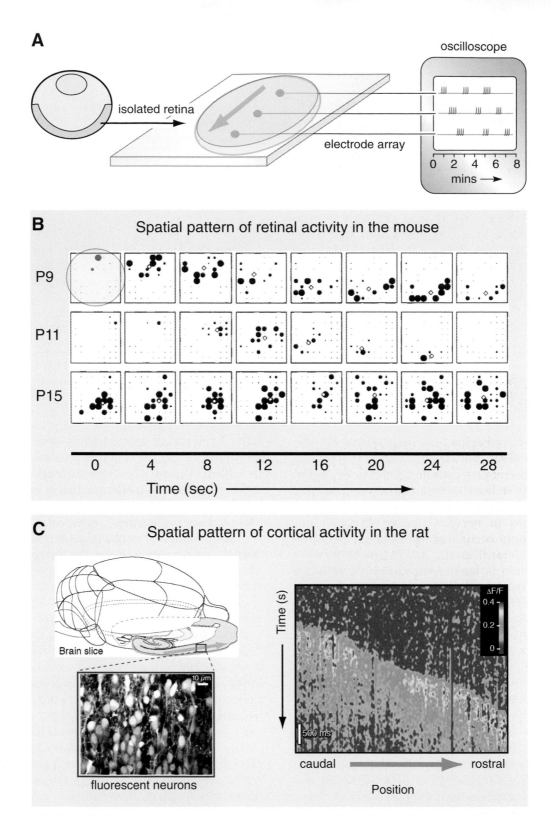

A

oscilloscope

isolated retina

electrode array

0 2 4 6 8
mins ⟶

B Spatial pattern of retinal activity in the mouse

P9

P11

P15

0   4   8   12   16   20   24   28

Time (sec) ⟶

C Spatial pattern of cortical activity in the rat

Brain slice

10 µm

fluorescent neurons

Time (s)

ΔF/F
0.4
0.2
0

500 ms

caudal ⟶ rostral

Position

FIGURE 9.21 Spontaneous activity in the developing visual pathway. Retinal explants are obtained from neonatal ferrets and placed on an array of electrodes to record bursts of spontaneous activity from several retinal locations at the same time. (For clarity, only three locations are shown.) These bursts of activity move across the retina, as shown by the increasing response latency in each oscilloscope trace (right). B. Electrode array recordings are plotted for mouse retina at three postnatal ages (P9, P11, and P15). Each square shows the spatial pattern of activity across the two dimensions of the flattened retinal, with circles representing individual neurons. By P15, the spontaneous activity no longer travels across the retina. The spatial activity patterns were acquired every 4 s. C. To observe spontaneous activity in the cortex, brain slices are obtained from rats during the first postnatal week, and intracellular calcium is imaged using a Ca-sensitive dye (left). Waves of spontaneous activity sweeps across the cortex from caudal to rostral. (Adapted from Meister et al., 1991; Demas et al., 2003; Garaschuk et al., 2000)

There seems to be a paradox between results showing that spontaneous retinal activity directs synapse development (Stellwagen and Shatz, 2002), and the observation that ocular dominance columns can form even when both eyes are removed (Crowley and Katz, 1999). Perhaps the spontaneous activity patterns that arise in the thalamus and cortex will resolve this question (Yuste et al., 1995; Weliky and Katz, 1999; Garaschuk et al., 2000; Chiu and Weliky, 2001). The influence of low levels of spontaneous activity may be supported by local inhibitory projections which further refine the spatial pattern. In fact, local GABAergic circuits within the cortex regulate the binocular competition between thalamic afferents, and establish the precise dimensions of ocular dominance columns (Iwai et al., 2003; Hensch and Stryker, 2004; Fagiolini et al., 2004).

MANY FORMS OF PLASTICITY HAVE A TIME LIMIT

Synaptic activity begins to exert an influence soon after synaptogenesis, but how long does this process continue? If we embrace learning and memory in our definition, then it lasts for our entire lifetime (Box: Remaining Flexible). However, there are certain significant changes in nervous system structure and function that only occur during a limited period of development (Berardi et al., 2000; Hensch, 2004). A common example is language acquisition, which is accomplished most easily before age 10 and becomes a grueling task for most of us if attempted as adults (see Chapter 10). At least some forms of neuronal plasticity only last for a limited period during developmental, often called a *critical period*. The influence of visual experience on ocular dominance has been explored in older animals to determine whether cortical neuron function is always dependent on vision (Hubel and Wiesel, 1970). Ocular dominance is most susceptible to monocular deprivation after several weeks of sight. The visual environment influences cortical neuron function for roughly the first three months of life in cats, but this critical period does not apply to connections forming throughout the brain.

The term *critical period* is a very general expression, and the neuronal property under discussion should always be taken into account. In primates, monocular deprivation affects the segregation of LGN afferents in Layer IV for about two to three months, but continues to produce severe weakening of intrinsic cortical synapses for up to a year (Figure 9.22). Similarly, when retinal activity is blocked with TTX after ganglion cell arbors have segregated into eye-specific layers in the cat LGN, the response properties of LGN neurons are nonetheless altered (Dubin et al., 1986).

The critical period may occur during a very narrow time window in some brain areas. One region of rodent somatosensory cortex receives afferents from each of the facial whiskers and contains an array of barrel-shaped cell clusters that are activated selectively by each of the whiskers. If the whiskers are destroyed before postnatal day 5, their associated barrels do not form, but if the whiskers are destroyed after that time, then the manipulation has no effect (Van der Loos and Woolsey, 1973). There are also extensive intracortical connections between each of the whisker barrels, and this connectivity is dependent on continued use. When the sensory nerve to the whiskers is cut on postnatal day 7, after the barrel fields are formed, there is a dramatic reduction in the number of local projections (McCasland et al., 1992). In fact, whisker trimming significantly decreases the motility of dendritic spines and filopodia in rat barrel cortex only from postnatal days 11 to 13 (Lendvai et al., 2000). This manipulation also affects the development of whisker-evoked receptive fields in cortical Layer 2/3 until day 14, but has no effect on Layer 4 receptive fields (Stern et al., 2001). Thus, the critical period for barrel cortex differs by layer. Even adult cortex can exhibit changes following deprivation. For example, monocular action potential blockade with TTX results in a decreased expression of $GABA_A$ receptor subunits in Layer IV of primate cortex (Hendry et al., 1994).

SYNAPSES INTERACT OVER A SHORT DISTANCE

The modification of synapse function involves both pre- and postsynaptic signaling mechanisms, and the neurotransmitter itself is likely to initiate the biochemical cascade. But is it really possible that there is a single mechanism to account for synaptic plasticity? After all, most central neurons are innervated by a wide variety of synapses: some release glutamate, others GABA; still others release a neuromodulator such as serotonin. Moreover, a single glutamate-releasing synapse can activate receptors that open ion channels (cf, ionotropic) and others that activate second messenger systems (cf, metabotropic). There are few generalizations that can cover the plasticity of all these systems. Therefore, we will focus on synapses that have been studied thoroughly, such as the nerve–muscle junction.

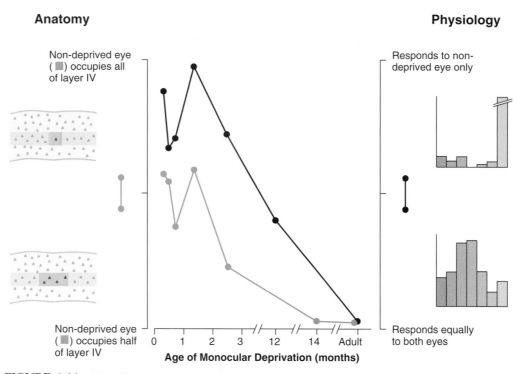

FIGURE 9.22 The effects of monocular deprivation are age-dependent. One eye was sutured shut in macaque monkeys at different postnatal ages. The sutures were removed after several months, and ocular dominance columns were examined with anatomical (left) and electrophysiological techniques (right). When deprivation began between 0 and 2 months of age, most neurons subsequently responded to the open eye only, and the open eye occupied much more of Layer IV. The effects of deprivation declined with age. The anatomical effect of deprivation on ocular dominance columns seemed to decline more rapidly than the physiological effect on ocular dominance histograms. (Adapted from LeVay et al., 1980)

Synaptic terminals can influence neighboring contacts on the same postsynaptic cell but only within a finite distance. When two motor nerves were grafted onto the same muscle in adult rats, both of them were able to maintain functional contacts over several months, even though muscles are normally innervated by one axon. The trick was to place the two motor terminals at least several millimeters from one another (Kuffler et al., 1977). If the terminals were placed within a millimeter or two, then one of the contacts was eliminated within about three weeks (Figure 9.23). In fact, some animals have muscle fibers that are normally innervated by more than one motor axon. In one polyneuronally innervated muscle in chicks, the distance between terminals can be reduced when synaptic transmission is blocked, presumably because competition between active terminals normally keeps them separated (Gordon et al., 1974).

The terminal endings from a single axonal arbor seem to innervate a continuous region of the postsynaptic cell, whether it is a primary dendrite or a small area of muscle cell (Forehand and Purves, 1984; Glanzman et al., 1991). For example, when sensory neurons from the sea slug, *Aplysia*, are grown in culture along with a common target motor neuron, their terminals come to occupy separate regions of the postsynaptic cell. However, if the two sensory neurons are grown without a target, then they grow extensively along one another. That is, the sensory neurons do not display contact inhibition, as discussed above (Chapter 4). Therefore, the antagonistic relationship between synapses must somehow be mediated by the postsynaptic neuron. Even if we know the amount and pattern of synaptic activity in a set of inputs, this may not be enough to predict whether one synapse will dislodge a second one. One must also know the spatial arrangement of these terminals.

HETEROSYNAPTIC DEPRESSION

Synapses that are capable of function can form in the absence of synaptic transmission (Cohen, 1972; Duxson, 1982). Presynaptic cholinergic terminals are even able to differentiate in a zebrafish mutant lacking

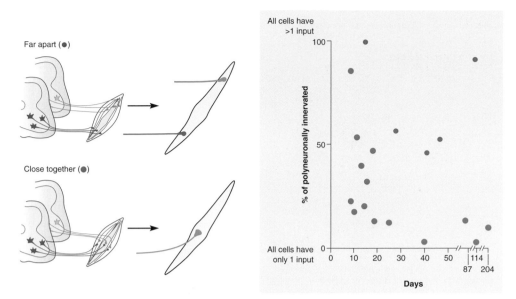

FIGURE 9.23 Synapse elimination depends on synapse distance. Two motor axons were positioned on the same muscle, either close to one another or at a distance. Intracellular recordings were made from muscle fibers to monitor polyneuronal innervation, as shown in Figure 9.2. When the synapses were close together (blue circles), synaptic elimination occurred within a few weeks. When the synapses were far apart (red circles), synaptic elimination failed to occur. (Adapted from Kuffler et al., 1977)

functional AChR clusters (Westerfield et al., 1990). In contrast, sensory deprivation studies suggest that the disuse of a synapse leads to its weakening or elimination, especially if other synapses are active. Synaptic terminals "compete" for the distinction of activating the postsynaptic neuron, and "stable" connections depend, in part, upon an active transmitter-receptor system. Therefore, the presence of synaptic transmission is not an absolute requirement for the initial formation of a synapse but influences its subsequent development and stability.

The activity of a neighboring synapse can be harmful to an inactive one. This mechanism was tested at an unusual muscle in the rat foot, called the lumbrical muscle, which receives its innervation via two separate peripheral nerves (Figure 9.24). It is possible to electrically stimulate one group of motor axons during the normal period of synapse elimination, and find out what happens to the synapses of unstimulated axons. The strength of the nerve–muscle connection was determined indirectly by measuring the size of nerve-evoked muscle contractions, where a larger contraction signifies a stronger connection. When one nerve is stimulated for about six days, the *unstimulated* axons are much less effective at producing a muscle contraction (Ridge and Betz, 1984). That is, the unstimulated axons are either eliminated from the muscle in disproportionate numbers, or their synapses become weaker.

How long does it take for one synapse to extinguish its neighbor? The onset of synapse elimination was studied using a very simple culture system that contained two presynaptic neurons and one postsynaptic cell, a myocyte (Figure 9.25A). Whole cell recordings were made from the myocyte, while the activity of one or both presynaptic neurons was controlled with stimulating electrodes. In the first experiment, the synaptic currents elicited by each neuron were first measured, and then one of the neurons was stimulated for several seconds. Within moments of this brief procedure, the unstimulated neuron produces much smaller synaptic currents, and the effect lasts for the duration of the experiment (Lo and Poo, 1991). When one synapse is able to modify the operation of a second one, the interaction is referred to as *heterosynaptic*.

Two additional observations seem to fit neatly into the puzzle. First, if the two synapses are separated from one another by >50 µm, then activation of one synapse is not able to suppress its neighbor. Two synapses can apparently compete for the right to activate a postsynaptic cell, but the interaction occurs over a very short distance, consistent with results from the adult NMJ and *Aplysia* cultures (Kuffler et al., 1977; Glanzman et al., 1991). Second, when both neurons are stimulated at the same time, there is also no change in the strength of either synapse. That is, synaptic activity is able to *protect* a synapse from the ill effects of an active neighbor.

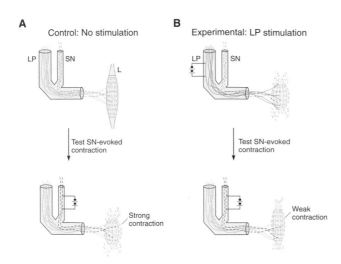

FIGURE 9.24 Synaptic activity depresses less active inputs in vivo. The lumbrical muscle in the rat foot is innervated by two physically separate motor nerve roots, LP and SN. A. In control animals, stimulation of the SN motor nerve root (red) elicits a robust contraction of the lumbrical muscle (bottom). B. If the LP motor nerve root (blue) is repetitively stimulated during development (top), then the SN nerve root stimulation elicits a much weaker contraction of the lumbrical muscle (bottom) when it is subsequently tested. (Adapted from Ridge and Betz, 1984)

Synaptic terminals do not battle each other directly but carry on their competition through an independent agent, the postsynaptic cell. It is a bit like choosing the winner of a boxing match by judging who punches the referee harder. The contribution of the postsynaptic cell to heterosynaptic depression can be demonstrated by activating it directly with transmitter. When a muscle cell is activated focally with a puff of ACh from the tip of a pipette, then the size of synaptic responses declines by about 50% (Dan and Poo, 1992).

Of course, this begs the question of whether a depressed synapse is necessarily an eliminated synapse. Furthermore, we have already learned that the rate of synapse elimination in vivo does not occur within minutes, but over a period of days (Figure 9.7). If synaptic depression really is the first hint of a dying connection, then one might expect to record a synaptic response that gradually becomes weaker during development. This can be observed at the mammalian NMJ by measuring the average number of ACh packets that are released by each synapse, called the *quantal content*. Even at birth, one synapse is about twice as strong as the second, but over the next week one of the synapses becomes about four times stronger than the other (Colman et al., 1997). That is, it releases four times as much transmitter. It has also been found that release probability varies dramatically between competing axons, and it is possible that presynaptic

terminals decrease their transmitter release as they are eliminated (Kopp et al., 2000). In summary, the strength of one synapse at the nerve–muscle junction apparently decreases with time, and it seems plausible that this must be the one that is eventually eliminated.

POSTSYNAPTIC RECEPTORS ARE ELIMINATED

The postsynaptic cell probably plays an active role in causing extra synapses to withdraw. At the NMJ, postsynaptic AChR clusters can disappear before nerve terminals withdraw from the muscle surface. Thus, presynaptic terminals may be forcibly evicted from their territory when the postsynaptic cell re-keys the locks (Role et al., 1987; Balice-Gordon and Lichtman, 1993). To demonstrate that the loss of functional AChRs is sufficient to produce synapse elimination, a small fraction of receptors were blocked by locally applying α-Btx. Over a number of days, the presynaptic terminal that is in contact with the inactivated region gradually withdraws (Balice-Gordon and Lichtman, 1994). However, it is essential that only a small fraction of neighboring receptors are blocked. If all the receptors are blocked, then the nerve terminal does not retract. It is also clear that motor axons can compete for existing receptors. One motor terminal will grow into the endplate region occupied by a second motor terminal as it is withdrawing (Walsh and Lichtman, 2003). These observations emphasize once again that synaptic competition occurs within very small dimensions.

To examine whether heterosynaptic depression and receptor turnover are related, simple dual innervated nerve–muscle cultures were examined (Figure 9.25B). Electrical stimulation of one neuron led to a depression of the excitatory response evoked by the unstimulated neuron. When the AChR density under each synapse was examined before and after stimulation, the unstimulated synapse was found to have lost more than 30% of its AChRs (Li et al., 2001). These experiments do not have the temporal resolution to decide whether receptor loss caused the reduction in synapse strength or whether either phenomenon leads to synapse withdrawal. However, they suggest how this process may proceed. Taken together with observations from in vivo recordings (Colman et al., 1997; Kopp et al., 2000), the higher release probability and greater quantal content of some competing axons may prevent the reduction of postsynaptic AChRs and maintain the synaptic area.

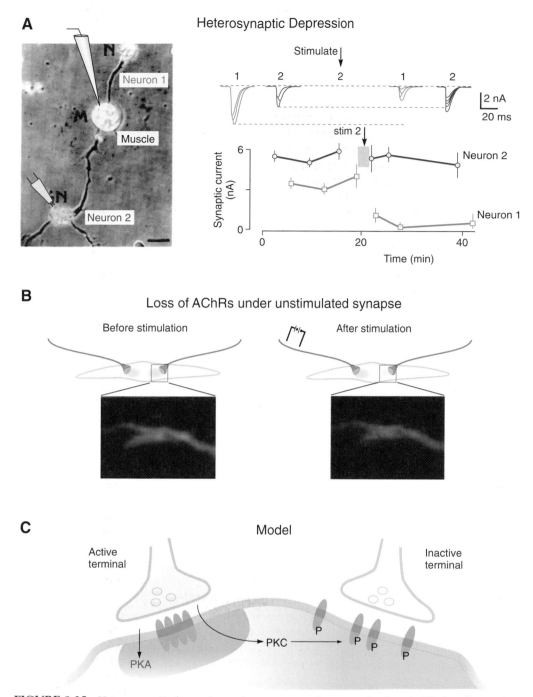

FIGURE 9.25 Heterosynaptic depression at the neuron-muscle synapse in vitro. A. Whole-cell recordings were made from *Xenopus* muscle cells that were co-innervated by two neurons (left). Evoked synaptic currents (ESCs) were measured in response to stimulation of each neuron. A strong stimulus was then applied to neuron 2 (red), and the strength of each neuron was tested again. This procedure leads to smaller evoked synaptic currents from the *unstimulated* neuron (green) but no change to neuron 2 (top right). The graph of ESC amplitude before and after stimulation of neuron 2 shows that the depression is long-lasting (bottom right). B. Heterosynaptic depression is associated with a loss of postsynaptic AChRs. Muscle cells are co-innervated by two neurons in vitro, as above. When one neuron is stimulated for 1–2 hours, the nonstimu-lated neuron displays heterosynaptic depression and a loss of AChRs beneath its synapse. The images show staining for nerve (green) and AChR clusters (red) beneath the unstimulated synapse before (left) and after (right) heterosynaptic depression. AChR aggregates are stained with rhodamine-bungarotoxin. C. Model of activity-dependent synapse elimination. The active terminal increases PKA and PKC locally. PKC phospho-rylates AChRs beneath the inactive terminal, leading to their loss. However, PKA activity can protect the AChRs beneath the active terminal. PKC may become activated by calcium influx, whereas PKA may become activated by a ligand that is released along with ACh. (From Lo and Poo, 1992; Li et al., 2001; Nelson et al., 2003)

INVOLVEMENT OF INTRACELLULAR CALCIUM

Clearly, the presynaptic terminals are carrying on a rather hostile conversation through the postsynaptic cell, and there must be some molecular pathway that conveys the signal from one contact to another through the cytoplasm. One hypothesis is that depolarizing synaptic potentials open voltage-gated Ca^{2+} channels, and Ca^{2+}-dependent proteolytic enzymes are recruited to demolish the nonactive terminals, ultimately leading to their withdrawal. A variety of proteolytic enzymes have been discovered in the nerve terminals and somata of developing neurons. This hypothesis was tested at the mammalian NMJ by decreasing extracellular Ca^{2+} or blocking specific Ca^{2+}-activated proteases, and both manipulations were able to slow down the process of synapse elimination (Connold et al., 1986).

Once again, there are important similarities between synapse elimination and heterosynaptic depression (above). First, it is possible to prevent heterosynaptic depression by injecting the muscle cell with a Ca^{2+} chelator that sops up free Ca^{2+}, suggesting that a rise in postsynaptic calcium is necessary for depression to occur. Second, it is possible to cause synaptic depression by momentarily raising postsynaptic calcium. This was accomplished by loading muscle cells with molecules of "caged" calcium, which can release the calcium into the cytoplasm when it is exposed to ultraviolet light (Figure 9.26A). Therefore, the synaptic responses at one muscle cell are recorded while a second neighboring muscle cell is exposed to a brief pulse of UV light, and synaptic transmission is depressed by 50% within seconds (Figure 9.26B) (Lo and Poo, 1994; Cash et al., 1996). Interestingly, this rise in postsynaptic calcium is only effective within 50 μm of the synapse, and stimulation of the synapse protects it from depression (Figure 9.26C).

Since synaptic activity leads to the depression and withdrawal of neighboring synapses, it follows that the active synapse must somehow be protected. Although the signaling pathways that lead to AChR loss and heterosynaptic depression are not fully understood, there is evidence that two kinases, PKA and PKC, are involved (Figure 9.25C). For example, PKC activators can produce synaptic depression in the absence of stimulation. In contrast, the PKC activator has no effect when the neuron is stimulated (Li et al., 2001). There are probably several molecular changes that attend synapse elimination at the nerve–muscle junction. The removal of AChRs, a decrease in ACh release, and loss of adhesion between pre- and postsynaptic cells (see below), all conspire to weaken the connection.

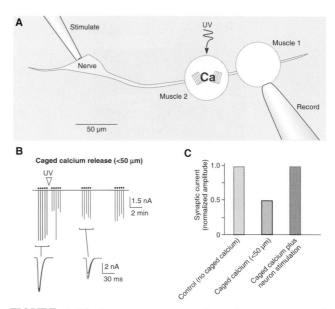

FIGURE 9.26 Synapse depression depends on postsynaptic calcium. A. A whole-cell recording is made from an innervated myocyte (Muscle 1), while the intracellular calcium was elevated in a second nearby myocyte (Muscle 2). Calcium was elevated by first filling Muscle 2 with caged calcium and then using UV light to release the calcium from its "cage". B. Baseline nerve-evoked synaptic currents (downward deflections beneath each dot which represent the stimuli) are first recorded from Muscle 1 (black). When intracellular calcium is elevated in Muscle 2 by exposure to UV light, the nerve-evoked synaptic currents (red) become depressed within seconds. Depression is greatest when the muscles are within 50 μm of one another. C. A summary of three conditions shows that synaptic currents decline by 50% when calcium is elevated (red bar), but the effect can be abolished by stimulating the nerve during UV-evoked uncaging (blue). (Adapted from Cash et al., 1996)

NMDA RECEPTORS AND CALCIUM SIGNALING

Calcium signaling seems to play an important role in the stabilization of developing synapses, but how does calcium get into the neuron? One important pathway is through neurotransmitter receptor-coupled channels. As we learned in Chapter 8, NMDA-sensitive glutamate receptors (NMDAR) are highly expressed in the central nervous system during synaptogenesis. These receptors become active when glutamate and membrane depolarization are present at the same instant (Figure 9.27A). When NMDARs are depolarized, a magnesium ion is expelled, and the open channel permits Ca^{2+} to rush into the postsynaptic neuron.

To find out whether NMDA receptors are involved in activity-dependent synapse plasticity, these receptors have been blocked in a number of developing

A

NMDA Receptor Activation and Calcium Influx

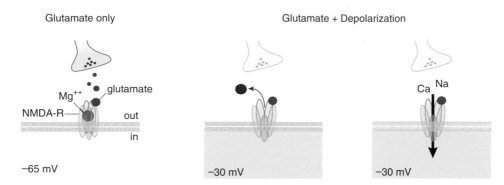

B

Segregation of Retinal Terminals in the Three-eyed Frog

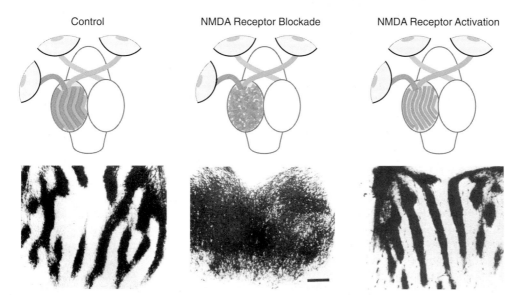

FIGURE 9.27 NMDA receptors are involved in synaptic plasticity. A. The NMDA receptor is activated by a combination of glutamate binding and membrane depolarization. A positively charged magnesium ion blocks the NMDA receptor channel at rest. Depolarization causes the magnesium ion to be ejected, and this permits sodium and calcium ions to flow in. B. When a third eye is implanted into a frog embryo, the tectum becomes co-innervated by two eyes. The afferents from each eye segregate into stripes, similar to primary visual cortex (left). When three-eyed frogs are treated with an NMDA receptor antagonist (either APV or MK801), the afferents do not segregate, and stripe formation is prevented (middle). When three-eyed frogs are treated with NMDA, stripe formation is enhanced (right), and the borders between stripes is sharper. (Adapted from Cline et al., 1987; Cline and Constantine-Paton, 1990)

systems. For example, we learned that monocular lid closure weakens synapses in the cortex that are driven by the deprived eye. However, when the same manipulation is performed during the chronic infusion of a NMDAR blocker, the strength of "deprived" synapses is preserved (Kleinschmidt et al., 1987). Thus, synapses from the open eye activate NMDARs in the cortex, allowing calcium to enter postsynaptic neurons. We will consider what calcium might be doing once it enters the postsynaptic neuron below.

NMDA receptors turn out to be broadly involved in the stability of developing excitatory synapses. In the cerebellum, where adult Purkinje cells are innervated by one climbing fiber, the chronic administration of a NMDAR blocker (AP5) results in 50% of Purkinje cells remaining multiply innervated (Rabacchi et al., 1992). In frogs, the innervation pattern of retinal afferents can be disrupted by NMDAR blockers. For example, when an extra eye is transplanted into a tadpole, the retinal axons project to the midbrain and form a "striped" innervation pattern, similar to the ocular dominance columns in mammalian visual cortex. The segregation of afferents into stripes was blocked completely in the presence of NMDAR antagonists (Figure 9.27B). This effect is reversible: when the NMDAR blocker is removed, the fibers from each eye became segregated. Interestingly, exposing the midbrain to NMDA leads to a sharpening of the borders between eye-specific stripes (Cline et al., 1987; Cline and Constantine-Paton, 1990).

In the mammalian superior colliculus (SC), NMDA blockade prevents the elimination of retinal projections that grow to the wrong topographic position during the first few postnatal weeks (Chapter 6). These studies, and many others like them, show that the NMDAR plays a fundamental role in activity-dependent maturation of synaptic connections (Scherer and Udin, 1989; Simon et al., 1992). One can observe the influence of NMDA receptor activity by recording the receptive field of individual SC neurons (i.e., the portion of visual space that activates the neuron). When the hamster SC is treated with an NMDAR blocker during development, the single neuron receptive field sizes are more than 60% larger than normal (Huang and Pallas, 2001). In fact, the receptive fields can attain normal values even when a portion of the superior colliculus is eliminated at birth, forcing the retinal axons to innervate fewer postsynaptic neurons. This ability is also prevented with a NMDAR antagonist. Therefore, NMDAR blockade commonly interferes with synapse elimination in the central nervous system, leading to less specific afferent projections. It is also significant that NMDARs play an important role in one neural analog of learning called *long-term potentiation* (see BOX: Remaining Flexible).

THE ROLE OF SECOND MESSENGER SYSTEMS

Since calcium plays an important role in developmental plasticity, it is reasonable to ask what calcium interacts with in the cytoplasm. There are a broad range of calcium-binding proteins in the nervous system. One of these, called *calmodulin*, is a major constituent of the postsynaptic density. Together with calcium, it serves to activate a cytoplasmic kinase called *Ca²⁺/calmodulin-dependent protein kinase II (CaMKII)*. Once activated, the CaMKII becomes autophosporylated, and its activity then becomes independent of Ca^{2+} and calmodulin binding.

To test the role of CaMKII, transgenic fruit flies were engineered to express an inhibitor of this enzyme during development. The motor nerve terminals of these transgenic animals have numerous sprouts, and

there is a greater number of presynaptic sites compared to wild-type flies (Wang et al., 1994). The opposite effect was produced in the frog optic tectum by causing CaMKII to be expressed constitutively in postsynaptic neurons (Zou and Cline, 1996). Retinal axons make simpler arborizations when CaMKII is highly expressed, suggesting that connections are being eliminated more rapidly than normal. In dissociated cultures of cortical pyramidal neurons, the number of connections per neuron can be modified by transfecting cells with a constitutively active form of CaMKII (i.e., catalytically active in the absence of calcium). The transfected neurons display a net loss of presynaptic contacts and a reduction in the number of presynaptic partners (Pratt et al., 2003). Thus, CaMKII participates in decisions about the number of connections to be made, making it a good candidate for the synapse elimination mechanism.

At the fly nerve muscle junction, there is an expansion of terminal boutons in the mutant, *ether-a-go-go*, which displays increased synaptic activity and an increase in bouton number (Figure 9.28A and B). As we learned in Chapter 8, *discs large* (DLG) is a clustering protein at the *Drosophila* nerve–muscle junction. One of the synaptic proteins that DLG regulates is the cell adhesion molecule, FasII. To test whether CaMKII signaling can influence the clustering of DLG and FasII, mutant flies expressing a constitutively active form of the kinase were examined (Koh et al., 1999). In these flies, DLG was apparently displaced from the synaptic region. Since DLG can be phosphorylated by CaMKII, these results lead to a model in which activity mediates synapse stability by modifying the level of adhesion (Figure 9.28C). An influx of calcium may activate CaMKII, leading to phosphorylation of DLG and declustering of FasII. Taken together with the results from flies in which CaMKII is inhibited, it appears that the activity of this enzyme controls synaptic sprouting and stabilization.

There are several other protein kinases that may play a similar role during synaptogenesis. One kinase that is partially dependent on intracellular calcium and phospholipid metabolites is protein kinase C (PKC). When mice are genetically engineered to be deficient in a neuron-specific form of PKC, synapse elimination is decreased in the cerebellum. About 40% of Purkinje neurons are innervated by more than one climbing fiber, while normal Purkinje cells are all innervated by a single axon (Kano et al., 1995). Other enzymes, such as protein kinase A (PKA), are activated by a rise in cAMP. A role for the cAMP and PKA signaling pathway has been suggested by single-gene mutant flies, named *dunce*, due to their poor learning abilities. The cAMP levels are persistently increased in *dunce* flies, leading to a decrease in FasII expression

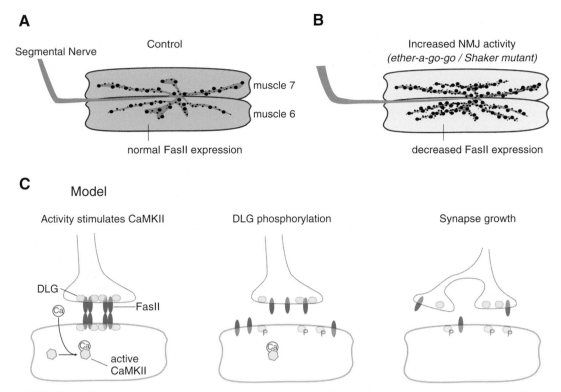

A Control

Segmental Nerve

muscle 7

muscle 6

normal FasII expression

B Increased NMJ activity
(ether-a-go-go / Shaker mutant)

decreased FasII expression

C Model

Activity stimulates CaMKII

DLG

Ca

FasII

Ca

active
CaMKII

DLG phosphorylation

P   P   P

Ca

Synapse growth

P   P   P

FIGURE 9.28 Activity-dependent synapse formation and cell adhesion molecules. A. In wild-type flies, Muscles 6 and 7 receive about 180 boutons, and the expression of FasII is relatively high (left). B. In a double mutant strain of flies (*ether-a-go-go*/Shaker), synaptic activity is elevated at the neuromuscular synapse, and FasII levels are lower. About 55% more boutonal endings are made on Muscles 6 and 7 in these flies (right). C. One model to explain these results is that synaptic activity leads to calcium influx and activation of the kinase, CaMKII (left). The active CaMKII phosphoyrylates a MAGUK protein that mediates clustering of synaptic proteins, including FasII (middle). The phosphorylated DLG is not as restricted to the synaptic complex, and FasII may no longer be located at the synapses, leading to less adhesion and sprouting of new boutons. (Adapted from Budnik et al., 1990; Schuster et al., 1996; Koh et al., 1999)

and a 70% increase in boutonal endings on muscle 6/7. The cAMP/PKA signal can lead to changes in protein synthesis by phosphorylating specific activators or repressors of transcription (see BOX: Remaining Flexible). When *dunce* flies are engineered to carry a transgene that maintains high levels of FasII expression, the sprouting of motor terminals does not occur (Schuster et al., 1996b). Thus, it appears that synaptic activity may also influence the adhesion of pre- and postsynaptic cell through cAMP signaling.

METABOTROPIC RECEPTORS

We have seen that mechanisms of synapse elimination are at least partially triggered by the primary transmitter and receptor. Nicotinic AChRs are likely to be involved in synapse elimination at the neuromuscular junction, and NMDARs appear to play an important role in the elimination of central glutamatergic synapses. Both of these signaling systems operate by

opening an ion channel that is permeable to sodium, potassium, and/or calcium. However, neurons typically express many different types of receptors, even within a single synapse. Are there other synaptic mechanisms involved besides ionotropic signaling? This question has been approached by examining synapse development in mice that are lacking specific receptor subunits. One interesting possibility is that glutamatergic transmission also activates metabotropic glutamate receptors (mGluRs). Rather than opening ion channels directly, mGluRs can activate either of two different signaling pathways: phospholipid metabolism or cAMP production.

One set of knockout mice that are deficient in a particular metabotropic glutamate receptor subunit, mGluR1, exhibits significantly less synapse elimination at cerebellar Purkinje cells compared to controls. Furthermore, normal synapse elimination can be rescued in these mice by expressing the mGluR1α transgene specifically in Purkinje cells (Kano et al., 1997; Ichise et al., 2000). Furthermore, the metabotropic glutamate receptor may well act through a phospholipase (PLC) signaling pathway. In mice that are deficient in a specific isoform, PLCβ4, climbing fiber elimination is impaired in the rostral portion of the cerebellum where the mRNA for PLCβ4 is predominantly expressed (Kano et al., 1998). Therefore, synapse elimination may involve the collaboration of more than one type of glutamate receptor and intracellular signaling cascade.

It is likely that transmitters besides ACh are released at the nerve–muscle junction and that they play an important role in synapse maturation and plasticity. For example, the neurotrophins, BDNF and NT-4 are expressed at the NMJ, and AChR clusters are rapidly lost when TrkB signaling is disrupted (Gonzalez et al., 1999; Wells et al., 1999). A second protein, calcitonin gene-related peptide (CGRP), is present at the developing nerve–muscle junction and may participate in the competition for postsynaptic space (Nelson et al., 2003). CGRP is known to upregulate AChR synthesis (New and Mudge, 1986). Furthermore, activation of the CGRP receptor can elevate postsynaptic cAMP, leading to activation of PKA and phosphorylation of AChRs (Miles et al., 1989; Lu et al., 1993). Therefore, it is possible that active motor nerve terminals use CGRP to recruit the protective influence of PKA signaling (Figure 9.25C).

Neuromodulatory transmitters have also been shown to affect synaptic plasticity during development. For example, the effects of monocular deprivation are markedly reduced when either cholinergic or noradrenergic or serotonergic terminals are eliminated from the developing nervous system, although none of these afferents mediates visually evoked activity in the cortex (Kasamatsu and Pettigrew, 1976; Bear and Singer, 1986; Gu and Singer, 1995). Apparently, the projections from certain brainstem nuclei (e.g., the raphe, the locus coeruleus, and the nucleus basalis) arborize widely in the brain, and can modify synaptic transmission in adult and developing animals. It is still unclear how these modulatory systems are activated during development or how they interact with the primary afferent transmitter system. However, such findings do suggest that our concept of synaptic plasticity in the central nervous system is still rather rudimentary.

GAIN CONTROL

To this point, our discussion has focused on the concept that synapses can be weakened and eliminated by the activity of neighboring connections. However, synaptic transmission can also undergo long-lasting alterations in strength during develop-ment. For example, bursts of high-frequency stimulation delivered to retinal afferents produce an enhancement of excitatory synaptic transmission in 40% of LGN neurons tested (Mooney et al., 1993). Moreover, a NMDAR antagonist is able to block this activity-dependent potentiation of synaptic transmission in most neurons. This phenomenon was called *long-term potentiation* (LTP) when it was first discovered in adult tissue (see BOX: Remaining Flexible).

LTP has been found at many other developing synapses, although it is most prominent in the cortex and hippocampus. One possibility is that LTP is an important mechanism in the adult nervous system and simply appears during development without playing any particular role in synaptogenesis. However, LTP is particularly prominent during development in some areas of the cortex (Figure 9.29A). For example, LTP has been studied at synapses between thalamic afferents and somatosensory "barrel" cortex, and it is only found before postnatal day 7 (Crair and Malenka, 1995). There are also suggestions that it determines which synaptic connections will drive the postsynaptic neuron most effectively. LTP is found in two different forebrain nuclei (LMAN and area X) that are required for vocal learning in developing zebra finches (Boetigger and Doupe, 2001; Ding and Perkel, 2004). Interestingly, each form of LTP is restricted to a particular developmental period, occurring earlier within LMAN and later within area X.

If an LTP mechanism helps to shape the connectivity between neurons during development, then it should selectively strengthen the active input, and not the inactive one. To test this idea, whole cell recordings were obtained from *Xenopus* tectal neurons in vivo. Small extracellular stimulating electrodes were placed at two positions in the retina that project retinal ganglion cell (RGC) axons to the recorded tectal neuron, and electrical stimulation evoked EPSCs (Tao et al., 2001). In young tadpoles, stimulation of one RGC leads to an enhanced excitatory postsynaptic response for *both* inputs. That is, there is no specificity. In contrast, when a single RGC is stimulated in slightly older tadpoles, it subsequently displays LTP, whereas the synaptic connection from the unstimulated RGC is unchanged. Dendritic growth and calcium regulation may account for this difference between ages. Synaptic stimulation elicits a rise in postsynaptic calcium throughout the younger tectal neurons, but the calcium rise is highly restricted in the older age group. These findings emphasize, once again, that synaptic plasticity is very restricted in the spatial domain (Figures 9.23 and 9.26).

Synaptic activity can also lead to a long-lasting decrease in EPSP amplitude, often called *long-term*

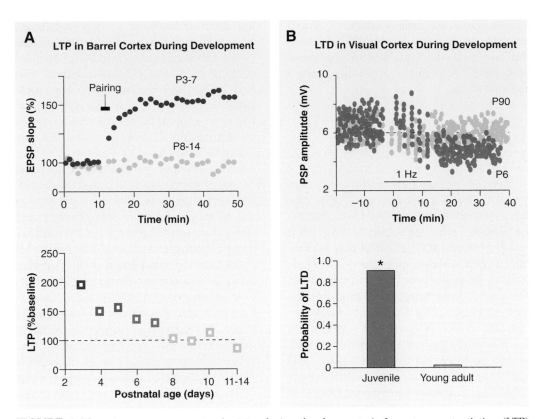

FIGURE 9.29 Alteration in synaptic plasticity during development. A. Long-term potentiation (LTP) declines with age in barrel cortex. During the first postnatal week, there is an enhancement of excitatory synaptic transmission when the neuron is depolarized during afferent stimulation at 1 Hz. The same stimulus does not enhance excitatory synaptic strength in tissue from P8–14 animals. LTP gradually declines in magnitude during the first postnatal week. B. Long-term depression (LTD) declines with age in Layer IV of visual cortex. At postnatal day 6, low-frequency stimulation of the thalamic afferents leads to a long-lasting depression of excitatory synaptic currents. By P90, the same treatment has no effect. The probability of detecting LTD is nearly 90% in tissue from juvenile animals but is negligible in young adults. (From Dudek and Friedlander, 1996; Crair and Malenka, 1995)

depression (LTD). In many areas of the brain, LTD is more prominent during early development (Figure 9.29B). For example, LTD is present in Layer IV of the visual cortex in juvenile cats and guinea pigs. However, it is virtually absent in adult animals (Dudek and Friedlander, 1996). Furthermore, LTD persists in other layers of the adult cortex. Therefore, LTD may play a role in eliminating thalamocortical synapses, possibly contributing to the formation of ocular dominance columns.

How can LTP and LTD play a role in the selective stabilization of inputs during development? To address this question, whole-cell recordings were again obtained from *Xenopus* tectal neurons in vivo, and stimulating electrodes were placed on two different RGCs that converged on the recorded neuron (Zhang et al., 1998). When the RGCs are stimulated synchronously for 20 s, the EPSCs evoked by stimulation at each site display LTP. The situation is a bit more

complicated when the two retinal sites are stimulated asynchronously. When stimulation of one retinal site produces a postsynaptic action potential, while the other site produces a subthreshold excitatory synaptic event, then the precise timing of the two inputs is crucial to the outcome. When the subthreshold event *leads* the retinally evoked action potential by <20 ms, then it displays LTP. When the subthreshold event *follows* the retinally evoked action potential by <20 ms, then it displays LTD. Thus, inputs that evoke subthreshold synaptic potentials can either be strengthened or weakened, depending on the timing of their activity with respect other inputs, particularly those that evoke suprathreshold activity. These result may help to explain how LTP and LTD contribute to the refinement of retinal projections. As visual stimuli (or spontaneous activity waves) sweep across the retina, a postsynaptic tectal neuron will be sequentially activated by a series of retinal ganglion cells. When the most effective retinal ganglion cell is activated, it will cause the tectal neuron to fire an action potential. Thus, any retinal ganglion cell that produces a subthreshold event within the next 20 ms may become depressed and subject to elimination.

SILENT SYNAPSES

An extreme case of synaptic potentiation occurs when the physical contacts between nerve cells display absolutely no transmission, yet can be turned on by using them. These "silent synapses" have been observed in both young and adult animals. For example, inactive contacts have been observed in cultures of chick ciliary ganglion neurons and myotubes. Only 58% of the contacts are functional, even though the myotubes are expressing plenty of AChRs (Dubinsky and Fischbach, 1990). When cAMP levels are increased in the cultures, 93% of contacts are functional, suggesting that "silent" presynaptic terminals are activated by a cAMP signaling pathway. Similarly, silent presynaptic terminals in the neonatal hippocampus can be turned on by activating AChRs (Maggi et al., 2003).

The enhancement of synaptic transmission that is observed soon after innervation may also occur if existing receptors on the postsynaptic membrane are modified. When embryonic rat medullary neurons are placed in dissociated culture, there is no indication of glycine-evoked currents during the first six days in vitro. However, glycine-evoked currents can be observed in excised patches of membrane within a day of plating. This result suggests that the glycine receptors are initially "silent" and are converted to an active state during development (Lewis et al., 1990).

It is possible that many glutamatergic synapses are silent during development because activation of the NMDAR requires membrane depolarization (Figure 9.27A). Intracellular recordings from the neonatal hippocampus show that most synapses have only NMDARs, and they do not respond when membrane potential is held at −60 mV. However, neural activity can enhance synaptic transmission by rapidly recruiting new functional AMPA-type glutamate receptors (Durand et al., 1996). The appearance of functional AMPARs may be due to either insertion or modification. For example, the phosphorylation state of AMPA-type glutamate receptors can be modified by NMDAR activity, leading to the dephosphorylation of a particular AMPA-type glutamate receptor, called GluR1. When this happens, excitatory synaptic transmission is depressed (Lee et al., 1998). Alternatively, glutamate receptors can be moved from the cytoplasm

and become inserted into the postsynaptic membrane within 10 minutes of stimulation (Shi et al., 1999).

Eye opening is associated with an increase in the percentage of "silent" NMDAR synapses in the rat superior colliculus. Furthermore, the AMPAR:NMDAR ratio increases during the next 24 hours of vision, and there is a commensurate reduction in the number of retinal inputs per postsynaptic neuron (Lu and Constantine-Paton, 2004). A similar observation was made in the optic tectum of developing tadpoles. When retinal afferents are stimulated electrically, most immature synapses do not exhibit a postsynaptic response if the cell is held near the resting potential. However, if the cell is depolarized to +55 mV, then the synapses immediately display a response. This is because the NMDAR is now able to open when bound by glutamate. As the synapses mature in this system, they begin to express a greater number of AMPA-type glutamate receptors. Furthermore, constitutive expression of CaMKII accelerates this transition (Wu et al., 1996). Thus, CaMKII may serve as an activity-dependent mechanism for strengthening some synapses by recruiting functional AMPARs, while eliminating others.

Two experiments suggest that glutamate receptor turnover and modification are controlled by electrical activity. When primary cultures of spinal neurons are grown in the presence of the glutamate receptor blockers, CNQX, a greater number of AMPA receptor subunits accumulate at synaptic contacts and spontaneous EPSCs are larger. Conversely, when excitatory synaptic activity is increased by growing the cultures in GABA and glycine receptor antagonists, the amount of synaptic AMPA receptors declines and spontaneous EPSCs are smaller (O'Brien et al., 1998). Apparently, synaptic transmission controls the number of synaptic AMPA receptors by regulating the half-life of the receptor subunits.

The ability of synapses to linger on, even though they provide little input to the postsynaptic cell, may be due to the presence of trophic substances. For example, when glial cell line-derived neurotrophic factor (GDNF) is overexpressed in the muscle cells of transgenic mice, the number of motor neuron terminals at each endplate is dramatically increased (Nguyen et al., 1998). The neurotrophin, BDNF, prolongs the multiple innervation of mammalian muscle cells in vivo, and this effect is far more prominent when assessed anatomically. At postnatal day 12, almost 80% of muscle cells are contacted by more than one nerve terminal, yet only about 25% of muscle cells display multiple EPSP amplitudes (as assessed with the procedure shown in Figure 9.2). This result suggests that many of the extramotor terminals must be functionally silent (Kwon and Gurney, 1996). Thus, synapse elimination may involve an intermediate state, one in which the synapse is anatomically present but cannot be detected with functional criteria.

HOMEOSTASIS: THE MORE THINGS CHANGE, THE MORE THEY STAY THE SAME

Experiments performed on simple invertebrate circuits and dissociated cortical neurons show that synapses and ion channels can each be regulated by the average level of postsynaptic activity (Marder and Prinz, 2002; Burrone and Murthy, 2003). This form of plasticity is homeostatic; its purpose is to keep the average postsynaptic discharge rate at about the same level, and it continues to operate in the adult nervous system (Royer and Pare, 2002). When postsynaptic activity is increased or decreased, voltage- and ligand-gated channels are adjusted to resist the manipulation. For example, cortical neurons that are cultured with the sodium channel blocker, TTX, increase their sodium channels and decrease their potassium channels. Similarly, excitatory synaptic currents increase and inhibitory currents decrease when cultures are grown in an assortment of activity blockers (Rao and Craig, 1997; Desai et al., 1999; Murthy et al., 2001; Kilman et al., 2002; Burrone et al., 2002). Conversely, when excitatory synaptic activity is increased by growing the cultures in GABA and glycine receptor antagonists, the amount of synaptic AMPA receptors declines and spontaneous EPSCs are smaller (O'Brien et al., 1998).

A change in postsynaptic activity can also affect the presynaptic terminal. In cultured hippocampal neurons, activity blockade leads to an increase in the size of the presynaptic terminal and docked vesicles as measured with electron microscopy. In one study, these changes were accompanied by an increase in presynaptic efficacy (Murthy et al., 2001). Homeostatic changes in presynaptic release have also been demonstrated in vivo. In congenitally deaf mice, there is an increase in release probability at the very first central synapse in the cochlear nucleus (Oleskevich and Walmsley, 2002). In an imaginative genetic manipulation, fruit fly muscle fibers were silenced by causing them to overexpress an inwardly rectifying potassium channel; this drives the resting membrane potential to the potassium equilibrium potential. The hyperpolarized muscle is no longer able to reach action potential threshold, yet motor neuron-evoked synaptic potentials are just as large as those recorded in wild-type flies. In this case, there is an increase in presynaptic release, with no change in quantal size (Paradis et al., 2001). Therefore, the balance

of ligand- and voltage-gated channels, as well as transmitter release, all depend on postsynaptic activity levels.

What is the evidence that synaptic strength is adjusted by a similar homeostatic mechanism in vivo? To answer this question, direct measures of EPSP and IPSP amplitudes have been made in a brain slice preparation following a period of sensory deprivation (e.g., blindness or deafness). When gerbils are surgically deafened before they would first experience sound, compensatory responses are observed for both excitatory and inhibitory synapses within the inferior colliculus. Inhibitory synaptic conductance declines, and the inhibitory reversal potential depolarizes. In contrast, afferent-evoked excitatory synaptic responses become larger and longer in duration (Vale and Sanes, 2000, 2002). Interestingly, the inhibitory reversal potential appears to become depolarized because chloride transport is downregulated (Vale et al., 2003). Similar observations have been made in the cortex. During normal development of the rat visual cortex, the amplitude of miniature EPSCs declines during the first three postnatal weeks. However, when rat pups are reared in complete darkness, this reduction in mEPSC amplitude is largely prevented (Desai et al., 2002), suggesting a compensatory response to the lost visual drive. Dark rearing also prevents the normal increase of inhibitory synaptic currents in Layer 2/3 cells (Morales et al., 2002). In the auditory cortex of deaf gerbils, there are three major compensatory responses that may sustain an operative level of cortical excitability in Layer 2/3 pyramidal neurons: the excitatory synaptic response becomes longer in duration, the inhibitory synaptic response becomes smaller in amplitude, and there is a modest depolarization of the resting membrane potential and increase in membrane resistance (Kotak et al., 2005).

The homeostatic mechanism does not explain some alterations in synaptic strength following a developmental manipulation. For example, one can selectively decrease inhibition to an auditory brainstem nucleus called the LSO (circuit shown in Figure 8.28) by ablating the contralateral ear. The strength of synaptic transmission was then measured with whole-cell recordings using an acute brain slice preparation. In normal animals, electrical stimulation of the inhibitory pathway produced large IPSPs, but following a short period of disuse the IPSPs were small or absent. Interestingly, the unmanipulated excitatory pathway became much stronger, displaying large NMDAR-dependent EPSPs (Kotak and Sanes, 1996). A homeostasis mechanism should have upregulated the inhibition and downregulated the excitation. The selective deprivation of an inhibitory pathway is not easily accomplished elsewhere in the nervous system, and it is possible that such a manipulation produces a different sort of neuronal response. Alternatively, the homeostatic response may depend on the state of maturation (Burrone et al., 2002).

PLASTICITY OF INHIBITORY CONNECTIONS

At the nerve–muscle junction, where there is only a single type of synapse, it may soon be possible to explain synaptic competition and elimination at the level of molecular pathways. However, the typical central neuron receives a variety of projections with distinct neurotransmitter systems. It is unlikely that these synaptic contacts form independently of one another. In fact, we have just learned that manipulations to one system inevitably affect the development of the others (see above: Homeostasis).

Inhibitory synapses contribute to neural processing in approximately equal numbers to synaptic excitation. Despite this, we are just now beginning to learn about their developmental plasticity. The development of one inhibitory projection nucleus in the auditory brainstem, called MNTB, has been relatively well studied. MNTB neurons are activated by the contralateral ear, and their axons project to auditory nuclei that encode binaural acoustic cues and contribute to sound localization. One of the postsynaptic targets to which MNTB projects to is the LSO (see schematic in Figure 8.28). The inhibitory terminals become refined in two phases during development, leading to a precise tonotopic map. During the first postnatal week, there is a large reduction of function contacts from the rat MNTB onto single LSO neurons. This was revealed by recording from individual LSO neuron while focally activating small areas within the MNTB. At birth, a 113 μm wide region of the MNTB functionally innervated a single LSO neuron, and this declined by almost 50% by postnatal day 9 (Kim and Kandler, 2003). During the second and third postnatal week, individual gerbil MNTB terminal arbors are pruned back by about 30% and come to occupy a narrow portion of the frequency map (Sanes and Siverls, 1990).

A complementary phenomenon has been described at a second target nucleus of the MNTB, the medial superior olivary nucleus (MSO). Inhibitory terminals are gradually eliminated from MSO dendrites during postnatal development and become restricted to the cell body in the adult (Figure 9.30). The staining pattern of glycine-containing boutons and glycine receptor clusters demonstrated that both pre- and postsynaptic elements of the inhibitory synapse were eliminated

A

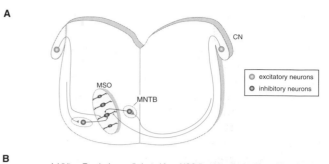

B

Inhibitory Terminals are eliminated from MSO Dendrites During Normal Development

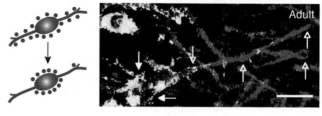

Inhibitory Terminals Remain on MSO Dendrites Following Neonatal Deafening

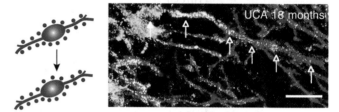

FIGURE 9.30 Elimination of inhibitory synapses during development. A. The schematic shows a nucleus in the ventral auditory brain stem (MSO) that receives inhibitory synapses from two nearby nuclei (LNTB and MNTB). B. At birth, inhibitory terminals are located on the soma and dendrites of MSO neurons. However, most of the dendritic synapses are eliminated during postnatal development (top left). The micrograph (top right) shows stained MSO neurons and glycine receptors from an adult animal. The glycine receptors (yellow) are largely restricted to the soma, and very few remain on the dendrites (blue). When animals are deafened unilaterally during development, the elimination of inhibitory synapses fails to occur (bottom left). The micrograph shows that significant glycine receptor staining (yellow) is now found on the dendrites (bottom right). Scale bars are 20 μm. (From Kapfer et al., 2002)

(Kapfer et al., 2002). Finally, inhibitory afferents can form a striping pattern in the rat auditory midbrain, reminiscent of the ocular dominance columns produced by thalamic afferents in the visual cortex. These stripes emerge from a diffuse projection pattern during the first two postnatal weeks (Gabriele et al., 2000).

To test whether the refinement of inhibitory synapses depends on their activity, MNTB neurons were deafferented by removal of the contralateral cochlea. This manipulation prevents the developmental refinement of MNTB arbors in both the LSO and MSO. In the LSO, single MNTB arbors develop the correct number of synaptic boutons, but these terminals remain spread over a larger portion of the tonotopic axis, compared to controls (Sanes and Takács, 1993). In the MSO, inhibitory terminals fail to be eliminated from the MSO dendrite (Figure 9.30). Interestingly, raising gerbils in omnidirectional white noise to reduce the binaural cues that MSO neurons respond to also leads to significantly higher density of glycine receptor puncta on the dendrites (Kapfer et al., 2002). Finally, the segregation of inhibitory afferents into stripes within the rat auditory midbrain is prevented by functionally denervating the inhibitory projection neurons (Gabriele et al., 2000). Thus, inhibitory arbors do not go through their normal period of anatomical refinement when they are deprived of synaptic input from the cochlea, suggesting that inhibitory activity also plays a role in the maintenance or stabilization of inhibitory synaptic contacts.

Since the refinement of excitatory terminals has been associated with their activity, a question that arises is whether the physical elimination of inhibitory synapses is associated with a weakening in the strength of inhibitory transmission. For example, we learned that inhibitory synapses elicit a depolarizing postsynaptic response in neonatal animals (Chapter 8). Therefore, it has been proposed that inhibitory terminals may use an "excitatory" calcium-depend mechanism, similar to glutamatergic terminals (Kullman et al., 2002; Kim and Kandler, 2003).

The gain of inhibitory synapses can also display activity-dependent adjustment (i.e., LTP and LTD), similar to glutamatergic synapses (Gaiarsa et al., 2002). Do MNTB synapses display an activity-dependent form of long-term depression that could support synapse elimination? In fact, low-frequency stimulation of MNTB afferents produces a profound depression of the evoked inhibitory synaptic response. Furthermore, this inhibitory LTD is age-dependent, being prominent during the period of synapse elimination and declining during the third postnatal week (Kotak et al., 2000).

Since IPSPs are no longer depolarizing at this time, it seems likely that alternative mechanisms mediate inhibitory LTD. One intriguing possibility for the signaling pathway is suggested by the observation that MNTB-evoked inhibition is primarily GABAergic at first and gradually becomes glycinergic during the first two postnatal weeks (Kotak et al., 1998). These results have been confirmed anatomically using immunohistochemical staining against GABA, glycine, the glycine receptor anchoring protein (gephyrin), and a $GABA_A$ receptor subunit (Kotak et al., 1998; Korada and Schwartz, 1999). Individual terminals actually display a developmental decrease in GABA content and a commensurate increase in glycine (Nabekura et al., 2004). Co-release of GABA and glycine has now been demonstrated directly in other developing systems and may be a general principle of inhibitory synapse maturation (Jonas et al., 1998; O'Brien and Berger, 1999; Russier et al., 2002; Keller et al., 2001: Dumoulin et al., 2001).

Whereas glycine receptors appear to be solely ionotropic, the presence of GABAergic transmission permits inhibitory terminals to communicate through a metabotropic pathway. LTD in the LSO can be blocked with a $GABA_B$ receptor antagonist (Kotak et al., 2001). Furthermore, focal application of GABA to the postsynaptic LSO neuron is sufficient to depress the inhibitory response; in contrast, glycine application has no effect (Chang et al., 2003). Thus, inhibitory synapses undergo a period of refinement, much as excitatory systems do, during which they attain a precise pattern of innervation. A growing number of studies suggest that spontaneous or sensory-evoked inhibitory transmission can influence this process.

SYNAPTIC INFLUENCE ON NEURON MORPHOLOGY

Synaptic activity plays an extremely important role in regulating postsynaptic neuron morphology. During early development, even if denervation does not result in cell death (see Chapter 7), then it certainly leads to shrinkage of cell body size, atrophy of dendritic processes, or loss of dendritic spines in most areas of the central nervous system (Globus and Scheibel, 1966; Valverde, 1968; Rakic, 1972; Harris and Woolsey, 1981; Vaughn et al., 1988). Furthermore, changes in the amount of sensory experience given to an animal during development (which presumably affects neural activity) also lead to measurable alterations in nerve cell morphology. Thus, young rats that are reared in an enriched, social environment have more dendritic branching than rats reared alone in an impoverished environment (Fiala et al., 1978). More precise manipulations of the sensory environment, such as sound attenuation or vertical stripe rearing, have also been associated with specific changes in auditory or visual neuron morphology, respectively.

Although it is common to perform a manipulation and then wait days or weeks to look for a change in the central nervous system, the effects of denervation occur at a surprising rate. In the chick *nucleus laminaris* (NL), it is possible to denervate the ventral dendrites while leaving the dorsal dendrites untouched (Figure 9.31). The entire NL dendritic arborization is then visualized with a Golgi stain, beginning one hour after the affer-

A

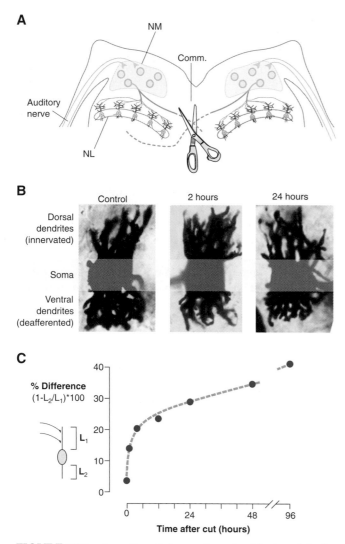

B

C

FIGURE 9.31 Synaptic activity regulates dendrite length in the chick auditory brain stem. A. Neurons of the *nucleus laminaris* (NL) receive afferents from ipsilateral NM to their dorsal dendrites and from the contralateral NM to their ventral dendrites. The ventral dendrites of NL neurons were denervated by cutting NM afferent axons at the midline. The dorsal dendrites remained fully innervated. B. When NL neurons were stained with the Golgi technique, the ventral dendrites were found to be significantly shorter than those on the dorsal side within one hour of the manipulation. C. The ventral dendrites shrunk by almost 40% by 96 hours after the lesion. (Adapted from Deitch and Rubel, 1984)

A   In vivo motility of dendritic filopodia in 1 month mice

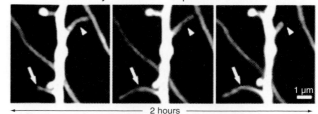

B   In vivo retraction and extension of dendritic spines

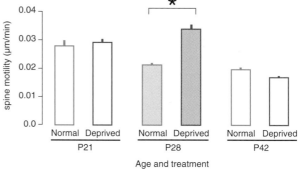

C

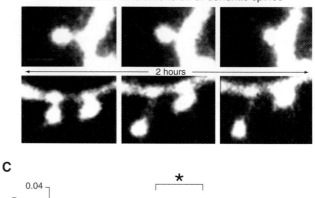

FIGURE 9.32 Two-photon imaging of dendritic spine and filopodia motility in vivo. A. A set of time-lapse images (one hour intervals) in a 1-month-old animal shows that filopodia undergo rapid extension (arrows) and retraction (arrowheads). B. Dendritic spines motility is more prominent at P21. Two time-lapse sequences are shown (acquired over two hours). In the first, a spine is retracted (top), and in the second a spine elongates (bottom). Binocular deprivation from P14 significantly increases spine motility at P28 by 60%. However, there is no change in motility when deprivation begins at P21. There is a slight reduction in motility in mice deprived at P42. (From Grutzendler et al., 2002; Majewska and Sur, 2003)

ents to the ventral dendrites are cut. Even at the earliest time point, there is a 14% decrease in the ventral dendrites compared to the dorsal set (Deitch and Rubel, 1984). Since this is well before terminals are actually removed from the postsynaptic neuron, the dendritic atrophy probably results from the sudden cessation of synaptic activity. Electron microscopy revealed that this change in morphology was correlated with a dramatic change in cytoskeletal structures, such as microtubules.

Dendrite morphology can be studied in vivo by imaging visual cortical neurons in transgenic mice that express a fluorescent protein. Images obtained over a two-hour period show that dendrites extend and retract filopodial processes and spines while they are developing (Figure 9.32A and B). However, the number of filopodia declines dramatically, and the spines become quite stable over the first two postnatal months (Grutzendler et al., 2002). When animals are visually deprived during the critical period, spine motility increases (Figure 9.32C). The results suggest that dendrites are maximally sensitive to synaptic activity during this period and that they participate in the activity-dependent formation and elimination of synaptic connections (Majewska and Sur, 2003).

As with cell death, the effect of synaptic transmission can be tested directly by manipulations that block glutamate receptors. When chick embryos are treated with an NMDA receptor antagonist, cerebellar Purkinje cell dendrites do not develop as many branches and occupy a smaller cross-sectional area (Vogel and Prittie, 1995). A similar effect has been observed in frog optic tectal neurons and in spinal motor neurons. In hippocampal cultures, the number of dendritic spines, but not the number of branches, is dependent on glutamatergic synaptic activity (Kossel et al., 1997). The effects of excitatory transmission can change during development as new signaling systems are added to the cytoplasm. For example, the decrease in the growth of optic tectum dendrites correlates with increased expression of CaMKII, and the rate of growth is experimentally increased when animals are treated with a CaMKII inhibitor (Wu and Cline, 1998). It is also likely that other growth-promoting factors are co-released with neurotransmitter. In organotypic cultures of visual cortex, function-blocking antibodies against the TrkB receptor decrease the amount of basal dendritic growth, indicating that endogenously released BDNF promotes postsynaptic growth (McAllister et al., 1997).

Spine morphology is controlled by the cytoskeletal component, actin, which is highly dynamic in developing systems (Star et al., 2002). Glutamatergic synaptic activity can influence actin polymerization and stabilize spines by raising intracellular calcium (Fischer et al., 2000). In fact, dendrite stability is very sensitive to local calcium levels. In the embryonic chick retina, dendritic retractions can be prevented by locally raising calcium levels within the process (Lohmann et al., 2002). A local rise in calcium may serve to stabilize actin filaments, perhaps by activating gelsolin, the calcium-dependent actin-binding protein. In general, GTP-binding proteins (Rho GTPases) have been implicated in regulating neuron morphology by affecting the stability and assembly of actin filaments (Luo, 2002).

Given that the effects of excitatory denervation are so dramatic, it would be surprising if inhibitory synapses did not have a trophic influence on postsynaptic maturation. In fact, inhibitory terminals appear to have the opposite effect of excitatory terminals in a central auditory nucleus, the lateral superior olive (LSO). Neurons in the LSO can be selectively deprived of functional inhibition by removal of the contralateral cochlea, and this manipulation leads to a significant increase in dendritic branching. Furthermore, when organotypic cultures of the LSO are grown in the presence of the inhibitory antagonist, strychnine, dendrites are twice as long as those grown in normal media (Sanes et al., 1992; Sanes and Hafidi, 1996). Thus, synaptic terminals do not provide a uniform signal to growing dendrites.

SUMMARY

At one level, the purpose of synapse elimination seems perfectly obvious: to create the optimal connections between neurons based upon their use. Perhaps the nervous system cannot take full advantage of the plasticity mechanisms unless it generates extra cell bodies, a surplus of dendritic branches, and a profusion of presynaptic arbors. In reality, the purpose of synapse elimination will remain enigmatic until we can produce a specific alteration in a single set of adult connections, let's say two climbing fibers per Purkinje cell, and determine the exact behavioral outcome. Therefore, our understanding of developmental plasticity is intimately tied to our insight into how the CNS encodes sensory information and controls movement. What is the optimal pattern of connectivity for running, singing, perfect pitch, speed reading, learning, and so forth? Is it even possible to have a nervous system that is optimized for diverse motor, sensory, and cognitive tasks? Unfortunately, we have yet to devise an experiment that tests whether extra cell bodies or small arborization errors actually affect animal behavior.

At present, we believe that synapses can be weakened or lost if they are not activated correctly during development. This might be particularly important for animals, such as humans, that inhabit a wide range of environments. For example, it is likely that our central auditory system is shaped by the spoken language(s) to which we are exposed as infants (see Chapter 10). However, the experiments that demonstrate an influence of environment or neural activity itself have remained rather flagrant. For example, it is unlikely that any animal sees only vertical stripes during development. Yet, we know that developing humans do experience many "extreme" rearing environments, such as blindness, deafness, malnutrition, and many others that result from genetic or epigenetic causes. Therefore, the clinical importance of understanding developmental plasticity is enormous.

10

Behavioral Development

BEHAVIORAL ONTOGENY

We often think of behavior in terms of the lives of postnatal animals, but behavior begins well before birth. As a fetus, you were making coordinated, though perhaps not goal-directed, movements. You were kicking, swallowing, putting your thumb in your mouth, and moving to songs. With ultrasound, we can now see the emergence of behaviors in humans, and even diagnose behavioral deficits prenatally. A bird embryo in its shell also moves and peeps. Even fly embryos wiggle and squirm before they hatch. The first movements and motor responses that an animal makes are far simpler than the sophisticated movements of an adult. Do these embryonic behaviors serve any strategic purpose, or are they merely the consequences of a nervous system that is wiring up and becoming electrically active? Which behaviors arise first in the embryo, how does the repertoire of behavior grow, is there any logic in the sequence, and how does behavior feed back onto the building of the nervous system? These are very intriguing questions for a developmental neurobiologist.

Early behaviors have often been classified as anticipatory, adaptive, or substractive (Oppenheim, 1981). Clearly, some behaviors develop in an *anticipatory* manner, with forward reference to actions that will be of value in later life (Carmichael, 1954). For instance, prematurely born humans can respond to light although they would normally not be exposed to light in the womb. Since the nervous system is constructed over a period of time, behaviors cannot be slapped together the moment they are needed, but arise with the developing circuitry. In the case of sensing light, one of the last things that happens in the nervous system is the addition of active photoreceptors; the rest

of the circuit from eye to brain and brain to motor output is ready and waiting for this input. Early behaviors tend to be imperfect at first and then fine-tuned in a manner that matches the fine-tuning of the development of synaptic connections, which goes on in humans through childhood and beyond. Some early behaviors are anticipatory in the sense that they reflect ancestral nervous systems and modes of behavior evolving toward later forms. There is evidence that mammalian embryonic spinal cords have rhythmic activity similar to swimming patterns of fish, and that at early stages even human embryos display swimming movements. The grasp reflex is an interesting example of this phenomenon (Figure 10.1). This reflex, present at birth in a human infant, disappears at about 3 months of age. When the palm of a baby is touched, it causes inward curling of the fingers and a forceful hold on any object. This grasp reflex is thought to be a relic of a more primitive primate nervous system in which clinging to branches or mother was important for survival. Perhaps these embryonic reflexes that are vestiges of ancestral behavioral traits, reflect important steps in the development of the neural control of our more derived behavioral repertoire.

Some embryonic and juvenile behaviors are useful and *adaptive*, and serve specific functions at particular stages of development. Hatching behaviors are a good example (Oppenheim, 1982). Quail embryos make clicking noises in the shell, which helps to synchronize hatching (Vince and Salter, 1967, Vince 1979). More frequent clicks accelerate hatching, and less frequent clicks retard it. In reptiles, birds, and insects, hatching is composed of repeated stereotypical movements that are transient, specific to that stage of life and clearly adaptive. In human babies, the rooting reflex, turning the face toward a touch on the cheek, is a transient reflex important for breast-feeding. It is important to

FIGURE 10.1   The grasp reflex demonstrated in a newborn baby.

recognize that embryonic and adult animals often live in very different environments. Each animal tends to exhibit stage-specific morphological, molecular, and behavioral adaptations to its specific environment. This is particularly obvious in the case of animals that go through metamorphosis, such as moths and frogs. Here, the larval and adult forms have a radically distinct appearance and behavior. The nervous systems, the substrates of these behaviors, undergo substantial modifications in response to metamorphic hormones (ecdysone for insects and thyroxine for amphibian), including death of larval neurons and genesis of new adult neurons.

Finally, we can consider embryonic behaviors as *substrates* for building more complex behaviors and thus for the continued maturation of the nervous system itself (Carpenter, 1874). Through function and feedback, neural circuits become finely tuned. The learning that we perform as adults may be little more than a continuation of the mechanisms used to adjust the embryonic nervous system. The behavioral patterns that we see in embryonic and juvenile animals are the integrated beginnings of more complex patterns that continue to develop. We must necessarily crawl before we can walk. If this view is correct, disruption of these early behaviors should have a significant impact on the development of later behaviors. A good example of this is the fine motor skills of the forearm. Experience is thought to be important for the development of these motor skills. Preventing normal forelimb use during early development not only causes defects in fine forelimb movements that last throughout life, but also produces defective development of the axonal terminals of the corticospinal neurons that control these movements (Martin et al., 2004).

In this chapter, we will examine the way behavior develops, especially as it reflects on the maturation of neural connections. We will look in detail at the development of various specific motor, sensory, and social behaviors, and try to understand from a neural, molecular, and genetic perspective how and why they develop in the order that they do. We will also examine the mechanisms by which behaviors become more precise and skillful, and the role of sensory and motor experience in the neural circuitry that controls them.

GENETIC AND ENVIRONMENTAL MECHANISMS

Donald Wilson, who worked on the neural basis of insect flight, was amazed that a locust, spreading its wings for the first time, could fly without practice and make appropriate adjustments to wind speed and visual signals. "How perfect is the motor score that is built into the thoracic ganglion?" he wondered. It seemed to him that the CNS is developmentally programmed to contain nearly everything that is necessary for flight before actual flight occurs (Wilson, 1968). Michael Bate, who works on the development of the *Drosophila* nervous system, expresses a similar surprise, but for a different reason (Bate, 1998). "How do we explain," Bate asks, "the remarkable fact that behavioral 'sense' of this kind is inherited and built into the nervous system as it develops?" Genes that affect behaviors function to control the differentiation and physiology of neurons. We have not found genes whose job it is to organize an entire neural circuit. It is difficult to comprehend the genetic basis of the neural circuits that underlie complicated behaviors, yet all of the mechanisms that we have discussed in Chapters 1–9 lead toward building a nervous system with functional circuits that orchestrate adaptive behavior.

Clearly, genetic factors are at the root of behavior; flies do not behave as humans, no matter how similar their rearing environment. Yet the nervous systems of both insects and mammals are built of the same type of neurons, synapses, and neurotransmitters. They also obey the same developmental rules often using homologous molecules. In the 1960s, Seymour Benzer began to address the question of behavior genetics by searching for single genes which, when mutated, would lead to aberrant behaviors in fruit flies (Benzer, 1971). Surprisingly, many of the genes he and his colleagues discovered by this process are conserved among different species. For instance, the genes found to govern the 24-hour circadian rhythm in flies have homologs that are involved in the same process in mammals (Ishida et al., 1999). Do these genetic homologies extend to neural development in the sense that genetically conserved programs construct similar neural circuits that control homologous behaviors? If so, how do the genes control how these circuits are built, and what mechanisms modify these circuits to give species-specific behaviors?

The nervous system is unique among organs in that it responds to a huge variety of environmental influences by changing itself structurally and functionally, even from an early age (Chapter 9). Neural activity modulates the expression of many of the same gene products that were used during neural embryogenesis. This interplay between the environment and the genome continues throughout life. And because of this, it is usually not reasonable to ascribe specific behaviors to purely genetic or environmental determinants. Genetic influences on behavior are as clear as experiential ones, and the two are interlinked. Identical twins who are separated at birth and reared in different families may show amazing similarities in attitude and taste, compared to nonidentical twins reared together. But these two human beings usually also show an enormous array of dissimilarities, reflecting lifelong interactions between the environment and the genome. Similarly, work in mice has shown that the different behavioral tendencies, like the willingness to explore, in different strains of mice are heavily influenced by epigeneitic mechanisms. A genetically B6 mouse that both develops in a BALB uterus and is reared by a BALB mother shows exploration behavior that is more similar to the BALB than it is to the B6 strain (Francis et al., 2003).

ENVIRONMENTAL DETERMINANTS OF BEHAVIORAL DEVELOPMENT

From early embryonic stages of life, genetic influences operate in the context of specific environments. These environments place selective pressures on the embryo and influence development. The most obvious examples for humans come from studies of environmental hazards such as drug use or alcohol consumption during pregnancy. Embryonic exposure to high levels of these substances can lead to mental retardation (Johnson and Leff, 1999). As we will see later in this chapter, embryonic exposure to sex-specific steroids can alter both physical and neural aspects of sexual maturation. In bird embryos, a brief exposure to a mother's call can imprint a preference for that call upon hatching. After hatching, song birds learn to produce the father's song, and vocalization centers in

its brain are strongly influenced by this process (Mooney, 1999).

Early deprivation of many kinds (visual, auditory, even emotional) is known to have permanent effects on behavioral development, just as early experiences can have profound affects on neural development. As we saw in Chapter 9, normal vision can never be restored to an adult mammal that has been blind from birth; the neural circuitry simply has not developed properly. Human infants who are blind also show delays in various motor skills, a fact that emphasizes the importance of vision as a sensory input for motor development (Levtzion-Korach et al., 2000). Lack of various aspects of education and emotional interactions early in life may also affect the ability to perform later in life because the neural substrates of these behaviors are wired in at particular stages. Thus, although we do not believe that human brains, or those of other animals, emerge as blank slates, it is nevertheless clear that experience shapes and adjusts the nervous system. An argument has been made that the brain is especially adaptive, in the Darwinian sense, simply because it is an organ that learns how to modify behavior in order to improve survival in a changing environment. In fact, learning is one of the main functions of the nervous system, and this process begins in the embryo. In dealing with the genetic and environmental influences on behavior, we must not only address how genes control cellular and molecular events to construct the neural substrate of behavior, but also how behavioral and sensory events feed back onto these molecular mechanisms.

The description of developing behavior is valuable because it provides a sensitive and fairly inclusive indicator of a successfully assembled nervous system. A multitude of human neural diseases have recognizable impacts on early behavior. For example, one of the earliest signs of fetal alcohol syndrome is the behavioral retardation of the fetus (Mulder et al., 1986). By affecting the way that neurons develop, both environmental insults and genetic mutations have an enormous impact on the emergence of the functional circuitry underlying behavior and may restrict an organism's ability to perceive the world and to respond to it with coordinated movements.

THE FIRST MOVEMENTS

The first simple twitch that an animal makes signals the beginning of its functional motor circuitry. In most species, the earliest skeletal movements are caused by spontaneous activity of motor neurons. If these motor neurons have already made a functional synapse onto a muscle, the result is a spontaneous movement. Reflexive movements, those elicited by sensory stimulation, also emerge early in development. All that is needed in addition to the above is a synapse between the sensory neuron and the motor neuron to complete a reflex arc. A question that has intrigued those who study the origins of behavior is whether the very first skeletal movements of an animal are *spontaneous* and involve no sensory input, or whether they are *reflexive* movements in response to sensory stimulation. Sometimes it is difficult to tell when a movement is truly spontaneous. For example, leech embryos show movements in their egg cases, and the frequency of these unprovoked movements increases during the latter half of embryonic leech development (Reynolds et al., 1998). This increase corresponds to the time of eye formation, suggesting that the embryos may actually be responding to light. Indeed, if the embryos are observed in red light, rather than white light, this increase in "spontaneous" behavior is not observed. If the first behaviors are always reflexive, then sensory input must have a dominant role in establishing behavior. If, however, truly spontaneous movements appear first, then the motor system is probably maturing independent of sensory input.

By shining light through chick eggs, candling them as it is called, it is possible to see the embryo moving inside of its shell. If one observes the later stages of development, chick embryos are very active, moving their wings and legs within their shell (Preyer, 1885). Careful studies of chick embryos raised in glass dishes, rather than shells, reveal that totally undisturbed animals exhibit a variety of behaviors (Figure 10.2). Neuroanatomical studies done at the same stages suggest that the earliest of these behaviors occurs prior to the establishment of any sensory input to the spinal cord. And indeed, for several days after their first occurrence, sensory stimulation does not change the frequency of these behaviors, nor is sensory stimulation able to evoke a motor response at all. One could argue, however, that there are subtle stimuli to which the chick can respond, or that the sensory inputs are themselves firing spontaneously. Perhaps, if all sensory stimuli were removed, the embryo would not move. To resolve this issue experimentally, Victor Hamburger, an extremely insightful developmental neurobiologist interested in the ontogeny of behavior (Oppenheim, 2001), surgically deafferented chick embryos by removing the neural crest cells that give rise to sensory DRG cells (Hamburger et al., 1966). In such embryos, which were sensory-deprived *ab initio*, he saw movements that began at the same stage as in unoperated controls. Moreover, these movements

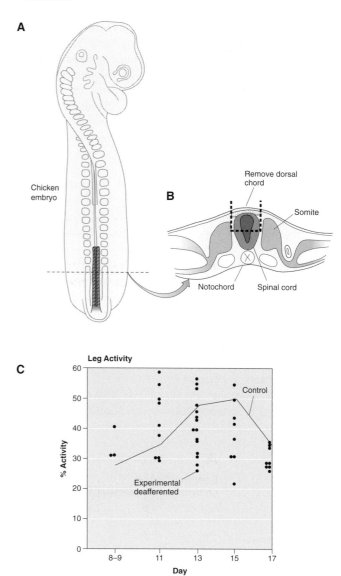

FIGURE 10.2 Spontaneous movements are generated in the absence of sensory input. A. The cross-hatched region of the lumbar spinal cord of this premotile chicken embryo is shown in cross section in (B), which shows the operation done in ovo, which removes the dorsal spinal cord and neural crest containing all the sensory neurons in this region. C. Activity monitor of average normal leg movements (line) and leg movements generated by legs without any sensory input in operated animals (circles). (Adapted from Hamburger et al., 1966)

were indistinguishable in frequency or quality for several days (Figure 10.2). These experimental findings are supported by electrophysiological studies of the spinal cord which reveal no detectable synaptic input onto motor neurons when sensory neurons are stimulated during this prereflexogenic period of behavior. The establishment of motor programming that is independent of sensory input is strongly reinforced by studies in *Drosophila* (Suster and Bate, 2002), which use a genetic technique that exclusively silences the entire sensory nervous system from the earliest stages. Such mutant larvae hatch and crawl in a coordinated manner in the complete absence of sensory input. The larvae often crawl backward instead of forward, but that may be because sensory input is needed to give directional purpose the movement. Thus, the motor circuit is assembled correctly and begins to drive behavior in the absence of such inputs.

THE MECHANISM OF SPONTANEOUS MOVEMENTS

If sensory stimulation is not involved in early motor behavior, then how are these spontaneous movements generated? Provine (1972) was the first to appreciate that the beginnings of electrical activity in the spinal cord of a chick were connected with the development of spontaneous behavior. To study this in detail, the chick embryonic spinal cord can be isolated and the activity patterns of particular nerve cells can be recorded (O'Donovan et al., 1998). The motor nerve roots of such isolated cords produce activity patterns that closely mimic their output in an intact animal that is walking or moving its wings (Figure 10.3). In both intact animals and isolated embryonic chick spinal cords, one observes alternating bursts of action potentials from extensor and flexor motor neurons. Electrophysiological records from the neurons themselves reveal that there are depolarizing synaptic inputs onto both flexor and extensor motor neurons. Both neuronal populations begin to depolarize toward threshold at the same time and begin to fire synchronously. However, at the peak of their depolarization phase, the flexor motor neurons receive synaptic input that causes a large shunt conductance, and as a result, these motor neurons stop firing just when the extensor motor neurons are firing most rapidly. This helps explain the mechanisms that lead to oscillations in flexor and extensor motor patterns.

But how do such bursts of activity first arise? Removing the dorsal half of the cord, where sensory axons travel and many interneurons reside, does not affect the rhythmic motor episodes. The episodes can also occur in bisected cords, implying that right-left communication is not essential. Local interneurons in the venral cord called *R-interneurons*, or *Renshaw cells*, receive cholinergic input from motor axon collaterals

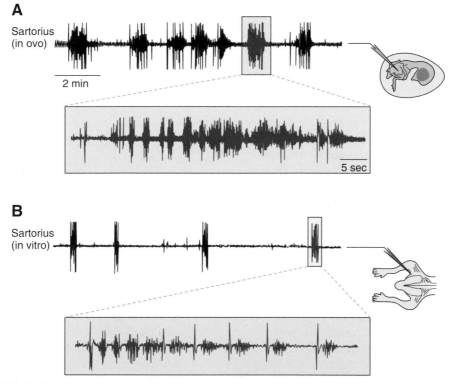

FIGURE 10.3 In vitro motor development physiology. A. Electromyographic recordings from the sortorius muscle of a chick embryo in ovo. B. A piece of the embryo kept in a culture dish. Very similar spontaneous movements begin to occur in these two situtations, although the bursts of activity in the in vitro preparation are shorter and less frequent. (Adapted from O'Donovan et al., 1998).

and feedback GABAergic input onto flexor motor neurons (Wenner and O'Donovan, 2001). In the adult, this is a negative feedback loop that prevents motor neurons from firing too much. But in the embryo, GABA is excitatory (Chapter 8), and any depolarization leads to massive excitatory activity among all the connected neurons, which gradually tapers off. Blocking both the GABAergic and glutaminergic synapses in an isolated cord preparation stops the rhythmic activity (Chub and O'Donovan, 1998). However, such experiments also show that when the activity is blocked, excitability gradually increases. This ensures that in normal spinal cords each quiescent period is followed by an active period. The rhythmic bursting pattern is thus an intrinsic property of the developing network.

EMBRYONIC MOVEMENTS: UNCOORDINATED OR INTEGRATED?

How coordinated is embryonic motor behavior? Do the episodes of trunk and limb movement represent an integrated behavior that needs only to be improved, or are the behaviors essentially spastic and random due to the immaturity of the circuit? The pioneering work of Coghill in the 1920s favored the notion that behavior develops in an integrated fashion (Coghill, 1929). Behaviors build upon each other as the circuitry matures and new components are added. Coghill began to study the first movements of the salamander embryo because, like all amphibian embryos, they grow from shell-less eggs in water and are accessible for observation from the earliest stages of development. Also, there was an extensive history of embryological and neuranatomical studies on these animals. By looking at large numbers of such embryos, Coghill found that slow bending of the head to one side was the very first movement executed by the salamander embryo (Figure 10.4). The movement involves the trunk muscles situated immediately behind the head. As development proceeds, muscles further and further down the body become involved so that the "bend," which started as a slow movement at the neck region, becomes a coiling of the entire body.

Careful examination of this fully developed coiling behavior reveals that movements start at the neck region and then proceed down the body (Figure 10.4), such that the sequence of each movement recapitulates the developmental progression of the movement as well as the progression of neuronal maturation (e.g., first in the hindbrain and then down the spinal cord). Both bending and coiling can be stimulated by a light touch of the skin on the side opposite the contraction.

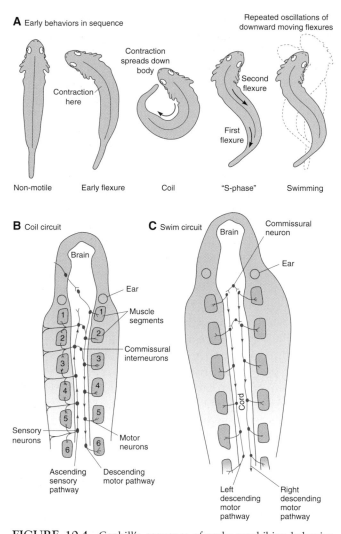

FIGURE 10.4 Coghill's sequence of early amphibian behavior and its proposed neural basis. A. An axolotl embryo at five stages of behavioral development. B. The neural circuit for the coil response. A stimulus anywhere on one side of the body is transmitted to the contralateral spinal motor pathway by commissural cells in the anterior cord or hindbrain where the neural signal descends, stimulating primary motorneurons of the cord. C. The early swim circuit (the sensory mechanism) is omitted but is the same as in B. Motor excitation travels down one side of the cord, but by this stage in development, some reciprocally exciting commissural neurons that cross the floorplate in the hindbrain have developed so that excitation on one side at the neck region can cross over after a delay to excite the contralateral motor pathway leading to coiling on one side being quickly followed by coiling on the other side. (Adapted from Coghill, 1929).

Coghill's anatomical explanation for this chronology is that sensory and motor neurons, which innervate the skin and muscles, are present in prereflexogenic stages. However, at the time that bending away from a light touch emerges, a set of interneurons appear to form the first commissural pathways from sensory neurons on one side to motor neurons on the other (Figure 10.4). Longitudinal ipsilateral tracts extend down the motor columns and up the sensory columns, but the commissural interneurons appear only in the rostral cord and hindbrain. Thus, a neural signal from a touch anywhere on the skin travels ipsilaterally up the sensory path, crosses over in the neck region, and stimulates motor neurons in that region of the other side, causing a bending of the neck away. The signal then proceeds down the motor pathway on that side of the spinal cord involving successively more posterior segments.

Although coiling movements do not propel the animal forward, the rostrocaudal propagation of a contraction along the body is a progression that is seen in swimming behavior, which develops next. So how does swimming arise? After the animal is capable of coiling, the next component of new behavior is the "S" phase. This arises, as the coiling movements become faster and alternate from side to side as new interneurons are added that communicate between the left and right sides. As the sustained swimming system develops in fish, more spinal neurons are added to the circuit such as the CiA interneurons in the spinal cord of zebrafish. These are an important component of the early rhythmic swimming circuit and provide all the ipsilateral glycinergic inhibition. The development of these cells can now be easily followed as they express the transcription factor *Engrailed-1* (Higashijima et al., 2004) (Figure 10.5). If a right-hand coiling movement proceeds only halfway down the body before a left-hand coiling movement starts at the neck region, the result is a "S"-shaped wave proceeding caudally, and the propagation of the animal forward: the beginning of swimming. Similarly in humans, the precursors to mature locomotion can be seen long before infants take their first steps. One can see left-right motor coordination in the crawling movements that babies make even at pre-crawling stages, when they are put on a gentle downhill slope.

Coghill saw many behaviors develop in this integrated way, and one of his other great contributions was his correlational study of nervous system anatomy with the behavior: he was able in some cases to attribute the origins of particular behaviors to newly added neuronal connections. An example of such a correlation occurs for the Mauthner neurons—very large cells in the hindbrain of tadpoles and fish. Mauthner neurons receive auditory input ipsilaterally and project posteriorly onto contralateral motor neurons. In the embryonic zebrafish, as early as 40 hours after fertilization, the Mauthner cells and homologous neurons in other hindbrain segments can initiate a directional escape response away from a stimulus followed by a series of strong tail flexures. This system is probably involved in evoked hatching behavior, as the response is sufficient to rupture the egg membrane and

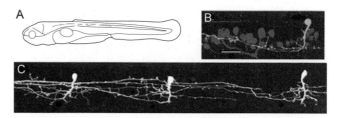

FIGURE 10.5 The morphology of En1 interneurons from live transgenic zebrafish expressing GFP under the control of the En1 promoter. A. The region of the spinal cord imaged below is shown in green. B. A CiA interneuron is shown in green, while in red retrogradely labeled motor neurons are shown. The ventral processes from the CiA interneuron extend among the motoneurons and appear to contact their somata. C. Three segmentally successive CiA internerons are shown, revealing the detail in which single neurons can be visualized in live fish. (From Higashijima et al., 2004)

allow the animal to emerge. The Mauthner cell sometimes fires spontaneously, which suggests that it might function also in spontaneous hatching behavior. The transparency of larval zebrafish has enabled physiologists to use calcium imaging in the intact fish to observe the activity of the Mauthner cell during behavior (Figure 10.6). Such work shows that during an escape, these cells are indeed activated in patterns that are exactly predicted by behavioral studies (O'Malley et al., 1996).

In the chick embryo, Hamburger (1963) describes the early movements as "uncoordinated twisting of the trunk, jerky flexions, extensions and kicking of the legs, gaping and later clapping of the beak, eye and eyelid movements and occasional wing-falling . . . performed in unpredictable combinations." The integration of movements between limbs, such that the left leg alternates with the right during walking, or the right and left wings beating synchronously during flight, do not emerge until later in development. Thus, the random thrashings and reflexes of individual parts of the chick embryo, as well as the mammalian embryo, are brought under control as more circuitry develops. After the chick hatches, circuits across the midline synchronize the right and left wings so that they beat together, and if one wing is weighted down so that it moves more slowly, the contralateral wing will follow the slower pattern (Provine, 1982). This imposed coordination of elemental movements may be carried out according to anatomically organized central pattern generators. Thus, if the brachial cord that drives wing movements in chick embryo is exchanged with the lumbrosacral cord that drives leg movements, the result is a very mixed up chicken in which the wings flap alternately and the legs hop synchronously (Figure 10.7) (Straznicky, 1967; Narayanan and Hamburger, 1971). The descending pathways from the brain clearly help integrate movements and bring

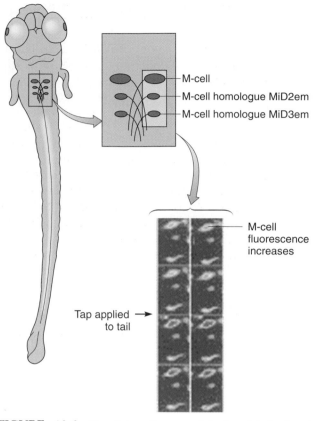

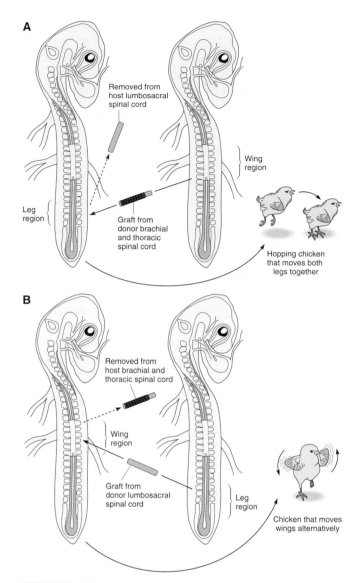

FIGURE 10.6 Visualizing the neural basis of behavior in zebrafish. In a newly post-hatched zebrafish larva, the descending cells including the giant Mauthner neuron (M-cell) and its two homologs in the hindbrain have been filled with a fluorescent calcium indicator. The panel below shows a trial consisting of a sequence of images during which an escape was elicited by an ipsilateral touch to the head or tail. The color scale represents fluorescence intensity (blue, lowest; red, highest). Simultaneous imaging of both the Mauthner cell (left) and one of its segmental homologs (far right). The top row shows that with a head stimulus, both cells respond. The bottom row shows the result from a tail stimulus. The Mauthner cell responds but not its homolog. (Adapted from O'Malley et al., 1996)

FIGURE 10.7 Flapping legs and walking arms. A. The lumbar cord of one chick embryo is replaced by the brachial cord of a donor embryo. When the chick hatches, instead of walking, it jumps in the rhythm of flapping wings. B. The reciprocal experiment leads to a chick that has alternating wing movements instead of normal flapping ones. (Adapted from Narayanan and Hamburger, 1971)

them under control. Voluntary movements are initiated in the cerebral cortex of primates, and so it is not until the corticospinal tract develops fully that macaque monkeys are able to make fine finger movements and exhibit mature manual dexterity (Armand et al., 1994).

THE ROLE OF ACTIVITY IN THE EMERGENCE OF COORDINATED BEHAVIOR

If the exchange of pieces of nervous system can lead to predictable abnormalities in behavior, then one wonders how much behavior is built into the nervous system. Are activity patterns and repetitive practice of simple movements important for the proper development of later movements? The deafferentation experiments mentioned above imply that sensory input is not necessary for the initiation of behavior. More remarkably, similar experiments show that neural activity altogether is not involved in the coordination and maturation of very early motor behaviors. Perhaps the most revealing experiments having to do with the role of activity in the maturation of early motor behaviors are the ones done almost a century ago by Harrison (1904). He raised some salamander embryos in an anesthetic solution throughout the period of bending, coiling, "S" movements, and early and sustained swimming. He then transferred the embryos to anesthetic-free solution. He found that the long-term anesthetized embryos were able to begin to swim as soon as the anesthetic was washed out. In a few minutes, they were behaviorally indistinguishable from the controls.

More recently, these types of experiments have been done with other drugs that block synaptic transmission or action potentials in combination with more quantitative behavioral measurements (Haverkamp and Oppenhein, 1986; Haverkamp 1986). The results are essentially the same (Figure 10.8). For the development of coordinated movements of the limbs in chicks or swimming movements in amphibians, activity in the nervous is not crucial. In other words, the earliest movements that an animal makes are not necessary stepping-stones to the development of at least some simple behaviors. Thus, it seems that many early motor patterns are determined by the molecular signals that direct the formation of neural connections (Chapters 5 and 6).

STAGE-SPECIFIC BEHAVIORS

If early motor behaviors serve no particular purpose in the building of the nervous system, then one might expect to see many such behaviors only in the embryo. Indeed, such embryonic specific behaviors are seen in the leech (Reynolds et al., 1998). For example, one behavior is called *lateral ridge formation* and is the result of the contraction of dorsoventral "flattener" muscle at a time in development when the embryo is still essentially a germinal plate. The contraction of these muscles lifts the boundary of the future dorsal and ventral territories. Another embryo-specific behavior is called *circumferential indentation* and occurs when an embryonic leech is prodded on one side (Figure 10.9). An adult leech, when presented with a similar stimulus, will usually exhibit a local bend away from the stimulus-contracting muscles on one side and relaxing those one the other. However, the embryo excites all the muscles in those segments causing a circumferential contraction. It is likely that circumferential indentation is a behavior that simply occurs at an incomplete stage of neural circuit formation.

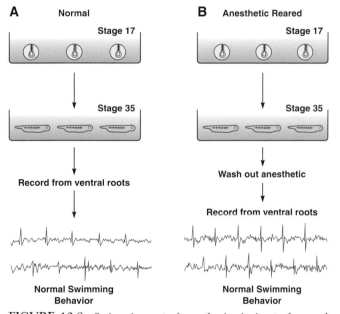

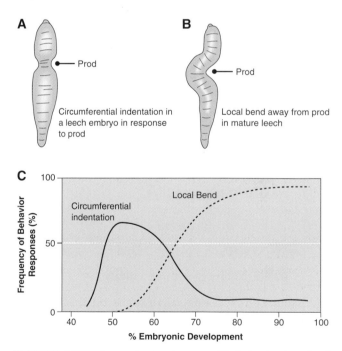

FIGURE 10.8 Swimming out of anesthesia. A. A set of normal *Xenopus* embryos put in a culture dish at the late neural plate stage and raised till the swimming stage in normal pond water. They swim normally, and electrophysiological records from ventral roots on opposite sides of the spinal cord show the expected alternating pattern of activity. B. A set of sister embryos raised in an anesthetic solution until stage 35, at which point some the embryos were immediately put into a recording chamber. The pattern of activity as the swimming behavior looks essentially normal. (Adapted from Haverkamp, 1986)

FIGURE 10.9 A transient embryo-specific behavior in a leech gets replaced. A. At 50% of embryonic development, most prods to one side of the mid-body of a leech embryo result in circumferential indentation of the body due to the contraction of muscles all around the body at the level of the poke. B. At 80% of development, a similar prod leads to local bending away from the prod, resulting from differential contraction of muscles on one side of the body versus the other. C. Graphic illustration of the transitory embryonic behavior with the more mature behavior. (Adapted from Reynolds et al., 1988)

But not all transitory embryonic motor behaviors are useless or simply the result of partially completed neural circuitry. One behavior that serves a clear purpose is the hatching movements of bird embryos. Chicks, like many other shell-bound embryos, go through a very specific motor behavior pattern just at hatching (Bekoff and Kauer, 1984). These behaviors allow the embryo to get into the appropriate position for breaking open the shell. If one places a post-hatch chick into the hatching position within a glass egg, the bird reinitiates its hatching behavior (Figure 10.10). If sensory input from the neck is eliminated with a local anesthetic, then hatching behavior is suppressed (Bekoff and Sabichi, 1987). Therefore, it appears that sensory receptors located in the neck provide a specific input signal for initiating hatching behavior. Another example of a transient embryonic behavior is the migration that marsupial embryos make from their womb to the mother's pouch. Born at an extremely early phase of development with their hind limbs little more than buds, these tiny embryos use their forelimbs to crawl tens of body lengths into the pouch where they attach onto a nipple and suckle (another transient but adaptive behavior of mammals) for several months. During this time they complete their embryonic development. Human infants that do not suckle can be fed through a tube, and the absence of these early suckling experiences does not impair the development of adult eating. This demonstrates, once again, that the juvenile behavior is not a prerequisite for the adult motor program, even though it is adaptive to the neonatal environment.

One of the most important transitory behaviors of a metamorphic insect such as a fly is eating. Fly larvae are eating machines, whereas adult flies are procreating machines. There are two styles of eating found in natural population of *Drosophila* larvae, and this difference can be genetically mapped to differences in the activity of a single gene, which encodes cGMP-regulated protein kinase (PKG) (Osborne et al., 1997). Larvae with the more active *rover* allele of PKG have significantly longer foraging path lengths than do those homozygous for the less active *sitter* allele. The two *Drosophila* foraging variants do not differ in their general activity in the absence of food, but when food is available the two variants may fare differentially based on the density of other animals feeding in the vicinity. If it is crowded, larvae that forage further may do better. Natural selection experiments done in the laboratory showed that under high-density rearing conditions for several generations, the *rovers* did better, whereas the short path (*sitter*) phenotype was selected for under low-density conditions (Sokolowski et al., 1997). Interestingly, the age-related transition by honeybees from hive work to foraging is also associated with an increase in the expression of the PKG gene and experimentally controlled cGMP treatment or elevated PKG activity cause premature foraging behavior in bees (Ben-Shahar et al., 2002).

Metamorphosis signals a dramatic change in lifestyle in insects, such as moths and flies, as well as amphibia, such as frogs (Harris, 1990, Truman, 1992). The larval behaviors of the swimming, filter-feeding tadpole are completely inappropriate for the land-dwelling, bug-eating frog. The transition from larval to adult state is activated by specific hormones: ecdysone for insects and thyroxine for frogs. Each of these hormones has a widespread effect on gene expression and

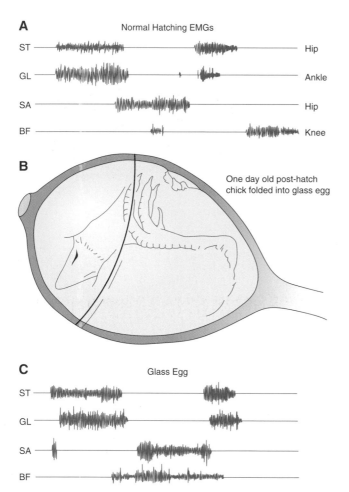

FIGURE 10.10 Chicken trying to hatch again in a glass egg. A. Electromyographic (EMG) recordings of the normal hatching motor program in the chick embryo. (ST and SA are hip muscles, GL is an ankle muscle, and BF is a knee muscle.) Note the alternation between the activity of hip and lower leg. B, A one-day-old chick is crammed back inside a glass egg. C. EMG records from such a chick show that it reinitiates the hatching motor program. (Adapted from Bekoff and Kauer, 1984)

cellular function. Many very adaptive larval behaviors are lost at this transition while new behaviors are gained. As would be expected, there are also dramatic changes in nervous system structure. In both insects and amphibia, larval neurons die upon exposure to metamorphic hormone, and some neuroblasts that have been quiescent throughout larval life begin to proliferate. Some larval neurons survive the transition to adult but are drastically reorganized. For example, the motor neurons that move the abdominal prolegs of the caterpillar do not die, even though these appendages are lost, but their axons and dendrites are remodeled, and old synapses are eliminated to support new behaviors (Figure 10.11).

In metamorphic insects, there are important transitional behaviors associated with building and emerging from the pupal state. In moths, the adult motor system is constructed primarily from remodeled larval components, whereas the adult sensory system is primarily composed of new neurons. Simple reflexes correlate these neuronal changes with the acquisition or loss of particular behaviors. The loss of the larval proleg retraction reflex is associated with the loss of the dendrites of the proleg motor neurons; the adult stretch receptor reflex begins when new adult-specific connections are added to new dendritic growth in an adult neuron (Levine and Weeks, 1990). Although humans do not go through metamorphosis in the same way as flies and frogs, the distinct behaviors of babies and adults must be largely due to changes in the nervous system that result not only from experience, but also from a variety of intrinsic influences, such as hormones and growth factors that are regulated throughout life.

BEGINNING TO MAKE SENSE OF THE WORLD

The nervous system becomes active well before animals experience the world around them or move about within it. In fact, this "spontaneous" activity may initially be necessary for the survival and maturation of synapses (Chapter 9). At some point, the nervous system begins to sense the world, and some of the information is of immediate use. Although neonatal mammals are helpless in many ways, their survival usually depends on perceptual abilities that lead them toward their mother's nipples and motor skills that allow them to ingest the milk. Many animals begin to hear, smell, see, taste, and feel well before the sensory epithelia (e.g., hair cells, photoreceptors, etc.) and central nervous system connections are mature.

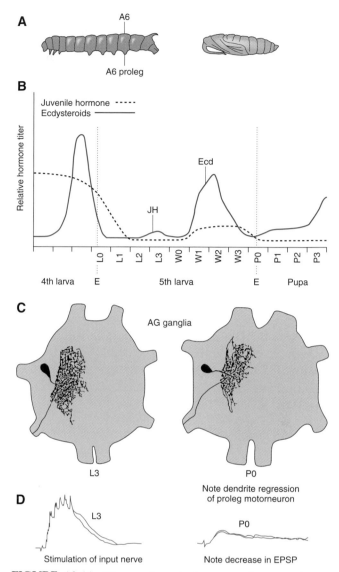

FIGURE 10.11 Changes in a single neuron during metamorphosis. A. The caterpillar of the moth (*Maduca sexta*) showing abdominal segment 6 and the proleg associated with this segment. To the right is shown the pupal stage. B. Profile of hormonal changes through larval life and the transition between the larval and pupal development where ecdysone rises without an increase in juvenile hormone. C. Remodeling of the A6 proleg motoneuron during metamorphoses. There is a dramatic pruning of the dendritic tree. D. Correlated with this dendritic remodeling, there is a decrease in the activation of this neuron when the sensory input nerve was stimulated. These changes reflect the changing role of this neuron in postmetamorphic life where there are no prolegs. (Adapted from Streichert and Weeks, 1995)

Therefore, it is important to understand the relationship between neural tissue development and perception. How is visual detection limited by neonatal retinal ganglion cell physiology and morphology? Why do infants rarely enjoy espresso or extra hot salsa? Why can it take several seconds for a baby to determine the location of a human voice?

Studies of sensory perception are among the most difficult experiments in the field of development. Baby animals (especially human infants) tend to be slow, sleepy, cranky, inattentive, and forgetful. These are generally referred to as nonsensory factors. For example, adults pay better attention to novel stimuli than young animals. When adult primates are presented with two images, one of which they have never seen before, they spend about 70% of the time staring at this novel object. In contrast, infants spend an equal amount of time staring at the familiar and novel objects (Bachevalier, 1990). Even though adults are more attentive to novel stimuli, they can also focus narrowly on a stimulus of interest and ignore novel stimuli that may be distracting. For example, when taking an examination, we tend to "block out" extraneous noise. This was demonstrated by asking people to detect a tone when it was presented on 75% of trials. Several other tones were presented on the other 25% of trials. Adults come to "expect" the tone that is presented 75% of the time, and they detect it quite well, whereas they are very poor at detecting the other tones (i.e., those presented on 25% of trials). Infant perception differs rather dramatically: they detect all of the tones equally well (Bargones and Werner, 1994). Thus, infants and adults experience the world in very different ways, and these non-sensory factors lurk in the background of all developmental studies of sensory perception.

ASKING BABIES QUESTIONS

How, then, can these uncooperative little animals tell us about their sensory experiences? Various experimental tricks have been devised to determine how sensory information is processed in young animals. In one scenario, the behavioral scientist watches for a motor reflex while a sensory stimulus is presented. For example, we often respond to an unexpected noise with a startle, and this rapid muscle twitch provides a reliable measure that sound has been detected. We can also take advantage of the fact that animals tend to stop responding, or habituate, to a stimulus when it is presented many times. After the animal has habituated, one can present a new stimulus and ask whether the animal responds. This is a good way of determining how well an animal notices the difference between two similar stimuli (e.g., middle C versus C sharp).

A young animal can also be trained to produce a stereotyped behavior, such as a head turn, when a stimulus is detected. Most animals will work for

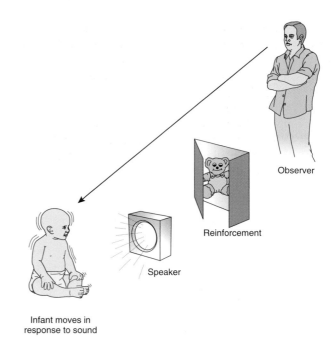

FIGURE 10.12 Determining an infant's sensitivity to sound. Infants will make small movements in response to a sound they can hear, and this natural tendency has been exploited to measure the infant's ability to hear different sound frequencies. To improve the sensitivity of this procedure, an adult observer watches the infant and judges whether the infant heard the stimulus based on any response that the infant makes. To increase the infant's responsiveness to sound, she can be reinforced for correct responses. In this case, the infant is rewarded with the appearance of a teddy bear.

a reward, even human infants. Thus, the head turn that an infant makes to a sound can be reinforced by showing her an interesting toy (Figure 10.12). The infant will then "work" for visual stimulation (that is, turn her head) when she hears a sound. This is somewhat below the minimum wage, but it serves the purpose. This procedure can be extended to very young infants (<6 months) by having an adult observer, who cannot hear the test sounds, watch the baby to determine when she makes a response to sound. The baby gets rewarded (with a viewing of the toy bear) whenever the observer determines that the baby has responded. In this manner, any possible response that a baby might make to sound (for example, an eye movement or a tongue wag) can be conditioned. Of course, we believe that the baby is oblivious to this process, but she nonetheless ends up working for a reward and providing valuable information about sensory development along the way. Such training techniques require far more of an animal than sensory skills, and one may well end up studying the development of attention or memory rather than the development of sensory perception.

Behavioral scientists tend to concentrate on two criteria: absolute sensitivity and discrimination. Absolute sensitivity is a measure of the minimum stimulus amplitude that can be detected: the softest touch, the finest line, the quietest sound. Discrimination is a measure of our ability to perceive a difference between two similar stimuli: sky-blue versus turquoise, middle C versus C sharp, margarine versus butter. Below, we explore how developing animals first comprehend their sensory world.

SHARP EYESIGHT

Human infants show clear evidence of being able to see at birth. For example, they stare for longer periods of time at a familiar face, such as their mother. However, their visual skills appear to be very poor as compared to those of an adult. Visual acuity, or the ability to detect fine detail, is almost entirely absent at birth (Figure 10.13). One measure of visual acuity is the number of black and white lines that can be observed per degree of visual space. (The "rule of thumb" states that, at arms distance, your thumb occupies about one degree of visual space.) Adults can see about 30 black and white lines per degree, but babies can only see about one. In the more common language of an eye doctor, the baby sees at 20 feet what a normal adult can see at 600 feet, and adult sensitivity is reached between 3 and 5 years of age (Birch et al., 1983). In principle, this level of acuity would permit an infant to distinguish the fingers of a hand, but their actual abilities remain somewhat of a mystery. Our best ideas come from "preferential looking" studies which tell us what babies prefer to look at, given a choice (e.g., faces, curved lines, complex patterns), but not what they see.

The modest visual acuity of primates at birth is partly due to an immature retina. Photoreceptors are relatively short and wide at birth, meaning that less light is absorbed and a greater piece of visual real estate is viewed (Yuodelis and Hendrickson, 1986). Thalamic and cortical neurons may also impose limits on visual acuity. If one compares the theoretical acuity of the retina (based on the density of cone photoreceptors) with the acuity of single cortex neurons, then the cortex neuron is found to be worse than expected in developing primates (Jacobs and Blakemore, 1988; Kiorpes and Movshon, 2004). Furthermore, the animal's behavioral acuity is worse than that of individual cortical neuron at first (Figure 10.13). Such results suggest that the development of accurate connections (Chapters 5 and 6) may only create a minimal operating system, and optimal performance is

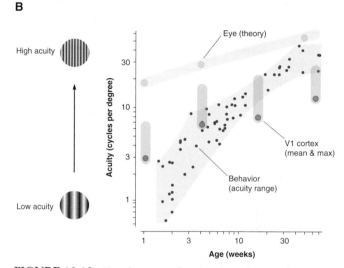

FIGURE 10.13 Development of acuity along the visual pathway in primates. A. These two images show how a visual scene appear to an adult (left) and a neonatal primate (right). The "infant view" is spatially lowpass filtered (blurred) and reduced in contrast to reflect the theoretical limitations of the retina. B. Acuity is measured by determining the number of visible bars per degree (left). The theoretical acuity of the eye is linked to the spacing of cones within the fovea (blue circles and shading). The behavioral acuity of individual animals is quite poor at birth and continues to mature until about 50 weeks postnatal (red circles and shading). This data suggests that acuity development is not limited by properties of the retina. Single neurons in the primary visual cortex respond to a broad range of spatial frequencies. (Green circle is the mean value recorded, and green bar extends to maximum value recorded at that age.) Behavioral acuity appears to track cortical neuron development until about 8 weeks, but there is a disparity in performance after this age. Therefore, it is possible that maturation of higher visual cortical areas is necessary for adult-like performance to emerge. (From Kiorpes and Movshon, 2004)

acquired through detailed changes in synaptic architecture or function (Chapter 9).

Binocular vision involves the coordinated use of both eyes to judge the distance of an object, and this requires both sensory and motor development. To look at an object close up, the eyes must be rotated toward one another (convergence), so that the visual image activates the correct portion of each retina. Similarly,

FIGURE 10.14 Investigating the process of depth perception in human infants. The testing device, called a *visual cliff*, consists of a sheet of plexiglas that covers a high-contrast checkerboard pattern. On one side of the device, the cloth is placed immediately beneath the plexiglas, and on the other side it is placed 4 feet below. The majority of infants would not crawl onto the seemingly unsupported surface, even when their mothers beckoned them from the other side. These results suggest that infants perceive depth by 6 months of age. (Gibson and Walk, 1960)

the eyes must be rotated away from one another (divergence) to look at a distant object. Cruel as it sounds, one of the simplest ways to determine whether depth perception is present in young animals is to ask whether they are willing to crawl off a "cliff." To do this safely, an infant is placed on an elevated glass surface that is patterned on one half and clear on the other. If an infant is willing to crawl out over the clear surface, off the "perceptual cliff," then one assumes poor depth perception (Figure 10.14). By the time that they crawl, most infants do avoid the "cliff," indicating that depth perception is present (Walk and Gibson, 1961). To determine whether infants can perceive depth before they crawl, 1- to 4-month-old subjects were equipped with a heart rate monitor and suspended either above the shallow side or the deep side. Interestingly, the heart rate was lower when the infants were suspended above the deep side, suggesting that they were interested but not fearful. An accelerated heart rate was measured after the infants began to crawl (Campos et al., 1970). More precise measurements of depth perception obtained in nonhuman species show a rather sudden improvement. For example, binocular perception in cats goes from being rather poor to almost adult-like between 4 and 6 weeks postnatal (Timney, 1981).

Why does binocular vision improve with age? Do neurons in the cortex suddenly become selective for binocular stimulation? In fact, several neural mechanisms may contribute to the maturation of binocular vision (Daw, 1995). First, since the visual system detects smaller objects with age (discussed above), it is likely that it can also resolve smaller differences between the two eyes. Detecting differences between the two eyes is fundamental to depth perception. We also know that ocular dominance columns in the kitten visual cortex continue to mature after eye opening (Chapter 9), during the period when binocular vision is improving. During the same period, individual neurons respond more selectively to visual stimuli. For example, individual cortex neurons respond to a smaller range of bar orientations during development in cats and ferrets (Bonds, 1979; Chapman and Stryker, 1993). Thus, some interesting candidate mechanisms have been identified, but we have yet to design the experiments that test their relationship to perception.

Color vision is another fascinating perceptual in that helps us categorize objects. Interestingly, human infants seem to use color in a different way than do adults. When 2- to 4-month-old infants are stimulated with a special stimulus that requires subjects to use color information to detect motion, they are found to perform better than adults when both groups were compared to a reference stimulus in which only luminance information was needed to perceive motion (Dobkins and Anderson, 2002). These results suggest that the visual pathways that carry information about an object's color have a relatively strong input to motion processing areas of the visual cortex at first, and this is reduced during postnatal maturation.

Even though acuity improves dramatically in human infants during the first two years, they can fail to make proper use of visual information. When 18- to 30-month-old children were provided with a large object, followed by exposure to a miniature replica, they often failed to understand the concept of size. For example, children were first permitted to use a child-sized chair on which they could sit comfortably. They were then escorted from the room, and when they returned a miniature chair replica had replaced the original object. In many instances, children would attempt to sit on the miniature chair; such "scale errors" reached a maximum at about 2 years of age (DeLoache et al., 2004). The children could easily discriminate between objects of different size and would choose to sit in the large chair when given a choice. The results suggest that visual information about object identity is not being integrated with information about its size.

ACUTE HEARING

As is true of the visual system, the auditory system displays improved sensitivity and discrimination as an animal matures. Although kittens can hear at birth, they only respond to extremely loud sounds, well above the level of city traffic (>100 dB sound pressure level), at 10 days postnatal. Their auditory thresholds gradually decrease so that they can detect sounds at the level of a whisper (≈30 db SPL) by one month, and by adulthood they become even more sensitive than humans. A similar change is found in all developing animals. In humans, auditory thresholds drop rapidly during the first 6 months of life and are virtually adult-like by 2 years of age. Improved behavioral thresholds are well correlated to a decrease in sound level needed to evoke an electrical potential from the cochlea, suggesting that the major factor limiting detection in young animals is the ear (Werner and Gray, 1998).

As thresholds decrease, most animals also respond to higher sound frequencies. This is probably due to a physical change in the cochlea because a single physical position along the basilar membrane responds to higher frequencies as the animal matures. Since the topographic projection from the cochlea to the central nervous system does not change significantly during this time, one might expect to find interesting changes in sound perception with development. In fact, 15-day-old rat pups can be trained to suppress activity when they hear a 8 kHz tone, but three days later they suppress activity to a higher frequency. That is, a higher sound frequency apparently sounds like the 8 kHz tone because the cochlear frequency map has shifted (Hyson and Rudy, 1987).

Although the basic sensitivity and frequency range mature rapidly, several features of sound remain difficult to detect. This is well illustrated for tasks in which one must detect a very brief event, often referred to as temporal processing (Figure 10.15). For example, adults are easily able to detect a 5 ms gap of silence in an ongoing sound. In contrast, one-year-old infants, who have already begun to process and produce speech sounds, can only detect gaps that are an order of magnitude longer (≈60 ms), and adult-like performance is not reached until about 5 years of age (Werner et al., 1992).

In humans, the maturation of adult-like performance on auditory perceptual tasks extends beyond the first decade of life (Stollman et al., 2004). The ability to detect small differences in the duration of a tone is more than an order of magnitude poorer in 4-year-olds, as compared to adults (Jensen and Neff, 1993). The ability to detect one sound in the presence of another may not mature until puberty. In one task, the listener is asked to recognize a long duration tone that is presented during an ongoing burst of noise. Even 10-year-old listeners cannot perform nearly as well as adults. Furthermore, children identified as learning

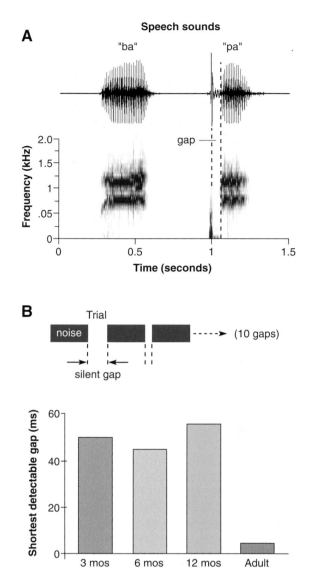

FIGURE 10.15 Development of temporal processing may affect speech perception. A. An oscilloscope record of the human speech phonemes, /ba/ and /pa/. Below each record is a spectrogram of the phoneme showing the sound frequencies that compose the phoneme and their relative intensity (darker is louder). Note that the /ba/ is a continuous sound, whereas /pa/ consists of a nonvoiced component (in this case, the p sound), followed by a brief gap, and then a voiced sound (the "a"). Perception of this brief gap is critical to word recognition. B. The minimum gap that humans can perceive was assessed with a brief silent period embedded in a white noise stimulus. The bar graph shows that even at 12 months of age, human infants are almost 10 times less sensitive at detecting a gap than adults. (Adapted from Werner et al., 1992)

disabled never reach the normal adult level of performance on this task, and it has been suggested that the onset of puberty may terminate the critical period during which neural maturation takes place (Wright and Zecker, 2004).

These behavioral measures of auditory processing are relevant to language development because human speech sounds are composed of rapid changes in frequency and intensity, including discrete periods of silence. In fact, children with learning disabilities that are due primarily to a difficulty with spoken language also perform poorly on simple auditory discrimination tasks that require temporal processing. For example, when normal children are exposed sequentially to two tones, they can report the correct sequence with delays as small as 8 ms. In contrast, the language-impaired group required a silent interval of 300 ms in order to report the correct sequence. Recently, it has been found that performance can be improved when language-impaired children are trained to recognize speech sounds that are slowed down. Apparently, once the nervous system has learned to recognize this slower speech, it is better able to recognize the rapid temporal variations in normal speech (Tallal and Piercy, 1973; Tallal et al., 1996).

Several mechanisms may explain poor temporal processing in young animals. For example, we have seen that synaptic potentials are usually of much longer duration in young animals (see Chapter 8). We might suppose that long PSPs effectively limit the "clock speed" of the organism, or the fastest rate at which information can be processed. Thus, it will be interesting to learn more about the neural basis of temporal processing, particularly in the auditory system.

Perhaps the most useful information that a developing animal gets from its ears is the location of significant objects, such as mother or a predator. Although infants can tell whether a sound source is coming from the left or the right (sound lateralization), they are not able to make fine discriminations. Adult humans can detect a 1° change in the position of a speaker (recall the "rule of thumb"), but newborns can only detect a change of about 25°. In fact, even sound lateralization is fairly challenging to a newborn infant (Figure 10.16). The sound stimulus must remain on for about one second if the infant is to make an appropriate head orientation response, whereas adults need only about a millisecond of sound, such as a finger snap (Clarkson et al., 1989). The ability of nonhuman mammals to lateralize sounds is also present even as the animal first experiences sound, yet we know little about the sensitivity of the system. For example, rat pups suddenly begin to turn their heads toward a noise at 14 days postnatal, a few days after the ear canal opens. However, the percentage of correct turns toward the sound continues to increase over the next seven days (Kelly et al., 1987).

The response of central neurons to sound is known to change during this period of development, and

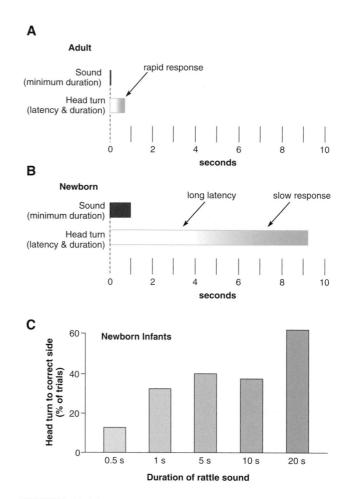

FIGURE 10.16 Infants are poor at sound lateralization. A. When presented with a sound located to one side, adult humans turn their head toward the sound source within a fraction of a second. B. When human infants are presented with a sound to one side, they may take several seconds to respond, and the head movement can be quite slow. C. For infants, a sound stimulus must be presented for a long time period in order to get accurate lateralization. Whereas adults can localize sounds that last only a millisecond, newborn infants require at least 1 s of sound. (Adapted from Clarkson and Clifton, 1991)

some of these alterations could help to explain immature sound localization. Maps of space are found in the superior colliculus (SC) of several mammals, and single SC neurons are selectively activated by sound (or visual and somatosensory stimuli) from a specific location. During the course of development, these SC neurons respond to a smaller part of the sensory world. In cats, the average size of a receptive field decreases about fourfold during the first two months after birth (Figure 10.17). Furthermore, the visual receptive fields become adult-like a few weeks earlier than the auditory receptive fields in kittens (Wallace and Stein, 1997). In the guinea pig, an orderly map of

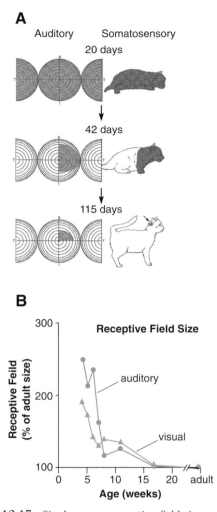

FIGURE 10.17 Single neuron receptive field sizes can decrease dramatically during development. A. Recordings were made from single neurons in the superior colliculus that respond to more than one sensory modality. The auditory (green) and somatosensory (red) receptive fields are shown for neurons from cats of increasing age. At 20 days, the neuron responded to auditory stimuli located anywhere in space, and to touch on any area of skin. At 115 days, the neuron responded to a small area of auditory space (green) and to touch on a small area of skin (red) under the right ear. B. The relative size of auditory (green) and visual (blue) receptive fields are plotted for neurons recorded throughout development. There is a dramatic decrease during the first eight weeks, and mature properties are attained by the seventeenth week. (Adapted from Wallace and Stein, 1997)

cortex neurons recorded in young animals are not very directional. In contrast, adult neurons respond to sounds from only a narrow range of auditory space, particularly if the sound is very quiet. To test whether this difference is due to maturation of the external ear, infant animals were provided with the sound cues that are only available with an adult pair of ears (i.e., the external ear filters sound arriving at your ear canal, and this depends on its size and shape). The adult ears produced a dramatic improvement in the spatial receptive fields of the infant cortical neurons, suggesting that the connections responsible for spatial tuning are quite mature at the outset of hearing. The result also raises an interesting prospect: Developmental plasticity may permit the central auditory system to remain properly sensitive to the changing sound cues as the ears grow.

Auditory processing may also be limited by the low discharge rates that are generally reported for young animals. Furthermore, the sound-evoked response fatigues rapidly during a period of stimulation in young neurons. What does this mean for the performance of a developing nervous system? First, neurons have a poorer resolution: they devote few action potentials to a given change in the stimulus. For example, adult LSO neurons can devote twice as many action potentials to a given change of interaural level compared to juvenile animals (Sanes and Rubel, 1988). Therefore, either young animals make decisions based on less neural information (e.g., fewer action potentials), or, more likely, they are not able to perform at adult levels because they have less neural information to work with. Striking as this result is, we still have no direct knowledge about the relationship between amount of neural activity and perception.

SEX-SPECIFIC BEHAVIOR

The many differences between male and female behavior are a popular subject of conversation. They are also a source of considerable controversy, and the political stakes can be quite high. We primates tend to debate whether sex-specific behaviors are due to our "biology" or to the social environment in which we are raised. While the debate is seductive, the relationship between brain development and sexual behavior varies tremendously from species to species. Since mating and maternity have been most thoroughly explored at the neural level, we will mostly focus on these behaviors. However, it is worth mentioning some complex behaviors that differ between male and female animals. These differences

auditory space is not apparent until postnatal day 32, even though hearing begins in utero (Withington-Wray et al., 1990).

It is essential to recognize how the development of peripheral structures, such as the external ears, influence the response properties of central neurons. An interesting demonstration of this comes from a study of spatial receptive fields in the developing ferret auditory cortex (Mrsic-Flogel et al., 2003). Single auditory in behavior are commonly referred to as sexual dimorphisms. Predatory behavior is sexually dimorphic in lions (females do more of it), urination posture is sexually dimorphic in dogs, and olfactory signaling is sexually dimorphic in moths. In both rats and monkeys, young animals engage in play behavior that differs between the sexes, at least in its frequency of occurrence. Males tend to have more play fights than females. These "fights" will typically begin with one animal jumping onto the other, and they end with one animal on top of the other. When testosterone is given to a pregnant monkey, the play behavior of her female offspring becomes more male-like (Abbott and Hearn, 1978).

Another behavioral sign of a sexually distinct nervous system comes from the prevalence of certain neurological and psychiatric diseases in males versus females. For example, both dyslexia and schizophrenia are more prominent in males (about 75% of cases), while anorexia nervosa is exhibited primarily by females (over 90% of cases). Many studies have also focused on the cognitive abilities of normal adult humans (Kimura, 1996). When presented with two figures drawn at different orientations, males are better able to "mentally rotate" the objects to determine whether the two figures are the same. In contrast, when presented with a picture containing many objects, females are better able to say which objects have been moved in a second picture. While these results tend to fascinate us, the challenge will be to understand what exactly is being measured and what its relevance is to behavior.

The male-female differences in complex behavior patterns do raise a host of interesting questions: Are these differences due to biology or environment? If there is a biological signal, then is it genetic or hormonal? Are the differences irretrievably established at birth, or are they modifiable throughout life? Certain sexual characteristics emerge during embryonic development, such as differentiation of the genitals and the motor neurons that innervate them (see Chapter 7). However, this is only the first step of a lifelong process. The nervous system continues to respond to steroid hormones throughout life, making it important to ask whether a behavior is determined by early exposure to a hormone or whether the behavior can be elicited in adults of either sex merely by adjusting the amount of circulating hormone. For example, one region of the amygdala has a greater volume in male rats than in females, but adult castration of males causes the volume to shrink to female values, and androgen treatment of adult females enlarges the structure to normal male size (Cooke et al, 1999). Therefore, we will begin by examining the early determinants of gender and then explore determinants of behavior and brain development.

GENETIC SEX

Animals seem to have two general ways of establishing gender. In fruit flies and other insects, the genetic sex of each cell is the key determinant. If a cell has a single X, then it is male. If it has two Xs, then the cell expresses a protein called sex-lethal and becomes female. The nematode, *C. elegans*, also comes in two categories, but they are male and hermaphrodite (i.e., an animal with both types of gonad). As with fruit flies, sex is determined by the ratio of X chromosomes to autosomes, and XO-lethal is the gene product that is activated in animals with a single X.

The genetic sex of each cell in a mammal is specified by the presence of either two X chromosomes (female) or one X and one Y (male). However, the genetic sex of most somatic cells is not thought to have an immediate influence on their development. It is the genetic sex of gonadal tissue that really matters. Primary sex determination refers to differentiation of the gonadal tissue, and this is determined by the *SRY* gene on the Y chromosome, which encodes a transcription factor (Goodfellow and Lovell-Badge, 1993). If *SRY* is present, the gonads develop into testes and secrete testosterone; if *SRY* is absent, the gonads develop into ovaries. The SRY gene product is a DNA-binding protein, and it probably controls the expression of downstream targets to prevent development along the female pathway. For example, a locus on the X chromosome, called Dax-1, is probably involved in ovary determination. Thus, it is thought that SRY represses Dax-1 in genetic males. Furthermore, SRY specifies the male gonads by activating the insulin receptor family of tyrosine kinases (Nef et al., 2003). After primary sex determination is complete, all sex differences, including those of the nervous system, are thought to originate from the gonads. The possibility remains that genetic sex of an individual somatic cell plays a role in maturation, as will be discussed below.

HORMONAL SIGNALS

In most vertebrates, as the gonads differentiate and begin to secrete hormones, tissues throughout the body respond by adopting a male or female phenotype. This is called *secondary sex determination*. The principal importance of gonadal hormones is powerfully demonstrated by removing the gonads before primary determination occurs (Jost, 1953). Without exception, animals develop as females (e.g., they have a vagina, a uterus, and oviducts). Furthermore, their sexual behavior is female-like, presumably because certain areas of the nervous system have developed female characteristics (Phoenix et al., 1959). When genetically female (XX) rats are treated with testosterone within a few days of birth, they will not display female sexual behaviors as adults. That is, they will not arch their back (lordosis) when approached by a male, and they will mount a female rat if given another shot of testosterone. When genetic males (XY) are castrated soon after birth, they will not mount a female as an adult, even if given a shot of testosterone.

The testes masculinize the body by releasing the steroid hormone, testosterone. The level first rises during the perinatal period, goes down after birth, and rises again at puberty. The testosterone must be converted to another compound in order to carry out some of its actions. For example, some members of a small community in the Dominican Republic carry a disrupted form of the 5α-reductase gene and cannot convert testosterone to 5α-dihydrotestosterone (DHT). Although affected genetic males (XY) have functional testes and plenty of circulating testosterone, their external genitals are female (Imperato-McGinley et al., 1979; Thigpen et al., 1992). Interestingly, most of the individuals who were unambiguously raised as girls nonetheless chose to adopt a male identity during or after puberty. The results suggest that testosterone has a potent influence in determining gender identity, even overcoming the prolonged "environmental" influence of being raised as a female. Since DHT is probably not involved in gender identity (although it is involved in differentiation of external genitalia), how does testosterone masculinize the brain?

In the brains of mammals, testosterone is also converted to the estrogen hormone, estradiol-17β, by an enzyme called aromatase. At first, this might seem puzzling because estradiol is secreted by the ovaries and promotes differentiation of the female reproductive organs. However, testosterone is also an intermediate metabolite of estradiol in the ovaries. Thus, we should probably not think of hormones as being "male" or "female." There are probably two factors that allow estradiol to act selectively on the brains of genetic males. First, aromatase activity is higher in the brains of male mice, particularly during the prenatal and neonatal periods (Hutchison, 1997). Second, the blood of young animals contains an estradiol-binding protein, called α-fetoprotein, that may prevent estrogen secreted by the ovaries from reaching the brain (Uriel et al., 1976). A direct masculinizing role for testosterone is revealed by examining androgen receptor (AR) null mice (Sato et al., 2004). Genetic males with the AR mutation do not display male-typical sexual and aggressive behaviors. Treatment with DHT does not restore normal sexual behavior but does partially rescue male aggressive behavior.

Since there are many different steroid hormones, and their actions are quite diverse, there must be specific transduction pathways. How do sex hormones influence neuron differentiation and function? Steroid receptors are cytoplasmic proteins with a steroid-binding domain and a DNA binding domain. That is, they provide a very direct pathway to the genome (Beato et al., 1995). When estradiol binds to its multi-subunit receptor, it dissociates, and the active DNA-binding complex enters the nucleus. Estradiol receptors are found in neurons of the hypothalamus and amygdala, and they are expressed transiently in the cortex and hypothalamus. Androgen receptors are also expressed at highest concentration in the hypothalamus and limbic structures.

HORMONAL CONTROL OF BRAIN GENDER

One might expect the hypothalamus to be a target of gonadal hormones during development. Lesion and stimulation studies show that some hypothalamic regions are involved directly in the production of sex-specific behaviors. For example, medial preoptic neurons fire rapidly just prior to male copulation, and copulatory behavior is disrupted when this area is lesioned. Medial preoptic neurons are also known to take up more testosterone than any other brain region in adult animals. One of the first studies to show that male and female brains actually differ in a measurable way was an ultrastructural study in the preoptic area (Raisman and Field, 1973). A few years later, it was found that one part of the preoptic area, aptly named the sexual dimorphic nucleus of the preoptic area (SDN-POA), is so much larger in male rats than females that one can actually see the difference in tissue sections without using a microscope (Figure 10.18A). A similar difference is found in the primate hypothalamus, including that of humans. Selective cell death may account for the sexual dimorphism in a human hypothalamic nucleus, called INAH 1. Until age 5, the number of INAH 1 neurons is about the same in males and females, but the number of neurons then declines more rapidly in females (Swaab and Hofman, 1988).

The sexual dimorphism of SDN-POA is an example of secondary sex determination in the nervous system.

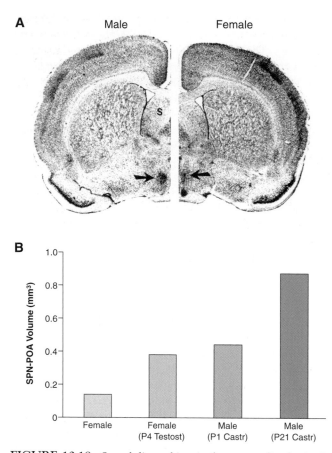

A

Male   Female

B

FIGURE 10.18   Sexual dimorphism in the mammalian brain. A. A hypothalamic structure called the sexually dimorphic nucleus of the preoptic area (SPN-POA, arrow) is almost six times larger in male than in female rats. B. In genetic females, the size of SPN-POA can be increased by treating with testosterone at postnatal day 4. In genetic males, the SPN-POA can be decreased in size by castrating at postnatal day 1. (Adapted from Gorski et al., 1978, 1980)

The hormonal environment of developing males yields a larger nucleus, and the dimorphism can be greatly reduced by castrating genetic males within a few days of birth (Figure 10.18B). Furthermore, this nucleus can be enlarged in genetic females when they are treated with testosterone as neonates (Gorski et al., 1978). Intracranial implants of estradiol turn out to be as effective as testosterone in masculinizing the SDN-POA, and such estradiol-treated females fail to lordose or ovulate. Presumably, testosterone is converted to estradiol in the male SDN-POA, whereas the circulating estradiol in females is bound by α-fetoprotein (Naftolin et al., 1975).

A second region of the hypothalamus, the ventromedial region (VMH), also participates in sexual behavior. Damage to this region disrupts female copulatory behaviors, such as lordosis in rats, and stimulation of the region seems to facilitate such behaviors.

Lesions also have more profound effects on food intake, particularly in females. Neurons of the VMH are selectively activated by the ovarian hormone, estrogen, and in female rats the cells respond by producing progesterone receptors. This does not occur in the male VMH. In primates, the hormonal signal may be somewhat different because loss of the adrenal glands, a source of androgen hormones, leads to reduction in copulatory behavior. Although there is little difference in the absolute size of the VMH, there is some reason to believe that it becomes sexually dimorphic during development (Sakuma, 1984). Estradiol and testosterone have a dramatic affect on both neurite outgrowth and dendritic branching in organotypic cultures of the mouse hypothalamus (Toran-Allerand, 1980; Toran-Allerand et al., 1983).

Sexual behavior in mice that have a disrupted estrogen receptor gene (ER-α and ER-β) is significantly attenuated. For example, females do not lordose. ER-α knockout males do initiate copulatory behavior, mounting females at a normal rate, but they rarely achieve an intromission or an ejaculation. Furthermore, the males are less aggressive than wild-type males, generally failing to attack an "'intruder" mouse when it is placed in the male's home cage (Ogawa et al., 1997). However, genetic male mice that lack both estrogen receptor genes display no sexual behavior, including mounting and ultrasonic vocalizations (Ogawa et al., 2000).

GENETIC CONTROL OF BRAIN GENDER

Since the control of gender is cell autonomous in insects, sexual behavior has been explored from a genetic perspective. Male fruit flies recognize females based on an olfactory cue, called a *contact pheromone*, and males will perform stereotyped courtship behavior when they receive this signal. The male will orient toward a female, tap her abdomen, flutter his wing in song, and place his proboscis (the mouthparts) on the female's genitals. If the female is receptive, the male will then mount her and copulate. How does the central nervous system create this complex set of behaviors? One approach to the problem is to create unusual flies, called mosaics, that have some cells that are genetically female (XX) and some cells that are genetically male (XO). By studying many of these animals, each one with a unique mosaic, it is possible to determine which brain cells must be male in order for male or female behaviors to occur (Hall, 1977).

In more recent studies, a genetic trick has been used to construct a line of animals in which a single part of

A   The P[GAL4] enhancer trap system

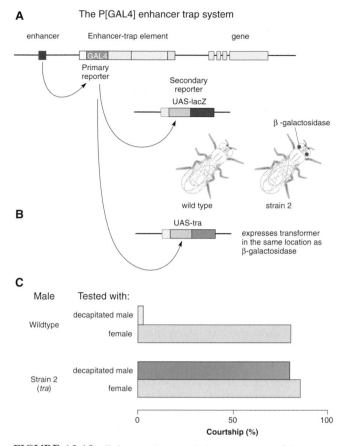

B

C

| Male | Tested with: | Courtship (%) |

FIGURE 10.19 Enhancer traps and the expression of the trans-former gene. A. An enhancer trap element inserts into the fly genome between an enhancer region and the gene that it normally controls. Whenever the enhancer is activated by a transcription factor, a reporter gene within the enhancer trap is expressed. In this example, a yeast transcription factor called *GAL-4* gene is expressed. To visualize the anatomical location of *GAL-4* expression, the enhancer trap flies are crossed to flies that have a *UAS-lacZ* gene. Since *GAL-4* is a transcription factor that activates UAS (blue), the *lacZ* gene (red) is expressed, and it encodes a protein (β-galactosi-dase) that can be stained for (red). Thus, labeled cells are known to have *GAL-4* expression. B. The enhancer trap line can also be used to drive the expression of native genes, such as *transformer* (green). Expression occurs only in cells with an activated enhancer. C. When an enhancer trap line was used to express transformer in olfactory neurons, the male flies courted males and females equally. Normal males only court females. (Adapted from Ferveur et al., 1995)

the brain is female (Figure 10.19). A piece of DNA, called PGAL4, is randomly inserted in the genome of many flies. By chance, it will occasionally insert next to an enhancer, and this enhancer will then activate the GAL4 gene in the enhancer trap element. If the enhancer is only active in one part of the body, then GAL4 will be expressed in that same part of the body. How does this help feminize the brain? It turns out that GAL4 can activate another promotor, called *upstream activating sequence* (UAS). If an experimenter

can hook up a gene of interest to the UAS promoter, then the gene of interest will be expressed wherever GAL4 is. This enhancer trap system was used to express the *transformer* gene, a feminizing signal, in olfactory neurons that might be processing the pheromonal signal (Ferveur et al., 1995).

Transformed males were presented with flies of either sex to see whether they would selectively court the female. Surprisingly, some strains of flies courted males with as much vigor as they did females (Figure 10.19C). The behavior of transformed animals may be due to their failure in discriminating the female pheromone. In fact, when the enhancer trap technique is used to make male flies that secrete only female pheromones, these flies are courted as if they are females (Ferveur et al., 1997). Thus, in flies, specific brain regions must have a gender if animals are to accurately interpret sensory information and produce sexually appropriate motor responses.

A separate tack has been used to explore what kinds of genes must be expressed in male or female nerve cells in order to produce correct sexual behaviors (Hall, 1994). For example, a gene product called *fruitless* is expressed in about 500 neurons of male flies only, and mutations of this gene also cause males to court one another. A mutation of the *dissatisfaction* gene leads virgin females to resist males during courtship, and they fail to lay mature eggs (Finley et al, 1997). Most mutations that affect sexual activity in flies are also found to affect other behaviors. Mutations of the *period* gene affects circadian rhythms, but they also change the temporal properties of the courtship song. Depending on the precise mutation, the interval between wing-beats can be shorter or longer than normal. That is, the song will have a lower or higher frequency, respectively.

SINGING IN THE BRAIN

One of the most striking correlations between sexual behavior and brain anatomy is found among several species of songbirds. Male birds attract a mate of the same species with vocalizations, or songs, that are commonly learned during juvenile development (see below). Zebra finches learn one song during the first 80 days after hatching, while canaries add new phrases to their song each breeding season. When scientists first looked at the brains of these animals, they were startled to find brain regions of remarkably different size in each sex (Nottebohm and Arnold, 1976). The sexual dimorphism occurs in brain nuclei that are known to participate in song production (RA, HVc),

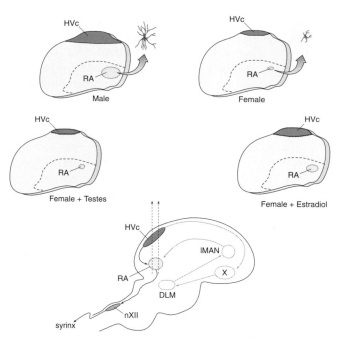

FIGURE 10.20 Sexual dimorphism in song production nuclei in birds. A sagittal section through the brain of song birds shows major nuclei involved in the learning and production of vocalizations. The pathway from HVc to RA to nXII is the primary output pathway to the song production apparatus (bottom). Both HVc (red) and RA (yellow) are much larger in adult male birds (top left) compared to adult females (top right). In addition, neurons in the male RA nucleus have a more elaborate dendritic architecture compared to those in females. Female zebra finches can be engineered to develop testes and little ovarian tissue, yet their HVc and RA nuclei do not become larger (middle left). In contrast, when female birds are treated with estradiol during development, the HVc and RA nuclei do become masculinized (middle right). (Adapted from Schlinger, 1998)

and these structures are much larger in males (Figure 10.20). Furthermore, when hatchling females are treated with estradiol, they can grow up to sing almost as adeptly as male birds (Gurney and Konishi, 1980; Simpson and Vicario, 1991). In male canaries, the size of vocal control nuclei changes during the course of a single breeding season, getting larger as testosterone levels rise (Nottebohm, 1981). Hormone treatment can apparently enhance the size of brain nuclei both by increasing afferent innervation and promoting dendritic growth.

It seems odd that females do not vocalize, if only to facilitate the mating process. In fact, female tropical wrens do sing a "duet" with the males. Furthermore, when the song repertoire of a female wren is relatively large, then the size of its song-control nuclei is similar to that of males (Brenowitz and Arnold, 1986). The vocal repertoire of the *Xenopus* females is also important in guaranteeing fertilized eggs. In this species, the male mating call has been well-characterized, and, like

birds, there is a sexual dimorphism of both neural and muscular components related to song production (Kelley, 1996). However, a female vocal behavior, termed *rapping*, is thought to trigger the entire copulatory repertoire (Tobias et al., 1998). When the female frog is unreceptive, it produces a ticking sound, but when it is ready to lay eggs, it begins to rap. This call stimulates males to vocalize even more vigorously and to attempt copulation. It is not yet known what the neural basis of ticking and rapping is, or whether the female brain becomes specialized for this behavior during development.

FROM GONADS TO BRAIN?

The simple hypothesis, then, is that testosterone is secreted by the testes, and this leads to a masculinized nervous system in male animals. A number of observations in birds and frogs suggest that other factors are involved (Wade and Arnold, 1996; Kelley, 1997). They raise the possibility that female and male brains differ from one another even in the absence of gonadal signals. First, the level of estradiol required to masculinize the nervous system of female birds is quite high, and even these high levels do not result in a fully masculinized phenotype. In frogs, the level of circulating androgen is quite similar in male and female animals during development. Second, it has not been possible to block masculine development of the nervous system in male birds by manipulations designed to decrease estrogen. Third, when genetic female zebra finches are pharmacologically engineered to develop with testes, and with little to no ovarian tissue, their vocal control nuclei continue to have a female phenotype (Figure 10.20). Finally, female frogs that receive transplanted testes have a larger larynx and more laryngeal motor neurons than do females that are treated with a single androgen (Watson et al., 1993).

These results suggest either that the gonads are a more complicated endocrine organ than we suspect, or that the nervous system contains intrinsic signals that bias its development in the absence of gonadal signals. For example, some rat diencephalic neurons express a sex-specific phenotype in vitro (Figure 10.21). When explanted at E14–17, before the initial surge of testosterone, tyrosine hydroxylase-expressing neurons are 30% larger in males, and the number of prolactin-expressing neurons is two to three times greater in female tissue, similar to adult animals (Kolbinger et al., 1991; Beyer et al., 1992). A similar experiment was performed in mice with a Y chromosome that lacks the *Sry*

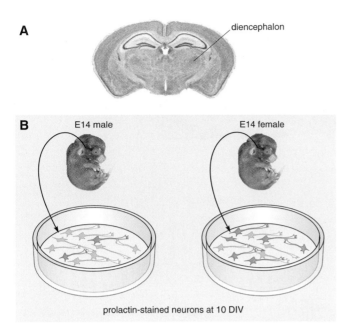

FIGURE 10.21 Development of sexual dimorphism before gonadal development. A. A transverse section through the adult mouse brain shows the location of the caudal diencephalon. B. When rat diencephalic neurons were explanted at embryonic day 14, the neurons from females embryos had two to three times more prolactin-expressing neurons (red) compared to cultures from E14 males. (Beyer et al., 1992)

gene. Testis development does not occur in these animals, and the gonadal signal is eliminated (Carruth et al., 2002). When mesencephalic neurons were explanted at embryonic day 14, the number of tyrosine hydroxylase-expressing cells was greater in genetic males than genetic females, even though both groups had female gonads. Therefore, it appears that the vertebrate nervous system may also develop some sex-specific characteristics independent of the gonads. In fact, the *Sry* gene is transcribed in the hypothalamus and midbrain of adult male mice, suggesting that these cells may be masculinized by a genetic signal.

A particularly compelling example of genetic determination of brain sexual dimorphism was discovered in a rare zebra finch gynandromorph (i.e., an animal that is a mosaic of male and female structures because some cells are chromosomal females while others are chromosomal males). In this instance, the animal was genetically male on one side and genetically female on the other (Figure 10.22A). Since only females carry a W chromosome, it was possible to stain for mRNA encoding a W chromosome gene and to show that expression was limited to one side of the brain (Figure 10.22B). As expected, one side of the animal developed a male-like gonad, and the other side developed a female-like gonad (Figure 10.22C). Thus, the brain was

exposed to an identical, if somewhat peculiar, gonadal hormone environment during development. In fact, the male side of the brain had a much larger HVc nucleus, as compared to the female side (Agate et al., 2003). Thus, the genetic sex of the brain cells plays a primary role in their differentiation.

LEARNING TO REMEMBER

Learning is often portrayed as an extension of neural development, and there are many similarities, particularly at the cellular level (Chapter 9). But this portrait does not capture an important fact: learning and memory themselves change during the course of development. Some forms of learning emerge during a limited period of development, and then disappear. The filial imprinting of a baby duckling on its mother occurs during a brief time interval after hatching. Other forms of learning are robust in young animals but gradually become less efficient. Humans retain the capacity to learn new vocabulary words throughout life, but there is a window of development when we learn words at a remarkable rate. Still other forms of learning seem to improve with time, perhaps owing to the wealth of information already stored in a mature nervous system. One clear line of evidence demonstrating the effect of environment on the developing nervous system comes from rearing rats in an enriched environment. Developing rats that are housed with many objects in the cage have almost 25% more synapses per neuron in the visual cortex (Turner and Greenough, 1985).

Memory is usually divided into two general types: recollection of facts (things that can be stated, or declarative memory) and recollection of skills (things that can be performed, or procedural memory). Humans with focal brain injuries are often found to have specific learning and memory deficits (Milner et al., 1998). For example, people with extensive damage to the limbic system are completely unable to recall new facts, yet they can learn and remember new motor tasks. Exceptionally rapid learning or memorization has sometimes been mistaken for intelligence. In fact, human brilliance is often specialized: a knack for game theory coupled to rapid learning of spatial patterns might allow one person to be a champion bridge player, while a taste for chewing tobacco coupled to robust motor learning can produce a major league pitcher. Therefore, it is not too surprising that clinical measures of learning and memory are often difficult to reconcile with the broad patterns of human behavior.

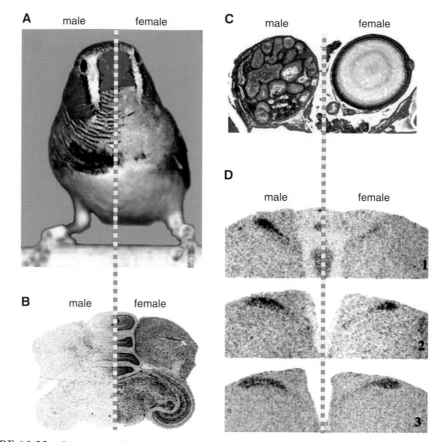

FIGURE 10.22 Genetic sex of brain determines the pattern of differentiation in the zebra finch. A. a zebra finch gynandromorph with male plumage on one side (left) and female plumage on the other side (right). B. The W chromosome is found normally only in females. The brain section shows *in situ* hybridization of mRNA encoding the W chromosome gene, *ASW*, to be ubiquitous on the female (right) side of the brain, but virtually absent on the male (left) side (dark areas show label). C. Histological sections of the gonads reveal dysmorphic testis on the genetically male side (left) and ovarian tissue containing a number of follicles on the female side (right). D. The HVC nucleus is normally larger in males than in females. The series of images shows *in situ* hybridization for androgen receptor mRNA (dark areas) to mark HVC at 3 caudal to rostral levels. The HVC is 82% larger on the male side of the brain (left) as compared to the female side (right). (From Agate et al., 2003)

The development of learning may also require a certain amount of practice, similar to many sensory and motor skills. For example, many animals build up a supply of provisions by hiding food in different locations. Of course, their spatial memory for the hiding places is crucial if they are to enjoy the fruits of their labor. When marsh tits are reared in captivity, they will continue to hide the sunflower seeds that they are fed. However, if the birds are given powdered seeds that cannot be stored, they develop a smaller hippocampus (Clayton and Krebs, 1994). Thus, learning and memory skills require practice, and this process may influence nervous system development. Since our immediate goal is to relate nervous system development to behavior, the following discussion focuses on reasonably simple forms of learning and procedural memory.

WHERE'S MAMMA?

Many vertebrates are born with an ability to obtain food and warmth from their mother, when offered. Nestling herring gulls peck at the tip of their mother's beak for food, neonatal rodents assume a specific position in order to suckle at a nipple, and newly hatched jewel fish have a natural tendency to approach objects that are colored like the broody adult. Although these innate motor behaviors are very sophisticated in the apparent absence of any experience, many animals learn to recognize and respond selectively to their mother (Lorenz, 1937). Konrad Lorenz, a co-recipient of the 1973 Nobel Prize, made the rather dramatic observation that hatchling ducks and geese will follow the first moving object that they see, forming a very stable attachment. Ordinarily, the mother goose fills this role, but hatchlings can also learn to follow inanimate objects, and even the experimenter himself. This learned behavior is termed *filial imprinting*. Filial imprinting has the immediate advantage of keeping offspring with the provider, and it can also have implications much later in life. When mature, the male birds will court a member of the species on which they imprinted, whether it is a bird, dog, or human.

Filial imprinting is not unique to birds. Tree shrew pups will imprint on the nursing mother during the second postnatal week. If removed from the nest during this period, a pup will not learn to follow its real mother, and it can be induced to follow a cloth permeated with the odor of a foster mother (Zippelius, 1972). Similarly, rat pups come to prefer their nest based on the mother's odor during the first few postnatal weeks, and this preference can be modified by providing a novel odorant during this period (Brunjes and Alberts, 1979). Subantarctic fur seal pups must learn their mother's vocalization within five days after birth. The mother seals set out on two to three week foraging trips, and the pups locate their mother within minutes of her return using the sound of her vocalization (Charrier et al., 2001). Newborn humans also display a preference for their mother. When infants are able to elicit either their mother's voice or the voice of another female by the rate at which they suck on a nipple, they preferentially activate their mother's voice (DeCasper and Fifer, 1980). What is the evidence that this preference is learned? When mothers read a story aloud during the last six weeks of pregnancy, their babies will subsequently prefer to hear that story over one that was not read aloud. Unexposed newborns display no preference between the two stories. Thus, even though an infant's hearing is quite limited in utero, he may already be forming certain auditory preferences (DeCasper and Spence, 1986).

What exactly is the nervous system learning during filial imprinting? Do infant animals simply learn their mother's smell or image? These questions have been explored in newly hatched ducklings, and the results suggest that several factors are necessary for filial imprinting to occur: visual cues, auditory cues, and social environment. When one-day-old mallard ducklings are allowed to follow a stuffed mallard hen for 30 minutes, they develop a preference for this replica, presumably based on its visual appearance (Johnston and Gottlieb, 1981). However, mother ducks also produce an "assembly" vocalization, and this auditory cue maintains filial imprinting as the ducklings begin to grow. The assembly call is such a powerful signal that ducklings will preferentially follow an unfamiliar red-and-white striped box that is producing this call rather than a familiar mallard hen model.

Why is the mother's assembly call such a powerful cue? One possibility is that the mother's call is necessary to maintain her duckling's attachment when the entire family leaves the nest and begins to move about the environment. Older ducklings become very attached to their siblings as they grow, and this "peer imprinting" can actually interfere with filial imprinting (Dyer et al., 1989). For example, socially reared ducklings will not preferentially follow a silent, familiar mallard model, although individually reared ducklings will do so. However, the mallard maternal call will induce socially reared ducklings to follow a familiar mallard or an unfamiliar pintail model (Dyer and Gottlieb, 1990).

This raises an important question: Do ducklings respond innately to their mother's call, or does it depend on sensory experience? Interestingly, ducklings have an inborn preference for the mother's call rate, 4 notes per second. However, to maintain preference through hatching, the duckling must either hear its own "contentment" call or that of its siblings (Figure 10.23). When ducklings are devocalized and reared in isolation with a "contentment" call that is slowed down to about 2 notes per second, they subsequently show no preference for the mother's "assembly" call (Gottlieb, 1980). Thus, imprinting is a far more elegant form of learning than was originally suspected. Although ducks, geese, and chicks can visually imprint on an object after walking behind it, many other factors regulate this learning. By studying the animal in its natural setting, it becomes clear that developing animals are "prepared" to learn certain cues they will likely encounter, such as the vocalization of its siblings.

Newly hatched domestic chicks also display filial imprinting, and the neural substrates have been studied in some depth. When chicks are presented with tones pulsed at about 3 Hz, they will develop a strong preference for this acoustic stimulus and selectively approach it (Wallhausser and Scheich, 1987). There is a dramatic reduction of spines in two different higher forebrain regions during the period of imprinting, and this is not observed in naive chicks. Furthermore, both imprinting and spine elimination depend on functional NMDARs within these forebrain regions (Bock et al., 1996; Bock and Braun, 1999a, 1999b). Chicks can also learn to imprint on the visual characteristics of an object and follow it around, just as duckings do. Destruction of a specific forebrain area impairs imprinting, and this same region displays an increase in NMDARs following imprinting (Horn, 2004). Therefore, some of the cellular mechanisms that have been implicated in synaptic plasticity (Chapter 9) have an important role in this early form of learning.

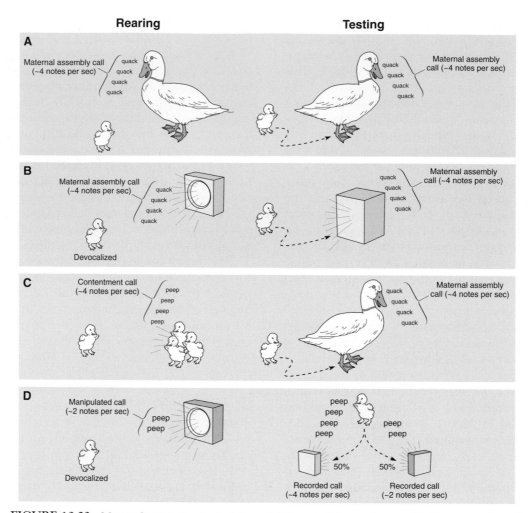

FIGURE 10.23 Maternal imprinting in ducklings. A. When ducklings are exposed to the maternal vocalization (~4 notes/s), they will subsequently approach an assembly call (~4 notes/s). B. Exposure to the assembly call alone is sufficient to promote auditory imprinting. C. Exposure to the duckling's own contentment call, which is also ~4 notes/s, is sufficient to promote auditory imprinting. D. When a duckling is exposed to an unnatural call (2 notes/s) during development, it is not able to recognize and respond to the assembly call when tested subsequently. (Adapted from Gottlieb, 1980)

The attachment of infant to mother requires an endogenous reward mechanism. Mice lacking the mu-opioid receptor (which mediates reward and analgesia) do not vocalize when they are removed from their mother, as wild-type animals do (Moles et al., 2004). Nor do they prefer their own bedding to that from another female. These results suggest that maternal attachment requires that both infant learning and an endogenous reward system validate maternal auditory, olfactory, and visual cues.

FEAR AND LOATHING

Beyond the procurement of food and water, most animals need behavioral mechanisms to avoid danger, such as a poisonous plant or a predator. Animals are born with the innate ability to avoid certain things. For example, several species of birds will run from a black hawk-shaped silhouette that is moved over their heads. This occurs even when the birds are reared in isolation with no chance to learn that the "hawk" image represents danger (Tinbergen, 1948). Many other dangers are not recognized at first, and animals must learn to avoid these situations through some sort of experience. A well-studied form of learning, called *fear conditioning*, is probably responsible for much of our skill at avoiding danger. During fear conditioning, an animal learns to associate an unconditioned stimulus and response (e.g., a snake bite and the pain or fear it produces) with a neutral stimulus (e.g., the image of a snake). Obviously, an image of a snake can do no harm, but the animal has learned that if he sees a snake, he may be bitten. Thus, the sight of a snake becomes a conditioned stimulus, and it produces a conditioned response (e.g., freezing). Although unethical by modern standards, a 9-month-old baby with no fear of animals was trained to fear rabbits by pairing the rabbit with a startling noise (a hammer striking metal just behind the baby's head). This conditioning eventually caused the baby to cry every time he saw a rabbit (Watson and Raynor, 1920).

In a typical fear conditioning experiment, an animal is exposed to a frightening stimulus such as a mild foot shock, and at the same time a pure tone is presented from a speaker. How do we know that the electric shock is frightening? Animals usually stop moving (i.e., they freeze) and their blood pressure goes up when they are in frightening situations, and this is precisely the response to mild foot shock. In contrast, the pure tone alone does not produce a change in move-

ment or blood pressure. By presenting these stimuli together several times, the sound alone is gradually able to elicit a fear response. Are developing animals able to form such associations? Actually, it seems to depend on the stimulus that the animal is asked to learn as well as the behavior that it is asked to perform. For example, rat pups at 15 days or older can learn to freeze in response to a tone that was paired previously with mild foot shock (Moye and Rudy, 1987).

For many learning tasks, animals improve with age. This was explored in rats by first pairing a brief loud sound that elicited a startle response with a long-lasting pure tone at moderate intensity (Figure 10.24). Adult animals learn quickly that the pure tone predicts the arrival of the loud sound. During test trials, they produce a much larger startle response when the pure tone is present, and this is referred to as fear-

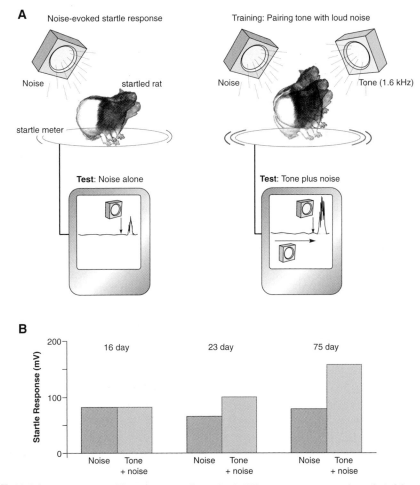

FIGURE 10.24 Emergence of fear-potentiated startle. A. When rats are exposed to a brief, loud noise (red speaker) they make a sudden movement, called a startle response (left). This can be recorded by a platform on which the animal stands and displayed on an oscilloscope. When a pure tone (green speaker) precedes the loud noise, the rats learn that the tone predicts the noise burst (right). In subsequent tests, they give a larger startle response to the paired tone plus noise, and this is called *fear-potentiated startle*. B. When trained in this paradigm, postnatal day 16 rat pups display no potentiation, indicating that they have not learned to associate the two signals. At 23 days, the animals do display a potentiation due to pairing, although the potentiation displayed by adults is greater still. (Adapted from Hunt et al., 1994)

potentiated startle. Thus, if the pure tone enhances the startle response, then one concludes that the animal learned about a dangerous situation. When 16-day rat pups are trained in the same paradigm, they do not show any sign of learning (Hunt et al., 1994). Their response to the tone plus noise is nearly identical to noise alone (Figure 10.24). A similar delay in learning is demonstrated when a light stimulus is paired with footshock (Hunt, 1999). For some learning tasks, neonates perform better than juvenile animals. Rat pups of 5–10 days can learn to avoid a sugar solution when it is paired with mild foot shock, yet 15-day pups fail to learn this task (Hoffman and Spear, 1988). Thus, learning is not simply poor in young animals and robust in adults. Rather, it is a complex function of age, sensory modality, and the motor response that is being modified by training.

Even extremely simple forms of learning, such as habituation and sensitization, can emerge at different times during development. This has been studied at the level of both behavior and neurophysiology in the sea slug, *Aplysia* (Rayport and Camardo, 1984; Rankin and Carew, 1988; Nolen and Carew, 1988). During habituation, animals produce a smaller reflexive response when they are exposed to repeated presentations of an identical stimulus. This form of learning can be demonstrated by squirting some seawater on the animal's siphon while monitoring its contraction. With each squirt of water the siphon withdrawal decreases, finally reaching about 30% of its initial amplitude. When stimulation ceases, the response gradually recovers over a few hours. Habituation can be observed in 5- to 10-day *Aplysia*, before most central neurons are born. However, the stimuli must be delivered with much shorter intervals to produce habituation in young animals. Consistent with these behavioral results, synaptic potentials that mediate the response decrease in size with repeated use in neurons from 5- to 10-day animals.

Do other simple forms of learning appear this early? This question was assessed for sensitization, a form of nonassociative learning in which an animal produces a larger reflexive response when it is preceded by a strong, usually noxious stimulus. For example, when an electric shock is delivered to the tail, the same squirt of seawater evokes a much larger siphon withdrawal response. Sensitization was found to emerge quite late in development, almost 60 days after the appearance of habituation. Once again, a neural analog of sensitization was first observed at roughly the same time as the behavior. In adult animals, stimuli to the siphon nerve produce synaptic potentials in a neuron called R2, and the size of these synaptic potentials can be increased by delivering stimuli to the nerve emanating from the tail. However, this synaptic facilitation is

observed only in animals >70 days. Thus, there is some reason to believe that specific forms of learning emerge at distinct periods of development owing to the maturation of explicit synaptic mechanisms.

Studies that target simple forms of learning will be critical for linking behavior with underlying neural mechanisms. However, it is also interesting to ask how developing animals learn complex, multistep tasks, such as how to write a sentence or make a peanut butter sandwich. Of course, even sophisticated learning tasks are studied with a formal paradigm. For example, in a delayed nonmatch to sample task, a primate is first shown an object that can be moved to reveal a reward, such as a food pellet (Bachevalier, 1990). After a delay, the animal is next presented with two objects, one of which it saw previously. In this case, the animal must learn to move the new object in order to obtain the reward (Figure 10.25). The task can be made more complicated by increasing the time between trials or by increasing the number of objects that must be memo-

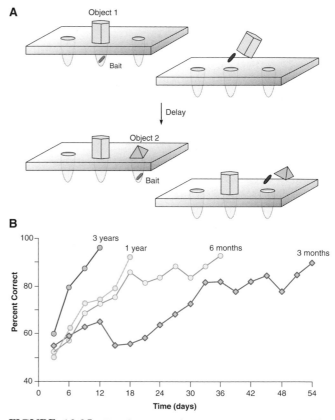

FIGURE 10.25 Development of memory in primates. A. The delayed nonmatch to sample task involves remembering the object presented first and, after a delay, choosing the new object on a test trial. B. The time to learn this task is shown for primates of different ages. At 2–3 years of age, animals learned the task within eight days of training. However, 3-month-old monkeys did not reach criterion (90% correct responses) until they had received 36 days of training. (Adapted from Bachevalier, 1990)

rized. Infant and adult primates were trained on a daily basis until they could perform the task correctly 90% of the time. Animals 2 to 3 years of age reached criterion after eight days of training, but 3-month-old monkeys required 36 days of training (Figure 10.25). One possible limitation for younger animals may be the amount of sensory activity entering the central nervous system (Bachevalier et al., 1991). In 3-month animals, visually evoked activity is significantly lower in regions of the cortex thought to mediate this form of learning, as measured with the 2-deoxyglucose technique (see Box: Watching Neurons Think in Chapter 9).

When humans of different ages were challenged with a similar delayed nonmatch to sample task, they also displayed a gradual improvement with age (Overman, 1990). Yet children nearly 3 years of age take about 10 times longer to learn the task compared to adults. Furthermore, they forget things more quickly. The duration of time that children can retain a simple associative learning task gradually increases from 2 to 18 months of age (Hartshorn et al., 1998). Complex learning tasks also emerge at different periods of development. As discussed above, memory is commonly divided into recollection of facts versus performance of skills. One type of factual ability is the recollection of the spatial environment. Spatial memory was tested in 2- and 4-year-old children by asking them to retrieve candy from eight different locations in an unfamiliar room. It was found that 2-year-olds revisited locations where they had already procured the candy more often than did 4-year-olds. That is, the younger subjects did not remember where they had been. In a different type of factual learning, children were asked to recall details of a story that they had been read, and there was significant improvement between 5 and 10 years of age. Finally, children were asked to learn a complex motor task (i.e., a skill), and their performance was equivalent at 5 and 10 years of age (Foreman et al., 1984; Hömberg et al., 1993). These studies point out the diversity of complex associative learning. Presumably, the improvements that are observed with age result from the maturation of sensory function, motor skills, and learning and memory systems themselves.

GETTING INFORMATION FROM ONE BRAIN TO ANOTHER

The development of animal communication is a fascinating mix of inherited traits and learning. For many of us animals, communication provides the foundation of our existence. Some might say that it forms the basis of our consciousness. Depending on our position in the food chain, it fetches us a mate, warns us of danger, informs of a food source, bonds us in society, and enriches us with artistry. Perhaps the best studied communication system is that of song birds, where adult males produce courtship vocalizations to attract conspecific females. While sex-specific behavior is explained in terms of genetic and epigenetic factors (above), the individual songs require learning and practice.

When juvenile birds are reared in isolation such that they do not hear a normal adult song, they develop abnormal vocalizations. The vocalizations are even more degraded by deafening the song bird soon after hatching (Marler and Sherman, 1983). Yet these vocalizations still retain a few species-specific characteristics, such as song duration. Even relatively boring vocalizations, such as those of the crow or rooster, may be affected by sensory experience. When a middle ear muscle is detached early in development, male chickens crow at a higher frequency than control animals, possibly because low-frequency sounds can no longer be damped by the middle ear mechanism (Grassi et al., 1990). These studies illustrate the role of learning, but suggest that there are intrinsic limitations on the song that any single species of bird is able to acquire.

Many neuroscientists have settled on the zebra finch as a model for studies of behavior and nervous system development. Juvenile birds leave the nest about 20 days after hatching, and they begin to sing a few days later. As with sparrows, male zebra finches must be exposed to the species-specific song, and they must be able to hear themselves sing if they are to produce an accurate rendition as adults. When males reach about 90 days of age, they produce a stereotyped song that remains unchanged throughout life, providing they continue to hear themselves sing. Lesions of the vocal control nuclei, HVc or RA, have a devastating effect on song production in adults (Nottebohm et al., 1976).

There is a second pathway from HVc to the anterior telencephalon, and this projection has been implicated in song learning. One indication of this special role in learning comes from the anatomy of the system (Hermann and Arnold, 1991; Johnson et al., 1995). The size of one of these telencephalic nuclei, lMAN, increases when birds first start to practice their tutor's song, and it eventually decreases in adulthood. The projection from lMAN to the motor output from the telencephalon, RA, is also greatest during the early stages of learning. Since degenerating nerve terminals can be stained within a few days of lesion, the lMAN was lesioned at three different posthatch ages, and the number of degenerating synapses within RA was assessed. The technique showed that the number of

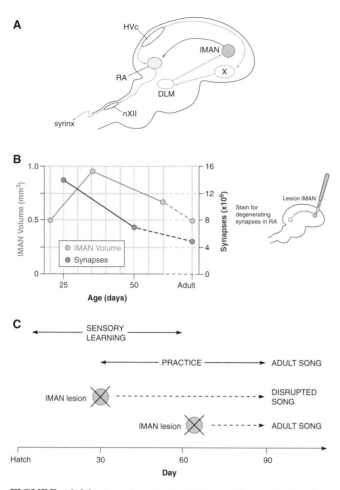

FIGURE 10.26 Song learning in birds. A. The sagittal section through the song bird brain shows the major nuclei involved in song learning and production. B. The projection from lMAN to RA was assessed by lesioning lMAN, waiting a day for the synapses to begin degenerating, and then staining the tissue degeneration. Thus, the greater the staining, the greater the innervation. The number of synapses begin to decline after day 25. For comparison, the size of lMAN is plotted, and it also begins to decline after day 35. C. The telencephalic nucleus, lMAN, has been implicated in song learning by lesioning it at different ages. If lesioned at 30 days, during sensory learning, the ability to produce song as an adult is disrupted. If lMAN is lesioned at around 60 days, then adult song is unaffected. (Adapted from Bottjer et al., 1984; Herman and Arnold, 1991)

lMAN synapses decreases almost threefold during development (Figure 10.26). The number of lMAN neurons that project to RA remains constant during this period, suggesting that terminals are being eliminated.

Interestingly, lesions to lMAN have no effect on song production in adult birds (Figure 10.26C). However, when lMAN is lesioned in animals before song learning has been completed, song learning and production is disrupted (Bottjer et al., 1984). Together, these experiments suggest that certain nuclei partici-

pate in the learning of song but not in its adult production. Does lMAN really play such a limited role in zebra finch behavior? In fact, lMAN may participate in song recognition by adult females. Lesions of HVc are known to disrupt song recognition such that the females perform a precopulatory behavior in response to the song of another species (Brenowitz, 1991). However, the role of lMAN in adult birds remains poorly understood.

Since lMAN seems to be essential for song learning, it would be interesting to know whether the cellular mechanisms are similar to other forms of plasticity. As discussed earlier, the NMDA receptor has a well-documented role in many forms of synaptic plasticity (see Chapter 9). The level of NMDA receptor expression is particularly high in lMAN during the period of song learning (Figure 10.27). When tissue sections are labeled with an NMDA receptor antagonist (^{3}H-MK-801), autoradiographic analysis shows that receptor number gradually declines over the first few months (Aamodt et al., 1995). To test whether these receptors mediate song learning, the NMDA receptor antagonist, AP5, was infused bilaterally into lMAN (Figure 10.27). Beginning on day 32, animals were presented with a tutor song every other day, and AP5 was either infused at the same time or on alternate days. Those animals receiving AP5 and training simultaneously performed very poorly at day 90, producing only 20% of the tutor song. In contrast, the animals that had active NMDA receptors while receiving auditory training learned about 50% of the tutor song (Basham et al., 1996). It is not yet known how AP5 affects synaptic activity in lMAN, but these results suggest that song learning in zebra finches shares one synaptic mechanism with other forms of plasticity.

Direct measures of synaptic plasticity can be made in brain slices through the forebrain nuclei involved in song learning. So far, long-term excitatory synaptic potentiation (LTP) has been described in two different nuclei (lMAN and area X) that are required for vocal learning in developing zebra finches (Boetigger and Doupe, 2001; Ding and Perkel, 2004). In lMAN, stimulation of intrinsic synapses led to their potentiation, while at the same time depressing the thalamic synapses. These forms of plasticity were gradually lost by posthatch day 60, when sensory learning comes to an end (Figure 10.26C). One proposal is that LTP of the lMAN synapses leads to the strong selectivity for a particular song structure, and the LTD of thalamic afferents refines the afferent input. A second locus of synaptic plasticity is found in area X, the second target of lMAN afferents. Stimulation of these glutamatertic afferents leads to a NMDAR-dependent increase in response amplitude. However, synaptic plasticity

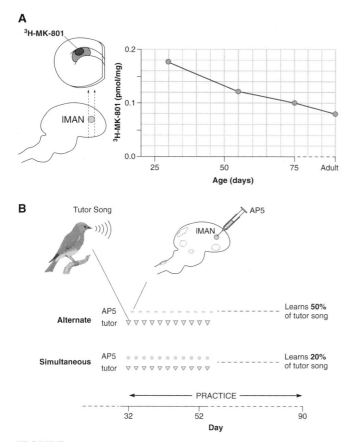

FIGURE 10.27 NMDA receptors are implicated in song learning. A. The number of NMDA receptors was assessed by measuring the amount of a receptor antagonist (^{3}H-MK-801) that bound to lMAN during development. NMDA receptors began to decrease after day 30. B. The influence of NMDA receptors on song learning was tested by injecting an antagonist (AP5) into lMAN while the bird was being exposed to a tutor song. In control experiments, AP5 was injected on days when the birds were not exposed to the tutor song. When AP5 and exposure to tutor song were delivered simultaneously, the birds performed very poorly at day 90, only producing 20% of the tutor song. (Adapted from Aamodt et al., 1995; Basham et al., 1996)

within area X is not observed until the period of sensory learning is nearly finished, and may be responsible for the phase of learning during which birds practice and adjust the memorized tutor song.

LANGUAGE

Although human communication is far more complicated than bird song, there are some similarities. Learning is certainly involved at every stage of development, from the production and perception of vowels to the syntax of a sentence. Humans generally speak their first words between 9 and 12 months and slowly acquire about 50 single words, mostly nouns, over the next eight months. As with birds, there is a period of development when communication skills are acquired most efficiently. From 2 to 6 years of age, children learn about eight words per day. One indication of a sensitive period for language development comes from studies of humans who learn to produce and understand a second language. When English-language skills were analyzed in native Korean or Chinese speakers who arrived in the United States as children or adults, the youngest subjects performed best (Johnson and Newport, 1989). A second indication of a sensitive period comes from studies of deaf individuals who were exposed to sign language from birth to one year of age. Those individuals who are exposed to sign language from birth are more skilled than infants who are exposed even as early as 6 months of age (Newport, 1990).

In contrast to bird song, human communication is performed with equal precision in three sensory modalities. Those born with profound hearing loss can learn to communicate perfectly with their hands and visual system using sign language. Those born without sight can learn to read with their somatosensory system using Braille. Furthermore, the development of language seems to be quite natural in any of these modalities. It has been known for some time that hearing infants begin to produce speech sounds well before they can understand words. These vocalizations, called vocal babbling, are commonly made up of repeated syllables (e.g., "dadadada"). Whereas deaf children are unable to produce perfect vocalizations as adults, similar to deafened songbirds, a remarkable thing happens: their language ability can be transferred to another sensory modality.

Early stages of language acquisition were studied in two infants who were deaf from birth but were continually exposed to American Sign Language (ASL) by their deaf parents (Petitto and Marentette, 1991). To determine whether the infants would "babble" with their hands, the manual activity of each infant was codified in some detail, and their production of ASL hand shapes was analyzed (Figure 10.28). Deaf children devote about 50% of their manual activity to ASL hand shapes, while hearing children only produce about 10% of this activity, presumably by chance. Interestingly, the disparity between deaf and hearing children increases from 10 to 14 months of age, suggesting that deaf children learn language at the same stage of development as hearing children when given the opportunity to use their visual system. Finally, 98% of the manual babbling was performed in front of the body, presumably within the infant's visual field. Thus, imitation of a "tutor" and sensory feedback are

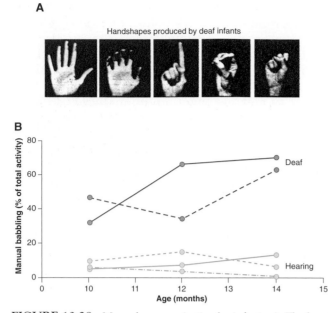

FIGURE 10.28   Manual communication by infants. A. The hand signals produced by deaf infants were studied and codified in order to detect hand shapes that correspond to American Sign Language (ASL). B. Deaf infants produced hand shapes corresponding to ASL (manual babbling) more often than hearing infants of the same age. (Adapted from Petitto and Marentette, 1991)

important when learning to "speak" with one's hands, in general agreement with studies of bird song development.

Given the complexity of language, it is not surprising that we are only beginning to understand the neural mechanisms that support human communication and how it develops. The ability of infants from two countries to recognize their native vowel sounds was studied to find out whether early experience effects perception. Six-month-old infants from Sweden and the United States were asked to judge two vowel sounds, one from their own country and one from the other country (Kuhl et al., 1992). The English vowel was a /i/ sound, as in the word "fee". The Swedish vowel was a front rounded vowel /y/ sound, as in the Swedish word "fy." Vowel sounds are composed of a unique set of frequencies, called *formants*, and the /i/ sound has slightly higher formants than the /y/ sound (Figure 10.29A). Most adult English-speaking listeners can categorize a sound as being like a /i/ sound if the first and second formants are reasonably close to the ideal. This ability to generalize is thought to prevent confusion since individual voice quality varies a good deal, particularly between children, adult females, and adult males. To see whether infants are able to categorize vowel sounds, ideal /i/ and /y/ vowels, called *prototypes*, were generated, and slight variations were

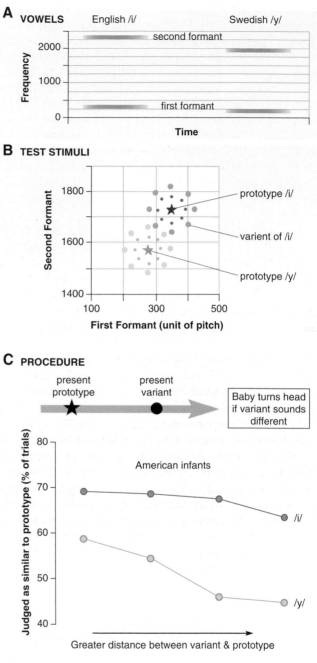

FIGURE 10.29   Recognition of language-specific phonemes. A. The frequency spectrum of an American vowel (/i/) and a Swedish vowel (/y/) are shown. Each vowel is composed of two major frequency bands, called *formants*. B. The ability of American and Swedish infants to recognize their native vowel sounds was examined with a range of computer-generated stimuli. An 'ideal' version of each vowel, called a *prototype*, was produced, along with vowels with small changes to one of the formant frequencies (called *variants*). C. Infants were trained to turn their head if the second of two vowels sounded different than the first. The data for American infants show that their native /i/ sound can be recognized even when formant frequency changes a good deal. However, the same frequency changes for the Swedish /y/ led to decreased recognition. (Adapted from Kuhl et al., 1992)

made to formant frequencies in order to produce variants (Figure 10.29B). Do infants treat the variants as a member of the group, as adults do? Infants were first exposed to the prototype vowel and then presented with a variant. If she perceived the variant to be different from the prototype, the infant was trained to turn her head. The data show that American infants are more likely to treat variants of /i/ as a member of that group, but are less likely to treat variants of /y/ as members of that group. The opposite result is found for Swedish infants. These results suggest that experience with one's own language in the first 6 months of life allow for improved perception of unique speech sounds.

Is it possible that language-specific activity is present in the brain as the infant is becoming sensitive to the unique attributes of human speech? A functional magnetic resonance imaging study in 3-month infants suggests that it is (Dehaene-Lambertz et al., 2002). Human adults exhibit the greatest activation along the left superior temporal sulcus when exposed to their native language, but the response is much smaller when the speech is played in reverse (i.e, "my dog has fleas" versus "saelf sah god ym"). When 3-month-old infants were tested with their native language (French), they also displayed greater activation of the left superior temporal gyrus (Figure 10.30). Forward speech elicited greater activation for one brain region on the left side, the angular gyrus, as compared to reversed speech. However, forward and reverse speech were equally effective at activating the temporal lobe, which is quite different than adults. Thus, left-hemisphere dominance for speech processing appears to be already present by 3 months of age, yet the sensitivity to phonological cues (e.g., forward vs. reverse speech) are immature.

SUMMARY

The great strides we have made in molecular and cellular neurobiology have underscored the importance of revisiting the behavior of animals, particularly during development. The maturation of neural processing, or the ability of a neuron to respond accurately to its synaptic inputs, depends on all the building blocks being in place. What can be gained from studying the system as a whole? If we learn the alphabet, are we not able to understand sentences? Of course, the blemish in this logic is simple to grasp: systems of mol-

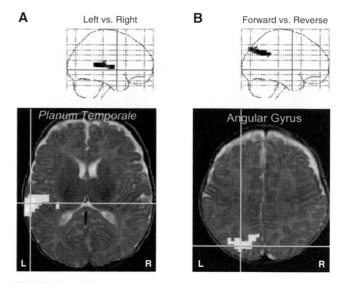

FIGURE 10.30 Speech activation of the human infant brain. fMRI images were obtained from 2- to 3-month-old infants during presentation of native speech. A. A transparent brain view (top) and an axial section (bottom) map the relative sound-evoked activity for left versus right temporal cortex. The activation was significantly greater on the left side. B. An activation map showing the relative sound-evoked activity in response to forward speech versus reverse speech. While there was no difference in the temporal cortex, there was an asymmetry within the angular gyrus. (From Dehaene-Lambertz et al., 2002).

ecules, or systems of nerve cells, take on new properties that were not expressed by the single molecule or nerve cell. While the genetic dissection of behavior is an important strategy, it should also be recognized that multiple gene products inevitably contribute to each phenotype, including behavior. Furthermore, the expression of many genes is influenced by the environment (i.e., neuronal activity). Therefore, a rich understanding of the relationship between brain and behavior is fundamental to interpreting all results. Studying animal behavior is one of the best ways to measure the properties of a system of nerve cells. It provides the most sensitive and universal indicator of a successfully assembled nervous system. All types of developmental errors (i.e., inappropriate fate, ion channel mutations, pathfinding errors, weak synapses) will affect the computational abilities of individual neurons. This is precisely why behavioral measures have long been used to tell clinicians when the nervous system is broken (e.g., schizophrenia, sleep apnea, delayed learning). It will, therefore, not be too great of a surprise when behavioral analyses reemerge as one of the most powerful tools available to developmental neuroscientists.

References

Aamodt SM, Nordeen EJ, Nordeen KW (1995) Early isolation from conspecific song does not affect the normal developmental decline of N-methyl-D-aspartate receptor binding in an avian song nucleus. J Neurobiol 27: 76–84.

Abbas L (2003) Synapse formation: let's stick together. Curr Biol 13: R25–27.

Abbott DH, Hearn JP (1978) The effects of neonatal exposure to testosterone on the development of behaviour in female marmoset monkeys. Ciba Found Symp 62: 299–327.

Acampora D, Mazan S, Lallemand Y, Avantaggiato V, Maury M, Simeone A, Brulet P. Forebrain and midbrain regions are deleted in Otx2-/- mutants due to a defective anterior neuroectoderm specification during gastrulation. Development. 1995 Oct;121(10): 3279–3290.

Adams RH, Sato K, Shimada S, Tohyama M, Puschel AW, Betz H (1995) Gene structure and glial expression of the glycine transporter GlyT1 embryonic and adult rodents. J Neurosci 15: 2524–2532.

Adler R, Hatlee M (1989) Plasticity and differentiation of embryonic retinal cells after terminal mitosis. Science 243: 391–393.

Agate RJ, Grisham W, Wade J, Mann S, Wingfield J, Schanen C, Palotie A, Arnold AP (2003) Neural not gonadal origin of brain sex differences in a gynandromorphic finch. Proc Natl Acad Sci USA 100: 4873–4878.

Agmon A, Hollrigel G, O'Dowd DK (1996) Functional GABAergic synaptic connection in neonatal mouse barrel cortex. J Neurosci 16: 4684–4695.

Aguado F, Carmona MA, Pozas E, Aguilo A, Martinez-Guijarro FJ, Alcantara S, Borrell V, Yuste R, Ibanez CF, Soriano E (2003) BDNF regulates spontaneous correlated activity at early developmental stages by increasing synaptogenesis and expression of the K+/Cl- co-transporter KCC2. Development 130: 1267–1280.

Aguayo AJ, Bray GM, Rasminsky M, Zwimpfer T, Carter D, Vidal-Sanz M (1990) Synaptic connections made by axons regenerating in the central nervous system of adult mammals. J Exp Biol 153: 199–224.

Ahmari SE, Buchanan J, Smith SJ (2000) Assembly of presynaptic active zones from cytoplasmic transport packets. Nat Neurosci 3: 445–451.

Airaksinen MS, Saarma M (2002) The GDNF family: Signaling, biological functions and therapeutic value. Nat Neuro Rev 3: 383–394.

Akai J, Storey K. Brain or brawn: how FGF signaling gives us both. Cell. 2003 Nov 26;115(5): 510–502.

Alcantara S, Frisen J, del Rio JA, Soriano E, Barbacic M, Silos-Santiago I (1997) TrkB signaling is required for postnatal survival of CNS neurons and protects hippocampal and motor neurons from axotomy-induced cell death. J Neurosci 17: 3623–3633.

Allendoerfer KL, Shatz CJ (1994) The subplate, a transient neocortical structure: its role in the development of connections between thalamus and cortex. Annu Rev Neurosci. 17: 185–218.

Alonso MC, Cabrera CV. The achaete-scute gene complex of Drosophila melanogaster comprises four homologous genes. EMBO J. 1988 Aug;7(8): 2585–2591.

Aloyz RS, Bamji SX, Pozniak CD, Toma JG, Atwal J, Kaplan DR, Miller FD (1998) P53 is essential for developmental neuron death as regulated by the TrkA and p75 neurotrophin receptors. J Cell Biol 143: 1691–1703.

Alsina B, Vu T, Cohen-Cory S (2001) Visualizing synapse formation in arborizing optic axons in vivo: dynamics and modulation by BDNF. Nat Neurosci 4: 1093–1101.

Altman J, Bayer SA. Embryonic development of the rat cerebellum. I. Delineation of the cerebellar primordium and early cell movements. J Comp Neurol. 1985 Jan 1;231(1): 1–26.

Altman J, Das GD (1965) Post-natal origin of microneurones in the rat brain. Nature 207: 953–956.

Alvarado-Mallart RM. Fate and potentialities of the avian mesencephalic/metencephalic neuroepithelium. J Neurobiol. 1993 Oct;24(10): 1341–1355.

Alzheimer C, Schwindt PC, Crill WE (1993) Postnatal development of a persistent Na+ current in pyramidal neurons from rat sensorimotor cortex. J Neurophysiol 69: 290–292.

Anderson DJ (1993) Molecular control of cell fate in the neural crest: the sympathoadrenal lineage. Annu Rev Neurosci 16: 129–158.

Anderson DJ, Groves A, Lo L, Ma Q, Rao M, Shah NM, Sommer L (1997) Cell lineage determination and the control of neuronal identity in the neural crest. Cold Spring Harb Symp Quant Biol 62: 493–504.

Anderson MJ, Cohen MW (1977) Nerve-induced and spontaneous redistribution of acetylcholine receptors on cultured muscle cells. J Physiol 268: 757–773.

Anderson SA, Eisenstat DD, Shi L, Rubenstein JL. Interneuron migration from basal forebrain to neocortex: dependence on Dlx genes. Science. 1997 Oct 17;278(5337): 474–476.

Angevine JB Jr. Time of neuron origin in the diencephalon of the mouse. An autoradiographic study. J Comp Neurol. 1970 Jun; 139(2): 129–187.

Antonini A, Stryker MP (1993) Development of individual geniculocortical arbors in cat striate cortex and effects of binocular impulse blockade. J Neurosci 13: 3549–3573.

Aoki C, Fujisawa S, Mahadomrongkul V, Shah PJ, Nader K, Erisir A (2003) NMDA receptor blockade in intact adult cortex increases trafficking of NR2A subunits into spines, postsynaptic densities, and axon terminals. Brain Res 963: 139–149.

Armand J, Edgley SA, Lemon RN, Olivier E (1994). Protracted postnatal development of corticospinal projections from the primary motor cortex to hand motoneurones in the macaque monkey. Exp Brain Res 101(1): 178–182.

Armstrong RC, Aja TJ, Hoang KD, Gaur S, Bai X, Alnemri ES, Litwack G, Karanewsky DS, Fritz LC, Tomaselli KJ (1997) Activation of the CED3/ICE-related protease CPP32 in cerebellar granule neurons undergoing apoptosis but not necrosis. J Neurosci 17: 553–562.

Asher RA, Morgenstern DA, Moon LD, Fawcett JW (2001) Chondroitin sulphate proteoglycans: inhibitory components of the glial scar. Prog Brain Res 132: 611–619.

Augsburger A, Schuchardt A, Hoskins S, Dodd J, Butler S (1999) BMPs as mediators of roof plate repulsion of commissural neurons. Neuron 24: 127–141.

Axelsson J, Thesleff F (1959) A study of supersensitivity in denervated mammalian skeletal muscle. J Physiol (Lond) 147: 178–193.

Baccaglini PI, Spitzer NC (1977) Developmental changes in the inward current of the action potential of Rohon-Beard neurones. J Physiol 271: 93–117.

Bachevalier J (1990) Ontogenetic development of habit and memory formation in primates. Ann New York Acad Sci 608: 457–477.

Bachiller D, Klingensmith J, Kemp C, Belo JA, Anderson RM, May SR, McMahon JA, McMahon AP, Harland RM, Rossant J, De Robertis EM (2000) The organizer factors Chordin and Noggin are required for mouse forebrain development. Nature 403(6770): 658–661.

Bagust J, Lewis DM, Westerman RA (1973) Polyneuronal innervation of kitten skeletal muscle. J Physiol 229: 241–255.

Bahr S, Wolff JR (1985) Postnatal development of axosomatic synapses in the rat visual cortex: morphogenesis and quantitative evaluation. J Comp Neurol 233: 405–420.

Baier H, Bonhoeffer F (1992) Axon guidance by gradients of a target-derived component. Science 255: 472–475.

Baier H, Korsching S (1994) Olfactory glomeruli in the zebrafish form an invariant pattern and are identifiable across animals. J Neurosci 14: 219–230.

Bailey AM, Posakony JW. Suppressor of hairless directly activates transcription of enhancer of split complex genes in response to Notch receptor activity. Genes Dev. 1995 Nov 1;9(21): 2609–2622.

Baird DH, Hatten ME, Mason CA (1992) Cerebellar target neurons provide a stop signal for afferent neurite extension in vitro. J Neurosci 12: 619–634.

Baker NE, Yu S-Y (2001) The EGF receptor defines domains of cell cycle progression and survival to regulate cell number in the developing drosophila eye. Cell 104: 699–708.

Balaban E (1997) Changes in multiple brain regions underlie species differences in a complex, congenital behavior. Proc Natl Acad Sci U S A 94(5): 2001–2006.

Balakrishnan V, Becker M, Lohrke S, Nothwang HG, Guresir E, Friauf E (2003) Expression and function of chloride transporters during development of inhibitory neurotransmission in the auditory brainstem. J Neurosci 23: 4134–4145.

Balice-Gordon RJ, Lichtman JW (1993) In vivo observations of pre- and postsynaptic changes during the transition from multiple to single innervation at developing neuromuscular junctions. J Neurosci 13: 834–855.

Balice-Gordon RJ, Lichtman JW (1994) Long-term synapse loss induced by focal blockade of postsynaptic receptors. Nature 372: 519–524.

Bamji SX, Majdan M, Pozniak CD, Belliveau DJ, Aloyz R, Kohn J, Causing CG, Miller FD (1998) The p75 neurotrophin receptor mediates neuronal apoptosis and is essential for naturally occurring sympathetic neuron death. J Cell Biol 140: 911–923.

Banerjee U, Zipursky SL (1990) The role of cell-cell interaction in the development of the Drosophila visual system. Neuron 4: 177–187.

Banker G, Goslin K (1991a) Culturing Nerve Cells. MIT Press: Cambridge.

Banker G, Goslin K (1991b) Rat hippocampal neurons in low density culture. In: Culturing Nerve Cells. (Banker G, Goslin K, eds), MIT Press: Cambridge, pp 251–281.

Barbacid M (1994) The Trk family of neurotrophin receptors. J Neurobiol 25: 1386–1403.

Bargones JY, Werner LA (1994) Adults listen selectively: infants do not. Psychol Sci 5: 170–174.

Barish ME (1986) Differentiation of voltage-gated potassium current and modulation of excitability in cultured amphibian spinal neurones. J Physiol 375: 229–250.

Barnea G, O'Donnell S, Mancia F, Sun X, Nemes A, Mendelsohn M, Axel R (2004) Odorant receptors on axon termini in the brain. Science 304: 1468.

Barres BA, Raff MC. Control of oligodendrocyte number in the developing rat optic nerve. Neuron. 1994 May; 12(5): 935–942.

Barres BA, Hart IK, Coles HS, Burne JF, Voyvodic JT, Richardson WD, Raff MC (1992) Cell death and control of cell survival in the oligodendrocyte lineage. Cell 70: 31–46.

Barres BA, Silverstein BE, Corey DP, Chun LL (1988) Immunological, morphological, and electrophysiological variation among retinal ganglion cells purified by panning. Neuron 1: 791–803.

Barres BA, MC Raff (1994). Control of oligodendrocyte number in the developing rat optic nerve. Neuron 12(5): 935–942.

Basham ME, Nordeen EJ, Nordeen KW (1996) Blockade of NMDA receptors in the anterior forebrain impairs sensory acquisition in the zebra finch (Poephila guttata). Neurobiol Learn Mem 66: 295–304.

Bashaw GJ, Goodman CS (1999) Chimeric axon guidance receptors: the cytoplasmic domains of slit and netrin receptors specify attraction versus repulsion. Cell 97: 917–926.

Bassell GJ, Zhang H, Byrd AL, Femino AM, Singer RH, Taneja KL, Lifshitz LM, Herman IM, Kosik KS (1998) Sorting of beta-actin mRNA and protein to neurites and growth cones in culture. J Neurosci 18: 251–265.

Bastiani MJ, Harrelson AL, Snow PM, Goodman CS (1987) Expression of fasciclin I and II glycoproteins on subsets of axon pathways during neuronal development in the grasshopper. Cell 48: 745–755.

Bastiani MJ, Raper JA, Goodman CS (1984) Pathfinding by neuronal growth cones in grasshopper embryos. III. Selective affinity of the G growth cone for the P cells within the A/P fascicle. J Neurosci 4: 2311–2328.

Bate M (1998). Making sense of behavior. Int J Dev Biol 42(3): 507–509.

Bayliss DA, Viana F, Bellingham MC, Berger AJ (1994) Characteristics and postnatal development of a hyperpolarization-activated inward current in rat hypoglossal motoneurons in vitro. J Neurophysiol 71: 119–128.

Bear MF, Singer W (1986) Modulation of visual cortical plasticity by acetylcholine and noradrenaline. Nature 320: 172–176.

Beato M, Herrlich P, Schütz G (1995) Steroid hormone receptors: many actors in search of a plot. Cell 83: 851–857.

Becker C-M, Hoch W, Betz H (1988) Glycine receptor heterogeneity in rat spinal cord during postnatal development. EMBO J 7: 3717–3726.

Beggs HE, Soriano P, Maness PF (1994) NCAM-dependent neurite outgrowth is inhibited in neurons from Fyn-minus mice. J Cell Biol 127: 825–833.

Bekoff A (1976). Ontogeny of leg motor output in the chick embryo: a neural analysis. Brain Res 106(2): 271–291.

Bekoff A, Kauer JA (1984). Neural control of hatching: fate of the pattern generator for the leg movements of hatching in post-hatching chicks. J Neurosci 4(11): 2659–2666.

Bekoff A, Sabichi AL (1987). Sensory control of the initiation of hatching in chicks: effects of a local anesthetic injected into the neck. Dev Psychobiol 20(5): 489–495.

Belluardo N, Westerblad H, Mudo G, Casabona A, Bruton J, Caniglia G, Pastoris O, Grassi F, Ibanez CF (2001) Neuromuscular junction disassembly and muscle fatigue in mice lacking neurotrophin-4. Mol Cell Neurosci 18: 56–67.

Belluscio L, Gold GH, Nemes A, Axel R (1998) Mice deficient in G(olf) are anosmic. Neuron 20: 69–81.

Ben-Ari Y, Cherubini E, Corradetti R, Gaiarsa JL (1989) Giant synaptic potentials in immature rat CA3 hippocampal neurones. J Physiol 416: 303–325.

Ben-Ari Y, Cherubini E, Krnjevic K (1988) Changes in voltage dependence of NMDA currents during development. Neurosci Lett 94: 88–92.

Bennett MR, Pettigrew AG (1974) The formation of synapses in reinnervated and cross-reinnervated striated muscle during development. J Physiol (Lond) 241: 547–573.

Ben-Shahar Y, Robichon A, Sokolowski MB, Robinson GE (2002) Influence of gene action across different time scales on behavior. Science 296: 741–744.

Bentley D, Caudy M (1983) Pioneer axons lose directed growth after selective killing of guidepost cells. Nature 304: 62–65.

Bentley D, Toroian-Raymond A (1986) Disoriented pathfinding by pioneer neurone growth cones deprived of filopodia by cytochalasin treatment. Nature 323: 712–715.

Benzer S (1971). From the gene to behavior. JAMA 218(7): 1015–1022.

Berardi N, Pizzorusso T, Maffei L (2000) Critical periods during sensory development. Curr Opin Neurobiol 10: 138–145.

Berg DK, Hall ZW (1975) Increased extrajunctional acetylcholine sensitivity produced by chronic postsynatpic neuromuscular blockade. J Physiol 244: 659–676.

Bevan S, Steinbach JH (1977) The distribution of alpha-bungarotoxin binding sites on mammalian skeletal muscle developing in vivo. J Physiol 267: 195–213.

Beyer C, Kolbinger W, Froehlich U, Pilgrim C, Reisert I (1992) Sex differences of hypothalamic prolactin cells develop independently of the presence of sex steroids. Brain Res 593: 253–256.

Bhat KM (1999) Segment polarity genes in neuroblast formation and identity specification during Drosophila neurogenesis. Bioessays 21: 472–485.

Bhattacharyya A, Watson F, Bradlee T, Pomeroy S, Stiles C, Segal R (1997) Trk receptors function as rapid retrograde signal carriers in the adult nervous system. J Neurosci 17: 7007–7016.

Biederer T, Sara Y, Mozhayeva M, Atasoy D, Liu X, Kavalali ET, Sudhof TC (2002) SynCAM, a synaptic adhesion molecule that drives synapse assembly. Science 297: 1525–1531.

Biggin MD, McGinnis W. Regulation of segmentation and segmental identity by Drosophila homeoproteins: the role of DNA binding in functional activity and specificity. Development. 1997 Nov; 124(22): 4425–4433.

Birch EE, Gwiazda J, Bauer Jr JA, Naegele J, Held R (1983) Visual acuity and its meridional variations in children aged 7–60 months. Vision Res 23: 1019–1024.

Birren SJ, Verdi JM, Anderson DJ (1992) Membrane depolarization induces p140trk and NGF responsiveness, but not p75LNGFR, in MAH cells. Science 257: 395–397.

Bishop KM, Goudreau G, O'Leary DD (2000). Regulation of area identity in mammalian neocortex by Emx2 and Pax6. Science 288: 344–349.

Bittman K, Owens DF, Kriegstein AR, LoTurco JJ. Cell coupling and uncoupling in the ventricular zone of developing neocortex. J Neurosci. 1997 Sep 15;17(18): 7037–7044.

Bixby JL, Harris WA (1991) Molecular mechanisms of axon growth and guidance. Annu Rev Cell Biol 7: 117–159.

Bixby JL, Pratt RS, Lilien J, Reichardt LF (1987) Neurite outgrowth on muscle cell surfaces involves extracellular matrix receptors as well as Ca2+-dependent and -independent cell adhesion molecules. Proc Natl Acad Sci U S A 84: 2555–2559.

Bixby JL, Spitzer NC (1982) The appearance and development of chemosensitivity in Rohon-Beard neurones of the Xenopus spinal cord. J Physiol 330: 513–536.

Blair LAC (1983) The timing of protein synthesis required for the development of the sodium action potential in embryonic spinal neurons. J Neurosci 3: 1430–1436.

Blakely RD, Clark JA, Pacholczyk T, Amara SG (1991) Distinct, developmentally regulated brain mRNAs direct the synthesis of neurotransmitter transporters. J Neurochem 56: 860–871.

Blakemore C, Cooper GF (1970) Development of the brain depends on the visual environment. Nature 228: 477–478.

Blakemore C, Van Sluyters RC (1975) Innate and environmental factors in the development of the kitten's visual cortex. J Physiol 248: 663–716.

Blaschke AJ, Weiner JA, Chun J (1998) Programmed cell death is a universal feature of embryonic and postnatal neuroproliferative regions throughout the central nervous system. J Comp Neurol 396: 39–50.

Blaschke AJ, Staley K, Chun J (1996) Widespread programmed cell death in proliferative and postmitotic regions of the fetal cerebral cortex. Development 122: 1165–1174.

Blasdel G, Obermayer K, Kiorpes L (1995) Organization of ocular dominance and orientation columns in the striate cortex of neonatal macaque monkeys. Vis Neurosci 12: 589–603.

Bliss TVP, Lømo T (1973) Long-lasting potentiation of synaptic transmission in the dentate area of the anesthtized rabbit following stimulation of the perforant path. J Physiol 232: 331–356.

Bloch RJ, Steinbach JH (1981) Reversible loss of acetylcholine receptor clusters at the developing rat neuromuscular junction. Dev Biol 81: 386–391.

Bock J, Braun K (1999a) Blockade of N-methyl-D-aspartate receptor activation suppresses learning-induced synaptic elimination. Proc Natl Acad Sci USA 96: 2485–2490.

Bock J, Braun K (1999b) Filial imprinting in domestic chicks is associated with spine pruning in the associative area, dorsocaudal neostriatum. Eur J Neurosci 11: 2566–2570.

Bock J, Wolf A, Braun K (1996) Influence of the N-methyl-D-aspartate receptor antagonist DL-2-amino-5-phosphonovaleric acid on auditory filial imprinting in the domestic chick. Neurobiol Learn Mem 65: 177–188.

Boettiger CA, Doupe AJ (2001) Developmentally restricted synaptic plasticity in a songbird nucleus required for song learning. Neuron 31: 809–818.

Bolz J, Castellani V, Mann F, Henke-Fahle S (1996) Specification of layer-specific connections in the developing cortex. Prog Brain Res 108: 41–54.

Bonds AB (1979) Development of orientation tuning in the visual cortex of kittens. In: Developmental Neurobiology of Vision (Freeman RD, ed), Plenum Press: New York, pp. 31–41.

Bonni A, Brunet A, West AE, Datta SR, Takasu MA, Greenberg ME (1999) Cell survival promoted by the Ras-MAPK signaling pathway by transcription-dependent and independent mechanisms. Science 286: 1358–1362.

Bonni A, Frank DA, Schindler C, Greenberg ME (1993) Characterization of a pathway for ciliary neurotrophic factor signaling to the nucleus. Science 262: 1575–1579.

Bonni A, Sun Y, Nadal-Vicens M, Bhatt A, Frank DA, Rozovsky I, Stahl N, Yancopoulos GD, Greenberg ME. Regulation of gliogenesis in the central nervous system by the JAK-STAT signaling pathway. Science. 1997 Oct 17;278(5337): 477–483.

Borasio GD, John J, Wittinghofer A, Barde YA, Sendtner M, Heumann R (1989) ras p21 protein promotes survival and fiber outgrowth of cultured embryonic neurons. Neuron 2: 1087–1096.

Born DE, Rubel EW (1985) Afferent influences on brain stem auditory nuclei of the chicken: neuron number and size following cochlea removal. J Comp Neurol 22: 435–445.

Born DE, Rubel EW (1988) Afferent influences on brain stem auditory nuclei of the chicken: presynaptic action potentials regulate protein synthesis in nucleus magnocellularis neurons. J. Neurosci. 8: 901–919.

Bose CM, Qiu D, Bergamaschi A, Gravante B, Bossi M, Villa A, Rupp F, Malgaroli A (2000) Agrin controls synaptic differentiation in hippocampal neurons. J Neurosci 20: 9086–9095.

Bottjer SW, Miesner EA, Arnold AP (1984) Forebrain lesions disrupt development but not maintenance of song in passerine birds. Science 224: 901–903.

Bouwmeester T, Kim S, Sasai Y, Lu B, De Robertis EM. Cerberus is a head-inducing secreted factor expressed in the anterior endoderm of Spemann's organizer. Nature. 1996 Aug 15;382(6592): 595–601.

Bovolenta P, Mason C (1987) Growth cone morphology varies with position in the developing mouse visual pathway from retina to first targets. J Neurosci 7: 1447–1460.

Bowe MA, Deyst KA, Leszyk JD, Fallon JR (1994) Identification and purification of an agrin receptor from Torpedo postsynaptic membranes: A heteromeric complex related to the dystroglycans. Neuron 12: 1173–1180.

Bowe MA, Nadler JV (1990) Developmental increase in the sensitivity to magnesium of NMDA receptors on CA1 hippocampal pyramidal cells. Dev Brain Res 56: 55–61.

Bradbury EJ, Moon LD, Popat RJ, King VR, Bennett GS, Patel PN, Fawcett JW, McMahon SB (2002) Chondroitinase ABC promotes functional recovery after spinal cord injury. Nature 416: 636–640.

Bradke F, Dotti CG (1999) The role of local actin instability in axon formation. Science 283: 1931–1934.

Bradley P, Berry M (1978) The Purkinje cell dendritic tree in mutant mouse cerebellum. A quantitative Golgi study of Weaver and Staggerer mice. Brain Res 142: 135–141.

Brainard MS, Knudsen EI (1993) Experience-dependent plasticity in the inferior colliculus: A site for visual calibration of the neural representation of auditory space in the barn owl. J Neurosci 13: 4589–4608.

Braisted JE, McLaughlin T, Wang HU, Friedman GC, Anderson DJ, O'Leary D D (1997) Graded and lamina-specific distributions of ligands of EphB receptor tyrosine kinases in the developing retinotectal system. Dev Biol 191: 14–28.

Brand M, Campos-Ortega JA (1988). Two groups of interrelated genes regulate early neurogenesis in Drosophila melanogaster. Roux's Arch Dev Biol 197, 457–470.

Brandt P, Neve RL (1992) Expression of plasm a membrane calcium pumping ATPase mRNAs in developing rat brain and adult brain subregions: evidence for stage-specific expression. J Neuroschem 59: 1566–1569.

Bray D (1979) Mechanical tension produced by nerve cells in tissue culture. J Cell Sci 37: 391–410.

Bray D, Thomas C, Shaw G (1978) Growth cone formation in ultures of sensory neurons. Proc Natl Acad Sci U S A 75: 5226–5229.

Breedlove SM, Arnold AP (1983) Hormonal control of a developing neuromuscular system. I. Complete Demasculinization of the male rat spinal nucleus of the bulbocavernosus using the antiandrogen flutamide. J Neurosci 3: 417–423.

Bregman BS, Goldberger ME (1983) Infant lesion effect: III. Anatomical correlates of sparing and recovery of function after spinal cord damage in newborn and adult cats. Brain Res 285: 137–154.

Bregman BS, Kunkel-Bagden E, Schnell L, Dai HN, Gao D, Schwab ME (1995) Recovery from spinal cord injury mediated by antibodies to neurite growth inhibitors. Nature 378: 498–501.

Brennan CA, Moses K (2000) Determination of Drosophila photoreceptors: timing is everything. Cell Mol Life Sci 57: 195–214.

Brenner S. The genetics of Caenorhabditis elegans. Genetics. 1974 May;77(1): 71–94.

Brenowitz EA (1991) Altered perception of species-specific song by female birds after lesions of a forebrain nucleus. Science 251: 303–305.

Brenowitz EA, Arnold AP (1986) Interspecific comparisons of the size of neural song control regions and song complexity in duetting birds: evolutionary implications. J Neurosci 6: 2875–2879.

Britsch S, Goerich DE, Riethmacher D, Peirano RI, Rossner M, Nave KA, Birchmeier C, Wegner M (2001) The transcription factor Sox10 is a key regulator of peripheral glial development. Genes Dev 15: 66–78.

Brittis PA, Lu Q, Flanagan JG (2002) Axonal protein synthesis provides a mechanism for localized regulation at an intermediate target. Cell 110: 223–235.

Broadie K, Sink H, Van Vactor D, Fambrough D, Whitington PM, Bate M, Goodman CS (1993) From growth cone to synapse: the life history of the RP3 motor neuron. Dev Suppl: 227–238.

Broadie KS, Bate M (1993a) Development of the embryonic neuromuscular synapse of Drosophila melanogaster. J Neurosci 13: 144–166.

Broadie KS, Bate M (1993b) Innervation directs receptor synthesis and localization in Drosophila embryo synaptogenesis. Nature 361: 350–353.

Broadie K, Sink H, Van Vactor D, Fambrough D, Whitington PM, Bate M, Goodman CS (1993). From growth cone to synapse: the life history of the RP3 motor neuron. Dev Suppl: 227–238.

Brodmann K (1909) Vergleichende Lokalisationslehre der Großhirnrinde in ihren Prinzipien dargestellt auf Grund des Zellenbaues. Barth, Leipzig.

Bronner-Fraser M, Fraser SE (1988) Cell lineage analysis reveals multipotency of some avian neural crest cells. Nature 335: 161–164.

Bronner-Fraser M, Fraser SE (1991) Cell lineage analysis of the avian neural crest. Development Suppl 2: 17–22.

Brose K, Bland KS, Wang KH, Arnott D, Henzel W, Goodman CS, Tessier-Lavigne M, Kidd T (1999) Slit proteins bind Robo receptors and have an evolutionarily conserved role in repulsive axon guidance. Cell 96: 795–806.

Brown A, Slaughter T, Black MM (1992) Newly assembled microtubules are concentrated in the proximal and distal regions of growing axons. J Cell Biol 119: 867–882.

Brown A, Yates PA, Burrola P, Ortuno D, Vaidya A, Jessell TM, Pfaff SL, O'Leary DD, Lemke G (2000) Topographic mapping from the retina to the midbrain is controlled by relative but not absolute levels of EphA receptor signaling. Cell 102: 77–88.

Brown NL, Patel S, Brzezinski J, Glaser T. Math5 is required for retinal ganglion cell and optic nerve formation. Development. 2001 Jul;128(13): 2497–2508.

Bruckenstein DA, Higgins D (1988a) Morphological differentiation of embryonic rat sympathetic neurons in tissue culture. II. Serum promotes dendritic growth. Dev Biol 128: 337–348.

Bruckenstein DA, Higgins D (1988b) Morphological differentiation of embryonic rat sympathetic neurons in tissue culture. I. Conditions under which neurons form axons but not dendrites. Dev Biol 128: 324–336.

Brunet A, Bonni A, Zigmond MJ, Lin MZ, Juo P, Hu LS, Anderson MJ, Arden KC, Blenis J, Greenberg ME (1999) Akt promotes cell survival by phosphorylating and inhibiting a Forkhead transcription factor. Cell 96: 857–868.

Brunjes PC, Alberts JR (1979) Olfactory stimulation induces filial preferences for huddling in rat pups. J Comp Physiol Psychol 93: 548–555.

Brunso-Bechtold JK, Henkel CK, Linville C (1992) Ultrastructural development of the medial superior olive (MSO) in the ferret. J Comp Neurol 324: 539–556.

Buchanan J, Sun Y-a, Poo M-m (1989) Studies of nerve-muslce interactions in Xenopus cell culture: fine structure of early functional contacts. J Neurosci 9: 1540–1554.

Buck KB, Zheng JQ (2002) Growth cone turning induced by direct local modification of microtubule dynamics. J Neurosci 22: 9358–9367.

Buck L, Axel R (1991) A novel multigene family may encode odorant receptors: a molecular basis for odor recognition. Cell 65: 175–187.

Budnik V, Koh Y-H, Guan B, Hartmann B, Hough C, Woods D, Gorczyca M (1996) Regulation of synapse structure and function by the Drosophila tumor suppressor gene dlg. Neuron 17: 627–640.

Bueker ED (1948) Implantation of tumors in the hind limb field of the embryonic chick and the developmental response of the lumbosacral nervous system. Anat Rec 102: 369–390.

Bunge RP (1975) Changing uses of nerve tissue culture. In: The Nervous System, Vol. 1: The Basic Neurosciences (Tower DB, ed), Raven Press: New York, pp. 31–42.

Burden S (1977a) Development of the neuromusclular junction in the chick embryo: The number, distribution, and stability of acetylcholine receptors. Dev Biol 57: 317–329.

Burden S (1977b) Acetylcholine receptors at the neuromuscular junction: Developmental change in receptor turnover. Dev Biol 61: 79–85.

Burden SJ, Sargent PB, McMahon UJ (1979) Acetylcholine receptors in regenerating muscle accumulate at original synaptic sites in the absence of the nerve. J Cell Biol 82: 412–425.

Burgard EC, Hablitz JJ (1993) Developmental changes in NMDA and Non-NMDA receptor-mediated synaptic potentials in rat neocortex. J Neurophysiol 69: 230–240.

Burr HS (1916) The effects of the removal of the nasal pits in Amblystoma embryos. J Exp Zool 20: 27–57.

Burrone J, O'Byrne M, Murthy VN (2002) Multiple forms of synaptic plasticity triggered by selective suppression of activity in individual neurons. Nature 420: 414–418.

Burrows RC, Wancio D, Levitt P, Lillien L. Response diversity and the timing of progenitor cell maturation are regulated by developmental changes in EGFR expression in the cortex. Neuron. 1997 Aug;19(2): 251–267.

Busetto G, Buffelli M, Tognana E, Bellico F, Cangiano A (2000) Hebbian mechanisms revealed by electrical stimulation at developing rat neuromuscular junctions. J Neurosci 20: 685–695.

Buszczak M, Segraves WA (2000) Insect metamorphosis: out with the old, in with the new. Curr Biol 10: R830–833.

Cabrera CV, Martinez-Arias A, Bate M. The expression of three members of the achaete-scute gene complex correlates with neuroblast segregation in Drosophila. Cell. 1987 Jul 31;50(3): 425–433.

Cai D, Qiu J, Cao Z, McAtee M, Bregman BS, Filbin MT (2001) Neuronal cyclic AMP controls the developmental loss in ability of axons to regenerate. J Neurosci 21: 4731–4739.

Callahan CA, Muralidhar MG, Lundgren SE, Scully AL, Thomas JB (1995) Control of neuronal pathway selection by a Drosophila receptor protein-tyrosine kinase family member. Nature 376: 171–174.

Calof AL, Reichardt LF (1984) Motoneurons purified by cell sorting respond to two distinct activities in myotube-conditioned medium. Dev Biol 106: 194–210.

Cameron JS, Dryer L, Dryer SE (1999) Regulation of neuronal K(+) currents by target-derived factors: opposing actions of two different isoforms of TGFbeta. Development 126: 4157–4164.

Cameron JS, Dryer L, Dryer SE (2001) beta-Neuregulin-1 is required for the in vivo development of functional Ca2+-activated K+ channels in parasympathetic neurons. Proc Natl Acad Sci USA 98: 2832–2836.

Cameron WE, Nunez-Abades PA, Kerman IA, Hodgson TM (2000) Role of potassium conductances in determining input resistance of developing brain stem motoneurons. J Neurophysiol 84: 2330–2339.

Campanelli JT, Roberds SL, Campbell KP, Scheller RH (1994) A role for dsytrophin-associated glycoproteins and utrophin in agrin-induced AchR clustering. Cell 77: 663–674.

Campbell DS, Holt CE (2001) Chemotropic responses of retinal growth cones mediated by rapid local protein synthesis and degradation. Neuron 32: 1013–1026.

Campbell DS, Holt CE (2003) Apoptotic pathway and MAPKs differentially regulate chemotropic responses of retinal growth cones. Neuron 37: 939–952.

Campbell DS, Regan AG, Lopez JS, Tannahill D, Harris WA, Holt CE (2001) Semaphorin 3A elicits stage-dependent collapse, turning, and branching in Xenopus retinal growth cones. J Neurosci 21: 8538–8547.

Campbell G, Shatz CJ (1992) Synapses formed by identified retinogeniculate axons during segregation of eye input. J Neurosci 12: 1847–1858.

Campenot RB (1977) Local control of neurite development by nerve growth factor. Proc Natl Acad Sci USA 74: 4516–4519.

Campenot RB (1982) Development of sympathetic neurons in compartmentalized cultures. II. Local control of neurite survival by nerve growth factor. Dev Biol 93: 13–21.

Campos JJ, Langer A, Krowitz A (1970) Cardiac responses on the visual cliff in prelocomotor human infants. Science 170: 196–197.

Carmichael L. (1954). The onset and early development of behavior. Manual of Child Psychology. L. Carmichael. New York, Wiley: 60–214.

Carpenter WB (1874) Principles of Mental Physiology. London, King.

Carr VM, Simpson SB (1978) Proliferative and degenerative events in the early development of chick dorsal root ganglia. II. Responses to altered peripheral fields. J Comp Neurol 182: 741–755.

Carruth LL, Reisert I, Arnold AP (2002) Sex chromosome genes directly affect brain sexual differentiation. Nat Neurosci 5: 933–934.

Casaccia-Bonnefil P, Carter BD, Dobrowsky RT, Chao MV (1996) Death of oligodendrocytes mediated by the interaction of nerve growth factor with its receptor p75. Nature 383: 716–719.

Casavant RH, Colbert CM, Dryer SE (2004) A-current expression is regulated by activity but not by target tissues in developing lumbar motoneurons of the chick embryo. J Neurophysiol. 2004 May 2 [Epub ahead of print].

Cash S, Zucker RS, Poo M-m (1996) Spread of synaptic depression mediated by presynaptic cytoplasmic signaling. Science 272: 998–1001.

Castellani V, Bolz J (1997) Membrane-associated molecules regulate the formation of layer-specific cortical circuits. Proc Natl Acad Sci U S A 94: 7030–7035.

Castellani V, Bolz J (1999) Opposing roles for neurotrophin-3 in targeting and collateral formation of distinct sets of developing cortical neurons. Development 126: 3335–3345.

Castellani V, Yue Y, Gao PP, Zhou R, Bolz J (1998) Dual action of a ligand for Eph receptor tyrosine kinases on specific populations of axons during the development of cortical circuits. J Neurosci 18: 4663–4672.

Casticas S, Thanos S, Clarke PGH (1987) Major role for neuronal death during brain development: refinement of topographic connections. Proc Natl Acad Sci USA 84: 8165–8168.

Catalano SM, Messersmith EK, Goodman CS, Shatz CJ, Chedotal A (1998) Many major CNS axon projections develop normally in the absence of semaphorin III. Mol Cell Neurosci 11: 173–182.

Catsicas M, Péquignot Y, Clarke PGH (1992) Rapid onset of neuronal death induced by blockade of either axoplasmic transport or action potentials in afferent fibers during brain development. J Neurosci 12: 4642–4650.

Caudy M, Bentley D (1986) Pioneer growth cone steering along a series of neuronal and non-neuronal cues of different affinities. J Neurosci 6: 1781–1795.

Caudy M, Vassin H, Brand M, Tuma R, Jan LY, Jan YN. daughterless, a Drosophila gene essential for both neurogenesis and sex determination, has sequence similarities to myc and the achaete-scute complex. Cell. 1988 Dec 23;55(6): 1061–1067.

Caviness VS Jr, Rakic P (1978) Mechanisms of cortical development: a view from mutations in mice. Annu Rev Neurosci 1: 297–326.

Cayouette M, Barres BA, Raff M (2003) Importance of intrinsic mechanisms in cell fate decisions in the developing rat retina. Neuron 40: 897–904.

Cepko CL, Austin CP, Yang X, Alexiades M, Ezzeddine D (1996) Cell fate determination in the vertebrate retina. Proc Natl Acad Sci U S A 93: 589–595.

Chalepakis G, Stoykova A, Wijnholds J, Tremblay P, Gruss P. Pax: gene regulators in the developing nervous system. J Neurobiol. 1993 Oct;24(10): 1367–1384.

Chalfie M (1993) Touch receptor development and function in Caenorhabditis elegans. J Neurobiol 24: 1433–1441.

Chalfie M, Au M (1989) Genetic control of differentiation of the Caenorhabditis elegans touch receptor neurons. Science 243: 1027–1033.

Chalfie M, Sulston J (1981) Developmental genetics of the mechanosensory neurons of Caenorhabditis elegans. Dev Biol 82: 358–370.

Chambon P (1995) The molecular and genetic dissection of the retinoid signaling pathway. Recent Prog Horm Res. 50: 317–332.

Chan SS, Zheng H, Su MW, Wilk R, Killeen MT, Hedgecock EM, Culotti JG (1996) UNC-40, a C. elegans homolog of DCC (Deleted in Colorectal Cancer), is required in motile cells responding to UNC-6 netrin cues. Cell 87: 187–195.

Chang EH, Kotak VC, Sanes DH (2003) Long-term depression of synaptic inhibition is expressed postsynaptically in the developing auditory system. J Neurophysiol 90: 1479–1488.

Chao MV (1994) The p75 neurotrophin receptor. J Neurobiol 25: 1373–1385.

Chapman B (2000) Necessity for afferent activity to maintain eye-specific segregation in ferret lateral geniculate nucleus. Science 287: 2479–2482.

Chapman B, Stryker MP (1993) Development of orientation selectivity in ferret visual cortex and effects of deprivation. J Neurosci 13: 5251–5262.

Charpier S, Behrends JC, Triller A, Faber DS, Korn H (1995) Latent inhibitory connections become functional during activity-dependent plasticity. Proc Natl Acad Scie USA 92: 117–120.

Charrier I, Mathevon N, Jouventin P (2001) Mother's voice recognition by seal pups. Nature 412: 873.

Charron F, Stein E, Jeong J, McMahon AP, Tessier-Lavigne M (2003) The morphogen sonic hedgehog is an axonal chemoattractant that collaborates with netrin-1 in midline axon guidance. Cell 113: 11–23.

Cheema ZF, Wade SB, Sata M, Walsh K, Sohrabji F, Miranda RC (1999) Fas/Apo [apoptosis]-1 and associated proteins in the differentiating cerebral cortex: induction of caspase-dependent cell death and activation of NF-kappaB. J Neurosci 19: 1754–1770.

Chen HH, Hippenmeyer S, Arber S, Frank E (2003) Development of the monosynaptic stretch reflex circuit. Curr Opin Neurobiol 13: 96–102.

Chen L, Wang H, Vicini S, Olsen RW (2000) The gamma-aminobutyric acid type A (GABAA) receptor-associated protein (GABARAP) promotes GABAA receptor clustering and modulates the channel kinetics. Proc Natl Acad Sci USA 97: 11557–11562.

Chen QX, Stelzner A, Kay AR, Wong RKS (1990) GABAA receptor function is regulated by phosphorylation in acutely dissociated guinea-pig hippocampal neurones. J Physiol 420: 207–221.

Cheng EH, Kirsch DG, Clem RJ, Ravi R, Kastan MB, Bedi A, Ueno K, Hardwick JM (1997) Conversion of Bcl-2 to a Bax-like death effector by caspases. Science 278: 1966–1968.

Cheng HJ, Nakamoto M, Bergemann AD, Flanagan JG (1995) Complementary gradients in expression and binding of ELF-1 and Mek4 in development of the topographic retinotectal projection map. Cell 82: 371–381.

Chiba A, Shepherd D, Murphey RK (1988) Synaptic rearrangement during postembryonic development in the cricket. Science 240: 901–905.

Chien CB, Rosenthal DE, Harris WA, Holt CE (1993) Navigational errors made by growth cones without filopodia in the embryonic Xenopus brain. Neuron 11: 237–251.

Chitnis A, Henrique D, Lewis J, Ish-Horowicz D, Kintner C (1995) Primary neurogenesis in Xenopus embryos regulated by a homologue of the Drosophila neurogenic gene Delta. Nature 375(6534): 761–766.

Chitnis AB, Patel CK, Kim S, Kuwada JY (1992) A specific brain tract guides follower growth cones in two regions of the zebrafish brain. J Neurobiol 23: 845–854.

Chiu C, Weliky M (2001) Spontaneous activity in developing ferret visual cortex in vivo. J Neurosci 21: 8906–8914.

Cho KO, Hunt CA, Kennedy MB (1992) The rat brain postsynaptic density fraction contains a homolog of the Drosophila discs-large tumor suppressor protein. Neuron 9: 929–942.

Chow I, Cohen MW (1983) Developmental changes in the distribution of acetylcholine receptors in the myotomes of Xenopus laevis. J Physiol 339: 553–571.

Chub N, O'Donovan MJ (1998) Blockade and recovery of spontaneous rhythmic activity after application of neurotransmitter antagonists to spinal networks of the chick embryo. J Neurosci 18(1): 294–306.

Chu-Wang IW, Oppenheim RW (1978) Cell death of motoneurons in the chick embryo spinal cord. II. A quantitative and qualitative analysis of degeneration in the ventral root, including evidence for axon outgrowth and limb innervation prior to cell death. J Comp Neurol 177: 59–85.

Clarke PG, Clarke S (1996) Nineteenth century research on naturally occurring cell death and related phenomena. Anat Embryol (Berl) 193: 81–99.

Clarkson MG, Clifton RK, Swain IU, Perris EE (1989) Stimulus duration and repetition rate influences newborns' head orientation towards sound. Dev Psychobiol 22: 683–705.

Clayton GH, Owens GC, Wolff JS, Smith RL (1998) Ontogeny of cation-Cl- cotransporter expression in rat neocortex. Brain Res Dev Brain Res 109: 281–292.

Clayton NC, Krebs JR (1994) Hippocampal growth and attrition in birds affected by experience. Proc Natl Acad Sci USA 91: 7410–7414.

Clendening B, Hume RI (1990) Cell interactions regulate dendritic morphology and responses to neurotransmitters in embryonic chick sympathetic preganglionic neurons in vitro. J Neurosci 10: 3992–4005.

Cline HT (2001) Dendritic arbor development and synaptogenesis. Curr Opin Neurobiol 11: 118–126.

Cline HT, Constantine-Paton M (1990) NMDA receptor agonist and antagonists alter retinal ganglion cell arbor structure in the developing frog retinotectal projection. J Neurosci 10: 1197–1216.

Cline HT, Debski EA, Constantine-Paton M (1987) N-methyl-D-aspartate receptor antagonist desegregates eye-specific stripes. Proc Natl Acad Sci USA 84: 4342–4345.

Coghill GE (1929) Anatomy and the Problem of Behaviour. London, Cambridge University Press.

Cohen J, Johnson AR (1991) Differential effects of laminin and merosin on neurite outgrowth by developing retinal ganglion cells. J Cell Sci Suppl 15: 1–7.

Cohen MW (1972) The development of neuromuscular connexions in the presence of D-tubocurarine. Brain Res 41: 457–463.

Cohen MW, Anderson MJ, Zorychta E, Weldon PR (1979) Accumulation of acetylcholine receptors at nerve-muscle contacts in culture. Prog Brain Res 49: 335–349.

Cohen MW, Weldon PR (1980) Localization of acetylcholine receptors and synaptic ultrastructure at nerve-muscle contacts in culture: dependence on nerve type. J Cell Biol 86: 388–401.

Cohen NA, Brenman JE, Snyder SH, Bredt DS (1996) Binding of the inward rectifier K+ channel Kir 2.3 to PSD-95 is regulated by protein kinase A phosphorylation. Neuron 17: 759–767.

Cohen S, Levi-Montalcini R (1956) A nerve growth-stimulating factor isolated from snake venom. Proc Natl Acad Sci USA 42: 571–574.

Cohen-Cory S, Fraser SE (1995) Effects of brain-derived neurotrophic factor on optic axon branching and remodelling in vivo. Nature 378: 192–196.

Colamarino SA, Tessier-Lavigne M (1995) The role of the floor plate in axon guidance. Annu Rev Neurosci 18: 497–529.

Colman H, Nabekura J, Lichtman JW (1997) Alterations in synaptic strength preceding axon withdrawal. Science 275: 356–361.

Connold AL, Evers JV, Vrbova G (1986) Effect of low calcium and protease inhibitors on synapse elimination during postnatal development in the rat soleus muscle. Dev Brain Res 28: 99–107.

Connors BW, Ransom BR, Kunis DM, Gutnick MJ (1982) Activity dependent K+ accumulation in the developing rat optic nerve. Science 216: 1341–1343.

Conradi S, Ronnevi L-O (1975) Spontaneous elimination of synapses on cat spinal motoneurons after birth: Do half of the synapses on the cell bodies disappear? Brain Res 92: 505–510.

Cook JE, Rankin EC (1986). Impaired refinement of the regenerated retinotectal projection of the goldfish in stroboscopic light: a quantitative WGA-HRP study. Exp Brain Res 63(2): 421–430.

Cooke BM, Tabibnia G, Breedlove SM (1999) A brain sexual dimorphism controlled by adult circulating androgens. Proc Natl Acad Sci USA 96: 7538–7540.

Corfas G, Fischbach GD (1993) The number of Na+ channels in cultured chick muscle is increased by ARIA, an acetylcholine receptor-inducing activity. J Neurosci 13: 2118–2125.

Cornell RA, Ohlen TV (2000) Vnd/nkx, ind/gsh, and msh/msx: conserved regulators of dorsoventral neural patterning? Curr Opin Neurobiol 10: 63–71.

Corriveau RA, Berg DK (1993) Coexpression of multiple acetylcholine receptor genes in neurons: quantification of transcripts during development. J Neurosci 13: 2662–2671.

Costa E, Auta J, Grayson DR, Matsumoto K, Pappas GD, Zhang X, Guidotti A (2002) GABAA receptors and benzodiazepines: a role for dendritic resident subunit mRNAs. Neuropharmacol 43: 925–937.

Cottrell JR, Dube GR, Egles C, Liu G (2000) Distribution, density, and clustering of functional glutamate receptors before and after

synaptogenesis in hippocampal neurons. J Neurophysiol 84: 1573–1587.

Covault J, Sanes JR (1985) Neural cell adhesion molecule (N-CAM) accumulates in denervated and paralyzed skeletal muscles. Proc Natl Acad Sci USA 82: 4544–4548.

Cragg BG (1975) The development of synapses in the visual system of the cat. J Comp Neurol 160: 147–166.

Craig AM, Banker G (1994) Neuronal polarity. Annu Rev Neurosci 17: 267–310.

Craig AM, Blackstone CD, Huganir RL, Banker G (1993) The distribution of glutamate receptors in cultured rat hippocampal neurons: postsynaptic clustering of AMPA-selective subunits. Neuron 10: 1055–1068.

Crair MC, Gillespie DC, Stryker MP (1998) The role of visual experience in the development of columns in cat visual cortex. Science 279: 566–570.

Crair MC, Horton JC, Antonini A, Stryker MP (2001) Emergence of ocular dominance columns in cat visual cortex by 2 weeks of age. J Comp Neurol 430: 235–249.

Crair MC, Malenka RC (1995) A critical period for long-term potentiation at thalamocortical synapses. Nature 375: 325–328.

Crossley PH, Martinez S, Martin GR. Midbrain development induced by FGF8 in the chick embryo. Nature. 1996 Mar 7;380(6569): 66–68.

Crowder RJ, Freeman RS (1998) Phosphatidylinositol 3-kinase and Akt protein kinase are necessary and sufficient for the survival of nerve growth factor-dependent sympathetic neurons. J Neurosci 18: 2933–2943.

Crowley JC, Katz LC (1999) Development of ocular dominance columns in the absence of retinal input. Nat Neurosci 2: 1125–1130.

Crowley JC, Katz LC (2000) Early development of ocular dominance columns. Science 290: 1321–1324.

Crowley C, Spencer SD, Nishimura MC, Chen KS, Pitts-Meek S, Armanini MP, Ling LH, MacMahon SB, Shelton DL, Levinson AD, Phillips HS (1994) Mice lacking nerve growth factor display perinatal loss of sensory and sympathetic neurons yet develop basal forebrain cholinergic neurons. Cell 76: 1001–1011.

Culotti JG (1994) Axon guidance mechanisms in Caenorhabditis elegans. Curr Opin Genet Dev 4: 587–595.

Cynader M, Berman N, Hein A (1973) Cats reared in stroboscopic illumination: Effects of receptive fields in visual cortex. Proc Natl Acad Sci 70: 1353–1354.

Dahm LM, Landmesser LT (1991) The regulation of synaptogenesis during normal development and following activity blockade. J Neurosci 11: 238–255.

Dai Z, Peng HB (1993) Elevation in presynaptic Ca2+ accompanying initial nerve-muscle contact in tissue culture. Neuron 10: 827–837.

Dalva MB, Takasu MA, Lin MZ, Shamah SM, Hu L, Gale NW, Greenberg ME (2000) EphB receptors interact with NMDA receptors and regulate excitatory synapse formation. Cell 103: 945–956.

Dan Y, Poo M-m (1992) Hebbian depression of isolated neuromuscular synapses in vitro. Science 256: 1570–1573.

D'Arcangelo G, Miao GG, Chen SC, Scares HD, Morgan JI, Curran T. A protein related to extracellular matrix proteins deleted in the mouse mutant reeler. Nature. 1995 Apr 20;374(6524): 719–723.

Dasen JS, Liu JP, Jessell TM (2003) Motor neuron columnar fate imposed by sequential phases of Hox-c activity. Nature 425: 926–933.

Davenport RW, Thies E, Cohen ML (1999) Neuronal growth cone collapse triggers lateral extensions along trailing axons. Nat Neurosci 2: 254–259.

Davenport RW, Dou P, Mills LR, Kater SB (1996). Distinct calcium signaling within neuronal growth cones and filopodia. J Neurobiol 31(1): 1–15.

David S, Aguayo AJ (1981) Axonal elongation into peripheral nervous system bridges after central nervous system injury in adult rats. Science 214: 931–933.

Davies AM, Bandtlow C, Heumann R, Korsching S, Rohrer H, Thoenen H (1987) Timing and site of nerve growth factor synthesis in developing skin in relation to innervation and expression of the receptor. Nature 326: 353–358.

Davies AM, Lee KF, Jaenisch R (1993) p75-deficient trigeminal sensory neurons have an altered response to NGF but not to other neurotrophins. Neuron 11: 565–574.

Davies KE, Tinsley JM, Blake DJ (1995) Molecular analysis of Duchenne muscular dystrophy: past, present, and future. Ann New York Acad Sci 758: 287–296.

Davis AA, Temple S. A self-renewing multipotential stem cell in embryonic rat cerebral cortex. Nature. 1994 Nov 17;372(6503): 263–266.

Davis GW, Schuster CM, Goodman CS (1996) Genetic dissection of structural and functional components of synaptic plasticity. III. CREB is necessary for presynaptic functional plasticity. Neuron 17: 669–679.

Daw NW (1995) Visual Development. Plenum Press: New York.

Daw NW, Fox K, Sato H, Czepita D (1992) Critical period for monocular deprivation in the cat visual cortex. J Neurophysiol 67: 197–202.

De Robertis EM, Sasai Y. A common plan for dorsoventral patterning in Bilateria. Nature. 1996 Mar 7;380(6569): 37–40.

Dean C, Scholl FG, Choih J, DeMaria S, Berger J, Isacoff E, Scheiffele P (2003) Neurexin mediates the assembly of presynaptic terminals. Nat Neurosci 6: 708–716.

Deardorff MA, Tan C, Saint-Jeannet JP, Klein PS. A role for frizzled 3 in neural crest development. Development. 2001 Oct;128(19): 3655–3663.

DeBello WM, Feldman DE, Knudsen EI (2001) Adaptive axonal remodeling in the midbrain auditory space map. J Neurosci 21: 3161–3174.

DeCasper AJ, Fifer WP (1980) Of human bonding: newborns prefer their mothers' voices. Science 208: 1174–1176.

DeCasper AJ, Spence MJ (1986) Prenatal maternal speech influences newborns' perception of speech sounds. Infant Behav Develop 9: 133–150.

DeChiara TM, Bowen DC, Valenzuela DM, Simmons MV, Poueymirou WT, Thomas S, Kinetz E, Compton DL, Rojas E, Park JS, Smith C, DiStefano PS, Glass DJ, Burden SJ, Yancopoulos GD (1996) The receptor tyrosine kinase MuSK is required for neuromuscular junction formation in vivo. Cell 85: 501–512.

DeChiara TM, Vejsada R, Poueymirou WT, Acheson A, Suri C, Conover JC, Friedman B, McClain J, Pan L, Stahl N, Ip NY, Kato A, Yancopoulos GD (1995) Mice lacking the CNTF receptor, unlike mice lacking CNTF, exhibit profound motor neuron deficits at birth. Cell 83: 313–322.

Decker RS (1976) Influence of thyroid hormones on neuronal death and differentiation in larval Rana pipiens. Dev Biol 49: 101–118.

Decker RS (1977) Lysosomal properties during thyroxine-induced lateral motor column neurogenesis. Brain Res 132: 407–422.

DeFazio RA, Keros S, Quick MW, Hablitz JJ (2000) Potassium-coupled chloride cotransport controls intracellular chloride in rat neocortical pyramidal neurons. J Neurosci 20: 8069–8076.

Dehaene-Lambertz G, Dehaene S, Hertz-Pannier L (2002) Functional neuroimaging of speech perception in infants. Science 298: 2013–2015.

Deiner MS, Kennedy TE, Fazeli A, Serafini T, Tessier-Lavigne M, Sretavan DW (1997) Netrin-1 and DCC mediate axon guidance locally at the optic disc: loss of function leads to optic nerve hypoplasia. Neuron 19: 575–589.

Deitch JS, Rubel EW (1984) Afferent influences on brain stem auditory nuclei of the chicken: Time course and specificity of dendritic atrophy following deafferentation. J Comp Neurol 229: 66–79.

Dekaban AS, Sadowsky D (1978) Changes in brain weights during the span of human life: relation of brain weights to body heights and body weights. Ann Neurology 4: 345–356.

del Barco Barrantes I, Davidson G, Grone HJ, Westphal H, Niehrs C. Dkk1 and noggin cooperate in mammalian head induction. Genes Dev. 2003 Sep 15;17(18): 2239–2244.

DeLoache JS, Uttal DH, Rosengren KS (2004) Scale errors offer evidence for a perception-action dissociation early in life. Science 304: 1027–1029.

Demas J, Eglen SJ, Wong RO (2003) Developmental loss of synchronous spontaneous activity in the mouse retina is independent of visual experience. J Neurosci 23: 2851–2860.

Dennis MJ, Yip JW (1978) Formation and elimination of foreign synapses on adult salamander muscle. J Physiol 274: 299–310.

Dent EW, Gertler FB (2003) Cytoskeletal dynamics and transport in growth cone motility and axon guidance. Neuron 40: 209–227.

Dent EW, Kalil K (2001) Axon branching requires interactions between dynamic microtubules and actin filaments. J Neurosci 21: 9757–9769.

Dent MA, Raisman G, Lai FA (1996) Expression of type 1 inositol 1,4,5-trisphosphate receptor during axogenesis and synaptic contact in the central and peripheral nervous system of developing rat. Dev 122: 1029–1039.

Desai AR, McConnell SK (2000) Progressive restriction in fate potential by neural progenitors during cerebral cortical development. Development 127: 2863–2872.

Desai C, Garriga G, McIntire SL, Horvitz HR (1988) A genetic pathway for the development of the Caenorhabditis elegans HSN motor neurons. Nature 336: 638–646.

Desai C, Horvitz HR (1989) Caenorhabditis elegans mutants defective in the functioning of the motor neurons responsible for egg laying. Genetics 121: 703–721.

Desai CJ, Gindhart JG Jr, Goldstein LS, Zinn K (1996) Receptor tyrosine phosphatases are required for motor axon guidance in the Drosophila embryo. Cell 84: 599–609.

Desai NS, Cudmore RH, Nelson SB, Turrigiano GG (2002) Critical periods for experience-dependent synaptic scaling in visual cortex. Nat Neurosci 5: 783–789.

Desai NS, Rutherford LC, Turrigiano GG (1999) Plasticity in the intrinsic excitability of cortical pyramidal neurons. Nat Neurosci 2: 515–520.

Desarmenien MG, Spitzer NC (1991) Role of calcium and protein kinase C in development of the delayed rectifier potassium current in Xenopus spinal neurons. Neuron 7: 797–805.

Deshmukh M, Du C, Wang X, Johnson EM Jr (2002) Exogenous smac induces competence and permits caspase activation in sympathetic neurons. J Neurosci 22: 8018–8027.

Detwiler SR (1936) Neuroembryology. An Experimental Study. MacMillan: New York.

Diamond J, Miledi R (1962) A study of fetal and new-born rat muscle fibres. J Physiol (Lond) 162: 393–408.

Dickson BJ (2001) Rho GTPases in growth cone guidance. Curr Opin Neurobiol 11: 103–110.

Dietrich WD, Durham D, Lowry OH, Woolsey TA (1981) Quantitative histochemical effects of whisker damage on single identified cortical barrels in the adult mouse. J Neurosci 1(9): 929–935.

Ding L, Perkel DJ (2004) Long-term potentiation in an avian basal ganglia nucleus essential for vocal learning. J Neurosci 24: 488–494.

Dingwell KS, Holt CE, Harris WA (2000) The multiple decisions made by growth cones of RGCs as they navigate from the retina to the tectum in Xenopus embryos. J Neurobiol 44: 246–259.

Dobkins KR, Anderson CM (2002) Color-based motion processing is stronger in infants than in adults. Psychol Sci 13: 76–80.

Doe CQ. Molecular markers for identified neuroblasts and ganglion mother cells in the Drosophila central nervous system. Development. 1992 Dec;116(4): 855–863.

Doe CQ, Smouse DT (1990) The origins of cell diversity in the insect central nervous system. Semin Cell Biol 1(3): 211–218. Review.

Doherty P, Ashton SV, Moore SE, Walsh FS (1991) Morphoregulatory activities of NCAM and N-cadherin can be accounted for by G protein-dependent activation of L-and N-type neuronal Ca2+ channels. Cell 67: 21–33.

Dorsky RI, Chang WS, Rapaport DH, Harris WA (1997) Regulation of neuronal diversity in the Xenopus retina by Delta signalling. Nature 385: 67–70.

Dorsky RI, Moon RT, Raible DW (2000) Environmental signals and cell fate specification in premigratory neural crest. Bioessays 22: 708–716.

Dourado MM, Brumwell C, Wisgirda ME, Jacob MH, Dryer SE (1994) Target tissues and innervation regulate the characteristics of K+ currents in chick ciliary ganglion neurons developing in situ. J Neurosci 14: 3156–3165.

Drager UC, Hubel DH (1975) Responses to visual stimulation and relationship between visual, auditory, and somatosensory inputs in mouse superior colliculus. J Neurophysiol 38(3): 690–713.

Drager UC, Hubel DH (1978) Studies of visual function and its decay in mice with hereditary retinal degeneration. J Comp Neurol 180(1): 85–114.

Drescher U, Kremoser C, Handwerker C, Loschinger J, Noda M, Bonhoeffer F (1995) In vitro guidance of retinal ganglion cell axons by RAGS, a 25 kDa tectal protein related to ligands for Eph receptor tyrosine kinases. Cell 82: 359–370.

Driever W, Nusslein-Volhard C. A gradient of bicoid protein in Drosophila embryos. Cell. 1988 Jul 1;54(1): 83–93.

Driever W, Nusslein-Volhard C. The bicoid protein determines position in the Drosophila embryo in a concentration-dependent manner. Cell. 1988 Jul 1;54(1): 95–104.

Driscoll M, Chalfie M (1992) Developmental and abnormal cell death in C. elegans. Trends Neurosci 15: 15–19.

Du C, Fang M, Li Y, Li L, Wang X (2000) Smac, a mitochondrial protein that promotes cytochrome c-dependent caspase activation by eliminating IAP proteins. Cell 102: 33–42.

Dubin MW, Stark LA, Archer SM (1986) A role for action potential activity in the developement of neuronal connections in the kitten retinogeniculate pathway. J Neurosci 6: 1021–1036.

Dubinsky JM, Fischbach GD (1990) A role for cAMP in the development of functional neuromuscular transmission. J Neurobiol 21: 414–426.

Dubinsky JM, Loftus DJ, Fischbach GD, Elson EL (1989) Formation of acetylcholine receptor clusters in chick myotubes: migration or new insertion? J Cell Biol 109: 1733–1743.

Dubnau J, Chiang AS, Grady L, Barditch J, Gossweiler S, McNeil J, Smith P, Buldoc F, Scott R, Certa U, Broger C, Tully T (2003) The staufen/pumilio pathway is involved in Drosophila long-term memory. Curr Biol 13: 286–296.

Dubnau J, Tully T (1998) Gene discovery in Drosophila: new insights for learning and memory. Annu Rev Neurosci 21: 407–444.

Duboule D, Morata G. Colinearity and functional hierarchy among genes of the homeotic complexes. Trends Genet. 1994 Oct;10(10): 358–364.

Duchen MR (2000) Mitochondria and calcium: from cell signalling to cell death. J Physiol 529: 57–68.

Dudek SM, Friedlander MJ (1996) Developmental down-regulation of LTD in cortical layer IV and its independence of modulation by inhibition. Neuron 16: 1097–1106.

Duffy FH, Burchfiel JL, Conway JL (1976) Bicuculline reversal of deprivation amblyopia in the cat. Nature 260: 256–257.

Dulabon L, Olson EC, Taglienti MG, Eisenhuth S, McGrath B, Walsh CA, Kreidberg JA, Anton ES. Reelin binds alpha3betal integrin and inhibits neuronal migration. Neuron. 2000 Jul 27(1): 33–44.

Dumoulin A, Triller A, Dieudonne S (2001) IPSC kinetics at identified GABAergic and mixed GABAergic and glycinergic synapses onto cerebellar Golgi cells. J Neurosci 21: 6045–6057.

Duorado MM, Dryer SE (1992) Changes in the electrical properties of chick ciliary ganglion neurones during embryonic development. J Physiol 449: 411–428.

Durand GM, Kovalchuk Y, Konnerth A (1996) Long-term potentiation and functional synapse induction in developing hippocampus. Nature 381: 71–75.

Duxson MJ (1982) The effect of post synaptic block on the development of the neuromuscular junction in postnatal rats. J Neurocytol 11: 395–408.

Dyer AB, Gottlieb G (1990) Auditory basis of maternal attachment in ducklings (Anas platyrhynchos) under simulated naturalistic imprinting conditions. J Comp Psychol 104: 190–194.

Dyer AB, Lickliter R, Gottlieb G (1989) Maternal and peer imprinting in mallard ducklings under experimentally simulated natural social conditions. Dev Psychobiol 22: 463–475.

Dyson SE, Jones DG (1980) Quantitation of terminal parameters and their interrelationships in maturing central synapses: A perspective for experimental studies. Brain Res 183: 43–59.

Eagleson GW, Harris WA. Mapping of the presumptive brain regions in the neural plate of Xenopus laevis. J Neurobiol. 1990 Apr;21(3): 427–440.

Eagleson KL, Pimenta AF, Burns MM, Fairfull LD, Cornuet PK, Zhang L, Levitt P (2003) Distinct domains of the limbic system-associated membrane protein (LAMP) mediate discrete effects on neurite outgrowth. Mol Cell Neurosci 24: 725–740.

Easter SS Jr., Burrill J, Marcus RC, Ross LS, Taylor JS, Wilson SW (1994) Initial tract formation in the vertebrate brain. Prog Brain Res 102: 79–93.

Easter SS, Jr., Stuermer CA (1984) An evaluation of the hypothesis of shifting terminals in goldfish optic tectum. J Neurosci 4: 1052–1063.

Ebens A, Brose K, Leonardo ED, Hanson MG Jr, Bladt F, Birchmeier C, Barres BA, Tessier-Lavigne M (1996) Hepatocyte growth factor/scatter factor is an axonal chemoattractant and a neurotrophic factor for spinal motor neurons. Neuron 17: 1157–1172.

Eccles JC, Krnjevic K, Miledi R (1959) Delayed effects of peripheral severance of afferent nerve fibres on the efficacy of their central synapses. J Physiol 145: 204–220.

Edelman GM (1984) Modulation of cell adhesion during induction, histogenesis, and perinatal development of the nervous system. Annu Rev Neurosci 7: 339–377.

Edmondson JC, Hatten ME. Glial-guided granule neuron migration in vitro: a high-resolution time-lapse video microscopic study. J Neurosci. 1987 Jun;7(6): 1928–1934.

Ehlers MD, Kaplan DR, Price DL, Koliatsos VE (1995) NGF-stimulated retrograde transport of TrkA in the mammalian nervous system. J Cell Biol 130: 149–156.

Ehrlich I, Lohrke S, Friauf E (1999) Shift from depolarizing to hyperpolarizing glycine action in rat auditory neurones is due to age-dependent Cl-regulation. J Physiol 520: 121–137.

Eisen JS (1991) Determination of primary motoneuron identity in developing zebrafish embryos. Science 252: 569–572.

Eisen JS, Melancon E (2001) Interactions with identified muscle cells break motoneuron equivalence in embryonic zebrafish. Nat Neurosci 4: 1065–1070.

Elkins T, Zinn K, McAllister L, Hoffmann FM, Goodman CS (1990) Genetic analysis of a Drosophila neural cell adhesion molecule: interaction of fasciclin I and Abelson tyrosine kinase mutations. Cell 60: 565–575.

Elmariah SB, Crumling MA, Parsons TD, Balice-Gordon RJ (2004) Postsynaptic TrkB-mediated signaling modulates excitatory and inhibitory neurotransmitter receptor clustering at hippocampal synapses. J Neurosci 24: 2380–2393.

ElShamy WM, Linnarsson S, Lee KF, Jaenisch R, Ernfors P (1996) Prenatal and postnatal requirements of NT-3 for sympathetic neuroblast survival and innervation of specific targets. Development 122(2): 491–500.

Ernfors P, Lee K-F, Jaenisch R (1994) Mice lacking brain-derived neurotrophic factor develop with sensory deficits. Nature 368: 147–150.

Ernfors P, Van De Water T, Loring J, Jaenisch R (1995) Complementary roles of BDNF and NT-3 in vestibular and auditory development. Neuron 14: 1153–1164.

Ernst M (1926) Ueber Untergang von Zellen während der normalen Entwicklung bei Wirbeltieren. Z Anat Entw Gesch 79: 228–262.

Ernstrom GG, Chalfie M (2002) Genetics of sensory mechanotransduction. Annu Rev Genet 36: 411–453.

Erskine L, Williams SE, Brose K, Kidd T, Rachel RA, Goodman CS, Tessier-Lavigne M, Mason CA (2000) Retinal ganglion cell axon guidance in the mouse optic chiasm: expression and function of robos and slits. J Neurosci 20: 4975–4982.

Eswaran H, Wilson J, Preissl H, Robinson S, Vrba J, Murphy P, Rose D, Lowery C (2002) Magnetoencephalographic recordings of visual evoked brain activity in the human fetus. Lancet 360: 779–780.

Everitt AB, Luu T, Cromer B, Tierney ML, Birnir B, Olsen RW, Gage PW (2004) Conductance of recombinant GABA (A) channels is increased in cells co-expressing GABA(A) receptor-associated protein. J Biol Chem 279: 21701–21706.

Evers J, Laser M, Sun Y-a, Xie Z-p, Poo M-m (1989) Studies of nerve-muscle interactions in Xenopus cell culture: analysis of early synaptic currents. J Neurosci 9: 1523–1539.

Fagiolini M, Fritschy JM, Low K, Mohler H, Rudolph U, Hensch TK (2004) Specific GABAA circuits for visual cortical plasticity. Science 303: 1681–1683.

Fahrbach SE, Choi MK, Truman JW (1994) Inhibitory effects of actinomycin D and cycloheximide on neuronal death in adult Manduca sexta. J Neurobiol 25: 59–69.

Fallon JR, Gelfman CE (1989) Agrin-related molecules are concentrated at acetylcholin receptor clusters in normal and aneural developing muscle. J Cell Biol 108: 1527–1535.

Falls DL, Rosen KM, Corfas G, Lane WS, Fischbach GD (1993) ARIA, a protein that stimulates acetylcholine receptor synthesis, is a member of the neu ligand family. Cell 72: 801–815.

Fambrough D, Goodman CS (1996) The Drosophila beaten path gene encodes a novel secreted protein that regulates defasciculation at motor axon choice points. Cell 87: 1049–1058.

Fannon AM, Colman DR (1996) A model for central synaptic junctional complex formation based on the differential adhesive specificities of the cadherins. Neuron 17: 423–434.

Farinas I, Yoshida CK, Backus C, Reichardt LF (1996) Lack of Neurotrophin-3 Results in Death of Spinal Sensory Neurons and Premature Differentiation of Their Precursors. Neuron 17: 1065–1078.

Fatt P, Katz B (1951) Spontaneous subthreshold activity at motor nerve endings. J Physiol (London) 117: 109–128.

Fehon RG, Kooh PJ, Rebay I, Regan CL, Xu T, Muskavitch MA, Artavanis-Tsakonas S. Molecular interactions between the protein products of the neurogenic loci Notch and Delta, two EGF-homologous genes in Drosophila. Cell. 1990 May 4;61(3): 523–534.

Feinstein P, Mombaerts P (2004) A contextual model for axonal sorting into glomeruli in the mouse olfactory system. Cell 117: 817–831.

Feldheim DA, Kim YI, Bergemann AD, Frisen J, Barbacid M, Flanagan JG (2000) Genetic analysis of ephrin-A2 and ephrin-A5 shows their requirement in multiple aspects of retinocollicular mapping. Neuron 25: 563–574.

Feldheim DA, Nakamoto M, Osterfield M, Gale NW, DeChiara TM, Rohatgi R, Yancopoulos GD, Flanagan JG (2004) Loss-of-function analysis of EphA receptors in retinotectal mapping. J Neurosci 24: 2542–2550.

Feldheim DA, Vanderhaeghen P, Hansen MJ, Frisen J, Lu Q, Barbacid M, Flanagan JG (1998) Topographic guidance labels in a sensory projection to the forebrain. Neuron 21: 1303–1313.

Feng G, Tintrup H, Kirsch J, Nichol MC, Kuhse J, Betz H, Sanes JR (1998) Dual requirement for gephyrin in glycine receptor clustering and molybdoenzyme activity. Science 282: 1321–1324.

Ferns M, Deiner M, Hall Z (1996) Agrin-induced acetylcholine receptor clustering in mammalian muscle requires tyrosine phosphorylation. J Cell Biol 132: 937–944.

Ferns MJ, Campanelli JT, Hoch W, Scheller RH, Hall Z (1993) The ability of agrin to cluster AChRs depends on alternative splicing and on cell surface proteoglycans. Neuron 11: 491–502.

Ferns MJ, Hoch W, Campinelli JT, Rupp F, Hall ZW, Scheller RH (1992) RNA spicing regulates agrin-mediated acetylcholine receptor clustering activity on cultured myotubes. Neuron 8: 1079–1086.

Ferreira A, Han H-Q, Greengard P, Kosik KS (1995) Suppression of synapsin II inhibits the formation and maintenance of synapses in hippocampal culture. Proc Natl Acad Sci US 92: 9225–9229.

Fertuck HC, Salpeter M (1976) Quantitation of junctional and extrajunctional acetylcholine receptors by electron microscopic autoradiography after 125I–alpha-bungarotoxin binding at mouse neuromuscular junctions. J Cell Biol 69: 144–158.

Ferveur JF, Savarit F, O'Kane CJ, Sureau G, Greenspan RJ, Jallon JM (1997) Genetic feminization of pheromones and its behavioral consequences in Drosophila males. Science 276: 1555–1558.

Ferveur JF, Störtkuhl KF, Stocker RF, Greenspan RJ (1995) Genetic feminization of brain structures and changed sexual orientation in male Drosophila. Science 267: 902–905.

Fiala BA, Joyce JN, Greenough WT (1978) Environmental complexity modulates growth of granule cell dendrites in developing but not adult hippocampus of rats. Exp Neurol 59: 372–383.

Fiala JC, Feinberg M, Popov V, Harris KM (1998) Synaptogenesis via dendritic filopodia in developing hippocampal area CA1. J Neurosci 18: 8900–8911.

Finlay BL, Slattery M (1983) Local differences in the amount of early cell death in neocortex predict adult local specializations. Science 219: 1349–1351.

Finley KD, Taylor BJ, Milstein M, McKeown M (1997) dissatisfaction, a gene involved in sex-specific behavior and neural development of Drosophila melanogaster. Proc Natl Acad Sci USA 94: 913–918.

Fischbach GD, Cohen SA (1973) The distribution of acetylcholine sensitivity over uninnervated and innervated muscle fibres grown in cell culture. Dev Biol 31: 147–162.

Fischer M, Kaech S, Wagner U, Brinkhaus H, Matus A (2000) Glutamate receptors regulate actin-based plasticity in dendritic spines. Nat Neurosci 3: 887–894.

Fishman MC (1996) GAP-43: putting constraints on neuronal plasticity. Perspect Dev Neurobiol 4(2–3): 193–198.

Flanagan JG, Vanderhaeghen P (1998) The ephrins and Eph receptors in neural development. Annu Rev Neurosci 21: 309–345.

Florence SL, Casagrande VA (1990) Development of geniculocortical axon arbors in a primate. Vis Neurosci 5: 291–309.

Florence SL, Taub HB, Kaas JH (1998) Large-scale sprouting of cortical connections after peripheral injury in adult macaque monkeys [see comments]. Science 282(5391): 1117–1121.

Forehand CJ, Purves D (1984) Regional innervation of rabbit ciliary ganglion cells by the terminals of preganglionic axons. J Neurosci 4: 1–12.

Foreman N, Arber M, Savage J (1984) Spatial memory in preschool infants. Dev Psychobiol 17: 129–137.

Forscher P, Kaczmarek LK, Buchanan JA, Smith SJ (1987) Cyclic AMP induces changes in distribution and transport of organelles within growth cones of Aplysia bag cell neurons. J Neurosci 7: 3600–3611.

Fortini ME, Artavanis-Tsakonas S. The suppressor of hairless protein participates in notch receptor signaling. Cell. 1994 Oct 21;79(2): 273–282.

Fournier AE, GrandPre T, Strittmatter SM (2001) Identification of a receptor mediating Nogo-66 inhibition of axonal regeneration. Nature 409: 341–346.

Fox K, Daw N, Sato H, Czepita D (1992) The effect of visual experience on development of NMDA receptor synaptic transmission in kitten visual cortex. J Neurosci 12: 2672–2684.

Frade JM, Barde YA (1999) Genetic evidence for cell death mediated by the nerve growth factor and the neurotrophin receptor p75 in the developing mouse retina and spinal cord. Dev 126: 683–690.

Frade JM, Rodríguez-Tébar A, Barde YA (1996) Induction of cell death by endogenous nerve growth factor through its p75 receptor. Nature 383: 166–168.

Francis DD, Szegda K, Campbell G, Martin WD, Insel T (2003) Epigenetic sources of behavioral differences in mice. Nature Neuroscience 6: 445–446.

Francis NJ, Landis SC (1999) Cellular and molecular determinants of sympathetic neuron development. Annu Rev Neurosci 22: 541–566.

Frank E, Wenner P (1993) Environmental specification of neuronal connectivity. Neuron 10(5): 779–785.

Freeman LM, Watson NV, Breedlove SM (1997) Androgen spares androgen-insensitive motoneurons from apoptosis in the spinal nucleus of the bulbocavernosus in rats. Horm Behav 30: 424–433.

Freeman M (1997) Cell determination strategies in the Drosophila eye. Development 124: 261–270.

Fricke C, Lee JS, Geiger-Rudolph S, Bonhoeffer F, Chien CB (2001) Astray, a zebrafish roundabout homolog required for retinal axon guidance. Science 292: 507–510.

Friedman HV, Bresler T, Garner CC, Ziv NE (2000) Assembly of new individual excitatory synapses: time course and temporal order of synaptic molecule recruitment. Neuron 27: 57–69.

Frisen J, Yates PA, McLaughlin T, Friedman GC, O'Leary DD, Barbacid M (1998) Ephrin-A5 (AL-1/RAGS) is essential for proper retinal axon guidance and topographic mapping in the mammalian visual system. Neuron 20: 235–243.

Fritzsch B, Silos-Santiago I, Bianchi LM, Farinas I (1997) The role of neurotrophic factors in regulating the development of inner ear innervation. Trends Neurosci 20: 159–164.

Fritzsch B, Tessarollo L, Coppola E, Reichardt LF (2004) Neurotrophins in the ear: their roles in sensory neuron survival and fiber guidance. Prog Brain Res 146: 265–278.

Frohner SC, Luetje CW, Scotland PB, Patrick J (1990) The postsynaptic 43K protein clusters muscle nicotinic acetylcholine receptors in Xenopus oocytes. Neuron 5: 403–410.

Fu W-m, Lin J-L (1993) Activation of protein kinase C potentiates postsynaptic acetylcholine response at developing neuromuscular synapses. Brit J Pharmacol 110: 707–712.

Fukata Y, Kimura T, Kaibuchi K (2002) Axon specification in hippocampal neurons. Neurosci Res 43: 305–315.

Fukuda A, Prince DA (1992) Postnatal development of electrogenic sodium pump activity in rat hippocampal pyramidal neurons. Dev Brain Res 65: 101–114.

Fulford J, Vadeyar SH, Dodampahala SH, Moore RJ, Young P, Baker PN, James DK, Gowland PA (2003) Fetal brain activity in response to a visual stimulus. Hum Brain Mapp. 20: 239–245.

Fulford J, Vadeyar SH, Dodampahala SH, Ong S, Moore RJ, Baker PN, James DK, Gowland P (2004) Fetal brain activity and hemodynamic response to a vibroacoustic stimulus. Hum Brain Mapp 22: 116–121.

Funakoshi H, Belluardo N, Arenas E, Yamamoto Y, Casabona A, Persson H, Ibanez CF (1995) Muscle-derived neurotrophin-4 as an activity-dependent trophic signal for adult motor neurons. Science 268: 1495–1499.

Funte LR, Haydon PG (1993) Synaptic target contact enhances presynaptic calcium influx by activating cAMP-dependent protein kinase during synaptogenesis. Neuron 10: 1069–1078.

Gagliardini V, Fernandez PA, Lee RK, Drexler HC, Rotello RJ, Fishman MC, Yuan J (1994) Prevention of vertebrate neuronal death by the crmA gene. Science 263: 826–828.

Gähwiler BH, Thompson SM, Audinat E, Robertson RT (1991) Organotypic slice cultures of nerual tissue. In: Culturing Nerve Cells. (Banker G, Goslin K, eds), MIT Press: Cambridge, pp. 379–411.

Gaiarsa JL, Caillard O, Ben-Ari Y (2002) Long-term plasticity at GABAergic and glycinergic synapses: mechanisms and functional significance. Trends Neurosci 25: 564–570.

Galli L, Maffei L (1988) Spontaneous impulse activity of rat retinal ganglion cells in prenatal life. Science 242: 90–91.

Gally C, Bessereau JL (2003) GABA is dispensable for the formation of junctional GABA receptor clusters in Caenorhabditis elegans. J Neurosci 23: 2591–2599.

Ganguly K, Schinder AF, Wong ST, Poo M-m (2001) GABA itself promotes the developmental switch of neuronal GABAergic responses from excitation to inhibition. Cell 105: 521–532.

Gao XB, van den Pol AN (2000) GABA release from mouse axonal growth cones. J Physiol 523: 629–637.

Garaschuk O, Linn J, Eilers J, Konnerth A (2000) Large-scale oscillatory calcium waves in the immature cortex. Nat Neurosci 3: 452–459.

Garcia-Castro MI, Marcelle C, Bronner-Fraser M. Ectodermal Wnt function as a neural crest inducer. Science. 2002 Aug 2;297(5582): 848–851.

Garcia-Verdugo JM, Doetsch F, Wichterle H, Lim DA, Alvarez-Buylla A. Architecture and cell types of the adult subventricular zone: in search of the stem cells. J Neurobiol. 1998 Aug;36(2): 234–248.

Gardette R, Listerud MD, Brussaard AB, Role LW (1991) Developmental changes in transmitter sensitivity and synaptic transmission in embryonic chicken sympathetic neurons innervated in vitro. Dev Biol 147: 83–95.

Garthwaite G, Yamini B Jr, Garthwaite J (1987) Selective loss of Purkinje and granule cell responsiveness to N-methyl-D-aspartate in rat cerebellum during development. Brain Res 433: 288–292.

Gautam M, Noakes PG, Moscoso L, Rupp F, Scheller RH, Merlie JP, Sanes JR (1996) Defective neuromuscular synaptogenesis in agrin-deficient mutant mice. Cell 85: 525–535.

Gautam M, Noakes PG, Mudd J, Nichol M, CHu GC, Sanes JR, Merlie JP (1995) Failure of postsynaptic specialization to develop at neruomuscular junctions of rapsyn-deficient mice. Nature 377: 232–236.

Gavalas A, Ruhrberg C, Livet J, Henderson CE, Krumlauf R. Neuronal defects in the hindbrain of Hoxa1, Hoxb1 and Hoxb2 mutants reflect regulatory interactions among these Hox genes. Development. 2003 Dec;130(23): 5663–5679.

Gaze RM, Keating MJ, Ostberg A, Chung SH (1979) The relationship between retinal and tectal growth in larval Xenopus: implications for the development of the retino-tectal projection. J Embryol Exp Morphol 53: 103–143.

Gee SH, Montanaro F, Lindenbaum MH, Carbonetto S (1994) Dystroglycin-alpha, a dystrophin-associated glycoprotein, is a functional agrin receptor. Cell 77: 675–686.

Gehring WJ. Exploring the homeobox. Gene. 1993 Dec 15;135(1–2): 215–221.

Geng Y, Yu Q, Sicinska E, Das M, Bronson RT, Sicinski P. Deletion of the p27Kip1 gene restores normal development in cyclin D1-deficient mice. Proc Natl Acad Sci U S A. 2001 Jan; 2;98(1): 194–199.

Ghosh A, Carnahan J, Greenberg ME (1994) Requirement for BDNF in activity-dependent survival of cortical neurons. Science 263: 1618–1623.

Ghosh A, Greenberg ME. Distinct roles for bFGF and NT-3 in the regulation of cortical neurogenesis. Neuron 1995 Jul;15(1): 89–103.

Gilbert SF, Raunio AM (1997) Embryology: Constructing the Organism. Sinauer Associates, Inc. Publishers, Sunderland, MA.

Gillespie SKH, Balasubramanian S, Fung ET, Huganir RL (1996) Rapsyn clusters and activates the synapse-specific receptor tyrosine kinase MuSK. Neuron 16: 953–962.

Glanzman DL, Kandel ER, Schacher S (1991) Target-dependent morphological segregation of Aplysia sensory outgrowth in vitro. Neuron 7: 903–913.

Glass DJ, Bowen DC, Stitt TN, Radziejewski C, Bruno J, Ryan TE, Gies DR, Shah S, Mattsson K, Burden SJ, DiStefano PS, Valenzuela DM, DeChiara TM, Yancopoulos GD (1996) Agrin acts via a MuSK receptor complex. Cell 85: 513–523.

Glavic A, Gomez-Skarmeta JL, Mayor R. The homeoprotein Xiro1 is required for midbrain-hindbrain boundary formation. Development. 2002 Apr;129(7): 1609–1621.

Glebova NO, Ginty DD (2004) Heterogeneous requirement of NGF for sympathetic target innervation in vivo. J Neurosci 24: 743–751.

Glinka A, Wu W, Delius H, Monaghan AP, Blumenstock C, Niehrs C. Dickkopf-1 is a member of a new family of secreted proteins and functions in head induction. Nature. 1998 Jan 22;391(6665): 357–362.

Glinka A, Wu W, Onichtchouk D, Blumenstock C, Niehrs C. Head induction by simultaneous repression of Bmp and Wnt signalling in Xenopus. Nature. 1997 Oct 2;389(6650): 517–519.

Globus A, Scheibel AB (1966) Loss of dendritic spines as an index of presynaptic terminal patterns. Nature (Lond) 212: 463–465.

Glover JC, Petursdottir G, Jansen JKS (1986) Fluorescent dextran-amines used as axonal tracers in the nervous system of the chicken embryo. J Neurosci Meth 18: 243–254.

Godfrey EW, Nitkin RM, Wallace BG, Rubin LL, McMahon UJ (1984) Components of Torpedo electric organ and muscle that cause aggregation of acetylcholine receptors in cultured muscle cells. J Cell Biol 99: 615–627.

Godsave SF, Slack JM. Clonal analysis of mesoderm induction in Xenopus laevis. Dev Biol. 1989 Aug;134(2): 486–490.

Goldberg DJ, Wu DY (1996) Tyrosine phosphorylation and protrusive structures of the growth cone. Perspect Dev Neurobiol 4: 183–192.

Goldman D, Carlson BM, Staple J (1991) Induction of adult-type nicotinic acetylcholine receptor gene expression in noninnervated regenerating muscle. Neuron 7: 649–6458.

Gomez TM, Robles E, Poo M, Spitzer NC (2001) Filopodial calcium transients promote substrate-dependent growth cone turning. Science 291: 1983–1987.

Gonzalez M, Ruggiero FP, Chang Q, Shi YJ, Rich MM, Kraner S, Balice-Gordon RJ (1999) Disruption of Trkb-mediated signaling induces disassembly of postsynaptic receptor clusters at neuromuscular junctions. Neuron 24: 567–583.

Goodearl AD, Yee AG, Sandrock AW Jr, Corfas G, Fischbach GD (1995) ARIA is concentrated in the synaptic basal lamina of the developing chick neuromuscular junction. J Cell Biol 130: 1423–1434.

Goodfellow PN, Lovell-Badge R (1993) SRY and sex determination in mammals. Ann Rev Genet 27: 71–92.

Goodman CS (1996) Mechanisms and molecules that control growth cone guidance. Annu Rev Neurosci 19: 341–377.

Goodman CS, Raper JA, Chang S, Ho R (1983) Grasshopper growth cones: divergent choices and labeled pathways. Prog Brain Res 58: 283–304.

Goodrich LV, Milenkovic L, Higgins KM, Scott MP. Altered neural cell fates and medulloblastoma in mouse patched mutants. Science. 1997 Aug 22;277(5329): 1109–1113.

Gordon T, Perry R, Tuffery AR, Vrbova G (1974) Possible mechanisms determining synapses formation in developing skeletal muscles of the chick. Cell Tissue Res 155: 13–25.

Gordon-Weeks PR (2004) Microtubules and growth cone function. J Neurobiol 58: 70–83.

Gorski RA, Gordon JH, Shryne JE, Southam AM (1978) Evidence for a morphological sex difference within the medial preoptic area of the rat brain. Brain Res 148: 333–346.

Gorski RA, Harlan RE, Jacobson CD, Shryne JE, Southam AM (1980) Evidence for the existence of a sexually dimorphic nucleus in the preoptic area of the rat. J Comp Neurol 193: 529–539.

Goslin K, Banker G (1989) Experimental observations on the development of polarity by hippocampal neurons in culture. J Cell Biol 108: 1507–1516.

Gottlieb G (1980) Development of species identification in ducklings: VI. Specific embryonic experience required to maintain species-typical perception in ducklings. J Comp Physiol Psychol 94: 579–587.

Gottmann K, Dietzel ID, Lux HD, Huck S, Rohrer H (1988) Development of inward currents in chick sensory and autonomic neuronal precurser cells in culture. J Neurosci 8: 3722–3732.

Gould AP, White RA (1992) Connectin, a target of homeotic gene control in Drosophila. Development 116: 1163–1174.

Grassi S, Ottaviani F, Bambagioni D (1990) Vocalization-related stapedius muscle activity in different age chickens (Gallus gallus), and its role in vocal development. Brain Res 529: 158–164.

Grenningloh G, Rehm EJ, Goodman CS (1991) Genetic analysis of growth cone guidance in Drosophila: fasciclin II functions as a neuronal recognition molecule. Cell 67: 45–57.

Grenningloh G, Soehrman S, Bondallaz P, Ruchti E, Cadas H (2004) Role of the microtubule destabilizing proteins SCG10 and stathmin in neuronal growth. J Neurobiol 58: 60–69.

Grens A, Mason E, Marsh JL, Bode HR. Evolutionary conservation of a cell fate specification gene: the Hydra achaete-scute homolog has proneural activity in Drosophila. Development. 1995 Dec;121(12): 4027–4035.

Grisham W, Lee J, McCormick ME, Yang-Stayner K, Arnold AP (2002) Antiandrogen blocks estrogen-induced masculinization of the song system in female zebra finches. J Neurobiol 51: 1–8.

Gross RE, Mehler MF, Mabie PC, Zang Z, Santschi L, Kessler JA. Bone morphogenetic proteins promote astroglial lineage commitment by mammalian.

Grove EA, Fukuchi-Shimogori T (2003) Generating the cerebral cortical area map. Annu Rev Neurosci. 26: 355–380.

Groves AK, Anderson DJ (1996) Role of environmental signals and transcriptional regulators in neural crest development. Dev Genet 18: 64–72.

Grubb MS, Rossi FM, Changeux JP, Thompson ID (2003) Abnormal functional organization in the dorsal lateral geniculate nucleus of mice lacking the beta 2 subunit of the nicotinic acetylcholine receptor. Neuron 40: 1161–1172.

Grunz H, Tacke L. Neural differentiation of Xenopus laevis ectoderm takes place after disaggregation and delayed reaggregation without inducer. Cell Differ Dev. 1989 Dec;28(3): 211–217.

Grutzendler J, Kasthuri N, Gan WB (2002) Long-term dendritic spine stability in the adult cortex. Nature 420: 812–816.

Gu Q, Singer W (1995) Involvement of serotonin in developmental plasticity of kitten visual cortex. Eur J Neurosci 7: 1146–1153.

Gu Y, Hall ZW (1988) Immunological evidence for a change in subunits of the acetylcholine receptor in developing and denervated rat muscle. Neuron 1: 117–125.

Gundersen RW, Barrett JN (1979) Neuronal chemotaxis: chick dorsal-root axons turn toward high concentrations of nerve growth factor. Science 206: 1079–1080.

Guo M, Jan LY, Jan YN (1996) Control of daughter cell fates during asymmetric division: interaction of Numb and Notch. Neuron 17: 27–41.

Guo X, Chandrasekaran V, Lein P, Kaplan PL, Higgins D (1999) Leukemia inhibitory factor and ciliary neurotrophic factor cause dendritic retraction in cultured rat sympathetic neurons. J Neurosci 19: 2113–2121.

Guo X, Metzler-Northrup J, Lein P, Rueger D, Higgins D (1997) Leukemia inhibitory factor and ciliary neurotrophic factor regulate dendritic growth in cultures of rat sympathetic neurons. Brain Res Dev Brain Res 104: 101–110.

Guo X, Rueger D, Higgins D (1998) Osteogenic protein-1 and related bone morphogenetic proteins regulate dendritic growth and the expression of microtubule-associated protein-2 in rat sympathetic neurons. Neurosci Lett 245: 131–134.

Guo Y, Udin SB (2000) The development of abnormal axon trajectories after rotation of one eye in Xenopus. J Neurosci 20: 4189–4197.

Gurney ME (1981) Hormonal control of cell form and number in the zebra finch song system. J Neurosci 1: 658–673.

Gurney ME, Konishi M (1980) Hormone-induced sexual differentiation of brain and behavior in zebra finches. Science 208: 1380–1383.

Hafen E, Dickson B, Brunner D, Raabe T (1994) Genetic dissection of signal transduction mediated by the sevenless receptor tyrosine kinase in Drosophila. Prog Neurobiol 42: 287–292.

Hafidi A, Moore T, Sanes DH (1996) Regional distribution of neurotrophin receptors in the developing auditory brainstem. J Comp Neurol 367: 454–464.

Halder G, Callaerts P, Gehring WJ. Induction of ectopic eyes by targeted expression of the eyeless gene in Drosophila. Science. 1995 Mar 24;267(5205): 1788–1792.

Hall AC, Lucas FR, Salinas PC (2000) Axonal remodeling and synaptic differentiation in the cerebellum is regulated by WNT-7a signaling. Ceil 100: 525–535.

Hall JC (1977) Portions of the central nervous system controlling reproductive behavior in Drosophila melanogaster. Behav Genet 7: 291–312.

Hall JC (1994) The mating of a fly. Science 264: 1702–1714.

Hallbook F, Fritzsch B (1997) Distribution of BDNF and trkB mRNA in the otic region of 3.5 and 4.5 day chick embryos as revealed with a combination of in situ hybridization and tract tracing. Int J Dev Biol 41: 725–732.

Halpern ME, Chiba A, Johansen J, Keshishian H (1991) Growth cone behavior underlying the development of stereotypic synaptic connections in Drosophila embryos. J Neurosci 11: 3227–3238.

Ham J, Babij C, Whitfield J, Pfarr CM, Lallemand D, Yaniv M, Rubin LL (1995) A c-Jun dominant negative mutant protects sympathetic neurons against programmed cell death. Neuron 14: 927–939.

Hamburger V (1934) The effects of wing bud extirpation on the development of the central nervous system in chick embryos. J Exp Zool 68: 449–494.

Hamburger V (1963) Some aspects of the embryology of behavior. Q. Rev. Biol. 38: 342–365.

Hamburger V (1969) Hans Spemann and the organizer concept. Experientia. Nov 15;25(11): 1121–1125.

Hamburger V, Levi-Montalcini R (1949) Proliferation, differentiation and degeneration in the spinal ganglia of the chick embryo under normal and experimental conditions. J Exp Zool 111: 457–501.

Hamburger V, Wenger E, Oppenhein RW (1966) Motility in the chick and embryo in the absence of sensory input. J. Exp. Zool. 162: 133–160.

Hamill OP, Marty A, Neher E, Sakmann B, Sigworth FJ (1981) Improved patch-clamp techniques for high-resolution current recording from cell and cell-free membrane patches. Pflügers Arch 391: 85–100.

Hammarback JA, Palm SL, Furcht LT, Letourneau PC (1985) Guidance of neurite outgrowth by pathways of substratum-adsorbed laminin. J Neurosci Res 13: 213–220.

Han H-Q, Nichols RA, Rubin MR, Bähler M, Greengard P (1991) Induction of formation of presynaptic terminals in neruoblastima cells by synapsin IIb. Nature 349: 697–700.

Hansen MJ, Dallal GE, Flanagan JG (2004) Retinal axon response to ephrin-as shows a graded, concentration-dependent transition from growth promotion to inhibition. Neuron 42: 717–730.

Hardingham GE, Fukunaga Y, Bading H (2002) Extrasynaptic NMDARs oppose synaptic NMDARs by triggering CREB shutoff and cell death pathways. Nat Neurosci 5: 405–414.

Harish OE, Poo M-m (1992) Retrograde modulation at developing neuromuscular synapses: Involvement of G protein and arachidonic acid cascade. Neuron 9: 1201–1209.

Harris DA, Falls DL, Dill-Devor RM, Fischbach GD (1988) Acetylcholine receptor-inducing factor from chicken brain increases the level of mRNA encoding the receptor alpha subunit. Proc Natl Acad Sci. 85: 1983–1987.

Harris L, McKenna WJ, Rowland E, Holt DW, Storey GC, Krikler DM (1983) Side effects of long-term amiodarone therapy. Circulation 67: 45–51.

Harris RM, Woolsey TA (1981) Dendritic plasticity in mouse barrel cortex following postnatal vibrissa follicle damage. J Comp Neurol 196: 357–376.

Harris W, Stark W, Walker J (1976) Genetic dissection of the photoreceptor system in the compound eye of Drosophila melanogaster. J Physiol 256: 415–439.

Harris WA (1989) Local positional cues in the neuroepithelium guide retinal axons in embryonic Xenopus brain. Nature 339: 218–221.

Harris WA (1990) Neurometamorphosis. J Neurobiol 21(7): 953–957.

Harris WA, Holt CE, Bonhoeffer F (1987) Retinal axons with and without their somata, growing to and arborizing in the tectum of Xenopus embryos: a time-lapse video study of single fibres in vivo. Development 101: 123–133.

Harrison R (1907) Observations on the living developing nerve fiber. Anat Rec 1: 116–118.

Harrison R (1910) The outgrowth of the nerve fiber as a mode of protoplasmic movement. J Exp Zool 9: 787–846.

Harrison RG (1904) An experimental study of the relation of the nervous system to the developing musculature in the embryo of the frog. Am. J. Anat. 3: 197–220.

Hartenstein V (1989) Early neurogenesis in Xenopus: the spatio-temporal pattern of proliferation and cell lineages in the embryonic spinal cord. Neuron 3: 399–411.

Hartley DA, Xu TA, Artavanis-Tsakonas S. The embryonic expression of the Notch locus of Drosophila melanogaster and the implications of point mutations in the extracellular EGF-like domain of the predicted protein. EMBO J. 1987 Nov;6(11): 3407–3417.

Hartshorn K, Rovee-Collier C, Gerhardstein P, Bhatt RS, Wondoloski TL, Klein P, Gilch J, Wurtzel N, Campos-de-Carvalho M (1998) The ontogeny of long-term memory over the first year-and-a-half of life. Dev Psychobiol 32: 69–89.

Hatten ME (1985) Neuronal regulation of astroglial morphology and proliferation in vitro. J Cell Biol 100: 384–396.

Hatten ME (1990) Riding the glial monorail: a common mechanism for glial-guided neuronal migration in different regions of the developing mammalian brain. Trends Neurosci. May;13(5): 179–184.

Hattori M, Osterfield M, Flanagan JG (2000) Regulated cleavage of a contact-mediated axon repellent. Science 289: 1360–1365.

Haverkamp LJ (1986) Anatomical and physiological development of the Xenopus embryonic motor system in the absence of neural activity. J Neurosci 6(5): 1338–1348.

Haverkamp LJ, Oppenheim RW (1986) Behavioral development in the absence of neural activity: effects of chronic immobilization on amphibian embryos. J Neurosci 6(5): 1332–1337.

Hay BA, Wassarman DA, Rubin GM (1995) Drosophila homologs of baculovirus inhibitor of apoptosis proteins function to block cell death. Cell 83: 1253–1262.

Haydon PG, McCobb DP, Kater SB (1984) Serotonin selectively inhibits growth cone motility and synaptogenesis of specific identified neurons. Science 226: 561–564.

Hayes BP, Roberts A (1973) Synaptic junction development in the spinal cord of an amphibian embryo: An electron microscope study. Z Zellforsch 137: 251–269.

Hedgecock EM, Culotti JG, Hall DH (1990) The unc-5, unc-6, and unc-40 genes guide circumferential migrations of pioneer axons and mesodermal cells on the epidermis in C. elegans. Neuron 4: 61–85.

Heidemann SR (1996) Cytoplasmic mechanisms of axonal and dendritic growth in neurons. Int Rev Cytol 165: 235–296.

Held H (1897) Beiträge zur Structur der Nervenzellen und ihrer Fortädtze. Arch Anat Physiol Anat Abt 21: 204–294.

Hemmati-Brivanlou A, Kelly OG, Melton DA. Follistatin, an antagonist of activin, is expressed in the Spemann organizer and displays direct neuralizing activity. Cell. 1994 Apr 22;77(2): 283–295.

Hemmati-Brivanlou A, Melton DA. Inhibition of activin receptor signaling promotes neuralization in Xenopus. Cell. 1994 Apr 22;77(2): 273–281.

Hendrikson CK, Vaughn JE (1974) Fine structural relationships between neurites and radial glial processes in developing mouse spinal cord. J Neuroscytol 3: 659–675.

Hendry IA, Stöckel K, Thoenen H, Iversen LL (1974) The retrograde axonal transport of nerve growth factor. Brain Res 68: 103–121.

Hendry SH, Huntsman MM, Vinuela A, Mohler H, de Blas AL, Jones EG (1994) GABAA receptor subunit immunoreactivity in primate visual cortex: distribution in macaques and humans and regulation by visual input in adulthood. J Neurosci 14: 2383–2401.

Hengartner MO, Ellis RE, Horvitz HR (1992) Caenorhabditis elegans gene ced-9 protects cells from programmed cell death. Nature 356: 494–499.

Henkemeyer M, Itkis OS, Ngo M, Hickmott PW, Ethell IM (2003) Multiple EphB receptor tyrosine kinases shape dendritic spines in the hippocampus. J Cell Biol 163: 1313–1326.

Henrique D, Hirsinger E, Adam J, Le Roux I, Pourquie O, Ish-Horowicz D, Lewis J (1997) Maintenance of neuroepithelial progenitor cells by Delta-Notch signalling in the embryonic chick retina. Curr Biol 7: 661–670.

Hensch TK (2004) Critical period regulation. Annu Rev Neurosci 27: 549–579.

Hensch TK, Stryker MP (2004) Columnar architecture sculpted by GABA circuits in developing cat visual cortex. Science 303: 1678–1681.

Herrmann K, Arnold AP (1991) The development of afferent projections to the robust archistriatal nucleus in male zebra finches: a quantitative electron microscopic study. J Neurosci 11: 2063–2074.

Hestrin S (1992) Developmental regulation of NMDA receptor-mediated synaptic currents at a central synapse. Nature 357: 686–689.

Heumann R, Schwab M, Merkl R, Thoenen H (1984) Nerve growth factor-mediated induction of choline acetyltransferase in PC12 cells: evaluation of the site of action of nerve growth factor and the involvement of lysosomal degradation products of nerve growth factor. J Neurosci 4: 3039–3050.

Hibbard E (1965) Orientation and directed growth of Mauthners cell axons form duplicated vestibular nerve roots. Exp Neurol 13: 289–301.

Higashijima S, Masino MA, Mandel G, Fetcho JR (2004) Engrailed-1 expression marks a primitive class of inhibitory spinal interneuron. J Neurosci 24: 5827–5839.

Hinds JW, Hinds PA (1976) Synapse formation in the mouse olfactory bulb. II. Morphogenesis. J Comp Neurol 169: 41–62.

Hindges R, McLaughlin T, Genoud N, Henkemeyer M, O'Leary DD (2002) EphB forward signaling controls directional branch extension and arborization required for dorsal-ventral retinotopic mapping. Neuron 35: 475–487.

His W (1868) Untersuchungen u"ber die erste Anlage des Wirbeltierleibes: die erste Entwickelung des Hu"hnchens im Ei, Vogel FCW, Leipzig 237 p.

Hoch W, McConville J, Helms S, Newsom-Davis J, Melms A, Vincent A (2001) Auto-antibodies to the receptor tyrosine kinase MuSK in patients with myasthenia gravis without acetylcholine receptor antibodies. Nat Med 7: 365–368.

Hoffman KL, Weeks JC (1998) Programmed cell death of an identified motoneuron in vitro: Temporal requirements for steroid exposure and protein synthesis. J Neurobiol 35: 300–322.

Hoffman KL, Weeks JC (2001) Role of caspases and mitochondria in the steroid-induced programmed cell ceath of a motoneuron during metamorphosis. Dev Biol 229: 517–536.

Hoffmann H, Spear NE (1988) Ontogenetic differences in conditioning of an aversion to a gustatory CS with a peripheral US. Behav Neural Biol 50: 16–23.

Holash JA, Soans C, Chong LD, Shao H, Dixit VM, Pasquale EB (1997) Reciprocal expression of the Eph receptor Cek5 and its ligand(s) in the early retina. Dev Biol 182: 256–269.

Holland RL, Brown MC (1980) Postsynaptic transmission block can cause terminal sprouting of a motor nerve. Science 207: 649–651.

Hollenbeck PJ, Bray D (1987) Rapidly transported organelles containing membrane and cytoskeletal components: their relation to axonal growth. J Cell Biol 105: 2827–2835.

Holley SA, Jackson PD, Sasai Y, Lu B, De Robertis EM, Hoffmann FM, Ferguson EL. A conserved system for dorsal-ventral patterning in insects and vertebrates involving sog and chordin. Nature. 1995 Jul 20;376(6537): 249–253.

Hollyday M, Hamburger V (1976) Reduction of the naturally occurring motor neuron loss by enlargement of the periphery. J Comp Neurol 170: 311–320.

Holt CE, Bertsch TW, Ellis HM, Harris WA (1988) Cellular determination in the Xenopus retina is independent of lineage and birth date. Neuron 1: 15–26.

Holt CE, Harris WA (1998) Target selection: invasion, mapping and cell choice. Curr Opin Neurobiol 8: 98–105.

Holtfreter J (1939a) Die totale exogastrulation, eine selbstablösung des ektoderms vom entomesoderm. Wilhem Roux' Arch. Entwicklungsmech. Org. 129, 669–793.

Holtfreter J (1939b) Studien zur Ermittlung der Gestaltlungsfaktoren in der Organentwicklung der Amphibien. I. Roux' Arch. Entw.-mech., 139; 227–273.

Hömberg V, Bickmann U, Müller K (1993) Ontogeny is different for explicit and implicit memory in humans. Neurosci Lett 150: 187–190.

Homma S, Yaginuma H, Vinsant S, Seino M, Kawata M, Gould T, Shimada T, Kobayashi N, Oppenheim RW (2003) Differential expression of the GDNF family receptors RET and GFRalpha1, 2, and 4 in subsets of motoneurons: a relationship between motoneuron birthdate and receptor expression. J Comp Neurol 456: 245–259.

Honig MG, Hume RI (1986) Fluorescent carbocyanine dyes allow living neurons of identified origin to be studied in long-term cultures. J Cell Biol 103: 171–187.

Hopf C, Hoch W (1996) Agrin binding to alpha-dystroglycan. Domains of agrin necessary to induce acetylcholine receptor clustering are overlapping but not identical to the alpha-dystroglycan-binding region. J Biol Chem 271: 5231–5236.

Hopker VH, Shewan D, Tessier-Lavigne M, Poo M, Holt C (1999) Growth-cone attraction to netrin-1 is converted to repulsion by laminin-1. Nature 401: 69–73.

Horn G (2004) Pathways of the past: the imprint of memory. Nat Rev Neurosci 5: 108–120.

Horton HL, Levitt P (1988) A unique membrane protein is expressed on early developing limbic system axons and cortical targets. J Neurosci 8: 4653–4661.

Hoyle GW, Mercer EH, Palmiter RD, Brinster RL (1993) Expression of NGF in sympathetic neurons leads to excessive axon outgrowth from ganglia but decreased terminal innervation within tissues. Neuron 10: 1019–1034.

Huang CF, Tong J, Schmidt J (1992) Protein kinase C couples membrane excitation to acetylcholine receptor gene inactivation in chick skeletal muscle. Neuron 9: 671–678.

Huang EJ, Reichardt LF (2001) Neurotrophins: Roles in neuronal development and function. Ann Rev Neurosci 24: 677–736.

Huang L, Pallas SL (2001) NMDA antagonists in the superior colliculus prevent developmental plasticity but not visual transmission or map compression. J Neurophysiol 86: 1179–1194.

Hubel DH, Wiesel TN (1965) Binocular interaction in striate cortex of kittens reared with artificial squint. J Neurophysiol 28: 1041–1059.

Hubel DH, Wiesel TN (1970) The period of susceptibility to the physiological effects of unilateral eye closure in kittens. J Physiol 206: 419–436.

Huberman AD, Wang GY, Liets LC, Collins OA, Chapman B, Chalupa LM (2003) Eye-specific retinogeniculate segregation independent of normal neuronal activity. Science 300: 994–998.

Hübner CA, Stein V, Hermans-Borgmeyer I, Meyer T, Ballanyi K, Jentsch TJ (2001) Disruption of KCC2 reveals an essential role of K-Cl cotransport already in early synaptic inhibition. Neuron 30: 515–524.

Huganir RL, Greengard P (1983) cAMP-dependent protein kinase phosphorylates the nicotinic acetylcholine receptor. Proc Natl Acad Sci 80: 1130–1134.

Huganir RL, Miles K, Greengard P (1984) Phosphorylation of the nicotinic acetylcholine receptor by an endogenous tyrosine-specific protein kinase. Proc Natl Acad Sci 81: 6968–6972.

Hughes AF (1961) Cell degeneration in the larval ventral horn of Xenopus laevis. J Embryol Exp Morphol 9: 269–284.

Huguenard JR, Hamill OP, Prince DA (1988) Developmental changes in Na+ conductances in rat neocortical neurons: appearance of a slowly inactivating component. J Neurophysiol 9: 778–795.

Hume RI, Role LW, Fischbach GD (1983) Acetylcholine release from growth cones detected with patches of acetylcholine receptor rich membranes. Nature 305: 632–634.

Hummel T, Vasconcelos ML, Clemens JC, Fishilevich Y, Vosshall LB, Zipursky SL (2003) Axonal targeting of olfactory receptor neurons in Drosophila is controlled by Dscam. Neuron 37: 221–231.

Hunt PS (1999) A further investigation of the developmental emergence of fear-potentiated startle in rats. Dev Psychobiol 34: 281–291.

Hunt PS, Richardson R, Campbell BA (1994) Delayed development of fear-potentiated startle in rats. Behav Neurosci 108: 69–80.

Hutchison JB (1997) Gender-specific steroid metabolism in neural differentiation. Cell Mol Neurobiol 17: 603–626.

Hutson LD, Chien CB (2002) Pathfinding and error correction by retinal axons: the role of astray/robo2. Neuron 33: 205–217.

Huttenlocher PR, de Courten C (1987) The development of synapses in striate cortex of man. Human Neurobiol 6: 1–9.

Hyson RL, Rubel EW (1989) Transneuronal regulation of protein synthesis in the brain-stem auditory system of the chick requires synaptic activation. J Neurosci 9: 2835–2845.

Hyson RL, Rudy JW (1987) Ontogenetic change in the analysis of sound frequency in the infant rat. Dev Psychobiol 20: 189–207.

Ibanez CF (1994) Structure-function relationships in the neurotrophin family. J Neurobiol 25: 1349–1361.

Ichise T, Kano M, Hashimoto K, Yanagihara D, Nakao K, Shigemoto R, Katsuki M, Aiba A (2000) mGluR1 in cerebellar Purkinje cells essential for long-term depression, synapse elimination, and motor coordination. Science 288: 1832–1835.

Imperato-McGinley J, Peterson RE, Gautier T, Sturla E (1979) Male pseudohermaphroditism secondary to 5 alpha-reductase deficiency^ model for the role of androgens in both the development of the male phenotype and the evolution of a male gender identity. J Steroid Biochem 11: 637–645.

Innocenti GM, Fiore L, Caminiti R (1977) Exuberant projection into the corpus callosum from the visual cortex of newborn cats. Neurosci Lett 4: 237–242.

Inoué S (1981) Video image processing greatly enhances contrast, quality, and speed in polarization-based micrscopy. J Cell Biol 89: 346–356.

Inoué S (1989) Foundations of confocal scanning imaging in light microscopy. In: The Handbook of Biological Confocal Micrscopy, (Pawley, ed). IMR Press: Madison, pp 1–14.

Irie A, Yates EA, Turnbull JE, Holt CE (2002) Specific heparan sulfate structures involved in retinal axon targeting. Development 129: 61–70.

Ishida N, Kaneko M, Allada R (1999) Biological clocks. Proc Natl Acad Sci U S A 96(16): 8819–8820.

Isshiki T, Pearson B, Holbrook S, Doe CQ (2001) Drosophila neuroblasts sequentially express transcription factors which specify the temporal identity of their neuronal progeny. Cell 106: 511–521.

Iwai Y, Fagiolini M, Obata K, Hensch TK (2003) Rapid critical period induction by tonic inhibition in visual cortex. J Neurosci 23: 6695–6702.

Iwasato T, Erzurumlu RS, Huerta PT, Chen DF, Sasaoka T, Ulupinar E and Tonegawa S (1997) NMDA receptor-dependent refinement of somatotopic maps. Neuron 19(6): 1201–1210.

Jablonska B, Gierdalski M, Kossut M and Skangiel-Kramska J (1999) Partial blocking of NMDA receptors reduces plastic changes induced by short-lasting classical conditioning in the SI barrel cortex of adult mice [In Process Citation]. Cereb Cortex 9(3): 222–231.

Jackson H, Parks TN (1982) Functional synapse elimination in the developing avian cochlear nucleus with simultaneous reduction in cochlear nerve axon branching. J Neurosci 2: 1736–1743.

Jacobs DS, Blakemore C (1988) Factors limiting the postnatal development of visual acuity in the monkey. Vision Res 8: 947–958.

Jacobson M. Mapping the developing retinotectal projection in frog tadpoles by a double label autoradiographic techinque. Brain Res. 1977 May 20;127(1): 55–67.

Jacobson M (1991) Developmental Neurobiology, Plenum Press.

James J, Das AV, Bhattacharya S, Chacko DM, Zhao X, Ahmad I (2003) In vitro generation of early-born neurons from late retinal progenitors. J Neurosci 23: 8193–8203.

Jan YN, Jan LY (2003) The control of dendrite development. Neuron 40: 229–242.

Jansen JKS, Lømo T, Nicolaysen K, Westgaard RH (1973) Hyperinnervation of skeletal muscle fibers: dependence on muscle activity. Science 181: 559–561.

Jarman AP, Grau Y, Jan LY, Jan YN. Atonal is a proneural gene that directs chordotonal organ formation in the Drosophila peripheral nervous system. Cel. 1993 Jul 2;73(7): 1307–1321.

Jendelová P, Syková E (1991) The role of glia in K+ and pH homeostasis in the neonatal rat spinal cord. Glia 4: 56–63.

Jensen JK, Neff DL (1993) Development of basic auditory discrimination in preschool children. Psychol Sci 4: 104–107.

Jessell TM (2000) Neuronal specification in the spinal cord: inductive signals and transcriptional codes. Nat Rev Genet 1: 20–29.

Jessell TM, Bovolenta P, Placzek M, Tessier-Lavigne M, Dodd J (1989) Polarity and patterning in the neural tube: the origin and function of the floor plate. Ciba Found Symp 144: 255–276; discussion 276–280, 290–255.

Jessell TM, Siegel RE, Fischbach GD (1979) Induction of acetylcholine receptors on cultured skeletal muscle by a factor extracted from brain and spinal cord. Proc Natl Acad Sci 76: 5397–5401.

Jo SA, Zhu X, Marchionni MA, Burden SJ (1995) Neuregulins are concentrated at nerve-muscle synapses and activate ACh-receptor gene expression. Nature 373: 158–161.

Johnson D, Lanahan A, Buck CR, Sehgal A, Morgan C, Mercer E, Bothwell M, Chao M (1986) Expression and structure of the human NGF receptor. Cell 47: 545–554.

Johnson EM Jr (1978) Destruction of the sympathetic nervous system in neonatal rats and hamsters by vinblastine: prevention by concomitant administration of nerve growth factor. Brain Res 141: 105–118.

Johnson EM Jr, Andres RY, Bradshaw RA (1978) Characterization of the retrograde transport of nerve growth factor (NGF) using high specific activity [125I] NGF. Brain Res 150: 319–331.

Johnson F, Sablan MM, Bottjer SW (1995) Topographic organization of a forebrain pathway involved with vocal learning in zebra finches. J Comp Neurol 358: 260–278.

Johnson JL, Leff M (1999) Children of substance abusers: overview of research findings. Pediatrics 103(5 Pt 2): 1085–1099.

Johnson JS, Newport EL (1989) Critical period effects in second language learning: the influence of maturational state on the acquisition of English as a second language. Cognit Psychol 21: 60–99.

Johnson KG, Harris WA (2000) Connecting the eye with the brain: the formation of the retinotectal pathway. Results Probl Cell Differ 31: 157–177.

Johnston TD, Gottlieb G (1981) Visual preferences of imprinted ducklings are altered by the maternal call. J Comp Physiol Psychol 95: 663–675.

Jonas P, Bischofberger J, Sandkuhler J (1998) Corelease of two fast neurotransmitters at a central synapse. Science 281: 419–424.

Jones EG, Pons TP (1998) Thalamic and brainstem contributions to large-scale plasticity of primate somatosensory cortex [see comments]. Science 282(5391): 1121–1125.

Joshi I, Wang LY (2002) Developmental profiles of glutamate receptors and synaptic transmission at a single synapse in the mouse auditory brainstem. J Physiol 540: 861–873.

Jossin Y, Bar I, Ignatova N, Tissir F, De Rouvroit CL, Goffinet AM. The reelin signaling pathway: some recent developments. Cereb Cortex. 2003 Jun;13(6): 627–633.

Jost A (1953) Problems of fetal endocrinology: The gonadal and hypophyseal hormones. Recent Prog Horm Res 8: 379–418.

Joyner AL, Liu A, Millet S (2000) Otx2, Gbx2 and Fgf8 interact to position and maintain a mid-hindbrain organizer. Curr Opin Cell Biol Dec;12(6): 736–741.

Ju W, Morishita W, Tsui J, Gaietta G, Deerinck TJ, Adams SR, Garner CC, Tsien RY, Ellisman MH, Malenka RC (2004) Activity-dependent regulation of dendritic synthesis and trafficking of AMPA receptors. Nat Neurosci 7: 244–253.

Kaang B-K, Kandel ER, Grant SGN (1993) Activation of cAMP-responsive genes by stimuli that produce long-term facilitation in Aplysia sensory neurons. Neuron 10: 427–435.

Kacharmina JE, Job C, Crino P, Eberwine J (2000) Stimulation of glutamate receptor protein synthesis and membrane insertion within isolated neuronal dendrites. Proc Natl Acad Sci USA 97: 11545–11550.

Kakazu Y, Akaike N, Komiyama S, Nabekura J (1999) Regulation of intracellular chloride by cotransporters in developing lateral superior olive neurons. J Neurosci 19: 2843–2851.

Kalil K, Reh T (1982) A light and electron microscopic study of regrowing pyramidal tract fibers. J Comp Neurol 211: 265–275.

Kalil K, Szebenyi G, Dent EW (2000) Common mechanisms underlying growth cone guidance and axon branching. J Neurobiol 44: 145–158.

Kaltschmidt JA, Brand AH (2002) Asymmetric cell division: microtubule dynamics and spindle asymmetry. J Cell Sci 115: 2257–2264.

Kanaka C, Ohno K, Okabe A, Kuriyama K, Itoh T, Fukuda A, Sato K (2001) The differential expression patterns of messenger RNAs encoding K-Cl cotransporters (KCC1,2) and Na-K-2Cl cotransporter (NKCC1) in the rat nervous system. Neuroscience 104: 933–946.

Kanamoto T, Mota MA, Takeda K, Rubin LL, Miyazopo K, Ichijo H, Bazenet CE (2000) Role of apoptosis signal-regulating kinase in regulation of the c-Jun N-terminal kinase pathway and apoptosis in sympathetic neurons. Mol Cell Biol 20: 196–204.

Kandel ER, Tauc L (1965b) Mechanism of heterosynaptic facilitation in the giant cell of the abdominal ganglion of Aplysia depilans. J Physiol 181: 28–47.

Kandler K, Friauf E (1995) Development of glycinergic and glutamatergic synaptic transmission in the auditory brainstem of perinatal rats. J Neurosci 15: 6890–6904.

Kano M, Hashimoto K, Chen C, Abeliovich A, Aiba A, Kurihara H, Tonegawa S (1995) Impaired synapse elimination during cerebellar development in PKC? mutant mice. Cell 83: 1223–1231.

Kano M, Hashimoto K, Kurihara H, Watanabe M, Inoue Y, Aiba A, Tonegawa S (1997) Persistent multiple climbing fiber innervation of cerebellar Purkinje cells in mice lacking mGluR1. Neuron 18: 71–79.

Kano M, Hashimoto K, Watanabe M, Kurihara H, Offermanns S, Jiang H, Wu Y, Jun K, Shin HS, Inoue Y, Simon MI, Wu D (1998) Phospholipase cbeta4 is specifically involved in climbing fiber synapse elimination in the developing cerebellum. Proc Natl Acad Sci 95: 15724–15729.

Kapfer C, Seidl AH, Schweizer H, Grothe B (2002) Experience-dependent refinement of inhibitory inputs to auditory coincidence-detector neurons. Nat Neurosci 5: 247–253.

Kapfhammer JP, Grunewald BE, Raper JA (1986) The selective inhibition of growth cone extension by specific neurites in culture. J Neurosci 6: 2527–2534.

Kapfhammer JP, Raper JA (1987a) Collapse of growth cone structure on contact with specific neurites in culture. J Neurosci 7: 201–212.

Kapfhammer JP, Raper JA (1987b) Interactions between growth cones and neurites growing from different neural tissues in culture. J Neurosci 7: 1595–1600.

Kaplan DR, Hempstead BL, Martin-Zanca D, Chao MV, Parada LF (1991a) The trk proto-oncogene product: a signal transducing receptor for nerve growth factor. Science 252: 554–558.

Kaplan DR, Martin-Zanca D, Parada LF (1991b) Tyrosine phosphorylation and tyrosine kinase activity of the trk proto-oncogene product induced by NGF. Nature 350: 158–160.

Kasamatsu T, Pettigrew JD (1976) Depletion of brain chatecholamines: failure of ocular dominance shift after monocular occlusion in kittens. Science 194: 206–209.

Kater SB, Mills LR (1991) Regulation of growth cone behavior by calcium. J Neurosci 11: 891–899.

Kay JN, Roeser T, Mumm JS, Godinho L, Mrejeru A, Wong RO, Baier H (2004) Transient requirement for ganglion cells during assembly of retinal synaptic layers. Development 131: 1331–1342.

Keating MJ, Feldman J (1975) Visual deprivation and intertectal neuronal connections in Xenopus laevis. Proc R Soc Lond Ser B 191: 467–474.

Keino-Masu K, Masu M, Hinck L, Leonardo ED, Chan SS, Culotti JG, Tessier-Lavigne M (1996) Deleted in Colorectal Cancer (DCC) encodes a netrin receptor. Cell 87: 175–185.

Keleman K, Rajagopalan S, Cleppien D, Teis D, Paiha K, Huber LA, Technau GM, Dickson BJ (2002) Comm sorts robo to control axon guidance at the Drosophila midline. Cell 110: 415–427.

Keller AF, Coull JA, Chery N, Poisbeau P, De Koninck Y (2001) Region-specific developmental specialization of GABA-glycine cosynapses in laminas I-II of the rat spinal dorsal horn. J Neurosci 21: 7871–7880.

Keller R, Shih J, Sater A (1992) The cellular basis of the convergence and extension of the Xenopus neural plate. Dev Dyn Mar;193(3): 199–217.

Kelley DB (1997) Generating sexually differentiated songs. Curr Opin Neurobiol 7: 839–843.

Kelly JB, Judge PW, Fraser IH (1987) Development of the auditory orientation response in the albino rat (Rattus norvegicus). J Comp Psychol 101: 60–66.

Kelsch W, Hormuzdi S, Straube E, Lewen A, Monyer H, Misgeld U (2001) Insulin-like growth factor 1 and a cytosolic tyrosine kinase activate chloride outward transport during maturation of hippocampal neurons. J Neurosci 21: 8339–8347.

Kennedy C, Des Rosiers MH, Jehle JW, Reivich M, Sharpe F, Sokoloff L (1975) Mapping of functional neural pathways by autoradiographic survey of local metabolic rate with (14C)deoxyglucose. Science 7: 850–853.

Kennedy TE, Serafini T, de la Torre JR, Tessier-Lavigne M (1994) Netrins are diffusible chemotropic factors for commissural axons in the embryonic spinal cord. Cell 78: 425–435.

Keshishian H, Bentley D (1983a) Embryogenesis of peripheral nerve pathways in grasshopper legs. II. The major nerve routes. Dev Biol 96: 103–115.

Keshishian H, Bentley D (1983b) Embryogenesis of peripheral nerve pathways in grasshopper legs. I. The initial nerve pathway to the CNS. Dev Biol 96: 89–102.

Keshishian H, Bentley D (1983c) Embryogenesis of peripheral nerve pathways in grasshopper legs. III. Development without pioneer neurons. Dev Biol 96: 116–124.

Kessaris N, Pringle N, Richardson WD (2001) Ventral neurogenesis and the neuron-glial switch. Neuron 31: 677–680.

Keynes R, Tannahill D, Morgenstern DA, Johnson AR, Cook GM, Pini A (1997) Surround repulsion of spinal sensory axons in higher vertebrate embryos. Neuron 18: 889–897.

Kidokoro Y, Anderson MJ, Gruener R (1980) Changes in synaptic potential properties during acetylcholine receptor accumulation and neurospecific interactions in Xenopus nerve-muscle cell cultures. Dev Biol 78: 464–483.

Kidokoro Y, Yeh E (1982) Initial synaptic transmission at the growth cone in Xenopus nerve-muscle cultures. Proc Natl Acad Sci 79: 6727–6731.

Kiecker C, Niehrs C (2001) A morphogen gradient of Wnt/beta-catenin signalling regulates anteroposterior neural patterning in Xenopus. Development Nov;128(21): 4189–4201.

Kil SH, Krull CE, Cann G, Clegg D, Bronner-Fraser M (1998) The alpha4 subunit of integrin is important for neural crest cell migration. Dev Biol 202(1): 29–42.

Killackey HP, Rhoades RW, Bennett-Clarke CA (1995) The formation of a cortical somatotopic map. Trends Neurosci 18(9): 402–407.

Killisch I, Dotti CG, Laurie DJ, Luddens H, Seeburg PH (1991) Expression patterns of GABAA receptor subtypes in developing hippocampal neurons. Neuron 7: 927–936.

Kilman V, van Rossum MC, Turrigiano GG (2002) Activity deprivation reduces miniature IPSC amplitude by decreasing the number of postsynaptic GABA(A) receptors clustered at neocortical synapses. J Neurosci 22: 1328–1337.

Kilpatrick TJ, Bartlett PF (1995) Cloned multipotential precursors from the mouse cerebrum require FGF-2, whereas glial restricted precursors are stimulated with either FGF-2 or EGF. J Neurosci. May;15(5 Pt 1): 3653–3661.

Kim E, Niethammer M, Rothschild A, Jan YN, Sheng M (1995) Clustering of Shaker-type K+ channels by interaction with a family of membrane-associated guanylate kinases. Nature 378: 85–88.

Kim E, Sheng M (1996) Differential K+ channel clustering activity of PSD-95 and SAP-97, two related membrane-associated putative guanylate kinases. Neuropharmacol 35: 993–1000.

Kim G, Kandler K (2003) Elimination and strengthening of glycinergic/GABAergic connections during tonotopic map formation. Nat Neurosci 6: 282–290.

Kimmel CB, Ballard WW, Kimmel SR, Ullmann B, Schilling TF (1995) Stages of embryonic development of the zebrafish. Dev Dyn 203(3): 253–310.

Kimura D (1996) Sex, sexual orientation and sex hormones influence human cognitive function. Curr Opin Neurobiol 6: 259–263.

Kinch G, Hoffman KL, Rodrigues EM, Zee MC, Weeks JC (2003) Steroid-triggered programmed cell death of a motoneuron is autophagic and involves structural changes in mitochondria. J Comp Neurol 457: 384–403.

King AJ, Hutchings ME, Moore DR, Blakemore C (1988) Developmental plasticity in the visual and auditory representations in the mammalian superior colliculus. Nature 332(6159): 73–76.

Kiorpes L, Movshon JA (2004) Neural limitations on visual development in primates. In: The Visual Neurosciences, ed. LM Chalupa, JS Werner. MIT Press.

Kirn JR, DeVoogd TJ (1989) Genesis and death of vocal control neurons during sexual differentiation in the zebra finch. J Neurosci 9: 3176–3187.

Kirsch J, Betz H (1998) Glycine-receptor activation is required for receptor clustering in spinal neurons. Nature 392: 717–720.

Kirsch J, Malosio M-L, Wolters I, Betz H (1993a) Distribution of gephyrin transcripts in the adult and developing rat brain. Eur J Neurosci 5: 1109–1117.

Kirsh J, Wolters I, Triller A, Betz H (1993b) Gephyrin antisense oligonucleotides prevent glycine receptor clustering in spinal neurons. Nature 366: 745–748.

Kirson ED, Schirra C, Konnerth A, Yaari Y (1999) Early postnatal switch in magnesium sensitivity of NMDA receptors in rat CA1 pyramidal cells. J Physiol 521: 99–111.

Kistner U, Wenzel BM, Veh RW, Cases-Langhoff C, Garner AM, Appeltauer U, Voss B, Gundelfinger ED, Garner CC (1993) SAP90, a rat presynaptic protein related to the product of the Drosophila tumor suppressor gene dlg-A. J Biol Chem 268: 4580–4583.

Klarsfeld A, Bessereau JL, Salmon AM, Triller A, Babinet C, Changeux JP (1991) An acetylcholine receptor alpha-subunit promoter conferring preferential synaptic expression in muscle of transgenic mice. EMBO J 10: 625–632.

Klarsfeld A, Laufer R, Fontaine B, Devillers-Thiery A, Dubreuil C, Changeux JP (1989) Regulation of muscle AChR alpha subunit gene expression by electrical activity: involvement of protein kinase C and Ca2+. Neuron 2: 1229–1236.

Klein R, Smeyne RJ, Wurst W, Long LK, Auerbach BA, Joyner AL, Barbacid M (1993) Targeted disruption of the trkB neurotrophin receptor gene results in nervous system lesions and neonatal death. Cell 8: 113–122.

Kleinschmidt A, Bear MF, Singer W (1987) Blockade of NMDA receptors disrupts experience-dependent plasticity of kitten striate cortex. Science 238: 355–358.

Kneussel M, Brandstatter JH, Laube B, Stahl S, Muller U, Betz H (1999) Loss of postsynaptic GABA(A) receptor clustering in gephyrin-deficient mice. J Neurosci 19: 9289–9297.

Knox RJ, Quattrocki EA, Connor JA, Kaczmarek LK (1992) Recruitment of Ca2+ channels by protein kinase C during rapid formation of putative neruopeptide release sites in isolated Aplysia neurons. Neuron 8: 883–889.

Knudsen EI, du Lac S, Esterly SD (1987) Computational maps in the brain. Annu Rev Neurosci 10: 41–65.

Kobayashi H, Koppel AM, Luo Y, Raper JA (1997) A role for collapsin-1 in olfactory and cranial sensory axon guidance. J Neurosci 17: 8339–8352.

Koh YH, Popova E, Thomas U, Griffith LC, Budnik V (1999) Regulation of DLG localization at synapses by CaMKII-dependent phosphorylation. Cell 98: 353–363.

Kolbinger W, Trepel M, Beyer C, Pilgrim C, Reisert I (1991) The influence of genetic sex on sexual differentiation of diencephalic dopaminergic neurons in vitro and in vivo. Brain Res 544: 349–352.

Kolodkin AL (1996) Growth cones and the cues that repel them. Trends Neurosci 19: 507–513.

Kolodkin AL, Matthes DJ, Goodman CS (1993) The semaphorin genes encode a family of transmembrane and secreted growth cone guidance molecules. Cell 75: 1389–1399.

Kolodziej PA, Timpe LC, Mitchell KJ, Fried SR, Goodman CS, Jan LY, Jan YN (1996) Frazzled encodes a Drosophila member of the DCC immunoglobulin subfamily and is required for CNS and motor axon guidance. Cell 87: 197–204.

Komatsu Y (1994) Age-dependent long-term potentiation of inibitory synaptic transmission in rat visual cortex. J Neurosci 14: 6488–6499.

Komatsu Y, Iwakiri M (1991) Postnatal development of neuronal connections in cat visual cortex studied by intracellular recording in slice preparation. Brain Res 540: 14–24.

Kontges G, Lumsden A (1996) Rhombencephalic neural crest segmentation is preserved throughout craniofacial ontogeny. Development 122: 3229–3242.

Kopp DM, Perkel DJ, Balice-Gordon RJ (2000) Disparity in neurotransmitter release probability among competing inputs during neuromuscular synapse elimination. J Neurosci 20: 8771–8779.

Kornau HC, Schenker LT, Kennedy MB, Seeburg PH (1995) Domain interaction between NMDA receptor subunits and the postsynaptic density protein PSD-95. Science 269: 1737–1740.

Kossel AH, Williams CV, Schweizer M, Kater SB (1997) Afferent innervation influences the development of dendritic branches and spines via both activity-dependent and non-activity-dependent mechanisms. J Neurosci 17: 6314–6324.

Kotak VC, DiMattina C, Sanes DH (2001) GABAB and Trk receptor signaling mediates long-lasting inhibitory synaptic depression. J Neurophysiol 86: 536–540.

Kotak VC, Fujisawa S, Lee FA, Karthikeyan O, Aoki C, Sanes DH (2005) Hearing loss raises excitability in the auditory cortex. J Neurosci 25: 3908–3918.

Kotak VC, Sanes DH (1995) Synaptically-evoked prolonged depolarizations in the developing auditory system. J Neurophysiol 74: 1611–1620.

Kotak VC, Sanes DH (1996) Developmental influence of glycinergic inhibition: Regulation of NMDA-mediated EPSPs. J Neurosci 16: 1836–1843.

Kotak VC, Sanes DH (1997) Deafferentation of glutamatergic afferents weakens synaptic strength in the developing auditory system. Eur J Neurosci 11: 2340–2347.

Kotak VK, Sanes DH (2000) Long-lasting inhibitory synaptic depression is age- and calcium-dependent. J Neurosci 20: 5820–5826.

Kratz KE, Spear PD (1976) Postcritical-period reversal of effects of monocular deprivation on striate cortex cells in the cat. J Neurophysiol 39: 501–511.

Krieglstein K, Richter S, Farkas L, Schuster N, Dunker N, Oppenheim RW, Unsicker K (2000) Reduction of endogenous transforming growth factors beta prevents ontogenetic neuron death. Nat Neurosci 3: 1085–1090.

Krull CE, Lansford R, Gale NW, Collazo A, Marcelle C, Yancopoulos GD, Fraser SE, Bronner-Fraser M (1997) Interactions of Eph–related receptors and ligands confer rostrocaudal pattern to trunk neural crest migration. Curr Biol 7(8): 571–580.

Krylova O, Herreros J, Cleverley KE, Ehler E, Henriquez JP, Hughes SM, Salinas PC (2002) WNT-3, expressed by motoneurons, regulates terminal arborization of neurotrophin-3-responsive spinal sensory neurons. Neuron 35: 1043–1056.

Kuffler D, Thompson W, Jansen JKS (1977) The elimination of synapses in multiply-innervated skeletal muslce fibres of the rat: dependence on distance between end-plates. Brain Res 138: 353–358.

Kuhl PK, Williams KA, Lacerda F, Stevens KN, Lindblom B (1992) Linguistic experience alters phonetic perception in infants by 6 months of age. Science 255: 606–608.

Kuhn HG, Winkler J, Kempermann G, Thal LJ, Gage FH (1997) Epidermal growth factor and fibroblast growth factor-2 have different effects on neural progenitors in the adult rat brain. J Neurosci Aug 1;17(15): 5820–5829.

Kuida K, Zheng TS, Na S, Kuan C, Yang D, Karasuyama H, Rakic P, Flavell RA (1996) Decreased apoptosis in the brain and premature lethality in CPP32-deficient mice. Nature 384: 368–372.

Kullberg RW, Lentz TL, Cohen MW (1977) Development of the myotomal neuromuscular junction in Xenopus laevis: An electrophysiological and fine-structural study. Dev Biol 60: 101–129.

Kullmann PHM, Ene FA, Kandler K (2002) Glycinergic and GABAergic calcium responses in the developing lateral superior olive. Eur J Neurosci 15: 1093–1104.

Kumar JP, Moses K (2000) Cell fate specification in the Drosophila retina. Results Probl Cell Differ 31: 93–114.

Kunda P, Paglini G, Quiroga S, Kosik K, Caceres A (2001) Evidence for the involvement of Tiam1 in axon formation. J Neurosci 21: 2361–2372.

Kurada P, White K (1998) Ras promotes cell survival in Drosophila by downregulating hid expression. Cell 95: 319–329.

Kuruvilla R, Zweifel LS, Glebova NO, Lonze BE, Valdez G, Ye H, Ginty DD (2004) A neurotrophin signaling cascade coordinates sympathetic neuron development through differential control of TrkA trafficking and retrograde signaling. Cell 118: 243–255.

Kwon YW, Gurney ME (1996) Brain–derived neurotrophic factor transiently stabilizes silent synapses on developing neuromuscular junctions. J Neurobiol 29: 503–516.

Lallier T, Deutzmann R, Ferris R, Bronner-Fraser M (1994) Neural crest cell interactions with laminin: structural requirements and localization of the binding site for alpha 1 beta 1 integrin. Dev Biol 162(2): 451–464.

Lamb TM, Knecht AK, Smith WC, Stachel SE, Economides AN, Stahl N, Yancopolous GD, Harland RM (1993) Neural induction by the secreted polypeptide noggin. Science Oct 29;262(5134): 713–718.

Landgraf M, Baylies M, Bate M (1999) Muscle founder cells regulate defasciculation and targeting of motor axons in the Drosophila embryo. Curr Biol 9: 589–592.

Landis SC (1992) Cellular and molecular mechanisms determining neurotransmitter phenotypes in sympathetic neurons. In: Determinants of Neural Identity (Shanland M, Macagno E, eds), pp 497–523. San Diego: Academic Press.

Landmesser L, Dahm L, Tang JC, Rutishauser U (1990) Polysialic acid as a regulator of intramuscular nerve branching during embryonic development. Neuron 4: 655–667.

Landmesser L, Pilar G (1972) The onset and development of transmission in the chick ciliary ganglion. J Physiol (Lond) 222: 691–713.

Landmesser L, Pilar G (1974) Synaptic transmission and cell death during normal ganglionic development. J Physiol (London) 241: 737–749.

Langley JN (1985) Note on regeneration of pre-ganglionic fibres of the sympathetic. J Physiol (Lond) 18: 280–284.

Langley JN (1897) On the regeneration of pre-ganglionic amd post-ganglionic visceral nerve fibres. J Physiol (Lond) 22: 215–230.

Lanier LM, Gertler FB (2000) From Abl to actin: Abl tyrosine kinase and associated proteins in growth cone motility. Curr Opin Neurobiol 10: 80–87.

Lankford KL, Letourneau PC (1991) Roles of actin filaments and three second-messenger systems in short-term regulation of chick dorsal root ganglion neurite outgrowth. Cell Motil Cytoskeleton 20: 7–29.

Larmet Y, Dolphin AC, Davies AM (1992) Intracellular calcium regulates the survival of early sensory neurons before they become dependent on neurotrophic factors. Neuron 9: 563–574.

Larrabee MG, Bronk DW (1947) Prolonged facilitation of synaptic excitation in sympathetic ganglia. J Neurophysiol 10: 139–154.

Larsell O (1931) The effect of experimental excision of one eye on the development of the optic lobe and opticus layer in larvae of the tree-frog. J Exp Zool 58: 1–20.

Laufer R, Klarsfeld A, Changeux JP (1991) Phorbol esters inhibit the activity of the chicken acetylcholine receptor alpha-subunit gene promoter. Role of myogenic regulators. Eur J Biochem 202: 813–818.

Laurie DJ, Wisden W, Seeburg PH (1992) The distribution of 13 GABAA receptor subunit mRNAs in the rat brain. III. Embryonic and postnatal development J Neurosci 12: 4151–4172.

Lazebnik YA, Kaufmann SH, Desnoyers S, Poirier GG, Earnshaw WC (1994) Cleavage of poly(ADP-ribose) polymerase by a proteinase with properties like ICE. Nature 371: 346–347.

Le Douarin NM (2004) The avian embryo as a model to study the development of the neural crest: a long and still ongoing story. Mech Dev 121(9): 1089–1102.

Le Douarin N (1982) The Neural Crest. New York: Cambridge University Press.

Le Douarin NM, Creuzet S, Couly G, Dupin E (2004) Neural crest cell plasticity and its limits. Development. Oct;131(19): 4637–4650.

Le Douarin NM, Dupin E (2003). Multipotentiality of the neural crest. Curr Opin Genet Dev 13: 529–536.

Le Douarin NM, Renaud D, Teillet MA, Le Douarin GH (1975) Cholinergic differentiation of presumptive adrenergic neuroblasts in interspecific chimeras after heterotopic transplantations. Proc Natl Acad Sci U S A 72: 728–732.

Le Roux PD, Reh TA (1994) Regional differences in glial-derived factors that promote dendritic outgrowth from mouse cortical neurons in vitro. J Neurosci 14: 4639–4655.

Leber SM, Breedlove SM, Sanes JR (1990) Lineage, arrangement, and death of clonally related motoneurons in chick spinal cord. J Neurosci Jul;10(7): 2451–2462.

Lebrand C, Dent EW, Strasser GA, Lanier LM, Krause M, Svitkina TM, Borisy GG, Gertler FB (2004) Critical Role of Ena/VASP Proteins for Filopodia Formation in Neurons and in Function Downstream of Netrin-1. Neuron 42: 37–49.

Ledda F, Paratcha G, Ibanez CF (2002) Target-derived GFRalpha1 as an attractive guidance signal for developing sensory and sympathetic axons via activation of Cdk5. Neuron 36: 387–401.

Lee FS, Chao MV (2001) Activation of Trk neurotrophin receptors in the absence of neurotrophins. Proc Natl Acad Sci USA 98: 3555–3560.

Lee H-K, Kameyama K, Huganir RL, Bear MF (1998) NMDA induces long-term synaptic depression and dephosphorylation of the Glur1 subunit of AMPA receptors in hippocampus. Neuron 21: 1151–1162.

Lee JE, Hollenberg SM, Snider L, Turner DL, Lipnick N, Weintraub H (1995) Conversion of Xenopus ectoderm into neurons by NeuroD, a basic helix-loop-helix protein. Science 268(5212): 836–844.

Lee KF, Li E, Huber LJ, Landis SC, Sharpe AH, Chao MV, Jaenisch R (1992) Targeted mutation of the gene encoding the low affinity NGF receptor p75 leads to deficits in the peripheral sensory nervous system. Cell 69: 737–749.

Lee R, Kermani P, Teng KK, Hempstead BL (2001) Regulation of cell survival by secreted proneurotrophins. Science 294: 1945–1948.

Leibrock J, Lottspeich F, Hohn A, Hofer M, Gengerer B, Masiakowski P, Thoenen H, Barde Y (1989) Molecular cloning and expression of brain-derived neurotrophic factor. Nature 341: 149–152.

Leimeroth R, Lobsiger C, Lussi A, Taylor V, Suter U, Sommer L (2002) Membrane-bound neuregulin1 type III actively promotes Schwann cell differentiation of multipotent Progenitor cells. Dev Biol 246: 245–258.

Lein P, Johnson M, Guo X, Rueger D, Higgins D (1995) Osteogenic protein-1 induces dendritic growth in rat sympathetic neurons. Neuron 15: 597–605.

Lemmon V, Burden SM, Payne HR, Elmslie GJ, Hlavin ML (1992) Neurite growth on different substrates: permissive versus instructive influences and the role of adhesive strength. J Neurosci 12: 818–826.

Lendvai B, Stern EA, Chen B, Svoboda K (2000) Experience-dependent plasticity of dendritic spines in the developing rat barrel cortex in vivo. Nature 404: 876–881.

Le-Niculescu H, Bonfoco E, Kasuya Y, Claret FX, Green DR, Karin M (1999) Withdrawal of survival factors results in activation of the JNK pathway in neuronal cells leading to Fas ligand induction and cell death. Mol Cell Biol 19: 751–763.

Letourneau PC (1975) Cell-to-substratum adhesion and guidance of axonal elongation. Dev Biol 44: 92–101.

Letourneau PC (1996) The cytoskeleton in nerve growth cone motility and axonal pathfinding. Perspect Dev Neurobiol 4: 111–123.

Leu M, Bellmunt E, Schwander M, Farinas I, Brenner HR, Muller U (2003) Erbb2 regulates neuromuscular synapse formation and is essential for muscle spindle development. Development 130: 2291–2301.

LeVay S, Stryker MP, Shatz CJ (1978) Ocular dominance columns and their development in layer IV of the cat's visual cortex. J Comp Neurol 179: 223–244.

Levi S, Logan SM, Tovar KR, Craig AM (2004) Gephyrin is critical for glycine receptor clustering but not for the formation of functional GABAergic synapses in hippocampal neurons. J Neurosci 24: 207–217.

Levi-Montalcini R (1949) The development of the acoustico-vestibular centers in the chick embryo in the absence of the afferent root fibers and of descending fiber tracts. J Comp Neurol 91: 209–241.

Levi-Montalcini R, Booker B (1960) Destruction of the sympathetic ganglia in mammals by an antiserum to a nerve growth protein. Proc Natl Acad Sci USA 46: 384–391.

Levi-Montalcini R, Cohen S (1956) In vitro and in vivo effects if a nerve growth-stimulating agent isolated from snake venom. Proc Natl Acad Sci USA 42: 695–699.

Levi-Montalcini R, Hamburger V (1951) Selective growth stimulating effects of mouse sarcoma on the sensory and sympathetic nervous system of the chick embryo. J Exp Zool 116: 321–361.

Levi-Montalcini R, Hamburger V (1953) A diffusible agent of mouse sarcoma, producing hyperplasia of sympathetic ganglia and hyperneurotization of viscera on the chick embryo. J Exp Zool 123: 233–287.

Levi-Montalcini R, Levi G (1942) Les consequences de la destruction d'un territoire d'innervation peripheique sur le développments des centres nerveux correspondents dans l'embryon de poulet. Arch Biol (Liege) 53: 537–545.

Levine RB, Weeks JC (1990) Hormonally mediated changes in simple reflex circuits during metamorphosis in Manduca. J Neurobiol 21(7): 1022–1036.

Levtzion-Korach O, Tennenbaum A, Schnitzer R, Ornoy A (2000) Early motor development of blind children. J Paediatr Child Health 36: 226–229.

Lewis CA, Ahmed Z, Faber DS (1990) Developmental changes in the regulation of glycine-activated Cl-channels of cultured rat medullary neurons. Dev Brain Res 51: 287–290.

Lewis EB (1978) A gene complex controlling segmentation in Drosophila. Nature. Dec 7;276(5688): 565–570.

Lewis JL, Bonner J, Modrell M, Ragland JW, Moon RT, Dorsky RI, Raible DW (2004) Reiterated Wnt signaling during zebrafish neural crest development. Development. Mar;131(6): 1299–1308.

Lhuillier L, Dryer SE (2000) Developmental regulation of neuronal KCa channels by TGFbeta 1: transcriptional and posttranscriptional effects mediated by Erk MAP kinase. J Neurosci 20: 5616–5622.

Li HS, Chen JH, Wu W, Fagaly T, Zhou L, Yuan W, Dupuis S, Jiang ZH, Nash W, Gick C, Ornitz DM, Wu JY, Rao Y (1999) Vertebrate slit, a secreted ligand for the transmembrane protein roundabout, is a repellent for olfactory bulb axons. Cell 96: 807–818.

Li M, Sendtner M, Smith A (1995) Essential function of LIF receptor in motor neurons. Nature 378: 724–727.

Li MX, Jia M, Jiang H, Dunlap V, Nelson PG (2001) Opposing actions of protein kinase A and C mediate Hebbian synaptic plasticity. Nat Neurosci 4: 871–872.

Li P, Nijhawan D, Budihardjo I, Srinivasula SM, Ahmad M, Alnemri ES, Wang X (1997) Cytochrome c and dATP-dependent formation of Apaf-1/caspase-9 complex initiates an apoptotic protease cascade. Cell 91: 479–489.

Li W, Cogswell CA, LoTurco JJ. Neuronal differentiation of precursors in the neocortical ventricular zone is triggered by BMP. J Neurosci 1998 Nov 1;18(21): 8853–8862.

Liao D, Scannevin RH, Huganir R (2001) Activation of silent synapses by rapid activity-dependent synaptic recruitment of AMPA receptors. J Neurosci 21: 6008–6017.

Lichtman JW (1977) The reorganization of synaptic connexions in the rat submandibular ganglion during post-natal development. J Physiol 273: 155–177.

Lichtman JW, Purves D (1980) The elimination of redundant preganglionic innervation to hamster sympathetic ganglion cells in early post-natal life. J Physiol 301: 213–228.

Liem KF Jr, Tremml G, Roelink H, Jessell TM (1995) Dorsal differentiation of neural plate cells induced by BMP-mediated signals from epidermal ectoderm. Cell Sep 22;82(6): 969–979.

Limb CJ, Ryugo DK (2000) Development of primary axosomatic endings in the anteroventral cochlear nucleus of mice. J Assoc Res Otolaryngol 1: 103–119.

Lin CH, Espreafico EM, Mooseker MS, Forscher P (1996) Myosin drives retrograde F-actin flow in neuronal growth cones. Neuron 16: 769–782.

Lin DM, Fetter RD, Kopczynski C, Grenningloh G, Goodman CS (1994) Genetic analysis of Fasciclin II in Drosophila: defasciculation, refasciculation, and altered fasciculation. Neuron 13: 1055–1069.

Lin DM, Wang F, Lowe G, Gold GH, Axel R, Ngai J, Brunet L (2000) Formation of precise connections in the olfactory bulb occurs in the absence of odorant-evoked neuronal activity. Neuron 26: 69–80.

Lin JH, Saito T, Anderson DJ, Lance-Jones C, Jessell TM, Arber S (1998) Functionally related motor neuron pool and muscle sensory afferent subtypes defined by coordinate ETS gene expression. Cell 95: 393–407.

Lin L-FH, Doherty DH, Lile JD, Bektesh S, Collins F (1993) GDNF: A glial cell line-derived neurotrophic factor for midbrain dopaminergic neurons. Science 260: 1130–1132.

Lin W, Burgess RW, Dominguez B, Pfaff SL, Sanes JR, Lee KF (2001) Distinct roles of nerve and muscle in postsynaptic differentiation of the neuromuscular synapse. Nature 410: 1057–1064.

Linden R (1994) The survival of developing neurons: A review of afferent control. Neurosci 58: 671–682.

Linsdell P, Moody WJ (1995) Electrical activity and calcium influx regulate ion channel development in embryonic Xenopus skeletal muscle. J Neurosci 15: 4507–4514.

Lippe WR (1994) Rhythmic spontaneous activity in the developing avian auditory system. J Neurosci 14: 1486–1495.

Lisberger SG (1988) The neural basis for learning of simple motor skills. Science 242(4879): 728–735.

Litingtung Y, Chiang C (2000) Specification of ventral neuron types is mediated by an antagonistic interaction between Shh and Gli3. Nat Neurosci Oct;3(10): 979–985.

Liu CN, Chambers WW (1958) Intraspinal sprouting of dorsal root axons; development of new collaterals and preterminals following partial denervation of the spinal cord in the cat. AMA Arch Neurol Psychiatry 79: 46–61.

Liu CW, Lee G, Jay DG (1999) Tau is required for neurite outgrowth and growth cone motility of chick sensory neurons. Cell Motil Cytoskeleton 43: 232–242.

Livesey FJ, Cepko CL (2001) Vertebrate neural cell-fate determination: lessons from the retina. Nat Rev Neurosci 2: 109–118.

Lloyd DPC (1949) Post-tetanic potentiation of response in monosynaptic reflex pathways of the spinal cord. J Gen Physiol 33: 147–170.

Lo L, Dormand E, Greenwood A, Anderson DJ (2002) Comparison of the generic neuronal differentiation and neuron subtype specification functions of mammalian achaete-scute and atonal homologs in cultured neural progenitor cells. Development 129: 1553–1567.

Lo Y-j, Poo M-m (1991) Activity-dependent synaptic competition in vitro: Heterosynaptic supression of developing synapses. Science 254: 1019–1022.

Lo Y-j, Poo M-m (1994) Heterosynaptic supression of developing neruomuscular synapses in culture. J Neurosci 14: 4684–4693.

Locksley RM, Killeen N, Lenardo MJ (2001) The TNF and TNF receptor superfamilies: integrating mammalian biology. Cell 104: 487–501.

Loeb DM, Maragos J, Martin-Zanca D, Chao MV, Parada LF, Greene LA (1991) The trk proto-oncogene rescues NGF responsiveness in mutant NGF-nonresponsive PC12 cell lines. Cell 66: 961–966.

Loetscher H, Deuschle U, Brockhaus M, Reinhardt D, Nelboeck P, Mous J, Grünberg J, Haass C, Jacobsen H (1997) Presenilins are processed by caspase-type proteases. J Biol Chem 272: 20655–20659.

Lohmann C, Myhr KL, Wong RO (2002) Transmitter-evoked local calcium release stabilizes developing dendrites. Nature 418: 177–181.

Lohof AM, Quillan M, Dan Y, Poo MM (1992) Asymmetric modulation of cytosolic cAMP activity induces growth cone turning. J Neurosci 12: 1253–1261.

Lois C, Alvarez-Buylla A (1993) Proliferating subventricular zone cells in the adult mammalian forebrain can differentiate into neurons and glia. Proc Natl Acad Sci U S A Mar 1;90(5): 2074–2077.

Lois C, Garcia-Verdugo JM, Alvarez-Buylla A. Chain migration of neuronal precursors. Science. 1996 Feb 16;271(5251): 978–981.

Lom B, Cogen J, Sanchez AL, Vu T, Cohen-Cory S (2002) Local and target-derived brain-derived neurotrophic factor exert opposing effects on the dendritic arborization of retinal ganglion cells in vivo. J Neurosci 22: 7639–7649.

Lom B, Cohen-Cory S (1999) Brain-derived neurotrophic factor differentially regulates retinal ganglion cell dendritic and axonal arborization in vivo. J Neurosci 19: 9928–9938.

Lømo T, Rosenthal J (1972) Control of ACh sensitivity by muscle activity in the rat. J Physiol 221: 493–513.

Lonze BE, Ginty DD (2002) Function and regulation of CREB family transcription factors in the nervous system. Neuron 35: 605–623.

Lorenz K (1937) The companion in the bird's world. Auk 54: 245–273.

Löwel S, Singer W (1992) Selection of intrinsic horizontal connections in the visual cortex by correlated neuronal activity. Science 255: 209–212.

Lu B, Fu WM, Greengard P, Poo MM (1993) Calcitonin gene-related peptide potentiates synaptic responses at developing neuromuscular junction. Nature 363: 76–79.

Lu B, Jan L, Jan YN (2000) Control of cell divisions in the nervous system: symmetry and asymmetry. Annu Rev Neurosci 23: 531–556.

Lu W, Constantine-Paton M (2004) Eye opening rapidly induces synaptic potentiation and refinement. Neuron 43: 237–249.

Lu W, Man H, Ju W, Trimble WS, MacDonald JF, Wang YT (2001) Activation of synaptic NMDA receptors induces membrane insertion of new AMPA receptors and LTP in cultured hippocampal neurons. Neuron 29: 243–254.

Luhmann HJ, Prince DA (1991) Postnatal maturation of the GABAergic system in rat neocortex. J Neurophysiol 65: 247–263.

Lumsden A, Keynes R (1989) Segmental patterns of neuronal development in the chick hindbrain. Nature Feb 2;337(6206): 424–428.

Lumsden AG, Davies AM (1986) Chemotropic effect of specific target epithelium in the developing mammalian nervous system. Nature 323: 538–539.

Luo L (2000) Rho GTPases in neuronal morphogenesis. Nat Rev Neurosci 1: 173–180.

Luo L (2002) Actin cytoskeleton regulation in neuronal morphogenesis and structural plasticity. Ann Rev Cell Dev Biol 18: 601–635.

Luo L, Jan LY, Jan YN (1997) Rho family GTP-binding proteins in growth cone signalling. Curr Opin Neurobiol 7: 81–86.

Luo Y, Raible D, Raper JA (1993) Collapsin: a protein in brain that induces the collapse and paralysis of neuronal growth cones. Cell 75: 217–227.

Luskin MB, Parnavelas JG, Barfield JA (1993) Neurons, astrocytes, and oligodendrocytes of the rat cerebral cortex originate from separate progenitor cells: an ultrastructural analysis of clonally related cells. J Neurosci Apr;13(4): 1730–1750.

Luskin MB (1993) Restricted proliferation and migration of postnatally generated neurons derived from the forebrain subventricular zone. Neuron Jul;11(1): 173–189.

Lyckman AW, Jhaveri S, Feldheim DA, Vanderhaeghen P, Flanagan JG, Sur M (2001) Enhanced plasticity of retinothalamic projections in an ephrin-A2/A5 double mutant. J Neurosci 21: 7684–7690.

Lytton WW, Kristan WB (1989) Localization of a leech inhibitory synapse by photo-ablation of individual dendrites. Brain Res 504(1): 43–48.

Lyuksyutova AI, Lu CC, Milanesio N, King LA, Guo N, Wang Y, Nathans J, Tessier-Lavigne M, Zou Y (2003) Anterior-posterior guidance of commissural axons by Wnt-frizzled signaling. Science 302: 1984–1988.

MacInnis BL, Campenot RB (2002) Retrograde support of neuronal survival without retrograde transport of nerve growth factor. Science 295: 1536–1539.

Maderdrut JL, Oppenheim RW, Prevette D (1988) Enhancement of naturally occurring cell death in the sympathetic and parasympathetic ganglia of the chicken embryo following blockade of ganglionic transmission. Brain Res 444: 189–194.

Magdaleno S, Keshvara L, Curran T. Rescue of ataxia and preplate splitting by ectopic expression of Reelin in reeler mice. Neuron. 2002 Feb 14;33(4): 573–586.

Maggi L, Le Magueresse C, Changeux J-P, Cherubini E (2003) Nicotine activates immature "silent" connections in the developing hippocampus. Proc Natl Acad Sci USA 100: 2059–2064.

Magill-Solc C, McMahon UJ (1988) Motor neurons contain agrin-like molecules. J Cell Biol 107: 1825–1833.

Maher PA (1988) Nerve growth factor induces protein tyrosine-phosphorylation. Proc Natl Acad Sci USA 85: 6788–6791.

Majewska A, Sur M (2003) Motility of dendritic spines in visual cortex in vivo: changes during the critical period and effects of visual deprivation. Proc Natl Acad Sci USA 100: 16024–16029.

Mallamaci A, Muzio L, Chan CH, Parnavelas J, Boncinelli E (2000) Area identity shifts in the early cerebral cortex of Emx2−/− mutant mice. Nat Neurosci Jul;3(7): 679–86.

Mann F, Peuckert C, Dehner F, Zhou R, Bolz J (2002b) Ephrins regulate the formation of terminal axonal arbors during the development of thalamocortical projections. Development 129: 3945–3955.

Mann F, Ray S, Harris W, Holt C (2002a) Topographic mapping in dorsoventral axis of the Xenopus retinotectal system depends on signaling through ephrin-B ligands. Neuron 35: 461–473.

Marder E, Prinz AA (2002) Modeling stability in neuron and network function: the role of activity in homeostasis. Bioessays 24: 1145–1154.

Mariani J, Changeux J-P (1981) Ontogenesis of olivocerebellar relationships. I. Studies by intracellular recordings of the multiple innervation of Pukinje cells by climbing fibers in the developing rat cerebellum. J Neurosci 1: 696–702.

Marin-Padilla M (1998) Cajal-Retzius cells and the development of the neocortex. Neurosci Feb;21(2): 64–71.

Marin-Padilla M (1978) Dual origin of the mammalian neocortex and evolution of the cortical plate. Anat Embryol (Berl) 152: 109–126.

Mark RF (1969) Matching muscles and motoneurones. A review of some experiments on motor nerve regeneration. Brain Res 14: 245–254.

Markus A, von Holst A, Rohrer H, Heumann R (1997) NGF-mediated survival depends on p21ras in chick sympathetic neurons from the superior cervical but not from lumbosacral ganglia. Dev Biol 191: 306–310.

Marler P, Sherman V (1983) Song structure without auditory feedback: emendations of the auditory template hypothesis. J Neurosci 3: 517–531.

Maroney AC, Glicksman MA, Basma AN, Walton KM, Knight E Jr, Murphy CA, Bartlett BA, Finn JP, Angeles T, Matsuda Y, Neff NT, Dionne CA (1998) Motoneuron apoptosis is blocked by CEP-1347 (KT 7515), a novel inhibitor of the JNK signaling pathway. J Neurosci 18: 104–111.

Marrus SB, DiAntonio A (2004) Preferential Localization of Glutamate Receptors Opposite Sites of High Presynaptic Release. Cur Biol 14: 924–931.

Marsh L, Letourneau PC (1984) Growth of neurites without filopodial or lamellipodial activity in the presence of cytochalasin B. J Cell Biol 99: 2041–2047.

Martin DP, Schmidt RE, DiStefano PS, Lowry OH, Carter JG, Johnson EM Jr (1988) Inhibitors of protein synthesis and RNA synthesis prevent neuronal death caused by nerve growth factor deprivation. J Cell Biol 106: 829–844.

Martin JH, Choy M, Pullman S, Meng Z (2004) Corticospinal system development depends on motor experience. J Neurosci 24: 2122–2132.

Martinou JC, Dubois-Dauphin M, Staple JK, Rodriguez I, Frankowski H, Missotten M, Albertini P, Talabot D, Catsicas S, Pietra C, Huarte J (1994) Overexpression of BCL-2 in transgenic mice protects neurons from naturally occurring cell death and experimental ischemia. Neuron 13: 1017–1030.

Martinou JC, Falls DL, Fischbach GD, Merlie JP (1991) Acetylcholine receptor-inducing acitivity stimulates expression of the epsilon-subunit gene of the muscle acetylcholine receptor. Proc Natl Acad Sci 88: 7669–7673.

Martin-Zanca D, Barbacid M, Parada LF (1990) Expression of the trk proto-oncogene is restricted to the sensory cranial and spinal ganglia of neural crest origin in mouse development. Genes Dev 4: 683–694.

Martin-Zanca D, Hughes SH, Barbacid M (1986) A human oncogene formed by the fusion of truncated tropomyosin and protein tyrosine kinase sequences. Nature 319: 743–748.

Marty S, Wehrle R, Sotelo C (2000) Neuronal activity and brain-derived neurotrophic factor regulate the density of inhibitory synapses in organotypic slice cultures of postnatal hippocampus. J Neurosci 20: 8087–8095.

Matsuo I, Kuratani S, Kimura C, Takeda N, Aizawa S (1995) Mouse Otx2 functions in the formation and patterning of rostral head. Genes Dev Nov 1;9(21): 2646–2658.

Mattson MP, Kater SB (1987) Calcium regulation of neurite elongation and growth cone motility. J Neurosci 7: 4034–4043.

Mattson MP, Kater SB (1989) Excitatory and inhibitory neurotransmitters in the generation and degeneration of hippocampal neuroarchitecture. Brain Res 478: 337–348.

Mattson MP, Taylor-Hunter A, Kater SB (1988a) Neurite outgrowth in individual neurons of a neuronal population is differentially regulated by calcium and cyclic AMP. J Neurosci 8: 1704–1711.

Mauceri D, Cattabeni F, Di Luca M, Gardoni F (2004) Calcium/calmodulin-dependent protein kinase II phosphorylation drives synapse-associated protein 97 into spines. J Biol Chem 279: 23813–23821.

Maynard TM, Wakamatsu Y, Weston JA (2000) Cell interactions within nascent neural crest cell populations transiently promote death of neurogenic precursors. Development 127: 4561–4572.

Mazzoni IE, Said FA, Aloyz R, Miller FD, Kaplan D (1999) Ras regulates sympathetic neuron survival by suppressing the p53-mediated cell death pathway. J Neurosci 19: 9716–9727.

McAllister AK (2000) Cellular and molecular mechanisms of dendrite growth. Cereb Cortex 10: 963–973.

McAllister AK, Katz LC, Lo DC (1997) Opposing roles for endogenous BDNF and NT-3 in regulating cortical dendritic growth. Neuron 18: 767–778.

McCasland JS, Bernardo KL, Probst KL, Woolsey TA (1992) Cortical local circuit axons do not mature after early deafferentation. Proc Natl Acad Sci USA 89: 1832–1836.

McCobb DP, Best PM, Beam KG (1989) Development alters the expression of calcium currents in chick limb motoneurons. Neuron 2: 1633–1643.

McCobb DP, Best PM, Beam KG (1990) The differentiation of excitability in embryonic chick limb motoneurons. J Neurosci 10: 2974–2984.

McCobb DP, Haydon PG, Kater SB (1988) Dopamine and serotonin inhibition of neurite elongation of different identified neurons. J Neurosci Res 19(1): 19–26.

McConnell SK (1988) Fates of visual cortical neurons in the ferret after isochronic and heterochronic transplantation. J Neurosci 8: 945–974.

McConnell SK (1995) Strategies for the generation of neuronal diversity in the developing central nervous system. J Neurosci 15: 6987–6998.

McDonald JA, Doe CQ (1997) Establishing neuroblast-specific gene expression in the Drosophila CNS: huckebein is activated by Wingless and Hedgehog and repressed by Engrailed and Gooseberry. Development 124: 1079–1087.

McFarlane S, Cornel E, Amaya E, Holt CE (1996) Inhibition of FGF receptor activity in retinal ganglion cell axons causes errors in target recognition. Neuron 17: 245–254.

McFarlane S, McNeill L, Holt CE (1995) FGF signaling and target recognition in the developing Xenopus visual system. Neuron 15: 1017–1028.

McKenna MP, Raper JA (1988) Growth cone behavior on gradients of substratum bound laminin. Dev Biol 130: 232–236.

McKerracher L, Chamoux M, Arregui CO (1996) Role of laminin and integrin interactions in growth cone guidance. Mol Neurobiol 12: 95–116.

McLaughlin T, Torborg CL, Feller MB, O'Leary DD (2003) Retinotopic map refinement requires spontaneous retinal waves during a brief critical period of development. Neuron 40: 1147–1160.

McMahon AP, Bradley A (1990) The Wnt-1 (int-1) proto-oncogene is required for development of a large region of the mouse brain. Cell Sep 21;62(6): 1073–1085.

Meadows LA, Gell D, Broadie K, Gould AP, White RA (1994) The cell adhesion molecule, connectin, and the development of the Drosophila neuromuscular system. J Cell Sci 107 (Pt 1): 321–328.

Mears SC, Frank E (1997) Formation of specific monosynaptic connections between muscle spindle afferents and motoneurons in the mouse. J Neurosci 17: 3128–3135.

Meier T, Gesemann M, Cavalli V, Ruegg MA, Wallace BG (1996) AChR phosphorylation and aggregation induced by an agrin fragment that lacks the binding domain for alpha-dystroglycan. EMBO Journal 15: 2625–2631.

Meier T, Masciulli F, Moore C, Schoumacher F, Eppenberger U, Denzer AJ, Jones G, Brenner HR (1998) Agrin can mediate acetylcholine receptor gene expression in muscle by aggregation of muscle derived neuregulins. J. Cell Biol 141: 715–726.

Meister M, Wong ROL, Baylor DA, Shatz CJ (1991) Synchronous bursts of action potentials in ganglion cells of the developing mammalian retina. Science 252: 939–943.

Merlie JP, Isenberg KE, Russell SD, Sanes JR (1984) Denervation supersensitivity in skeletal muscle: analysis with a cloned cDNA probe. J Cell Biol 99: 332–335.

Merlie JP, Sanes JR (1985) Concentration of acetylcholine receptor mRNA in synaptic regions of adult muscle fibres. Nature 317: 66–68.

Merzenich M (1998) Long-term change of mind. Science 282: 1062–1063.

Merzenich MM, Jenkins WM (1993) Reorganization of cortical representations of the hand following alterations of skin inputs induced by nerve injury, skin island transfers, and experience. J Hand Ther 6: 89–104.

Messersmith EK, Leonardo ED, Shatz CJ, Tessier-Lavigne M, Goodman CS, Kolodkin AL (1995) Semaphorin III can function as a selective chemorepellent to pattern sensory projections in the spinal cord. Neuron 14: 949–959.

Metin C, Frost DO (1989) Visual responses of neurons in somatosensory cortex of hamsters with experimentally induced retinal projections to somatosensory thalamus. Proc Natl Acad Sci U S A 86: 357–361.

Metzger F, Wiese S, Sendtner M (1998) Effect of glutamate on dendritic growth in embryonic rat motoneurons. J Neurosci 18: 1735–1742.

Meyer RL, Wolcott LL (1987) Compression and expansion without impulse activity in the retinotectal projection of goldfish. J Neurobiol 18: 549–567.

Meyer RL, Wolcott LL (1987) Compression and expansion without impulse activity in the retinotectal projection of goldfish. J Neurobiol 18(6): 549–567.

Meyer-Franke A, Kaplan MR, Pfrieger FW, Barres BA (1995) Characterization of the signaling interactions that promote the survival and growth of developing retinal ganglion cells in culture. Neuron 15: 805–819.

Meyers EN, Lewandoski M, Martin GR (1998) An Fgf8 mutant allelic series generated by Cre- and Flp-mediated recombination. Nat Genet Feb;18(2): 136–141.

Migaud M, Charlesworth P, Dempster M, Webster LC, Watabe AM, Makhinson M, He Y, Ramsay MF, Morris RG, Morrison JH, O'Dell TJ, Grant SG (1998) Enhanced long-term potentiation and impaired learning in mice with mutant postsynaptic density-95 protein. Nature 396: 433–439.

Miles K, Greengard P, Huganir RL (1989) Calcitonin gene-related peptide regulates phosphorylation of the nicotinic acetylcholine receptor in rat myotubes. Neuron 2: 1517–1524.

Miller MW (1995) Relationship of the time of origin and death of neurons in rat somatosensory cortex: barrel versus septal cortex and projection versus local circuit neurons. J Comp Neurol 355: 6–14.

Millet S, Bloch-Gallego E, Simeone A, Alvarado-Mallart RM (1996) The caudal limit of Otx2 gene expression as a marker of the midbrain/hindbrain boundary: a study using in situ hybridisation and chick/quail homotopic grafts. Development Dec;122(12): 3785–3797.

Millet S, Campbell K, Epstein DJ, Losos K, Harris E, Joyner AL (1999) A role for Gbx2 in repression of Otx2 and positioning the mid/hindbrain organizer. Nature Sep 9;401(6749): 161–164.

Milligan CE, Prevette D, Yaginuma H, Homma S, Cardwell C, Fritz LC, Tomaselli KJ, Oppenheim RW, Schwartz LM (1995) Peptide inhibitors of the ICE protease family arrest programmed cell death of motoneurons in vivo and in vitro. Neuron 15: 385–393.

Ming GL, Song HJ, Berninger B, Holt CE, Tessier-Lavigne M, Poo MM (1997) cAMP-dependent growth cone guidance by netrin-1. Neuron 19: 1225–1235.

Ming GL, Wong ST, Henley J, Yuan XB, Song HJ, Spitzer NC, Poo MM (2002) Adaptation in the chemotactic guidance of nerve growth cones. Nature 417: 411–418.

Minichiello L, Casagranda F, Tatche RS, Stucky CL, Postigo A, Lewin GR, Davies AM, Klein R (1998) Point mutation in trkB causes loss of NT4-dependent neurons without major effects on diverse BDNF responses. Neuron 21: 335–345.

Minichiello L, Klein R (1996) TrkB and TrkC neurotrophin receptors cooperate in promoting survival of hippocampal and cerebellar granule neurons. Genes Dev 10: 2849–2858.

Miskevich F, Zhu Y, Ranscht B, Sanes JR (1998) Expression of multiple cadherins and catenins in the chick optic tectum. Mol Cell Neurosci 12: 240–255.

Mitchell KJ, Doyle JL, Serafini T, Kennedy TE, Tessier-Lavigne M, Goodman CS, Dickson BJ (1996) Genetic analysis of Netrin genes in Drosophila: Netrins guide CNS commissural axons and peripheral motor axons. Neuron 17: 203–215.

Mitchison T, Kirschner M (1988) Cytoskeletal dynamics and nerve growth. Neuron 1: 761–772.

Mittaud P, Marangi PA, Erb-Vogtli S, Fuhrer C (2001) Agrin-induced activation of acetylcholine receptor-bound Src family kinases requires Rapsyn and correlates with acetylcholine receptor clustering. J Biol Chem 276: 14505–14513.

Miyashiro K, Dichter M, Eberwine J (1994) On the nature and differential distribution of mRNAs in hippocampal neurites: implications for neuronal functioning. Proc Natl Acad Sci USA 91: 10800–10804.

Miyata M, Miyata H, Mikoshiba K, Ohama E (1999) Development of Purkinje cells in humans: an immunohistochemical study using a monoclonal antibody against the inositol 1,4,5-triphosphate type 1 receptor (IP3R1). Acta Neuropathol (Berl) Sep;98(3): 226–232.

Mizoguchi A, Nakanishi H, Kimura K, Matsubara K, Ozaki-Kuroda K, Katata T, Honda T, Kiyohara Y, Heo K, Higashi M, Tsutsumi T, Sonoda S, Ide C, Takai Y (2002) Nectin: an adhesion molecule involved in formation of synapses. J Cell Biol 156: 555–565.

Mizuseki K, Kishi M, Shiota K, Nakanishi S, Sasai Y (1998) SoxD: an essential mediator of induction of anterior neural tissues in Xenopus embryos. Neuron 21(1): 77–85.

Moles A, Kieffer BL, D'Amato FR (2004) Deficit in attachment behavior in mice lacking the mu-opioid receptor gene. Science 304: 1983–1986.

Mombaerts P (1996) Targeting olfaction. Curr Opin Neurobiol 6: 481–486.

Mombaerts P, Wang F, Dulac C, Chao SK, Nemes A, Mendelsohn M, Edmondson J, Axel R (1996) Visualizing an olfactory sensory map. Cell 87: 675–686.

Mooney R, Madison DV, Shatz CJ (1993) Enhancement of transmission at the developing retinogeniculate synapse. Neuron 10: 815–825.

Mooney R, Penn AA, Gallego R, Shatz CJ (1996) Thalamic relay of spontaneous retinal activity prior to vision. Neuron 17: 863–874.

Mooney R (1999) Sensitive periods and circuits for learned birdsong. Curr Opin Neurobiol 9(1): 121–127.

Moore C, Leu M, Muller U, Brenner HR (2001) Induction of multiple signaling loops by MuSK during neuromuscular synapse formation. Proc Natl Acad Sci USA 98: 14655–14660.

Morales B, Choi SY, Kirkwood A (2002) Dark rearing alters the development of GABAergic transmission in visual cortex. J Neurosci 22: 8084–8090.

Mori-Okamoto J, Ashida H, Maru E, Tatsuno J (1983) The development of action potentials in cultures of explanted cortical neurons from chick embryos. Dev Biol 97: 408–416.

Morrison SJ (2001) Neuronal potential and lineage determination by neural stem cells. Curr Opin Cell Biol 13: 666–672.

Moscoso LM, Chu GC, Gautam M, Noakes PG, Merlie JP, Sanes JR (1995) Synapse-associated expression of an acetylcholine receptor-inducing protein, ARIA/heregulin, and its putative receptors, ErbB2 and ErbB3, in developing mammalian muscle. Dev Biol 172: 158–169.

Moscoso LM, Cremer H, Sanes JR (1998) Organization and reorganization of neuromuscular junctions in mice lacking neural cell adhesion molecule, tenascin-C, or fibroblast growth factor-5. J Neurosci 18: 1465–1477.

Moss BL, Role LW (1993) Enhanced ACh sensitivity is accompanied by changes in ACh receptor channel properties and segregation of ACh receptor subtypes on sympathetic neurons during innervation in vivo. J Neurosci 13: 13–28.

Motoyama N, Wang F, Roth KA, Sawa H, Nakayama K, Nakayama K, Negishi I, Senju S, Zhang Q, Fujii S, Loh DY (1995) Massive cell death of immature hematopoietic cells and neurons in Bcl-x-deficient mice. Science 267: 1506–1510.

Moury JD, Jacobson AG. The origins of neural crest cells in the axolotl. Dev Biol. 1990 Oct;141(2): 243–253.

Moye TB, Rudy JW (1987) Ontogenesis of trace conditioning in young rats: Dissociation of associative and memory processes. Dev Psychobiol 20: 405–414.

Mrsic-Flogel TD, Schnupp JW, King AJ (2003) Acoustic factors govern developmental sharpening of spatial tuning in the auditory cortex. Nat Neurosci 6: 981–988.

Mueller AL, Chesnut RM, Schwartzkroin PA (1983) Actions of GABA in developing rabbit hippocampus: An in vitro study. Neurosci Lett 39: 193–198.

Mueller AL, Taube JS, Schwartzkroin PA (1984) Development of hyperpolarizing inhibitory postsynaptic potentials and hyperpolarizing responses to gamma-aminobutyric acid in rabbit hippocampus studied in vitro. J Neurosci 4: 860–867.

Mukhopadhyay M, Shtrom S, Rodriguez-Esteban C, Chen L, Tsukui T, Gomer L, Dorward DW, Glinka A, Grinberg A, Huang SP, Niehrs C, Belmonte JC, Westphal H. Dickkopf1 is required for embryonic head induction and limb morphogenesis in the mouse. Dev Cell. 2001 Sep;1(3): 423–434.

Mulder EJ, Kamstra A, O'Brien MJ, Visser GH, Prechtl HF (1986) Abnormal fetal behavioural state regulation in a case of high maternal alcohol intake during pregnancy. Early Hum Dev 14(3–4): 321–326.

Muller BK, Jay DG, Bonhoeffer F (1996) Chromophore-assisted laser inactivation of a repulsive axonal guidance molecule. Curr Biol 6(11): 1497–1502.

Murphy GG, Glanzman DL (1999) Cellular analog of differential classical conditioning in Aplysia: disruption by the NMDA receptor antagonist DL-2-amino-5-phosphonovalerate. J Neurosci 19: 10595–10602.

Murthy VN, Schikorski T, Stevens CF, Zhu Y (2001) Inactivity produces increases in neurotransmitter release and synapse size. Neuron 32: 673–682.

Mutani R, Futamachi K, Prince DA (1974) Potassium activity in immature cortex. Brain Res 75: 27–39.

Muzio L, DiBenedetto B, Stoykova A, Boncinelli E, Gruss P, Mallamaci A. Conversion of cerebral cortex into basal ganglia in Emx2(−/−) Pax6(Sey/Sey) double-mutant mice. Nat Neurosci. 2002 Aug;5(8): 737–745.

Muzio L, DiBenedetto B, Stoykova A, Boncinelli E, Gruss P, Mallamaci A. Emx2 and Pax6 control regionalization of the pre-neuronogenic cortical primordium. Cereb Cortex. 2002 Feb;12(2): 129–139.

Muzio L, Mallamaci A. Emx1, emx2 and pax6 in specification, regionalization and arealization of the cerebral cortex. Cereb Cortex. 2003 Jun;13(6): 641–647.

Myat A, Henry P, McCabe V, Flintoft L, Rotin D, Tear G (2002) Drosophila Nedd4, a ubiquitin ligase, is recruited by Commissureless to control cell surface levels of the roundabout receptor. Neuron 35: 447–459.

Nabekura J, Katsurabayashi S, Kakazu Y, Shibata S, Matsubara A, Jinno S, Mizoguchi Y, Sasaki A, Ishibashi H (2004) Developmental switch from GABA to glycine release in single central synaptic terminals. Nat Neurosci 7: 17–23.

Nadarajah B, Brunstrom JE, Grutzendler J, Wong RO, Pearlraan AL. Two modes of radial migration in early development of the cerebral cortex. Nat Neurosci. 2001 Feb;4(2): 143–150.

Naftolin F, Ryan KJ, Davies IJ, Reddy VV, Flores F, Petro Z, Kuhn M, White RJ, Takaoka Y, Wolin L (1975) The formation of estrogens by central neuroendocrine tissues. Recent Prog Horm Res 31: 295–319.

Nakagawa S, Brennan C, Johnson KG, Shewan D, Harris WA, Holt CE (2000) Ephrin-B regulates the Ipsilateral routing of retinal axons at the optic chiasm. Neuron 25: 599–610.

Nakamura H, O'Leary DD (1989) Inaccuracies in initial growth and arborization of chick retinotectal axons followed by course corrections and axon remodeling to develop topographic order. J Neurosci 9: 3776–3795.

Nakashima K, Wiese S, Yanagisawa M, Arakawa H, Kimura N, Hisatsune T, Yoshida K, Kishimoto T, Sendtner M, Taga T (1999) Developmental requirement of gp130 signaling in neuronal survival and astrocyte differentiation. J Neurosci 19: 5429–5434.

Narayanan CH, Hamburger V (1971) Motility in chick embryos with substitution of lumbosacral by brachial and brachial by lumbosacral spinal cord segments. J Exp Zool 178(4): 415–431.

Nardi JB (1983) Neuronal pathfinding in developing wings of the moth Manduca sexta. Dev Biol 95: 163–174.

Nardi JB, Vernon RA (1990) Topographical features of the substratum for growth of pioneering neurons in the Manduca wing disc. J Neurobiol 21: 1189–1201.

Naumann T, Casademunt E, Hollerbach E, Hofmann J, Dechant G, Frotscher M, Barde YA (2002) Complete deletion of the neurotrophin receptor p75NTR leads to long-lasting increases in the number of basal forebrain cholinergic neurons. J Neurosci 22: 2409–2418.

Nelson PG, Lanuza MA, Jia M, Li MX, Tomas J (2003) Phosphorylation reactions in activity-dependent synapse modification at the neuromuscular junction during development. J Neurocytol 32: 803–816.

New HV, Mudge AW (1986) Calcitonin gene-related peptide regulates muscle acetylcholine receptor synthesis. Nature 323: 809–811.

Newman-Gage H, Westrum LE, Bertrum JF (1987) Stereological analysis of synaptogenesis in the molecular layer of piriform cortex in the prenatal rat. J Comp Neurol 261: 295–305.

Newport E (1990) Maturational constraints on language learning. Cognitive Sci 14: 11–28.

Nguyen QT, Parsadanian AS, Snider WD, Lichtman JW (1998) Hyperinnervation of neuromuscular junctions caused by GDNF overexpression in muscle. Science 279: 1725–1729.

Nicholls DG, Ward MW (2000) Mitochondrial membrane potential and neuronal glutamate excitotoxity: mortality and millivolts. Trends Neurosci 23: 166–174.

Nicholson DW, Thornberry NA (1997) Caspases: killer proteases. Trends Biochem 22: 299–306.

Niell CM, Meyer MP, Smith SJ (2004) In vivo imaging of synapse formation on a growing dendritic arbor. Nat Neurosci 7: 254–260.

Niethammer M, Kim E, Sheng M (1996) Interaction between the C terminus of NMDA receptor subunits and multiple members of the PSD-95 family of membrane-associated guanylate kinases. J Neurosci 16: 2157–2163.

Nieuwkoop PD. Inductive interactions in early amphibian development and their general nature. J Embryol Exp Morphol. 1985 Nov;89 Suppl: 333–347.

Nieuwkoop PD. The organization center of the amphibian embryo: its origin, spatial organization, and morphogenetic action. Adv Morphog. 1973;10: 1–39.

Nitkin RM, Smith MA, Magill C, Fallon JR, Yao M, Wallace BG, McMahon UJ (1987) Identification of agrin, a synaptic organizing protein from Torpedo electric organ. J Cell Biol 105: 2471–2478.

Njå A, Purves D (1977) Specific innervation of guinea-pig superior cervical ganglion cells by preganglionic fibres arising from different levels of the spinal cord. J Physiol 264: 565–583.

Noakes PG, Gautam M, Mudd J, Sanes JR, Merlie JP (1995) Aberrant differentiation of neuromuscular junctions in mice lacking s-laminin/laminin beta 2. Nature 374: 258–262.

Noakes PG, Phillips WD, Hanley TA, Sanes JR, Merlie JP (1993) 43K protein and acetylcholine receptors colocalize during the initial stages of neuromuscular synapse formation in vivo. Dev Biol 155: 275–280.

Noctor SC, Flint AC, Weissman TA, Wong WS, Clinton BK, Kriegstein AR. Dividing precursor cells of the embryonic cortical ventricular zone have morphological and molecular characteristics of radial glia. J Neurosci. 2002 Apr 15;22(8): 3161–3173.

Noctor SC, Flint AC, Weissman TA, Dammerman RS, Kriegstein AR. Neurons derived from radial glial cells establish radial units in neocortex. Nature. 2001 Feb 8;409(6821): 714–720.

Noden DM (1975) An analysis of migratory behavior of avian cephalic neural crest cells. Dev Biol 42: 106–130.

Nordeen EJ, Nordeen KW (1988) Sex and regional differences in the incorporation of neurons born during song learning in zebra finches. J Neurosci 8: 2869–2874.

Nordeen EJ, Nordeen KW, Sengelaub DR, Arnold AP (1985) Androgens prevent normally occurring cell death in a sexually dimorphic spinal nucleus. Science 229: 671–673.

Nose A, Mahajan VB, Goodman CS (1992a) Connectin: a homophilic cell adhesion molecule expressed on a subset of muscles and the motoneurons that innervate them in Drosophila. Cell 70: 553–567.

Nose A, Takeichi M, Goodman CS (1994) Ectopic expression of connectin reveals a repulsive function during growth cone guidance and synapse formation. Neuron 13: 525–539.

Nose A, Umeda T, Takeichi M (1997) Neuromuscular target recognition by a homophilic interaction of connectin cell adhesion molecules in Drosophila. Development 124: 1433–1441.

Nose A, Van Vactor D, Auld V, Goodman CS (1992b) Development of neuromuscular specificity in Drosophila. Cold Spring Harb Symp Quant Biol 57: 441–449.

Nottebohm F (1980) Testosterone triggers growth of brain vocal control nuclei in adult female canaries. Brain Res May 12;189(2): 429—436.

Nottebohm F (1981) A brain for all seasons: cyclical anatomical changes in song control nuclei of the canary brain. Science 214: 1368–1370.

Nottebohm F (1985) Neuronal replacement in adulthood. Ann N Y Acad Sci 457: 143–161.

Nottebohm F, Arnold AP (1976) Sexual dimorphism in vocal control areas of the songbird brain. Science 194: 211–213.

Novak KD, Prevette D, Wang S, Gould TW, Oppenheim RW (2000) Hepatocyte growth factor/scatter factor is a neurotrophic survival factor for lumbar but not for other somatic motoneurons in the chick embryo. J Neurosci 20: 326–337.

Novitch BG, Wichterle H, Jessell TM, Sockanathan S. A requirement for retinoic acid-mediated transcriptional activation in ventral neural patterning and motor neuron specification. Neuron. 2003 Sep 25;40(1): 81–95.

Nusslein-Volhard C, Wieschaus E (1980) Mutations affecting segment number and polarity in Drosophila. Nature Oct 30;287(5785): 795–801.

Obata K, Oide M, Tanaka H (1978) Excitatory and inhibitory actions of GABA and glycine on embryonic chick spinal neurons in culture. Brain Res 144: 179–184.

O'Brien JA, Berger AJ (1999) Cotransmission of GABA and glycine to brain stem motoneurons. J Neurophysiol 82: 1638–1641.

O'Brien R, Xu D, Mi R, Tang X, Hopf C, Worley P (2002) Synaptically targeted narp plays an essential role in the aggregation of AMPA receptors at excitatory synapses in cultured spinal neurons. J Neurosci 22: 4487–4498.

O'Brien RAD, Ostberg AJC, Vrbova G (1984) Protease inhibitors reduce the loss of nerve terminals induced by activity and calcium in developing rat soleus muscles in vitro. Neurosci 12: 637–646.

O'Brien RJ, Fischbach GD (1986a) Isolation of embryonic chick motoneurons and their survival in vitro. J Neurosci 6: 3265–3274.

O'Brien RJ, Fischbach GD (1986b) Modulation of embryonic chick motoneuron glutamate sensitivity by interneurons and agonists. J Neurosci 6: 3290–3296.

O'Brien RJ, Kamboz S, Ehlers MD, Rosen KR, Fischbach GD, Huganir RL (1998) Activity-dependent modulation of synaptic AMPA receptor accumulation. Neuron 21: 1067–1078.

O'Brien RJ, Xu D, Petralia RS, Steward O, Huganir RL, Worley P (1999) Synaptic clustering of AMPA receptors by the extracellular immediate-early gene product Narp. Neuron 23: 309–323.

Obrietan K, van den Pol AN (1995) GABA neurotransmission in the hypothalamus: developmental reversal from Ca2+ elevating to depressing. J Neurosci 15: 5065–5077.

O'Connor LT, Lauterborn JC, Gall CM, Smith MA (1994) Localization and alternative splicing of agrin mRNA in adult rat brain: transcripts encoding isoforms that aggregate acetylcholine receptors are not restricted to cholinergic regions. J Neurosci 14: 1141–1152.

O'Connor R, Tessier-Lavigne M (1999) Identification of maxillary factor, a maxillary process-derived chemoattractant for developing trigeminal sensory axons. Neuron 24: 165–178.

O'Connor TP, Duerr JS, Bentley D (1990) Pioneer growth cone steering decisions mediated by single filopodial contacts in situ. J Neurosci 10: 3935–3946.

O'Donovan M, Ho S, Yee W (1994) Calcium imaging of rhythmic network activity in the developing spinal cord of the chick embryo. J Neurosci 14: 6354–6369.

O'Donovan MJ, Wenner P, Chub N, Tabak J, Rinzel J (1998) Mechanisms of spontaneous activity in the developing spinal cord and their relevance to locomotion. Ann N Y Acad Sci 860: 130–141.

O'Dowd DK (1983) RNA synthesis dependence of action potential development in spinal cord neurones. Nature 303: 619–621.

O'Dowd DK, Ribera AB, Spitzer NC (1988) Development of voltage-dependent calcium, sodium, and potassium currents in Xenopus spinal neurons. J Neurosci 8: 792–805.

Ogawa S, Chester AE, Hewitt SC, Walker VR, Gustafsson JA, Smithies O, Korach KS, Pfaff DW (200) Abolition of male sexual behaviors in mice lacking estrogen receptors alpha and beta (alpha beta ERKO). Proc Natl Acad Sci USA 97: 14737–14741.

Ogawa S, Lubahn DB, Korach KS, Pfaff DW (1997) Behavioral effects of estrogen receptor gene disruption in male mice. Proc Natl Acad Sci USA 94: 1476–1481.

Ohnuma S, Philpott A, Wang K, Holt CE, Harris WA (1999) p27Xic1, a Cdk inhibitor, promotes the determination of glial cells in Xenopus retina. Cell 99: 499–510.

Okada M, Corfas G (2004) Neuregulin1 downregulates postsynaptic GABAA receptors at the hippocampal inhibitory synapse. Hippocampus 14: 337–344.

O'Leary DDM, Stanfield BB, Cowan WM (1981) Evidence that the early postnatal restriction of the cells or orignin of the callosal projection is due to the elimination of axonal collaterals rather than to the death of the neurons. Dev Brain Res 1: 607–617.

Oleskevich S, Walmsley B (2002) Synaptic transmission in the auditory brainstem of normal and congenitally deaf mice. J Physiol 540: 447–455.

Olsen CR, Pettigrew JD (1974) Single units in the visual cortex of kittens reared in stroboscopic illumination. Brain Res 70: 189–204.

O'Malley DM, Kao YH, Fetcho JR (1996) Imaging the functional organization of zebrafish hindbrain segments during escape behaviors. Neuron 17(6): 1145–1155.

Oppenheim RW (1981) Ontogenetic adaptation and regressive processes in the development of the nervous system and behavior: a neuro-embryological perspective. In Development and Maturation, K Connolly and H Prechtl (eds), J. Lippincott, Philadelphia pp 73–109.

Oppenheim RW (1982) The neuroembryological study of behavior: progress, problems, perspectives. Current Topics in Devel. Biol 17: 257–309.

Oppenheim RW (1989) The neurotrophic theory and naturally occurring motoneuron death. Trends Neurosci 12: 252–255.

Oppenheim RW (1991) Cell death during development in the nervous system. Ann Rev Neurosci 14: 453–501.

Oppenheim RW (1992). Pathways in the emergence of developmental neuroethology: antecedents to current views of neurobehavioral ontogeny. J Neurobiol 23(10): 1370–1403.

Oppenheim, RW (2001) Viktor Hamburger (1900–2001): Journey of a neuroembryologist to the end of the millennium and beyond. Neuron 31: 179–190.

Oppenheim RW, Chu-Wang I-W, Foelix RF (1975) Some aspects of synaptogenesis in the spinal cord of the chick embryo: A quantitative electron microscopic study. J Comp Neurol 161: 383–418.

Oppenheim RW, Flavell RA, Vinsant S, Prevette D, Kuan C-Y, Rakic P (2001) Programmed Cell Death of Developing Mammalian Neurons after Genetic Deletion of Caspases. J Neurosci 21: 4752–4760.

Oppenheim RW, Houenou LJ, Johnson JE, Lin LF, Li L, Lo AC, Newsome AL, Prevette DM, Wang S (1995) Developing motor neurons rescued from programmed and axotomy-induced cell death by GDNF. Nature 373: 344–346.

Oppenheim RW, Houenou LJ, Parsadanian AS, Prevette D, Snider WD, Shen L (2000) Glial cell line-derived neurotrophic factor and developing mammalian motoneurons: regulation of programmed cell death among motoneuron subtypes. J Neurosci 20: 5001–5011.

Oppenheim RW, Johnson JE (2003) Programmed Cell Death and Neurotrophic Factors. In: Fundamental Neuroscience (Squire LE, Bloom FE, McConnell SK, Roberts JL, Spitzer NC, Zigmond MJ, eds), Academic Press, New York).

Oppenheim RW, Prevette D, D'Costa A, Wang S, Houenou LJ, McIntosh JM (2000) Reduction of neuromuscular activity is required for the rescue of motoneurons from naturally occurring cell death by nicotinic-blocking agents. J Neurosci 20: 6117–6124.

Oppenheim RW, Prevette D, Tytell M, Homma S (1990) Naturally occurring and induced neuronal death in the chick embryo in vivo requires protein and RNA synthesis: evidence for the role of cell death genes. Dev Biol 138: 104–113.

Oppenheim RW, Prevette D, Yin QW, Collins F, MacDonald J (1991) Control of embryonic motoneuron survival in vivo by ciliary neurotrophic factor. Science 29: 1616–1618.

Oppenheim RW, Reitzel J (1975) Ontogeny of behavioral sensitivity to strychnine in the chick embryo: Evidence for the early onset of CNS inhibition. Brain Behav Evol 11: 130–159.

Oppenheim RW, Wiese S, Prevette D, Armanini M, Wang S, Houenou LJ, Holtmann B, Gotz R, Pennica D, Sendtner M (2001) Cardiotrophin-1, a Muscle-Derived Cytokine, Is Required for the Survival of Subpopulations of Developing Motoneurons. J Neurosci 21: 1283–1291.

Oppenheim RW, Yin QW, Prevette D, Yan Q (1992) Brain-derived neurotrophic factor rescues developing avian motoneurons from cell death. Nature 360: 755–757.

O'Rourke NA, Sullivan DP, Kaznowski CE, Jacobs AA, McConnell SK. Tangential migration of neurons in the developing cerebral cortex. Development. 1995 Jul;121(7): 2165–2176.

Osborne KA, Robichon A, Burgess E, Butland S, Shaw RA, Coulthard A, Pereira HS, Greenspan RJ, Sokolowski MB (1997) Natural behavior polymorphism due to a cGMP-dependent protein kinase of Drosophila. Science 277: 834–836.

Overman WH (1990) Performance on traditional matching to sample, non-matching to sample, and object discrimination tasks by 12- and 32-month-old children. New York Acad Sci 608: 365–385.

Owens DF, Boyce LH, Davis MBE, Kriegstein AR (1996) Excitatory GABA responses in embryonic and neonatal cortical slices demonstrated by gramicidin perforated-patch recordings and calcium imaging. J Neurosci 16: 6414–6423.

Ozaki M, Sasner M, Yano R, Lu HS, Buonanno A (1997) Neuregulin-beta induces expression of an NMDA-receptor subunit. Nature 390: 691–694.

Ozaki S, Snider WD (1997) Initial trajectories of sensory axons toward laminar targets in the developing mouse spinal cord. J Comp Neurol 380: 215–229.

Packard M, Koo ES, Gorczyca M, Sharpe J, Cumberledge S, Budnik V (2002) The Drosophila Wnt, wingless, provides an essential signal for pre- and postsynaptic differentiation. Cell 111: 319–330.

Palade GE, Palay SL (1954) Electron microscope observations of interneuronal and neuromuscular synapses. Anat Rec 118: 335–336.

Palca J (1991) Famous monkeys provide surprising results [news]. Science 252(5014): 1789.

Pappas GD, Purpura DP (1964) Electron microscopy of immature human and feline hippocampus. Prog Brain Res 4: 176–186.

Paradis S, Sweeney ST, Davis GW (2001) Homeostatic control of presynaptic release is triggered by postsynaptic membrane depolarization. Neuron 30: 737–749.

Parks TN (1979) Afferent influences on the development of the brain stem auditory nuclei of the chicken: otocyst ablation. J Comp Neurol 183: 665–678.

Parnavelas JG, Barfield JA, Franke E, Luskin MB Separate progenitor cells give rise to pyramidal and nonpyramidal neurons in the rat telencephalon. Cereb Cortex 1991 Nov–Dec;1(6): 463–468.

Pattyn A, Morin X, Cremer H, Goridis C, Brunet JF (1999) The homeobox gene Phox2b is essential for the development of autonomic neural crest derivatives. Nature 399: 366–370.

Payne JA, Rivera C, Voipio J, Kaila K (2003) Cation-chloride cotransporters in neuronal communication, development, and trauma. Trends Neurosci 26: 199–206.

Penfield W (1954) The excitable cortex in conscious man. Liverpool: Liverpool University Press.

Penfield W, Rasmussen T (1950) The Cerebral Cortex of Man: A Clinical Study of Localization of Function. Macmillan: New York.

Péquignot Y, Clarke PG (1992) Changes in lamination and neuronal survival in the isthmo-optic nucleus following the intraocular injection of tetrodotoxin in chick embryos. J Comp Neurol 321: 336–350.

Person AL, Cerretti DP, Pasquale EB, Rubel EW, Cramer KS (2004) Tonotopic gradients of Eph family proteins in the chick nucleus laminaris during synaptogenesis. J Neurobiol 60: 28–39.

Personius KE, Balice-Gordon RJ (2001) Loss of correlated motor neuron activity during synaptic competition at developing neuromuscular synapses. Neuron 31: 395–408.

Peterson ER, Crain SM (1981) Preferential growth of neurites from isolated fetal mouse dorsal root ganglia in relation to specific regions of co-cultured spinal cord explants. Brain Res 254: 363–382.

Petitto LA, Marentette PF (1991) Babbling in the manual mode: evidence for the ontogeny of language. Science 251: 1493–1496.

Phelan P, Nakagawa M, Wilkin MB, Moffat KG, O'Kane CJ, Davies JA, Bacon JP (1996) Mutations in shaking-B prevent electrical synapse formation in the Drosophila giant fiber system. J Neurosci 16: 1101–1113.

Phillips GR, Tanaka H, Frank M, Elste A, Fidler L, Benson DL, Colman DR (2003) Gamma-protocadherins are targeted to subsets of synapses and intracellular organelles in neurons. J Neurosci 23: 5096–5104.

Phillips WD, Kopta C, Blount P, Gardner PD, Steinbach JH, Merlie JP (1991) ACh receptor-rich domains organized in fibroblasts by recombinant 43-Kilodalton protein. Science 251: 568–570.

Phoenix CH, Goy RW, Gerall AA, Young AC (1959) Organizing action of prenatally administered testosterone propionate on the tissues mediating mating behavior in the female guinea pig. Endocrinol 65: 369–382.

Piatt J (1955) Regeneration in the spinal cord of the salamander. J Exp Zool 129: 177–207.

Piccolo S, Sasai Y, Lu B, De Robertis EM. Dorsoventral patterning in Xenopus: inhibition of ventral signals by direct binding of chordin to BMP-4. Cell. 1996 Aug 23;86(4): 589–598.

Picken Bahrey HL, Moody WJ (2003) Early development of voltage-gated ion currents and firing properties in neurons of the mouse cerebral cortex. J Neurophysiol 89: 1761–1773.

Pilar G, Landmesser L, Burstein L (1980) Competition for survival among developing ciliary ganglion cells. J Neurophysiol 43: 233–254.

Pimenta AF, Zhukareva V, Barbe MF, Reinoso BS, Grimley C, Henzel W, Fischer I, Levitt P (1995) The limbic system-associated membrane protein is an Ig superfamily member that mediates selective neuronal growth and axon targeting. Neuron 15: 287–297.

Pini A (1993) Chemorepulsion of axons in the developing mammalian central nervous system. Science 261: 95–98.

Piper M, Salih S, Weinl C, Holt C, Harris WA (2004) Rapid desensitisation and resensitisation of retinal growth cones. Neuron.

Pipes GC, Lin Q, Riley SE, Goodman CS (2001) The Beat generation: a multigene family encoding IgSF proteins related to the Beat axon guidance molecule in Drosophila. Development 128: 4545–4552.

Pittman A, Chien CB (2002) Understanding dorsoventral topography: backwards and forwards. Neuron 35: 409–411.

Pittman R, Oppenheim RW (1979) Cell death of motoneurons in the chick embryo spinal cord. IV. Evidence that a functional neuromuscular interaction is involved in the regulation of naturally occurring cell death and the stabilization of synapses. J Comp Neurol 187: 425–446

Pittman RN (1985) Release of plasminogen activator and a calcium-dependent metalloprotease from cultured sympathetic and sensory neurons. Dev Biol 110(1): 91–101.

Plunkett JA, Simmons RB, Walthall WW (1996) Dynamic interactions between nerve and muscle in Caenorhabditis elegans. Dev Biol 175: 154–165.

Pompeiano M, Blaschke AJ, Flavell RA, Srinivasan A, Chun J (2000) Decreased Apoptosis in Proliferative and Postmitotic Regions of the Caspase 3-Deficient Embryonic Central Nervous System. J Comp Neurol 423: 1–12.

Pons TP, Garraghty PE, Ommaya AK, Kaas JH, Taub E, Mishkin M (1991) Massive cortical reorganization after sensory deafferentation in adult macaques. Science 252(5014): 1857–1860.

Poo MM (1982) Rapid lateral diffusion of functional ACh receptors in embryonic muscle cell membrane. Nature 295: 332–334.

Porter BE, Weis J, Sanes JR (1995) A motoneuron-selective stop signal in the synaptic protein s-laminin. Neuron 14: 549–559.

Potter SM, Zheng C, Koos DS, Feinstein P, Fraser SE, Mombaerts P (2001) Structure and emergence of specific olfactory glomeruli in the mouse. J Neurosci 21: 9713–9723.

Potts RA, Dreher B, Bennett MR (1982) The loss of ganglion cells in the developing retina of the rat. Dev Brain Res 3: 481–486.

Pratt KG, Watt AJ, Griffith LC, Nelson SB, Turrigiano GG (2003) Activity-dependent remodeling of presynaptic inputs by postsynaptic expression of activated CaMKII. Neuron 39: 269–281.

Preyer, W (1885) Specielle physiologie des Embryo. Leipzig, Grieben.

Price SR, De Marco Garcia NV, Ranscht B, Jessell TM (2002) Regulation of motor neuron pool sorting by differential expression of type II cadherins. Cell 109: 205–216.

Prokop A, Landgraf M, Rushton E, Broadie K, Bate M (1996) Presynaptic development at the neruomuscular junction: Assembly

and localization of presynaptic active zones. Neuron 17: 617–626.

Provine, RR (1982) Preflight development of bilateral wing coordination in the chick (Gallus domesticus): effects of induced bilateral wing asymmetry. Dev Psychobiol 15: 245–255.

Provine, RR (1972) Ontogeny of bioelectric activity in the spinal cord of the chick embryo and its behavioral implications. Brain Res 41: 365–378.

Puelles L, Rubenstein JL. Expression patterns of homeobox and other putative regulatory genes in the embryonic mouse forebrain suggest a neuromeric organization. Trends Neurosci. 1993 Nov;16(11): 472–479.

Purves D, Lichtman JW (1985) Geometrical differences among homologous neurons in mammals. Science 228: 298–302.

Purves D, Thompson W, Yip JW (1981) Re-innervation of ganglia transplanted to the neck from different levels of the guinea-pig sympathetic chain. J Physiol 313: 49–63.

Qian X, Davis AA, Goderie SK, Temple S. FGF2 concentration regulates the generation of neurons and glia from multipotent cortical stem cells. Neuron. 1997 Jan;18(1): 81–93.

Qiu J, Cai D, Dai H, McAtee M, Hoffman PN, Bregman BS, Filbin MT (2002) Spinal axon regeneration induced by elevation of cyclic AMP. Neuron 34: 895–903.

Qu Z, Huganir RL (1994) Comparison of innervation and agrin-induced tyrosine phosphorylation of the nicotinic acetylcholine receptor. J Neurosci 14: 6834–6841.

Rabacchi S, Bailly Y, Delhaye-Bouchaud N, Mariani J (1992) Involvement of the N-methyl D-aspartate receptor in synapse elimination during cerebellar development. Science 256: 1823–1825.

Radel JD, Hankin MH, Lund RD (1990) Proximity as a factor in the innervation of host brain regions by retinal transplants. J Comp Neurol 300: 211–229.

Raff MC, Lillien LE, Richardson WD, Burne JF, Noble MD. Platelet-derived growth factor from astrocytes drives the clock that times oligodendrocyte development in culture. Nature. 1988 Jun 9;333(6173): 562–565.

Raff MC, Miller RH, Noble M. A glial progenitor cell that develops in vitro into an astrocyte or an oligodendrocyte depending on culture medium. Nature. 1983 Jun 2–8;303(5916): 390–396.

Raghavan S, White RA (1997) Connectin mediates adhesion in Drosophila. Neuron 18: 873–880.

Raisman G, Field PM (1973a) A quantitative investigation of the development of collateral reinnervation after partial deafferentation of the septal nuclei. Brain Res 50: 241–264.

Raisman G, Field PM (1973b) Sexual dimorphism in the neuropil of the preoptic area of the rat and its dependence on neonatal androgen. Brain Res 54: 1–29.

Rajagopalan S, Vivancos V, Nicolas E, Dickson BJ (2000) Selecting a longitudinal pathway: Robo receptors specify the lateral position of axons in the Drosophila CNS. Cell 103: 1033–1045.

Rakic P. Neuron-glia relationship during granule cell migration in developing cerebellar cortex. A Golgi and electronmicroscopic study in Macacus Rhesus. J Comp Neurol. 1971 Mar;141(3): 283–312.

Rakic P (1972) Mode of cell migration to the superficial layers of fetal monkey neocortex. J Comp Neurol 145: 61–84.

Rakic S, Zecevic N (2000) Programmed cell death in the developing human telencephalon. Eur J Neurosci 12: 2721–2734.

Ramachandran VS, Rogers-Ramachandran D (2000) Phantom limbs and neural plasticity. Arch Neurol 57: 317–320.

Ramarao MK, Bianchetta MJ, Lanken J, Cohen JB (2001) Role of rapsyn tetratricopeptide repeat and coiled-coil domains in self-association and nicotinic acetylcholine receptor clustering. J Biol Chem 276: 7475–7483.

Ramoa AS, McCormick DA (1994) Developmental changes in electrophysiological properties of LGNd neurons during reorganization of retinogeniculate connections. J Neurosci 14: 2089–2097.

Ramón y Cajal S (1890) A quelle epoque aparaissent les expansions des cellule neurveuses de la moelle epinere du poulet. Anat Anz 5: 609–613.

Ramón y Cajal S (1905) Genèse des fibres nerveuses de l'embryon et observations contraires à la thérie catenaire. Trab Lab Invest Biol, Univ Madrid 4: 219–284.

Ramón y Cajal S (1928) Degeneration and Regeneration of the Nervous System. New York: Hafner.

Ramón y Cajal, S Histologie du systeme nerveux de 1'homme et des vertebres. In: Consejo Sup Invest Cient, Madrid (1952), p. 589.

Ramos RG, Igloi GL, Lichte B, Baumann U, Maier D, Schneider T, Brandstatter JH, Frohlich A, Fischbach KF (1993) The irregular chiasm C-roughest locus of Drosophila, which affects axonal projections and programmed cell death, encodes a novel immunoglobulin-like protein. Genes Dev 7: 2533–2547.

Rankin CH, Carew TJ (1988) Dishabituation and sensitization emerge as separate processes during development in Aplysia. J Neurosci 8: 197–211.

Rao A, Craig AM (1997) Activity regulates the synaptic localization of the NMDA receptor in hippocampal neurons. Neuron 19: 801–812.

Raoul C, Henderson CE, Pettmann B (1999) Programmed cell death of embryonic motoneurons triggered through the Fas death receptor. J Cell Biol 147: 1049–1062.

Raper JA, Bastiani MJ, Goodman CS (1984) Pathfinding by neuronal growth cones in grasshopper embryos. IV. The effects of ablating the A and P axons upon the behavior of the G growth cone. J Neurosci 4: 2329–2345.

Raven CP, Kloos J (1945) Induction by medial and lateral pieces of the archenteron roof with special reference to the determination of the neural crest. Acta Neerl Morphol 5: 348–362.

Raynaud A, Clairambault P, Renous S, Gasc JP (1977) Organisation des cornes ventrales de la moelle epiniere, dans les regions brachiale et lombiare, chez les embryons de Reptiles serpentiformes et de Reptiles a membres bien developpes. C R Acad Sci Hebd Seances Acad Sci D 19: 1507–1509.

Raymond PA, Rivlin PK. Germinal cells in the goldfish retina that produce rod photoreceptors. Dev Biol. 1987 Jul;122(l): 120–138.

Ready DF (1989) A multifaceted approach to neural development. Trends Neurosci 12: 102–110.

Ready DF, Hanson TE, Benzer S (1976) Development of the Drosophila retina, a neurocrystalline lattice. Dev Biol 53: 217–240.

Recanzone GH, Jenkins WM, Hradek GT, Merzenich MM (1992) Progressive improvement in discriminative abilities in adult owl monkeys performing a tactile frequency discrimination task. J Neurophysiol 67: 1015–1030.

Recanzone GH, Merzenich MM, Jenkins WM (1992). Frequency discrimination training engaging a restricted skin surface results in an emergence of a cutaneous response zone in cortical area 3a. J Neurophysiol 67(5): 1057–1070.

Redfern PA (1970) Neuromuscular transmission in new-born rats. J Physiol 209: 701–709.

Redies, C (1997). Cadherins and the formation of neural circuitry in the vertebrate CNS. Cell Tissue Res 290(2): 405–413.

Redies C, Takeichi M (1996) Cadherins in the developing central nervous system: an adhesive code for segmental and functional subdivisions. Dev Biol 180: 413–423.

Rees RP, Bunge MB, Bunge RP (1976) Morphological changes in the neuritic growth cone and target neuron during synaptic junction development in culture. J Cell Biol 68: 240–263.

Reh TA, Kljavin IJ (1989) Age of differentiation determines rat retinal germinal cell phenotype: induction of differentiation by dissociation. J Neurosci 9: 4179–4189.

Reichling DB, Kyrozis A, Wang J, MacDermott AB (1994) Mechanisms of GABA and glycine depolarization-induced calcium transients in rat dorsal horn neurons. J Physiol 476: 411–421.

Reimer MK, Mokshagundam SP, Wyler K, Sundler F, Ahren B, Stagner JI (2003) Local growth factors are beneficial for the autonomic reinnervation of transplanted islets in rats. Pancreas 26: 392–397.

Reiness CG, Hall ZW (1981) The developmental change in immunological properties of the acetylcholine receptor in rat muscle. Dev Biol 81: 324–331.

Reiness CG, Weinberg CB (1981) Metabolic stabilization of acetylcholine receptors at newly formed neuromuscular junctions in rat. Dev Biol 84: 247–254.

Reinke R, Zipursky SL (1988) Cell-cell interaction in the Drosophila retina: the bride of sevenless gene is required in photoreceptor cell R8 for R7 cell development. Cell 55: 321–330.

Reissmann E, Ernsberger U, Francis-West PH, Rueger D, Brickell PM, Rohrer H (1996) Involvement of bone morphogenetic protein-4 and bone morphogenetic protein-7 in the differentiation of the adrenergic phenotype in developing sympathetic neurons. Development 122: 2079–2088.

Reist NE, Magill C, McMahon UJ (1987) Agrin-like molecules at synaptic sites in normal, denervated, and damaged skeletal muscles. J Cell Biol 105: 2457–2469.

Reist NE, Werle MJ, McMahon UJ (1992) Agrin released by motor neurons induces the aggregation of acetylcholine receptors at neuromuscular junctions. Neuron 8: 865–868.

Reyes A, Sakmann B (1999) Developmental switch in the short-term modification of unitary EPSPs evoked in layer 2/3 and layer 5 pyramidal neurons of rat neocortex. J Neurosci 19: 3827–3835.

Reynolds BA, Weiss S. Generation of neurons and astrocytes from isolated cells of the adult mammalian central nervous system. Science. 1992 Mar 27;255(5052): 1707–1710.

Reynolds SA, French KA, Baader A, Kristan WB, Jr (1998) Development of spontaneous and evoked behaviors in the medicinal leech. J Comp Neurol 402(2): 168–180.

Rhinn M, Brand M. The midbrain–hindbrain boundary organizer. Curr Opin Neurobiol. 2001 Feb;11(1): 34–42.

Ribera AB, Spitzer NC (1989) A critical period of transcription required for differentiation of the action potential of spinal neurons. Neuron 2: 1055–1062.

Riccio A, Ahn S, Davenport CM, Blendy JA, Ginty DD (1999) Mediation by a CREB family transcription factor of NGF-dependent survival of sympathetic neurons. Science 286: 2358–2361.

Riccio A, Pierchala BA, Ciarallo CL, Ginty DD (1997) An NGF-TrkA–Mediated Retrograde Signal to Transcription Factor CREB in Sympathetic Neurons. Science 277: 1097–1100.

Richardson PM, McGuinness UM, Aguayo AJ (1980) Axons from CNS neurons regenerate into PNS grafts. Nature 284: 264–265.

Ridge RMAP, Betz WJ (1984) The effect of selective, chronic stimulation on motor unit size in developing rat muscle. J Neurosci 4: 2614–2620.

Rieff HI, Raetzman LT, Sapp DW, Yeh HH, Siegel RE, Corfas G (1999) Neuregulin induces GABAA receptor subunit expression and neurite outgrowth in cerebellar granule cells. J Neurosci 19: 10757–10766.

Rimer M, Cohen I, Lømo T, Burden SJ, McMahan UJ (1998) Neuregulins and ErbB receptors at neuromuscular junctions and at agrin-induced postsynaptic-like apparatus in skeletal muscle. Mol Cell Neurosci 12: 1–15.

Rimer M, Prieto AL, Weber JL, Colasante C, Ponomareva O, Fromm L, Schwab MH, Lai C, Burden SJ (2004) Neuregulin-2 is synthesized by motor neurons and terminal Schwann cells and activates acetylcholine receptor transcription in muscle cells expressing ErbB4. Mol Cell Neurosci 26: 271–281.

Riquimaroux H, Gaioni SJ, Suga N (1991) Cortical computational maps control auditory perception. Science 1: 565–568

Ritzenthaler S, Suzuki E, Chiba A (2000) Postsynaptic filopodia in muscle cells interact with innervating motoneuron axons. Nat Neurosci 3: 1012–1017.

Rivera C, Li H, Thomas-Crusells J, Lahtinen H, Viitanen T, Nanobashvili A, Kokaia Z, Airaksinen MS, Voipio J, Kaila K, Saarma M (2002) BDNF-induced TrkB activation down-regulates the K+-Cl-cotransporter KCC2 and impairs neuronal Cl-extrusion. J Cell Biol 159: 747–752.

Robinow S, Talbot WS, Hogness DS, Truman JW (1993) Programmed cell death in the Drosophila CNS is ecdysone-regulated and coupled with a specific ecdysone receptor isoform. Development 119: 1251–1259.

Roche KW, Standley S, McCallum J, Dune Ly C, Ehlers MD, Wenthold RJ (2001) Molecular determinants of NMDA receptor internalization. Nat Neurosci 4: 794–802.

Roe AW, Pallas SL, Kwon YH, Sur M (1992) Visual projections routed to the auditory pathway in ferrets: receptive fields of visual neurons in primary auditory cortex. J Neurosci 12: 3651–3664.

Roelink H, Augsburger A, Heemskerk J, Korzh V, Norlin S, Ruiz i Altaba A, Tanabe Y, Placzek M, Edlund T, Jessell TM, et al. Floor plate and motor neuron induction by vhh-1, a vertebrate homolog of hedgehog expressed by the notochord. Cell. 1994 Feb 25;76(4): 761–775.

Roessler E, Belloni E, Gaudenz K, Vargas F, Scherer SW, Tsui LC, Muenke M. Mutations in the C-terminal domain of Sonic Hedgehog cause holoprosencephaly. Hum Mol Genet. 1997 Oct;6(11): 1847–1853.

Roessler E, Ward DE, Gaudenz K, Belloni E, Scherer SW, Donnai D, Siegel-Bartelt J, Tsui LC, Muenke M. Cytogenetic rearrangements involving the loss of the Sonic Hedgehog gene at 7q36 cause holoprosencephaly. Hum Genet. 1997 Aug;100(2): 172–181.

Rohrbough J, Spitzer NC (1996) Regulation of intracellular Cl- levels by Na(+)-dependent Cl- cotransport distinguishes depolarizing from hyperpolarizing GABAA receptor-mediated responses in spinal neurons. J Neurosci 16: 82–91.

Role LW (1985) Neural regulation of acetylcholine sensitivity in embryonic sympathetic neurons. Proc Natl Acad Soc 85: 2825–2829.

Role LW, Matossian VR, O'Brien RJ, Fischbach GD (1985) On the mechanism of acetylcholine receptor accumulation at newly formed synapses on chick myotubes. J Neurosci 5: 2197–2204.

Role LW, Roufa DG, Fischbach GD (1987) The distribution of acetylcholine receptor clusters and sites of transmitter release along chick ciliary ganglion neurite-myotube contacts in culture. J Cell Biol 104: 371–379.

Rosentreter SM, Davenport RW, Loschinger J, Huf J, Jung J, Bonhoeffer F (1998) Response of retinal ganglion cell axons to striped linear gradients of repellent guidance molecules. J Neurobiol 37: 541–562.

Rosoff WJ, Urbach JS, Estrick MA, McAllister RG, Richards LJ, Goodhill GJ (2004) A novel chemotaxis assay reveals the extreme sensitivity of axons to molecular gradients. Nature Neurosci 7: 678–682.

Ross LS, Parrett T, Easter SS, Jr (1992) Axonogenesis and morphogenesis in the embryonic zebrafish brain. J Neurosci 12: 467–482.

Rotzler S, Schramek H, Brenner HR (1991) Metabolic stabilization of endplate acetylcholine receptors regulated by Ca2+ influx associated with muscle activity. Nature 349: 337–339.

Roy N, Mahadevan MS, McLean M, Shutler G, Yaraghi Z, Farahani R, Baird S, Besner-Johnston A, Lefebvre C, Kang X, Salih M, Arbry H, Tamai K, Guan X, Ioannon P, Crawford TO, deJong PJ, Surh L, Ikeda J-E, Korneluk RG, MacKenzie RG (1995) The gene for neuronal apoptosis inhibitory protein is partially deleted in individuals with spinal muscular atrophy. Cell 80: 167–178.

Royer S, Pare D (2003) Conservation of total synaptic weight through balanced synaptic depression and potentiation. Nature 422: 518–522.

Rubel EW, Smith DJ, Miller LC (1976) Organization and development of brain stem auditory nuclei of the chicken: Ontogeny of n. magnocellularis and n laminaris.

Rubin GM (1991) Signal transduction and the fate of the R7 photoreceptor in Drosophila. Trends Genet 7: 372–377.

Rudhard Y, Kneussel M, Nassar MA, Rast GF, Annala AJ, Chen PE, Tigaret CM, Dean I, Roes J, Gibb AJ, Hunt SP, Schoepfer R (2003) Absence of Whisker-related pattern formation in mice with NMDA receptors lacking coincidence detection properties and calcium signaling. J Neurosci 23: 2323–2332.

Ruegg MA, Tsim KWK, Horton SE, Kroger S, Escher G, Gensch EM, McMahan UJ (1992) The agrin gene codes for a family of basal lamina proteins that differ in function and distribution. Neuron 8: 691–699.

Ruiz i Altaba A. Planar and vertical signals in the induction and patterning of the Xenopus nervous system. Development. 1992 Sep;116(1): 67–80.

Rupp F, Payan DG, Magill-Solc C, Cowan DM, Scheller RH (1991) Structure and expression of a rat agrin. Neuron 6: 811–823.

Russier M, Kopysova IL, Ankri N, Ferrand N, Debanne D (2002) GABA and glycine co-release optimizes functional inhibition in rat brainstem motoneurons in vitro. J Physiol 541: 123–137.

Ruthazer ES, Akerman CJ, Cline HT (2003) Control of axon branch dynamics by correlated activity in vivo. Science 301: 66–70.

Ruthel G, Hollenbeck PJ (2003) Response of mitochondrial traffic to axon determination and differential branch growth. J Neurosci 23: 8618–8624.

Sabo SL, McAllister AK (2003) Mobility and cycling of synaptic protein-containing vesicles in axonal growth cone filopodia. Nat Neurosci 6: 1264–1269.

Saitoe M, Schwarz TL, Umbach JA, Gundersen CB, Kidokoro Y (2001) Absence of junctional glutamate receptor clusters in Drosophila mutants lacking spontaneous transmitter release. Science 293: 514–517.

Sakuma Y (1984) Influences of neonatal gonadectomy or androgen exposure on the sexual differentiation of the rat ventromedial hypothalamus. J Physiol (Lond) 349: 273–286.

Salmelin R, Service E, Kiesila P, Uutela K, Salonen O (1996) Impaired visual word processing in dyslexia revealed with magnetoencephalography. Ann Neurol 40: 157–162.

Salpeter MM, Harris R (1983) Distribution and turnover rate of acetylcholine receptors throughout the junction folds at a vertebrate neuromuscular junction. J Cell Biol 96: 1781–1785.

Sandrock AW Jr, Dryer SE, Rosen KM, Gozani SN, Kramer R, Theill LE, Fischbach GD (1997) Maintenance of acetylcholine receptor number by neuregulins at the neuromuscular junction in vivo. Science 276: 599–603.

Sandrock AW Jr, Goodearl AD, Yin QW, Chang D, Fischbach GD (1995) ARIA is concentrated in nerve terminals at neuromuscular junctions and at other synapses. J Neurosci 15: 6124–6136.

Sanes DH (1993) The development of synaptic function and integration in the central auditory system. J Neurosci 13: 2627–2637.

Sanes DH, Constantine-Paton M (1985b) The sharpening of frequency tuning curves requires patterned activity during development in the mouse, Mus musculus. J Neurosci 5: 1152–1166.

Sanes DH, Hafidi A (1996) Glycinergic transmission regulates dendrite size in organotypic culture. J Neurobiol 4: 503–511.

Sanes DH, Markowitz S, Bernstein J, Wardlow J (1992) The influence of inhibitory afferents on the development of postsynaptic dendritic arbors. J Comp Neurol 321: 637–644.

Sanes DH, Siverls V (1991) Development and specificity of inhibitory terminal arborizations in the central nervous system. J Neurobiol 22: 837–854.

Sanes DH, Takacs C (1993) Activity-dependent refinement of inhibitory connections. Eur J Neurosci 5: 570–574.

Sanes JR, Johnson YR, Kotzbauer PT, Mudd J, Hanley T, Martinou JC, Merlie JP (1991) Selective expression of an acetylcholine receptor-lacZ transgene in synaptic nuclei of adult muscle fibers. Dev 113: 1181–1191.

Sanes JR, Yamagata M (1999) Formation of lamina-specific synaptic connections. Curr Opin Neurobiol 9: 79–87.

Sasai Y, Lu B, Steinbeisser H, De Robertis EM. Regulation of neural induction by the Chd and Bmp-4 antagonistic patterning signals in Xenopus. Nature. 1995 Jul 27;376(6538): 333–336.

Sasai Y, Lu B, Steinbeisser H, Geissert D, Gont LK, De Robertis EM. Xenopus chordin: a novel dorsalizing factor activated by organizer-specific homeobox genes. Cell. 1994 Dec 2;79(5): 779–790.

Sato T, Matsumoto T, Kawano H, Watanabe T, Uematsu Y, Sekine K, Fukuda T, Aihara K, Krust A, Yamada T, Nakamichi Y, Yamamoto Y, Nakamura T, Yoshimura K, Yoshizawa T, Metzger D, Chambon P, Kato S (2004) Brain masculinization requires androgen receptor function. Proc Natl Acad Sci USA 101: 1673–1678.

Sattler R, Charlton MP, Hafner M, Tymianski M (1998) Distinct influx pathways, not calcium load, determine neuronal vulnerability to calcium neurotoxicity. J Neurochem 71: 2349–2364.

Sauer FC (1953) Mitosis in the neural tube. J Comp Neurol 62: 377–405.

Scheetz AJ, Nairn AC, Constantine-Paton M (1997) N-methyl-D-aspartate receptor activation and visual activity induce elongation factor-2 phosphorylation in amphibian tecta: a role for N-methyl-D-aspartate receptors in controlling protein synthesis. Proc Natl Acad Sci USA 94: 14770–14775.

Scheiffele P, Fan J, Choih J, Fetter R, Serafini T (2000) Neuroligin expressed in nonneuronal cells triggers presynaptic development in contacting axons. Cell 101: 657–669.

Scherer WJ, Udin SB (1989) N-methyl-D-aspartate antagonists prevent interaction of binocular maps in Xenopus tectum. J Neurosci 9: 3837–3843.

Schlinger BA (1998) Sexual differentiation of avian brain and behavior: Current views on gonadal hormone-dependent and independent mechanisms. Ann Rev Physiol 60: 407–429.

Schmidt J, Coen T (1995) Changes in retinal arbors in compressed projections to half tecta in goldfish. J Neurobiol 28: 409–418.

Schmidt JT, Buzzard M (1990) Activity-driven sharpening of the regenerating retinotectal projection: effects of blocking or synchronizing activity on the morphology of individual regenerating arbors. J Neurobiol 21: 900–917.

Schmidt JT, Easter SS (1978) Independent biaxial reorganization of the retinotectal projection: a reassessment. Exp Brain Res 31: 155–162.

Schmidt TA, Larsen JS, Kjeldsen (1992) Quantification of rat cerebral cortex Na+,K(+)-ATPase: effect of age and potassium depletion. J Neuroschem 59: 2094–2104.

Schmucker D, Clemens JC, Shu H, Worby CA, Xiao J, Muda M, Dixon JE, Zipursky SL (2000) Drosophila Dscam is an axon guidance receptor exhibiting extraordinary molecular diversity. Cell 101: 671–684.

Schneider C, Wicht H, Enderich J, Wegner M, Rohrer H (1999) Bone morphogenetic proteins are required in vivo for the generation of sympathetic neurons. Neuron 24: 861–870.

Schneider MB, Standop J, Ulrich A, Wittel U, Friess H, Andren-Sandberg A, Pour PM (2001) Expression of nerve growth factors in pancreatic neural tissue and pancreatic cancer. J Histochem Cytochem 49: 1205–1210.

Schneider U, Schleussner E, Haueisen J, Nowak H, Seewald HJ (2001) Signal analysis of auditory evoked cortical fields in fetal magnetoencephalography. Brain Topogr 14: 69–80.

Schneiderman AM, Hildebrand JG, Brennan MM,Tumlinson JH (1986) Trans-sexually grafted antennae alter pheromone-directed behaviour in a moth. Nature 323(6091): 801–803.

Schnell L, Schwab ME (1990) Axonal regeneration in the rat spinal cord produced by an antibody against myelin-associated neurite growth inhibitors. Nature 343: 269–272.

Schrader N, Kim EY, Winking J, Paulukat J, Schindelin H, Schwarz G (2004) Biochemical characterization of the high affinity binding between the glycine receptor and gephyrin. J Biol Chem 279: 18733–18741.

Schulte-Merker S, Lee KJ, McMahon AP, Hammerschmidt M. The zebrafish organizer requires chordino. Nature. 1997 Jun 26;387(6636): 862–863.

Schuster CM, Davis GW, Fetter RD, Goodman CS (1996a) Genetic dissection of structural and functional components of synaptic plasticity. I. Fasciclin II controls synaptic stabilization and growth. Neuron 17: 641–654.

Schuster CM, Davis GW, Fetter RD, Goodman CS (1996b) Genetic dissection of structural and functional components of synaptic plasticity. II. Fasciclin II controls presynaptic structural plasticity. Neuron 17: 655–667.

Schwab ME, Bartholdi D (1996) Degeneration and regeneration of axons in the lesioned spinal cord. Physiol Rev 76: 319–370.

Schwartz IR, Pappas GD, Purpura DP (1968) Fine structure of neruons and synapses in the feline hippocampus during postnatal ontogenesis. Exp Neurol 22: 394–407.

Schweisguth F, Gho M, Lecourtois M (1996) Control of cell fate choices by lateral signaling in the adult peripheral nervous system of Drosophila melanogaster. Dev Genet 18: 28–39.

Sealock R, Wray BE, Froehner SC (1984) Ultrastructural localization of the Mr 43,000 protein and the acetylcholine receptor in Torpedo postsynaptic membranes using monoclonal antibodies. J Cell Biol 98: 2239–2244.

Seebach BS, Ziskind-Conhaim L (1994) Formation of transient inappropriate sensorimotor synapses in developing rat spinal cords. J Neurosci 14: 4520–4528.

Seeger M, Tear G, Ferres-Marco D, Goodman CS (1993) Mutations affecting growth cone guidance in Drosophila: genes necessary for guidance toward or away from the midline. Neuron 10: 409–426.

Selleck MA, Bronner-Fraser M (1995) Origins of the avian neural crest: the role of neural plate-epidermal interactions. Development. Feb; 121(2): 525–538.

Serafini T, Kennedy TE, Galko MJ, Mirzayan C, Jessell TM, Tessier-Lavigne M (1994) The netrins define a family of axon outgrowth-promoting proteins homologous to C. elegans UNC-6. Cell 78: 409–424.

Serpinskaya AS, Feng G, Sanes JR, Craig AM (1999) Synapse formation by hippocampal neurons from agrin-deficient mice. Dev Biol 205: 65–78.

Shainberg A, Burstein M (1976) Decrease of acetylcholine receptor synthesis in muscle cultures by electrical stimulation. Nature 264: 368–369.

Shapira M, Zhai RG, Dresbach T, Bresler T, Torres VI, Gundelfinger ED, Ziv NE, Garner CC (2003) Unitary assembly of presynaptic active zones from Piccolo-Bassoon transport vesicles. Neuron 38: 237–252.

Shen K, Bargmann CI (2003) The immunoglobulin superfamily protein SYG-1 determines the location of specific synapses in C. elegans. Cell 112: 619–630.

Shen K, Fetter RD, Bargmann CI (2004) Synaptic specificity is generated by the synaptic guidepost protein SYG-2 and its receptor, SYG-1. Cell 116: 869–881.

Sherman SM, Spear PD (1982) Organization of visual pathways in normal and visually deprived cats. Physiol Rev 62: 738–855.

Sherrington CS (1906) The Integrative Action of the Nervous System. New York: Charles Scribner's Sons.

Shewan D, Dwivedy A, Anderson R, Holt CE (2002) Age-related changes underlie switch in netrin-1 responsiveness as growth cones advance along visual pathway. Nat Neurosci 5: 955–962.

Shi SH, Hayashi Y, Petralia RS, Zaman SH, Wenthold RJ, Svoboda K, Malinow R (1999) Rapid spine delivery and redistribution of AMPA receptors after synaptic NMDA receptor activation. Science 284: 1811–1816.

Shi SH, Jan LY, Jan YN (2003) Hippocampal neuronal polarity specified by spatially localized mPar3/mPar6 and PI 3-kinase activity. Cell 112: 63–75.

Shimizu S, Narita M, Tsujimoto Y (1999) Bcl-2 family proteins regulate the release of apoptogenic cytochrome c by the mitochondrial channel VDAC. Nature 399: 483–487.

Shindler KS, Latham CB, Roth KA (1997) Bax deficiency prevents the increased cell death of immature neurons in bcl-x-deficient mice. J Neurosci 17: 3112–3119.

Shirasaki R, Katsumata R, Murakami F (1998) Change in chemoattractant responsiveness of developing axons at an intermediate target. Science 279: 105–107.

Shitaka Y, Matsuki N, Saito H, Katsuki H (1996) Basic fibroblast growth factor increases functional L-type Ca2+ channels in fetal hippocampal neurons: Implications for neurite morphogenesis in vitro. J Neurosci 16: 6476–6489.

Shyng S-L, Xu R, Salpeter MM (1991) cAMP stabilizes the degradation of original junctional acetylcholine receptors in denervated muscle. Neuron 6: 469–475.

Si J, Luo Z, Mei L (1996) Induction of acetylcholine receptor gene expression by ARIA requires activation of mitogen-activated protein kinase. J Biol Chem 16;19752–19759.

Sidman R (1961) Histogenesis of the mouse retina studied with thymidine 3H. In: The Structure of the Eye (Smelse G, ed), pp 487–506. New York: Academic Press.

Sidman RL (1961) Histogenesis of the mouse retina studied with tritiated thymidine. In "The structure of the eye" (GK Smelser, Ed.). Academic Press, New York.

Siegelbaum SA, Trautmann A, Koenig J (1984) Single acetylcholine-activated channel currents indeveloping muscle cells. Dev Biol 104: 366–379.

Silver J, Ogawa MY (1983) Postnatally induced formation of the corpus callosum in acallosal mice on glia-coated cellulose bridges. Science 220: 1067–1069.

Simeone A, Acampora D, Nigro V, Faiella A, D'Esposito M, Stornaiuolo A, Mavilio F, Boncinelli E. Differential regulation by retinoic acid of the homeobox genes of the four HOX loci in human embryonal carcinoma cells. Mech Dev. 1991 Mar; 33(3): 215–227.

Simon AM, Hoppe P, Burden SJ (1992) Spatial restriction of AChR gene expression to subsynaptic nuclei. Dev 114: 545–553.

Simon DK, O'Leary DD (1990) Limited topographic specificity in the targeting and branching of mammalian retinal axons. Dev Biol 137: 125–134.

Simon DK, O'Leary DD (1992) Development of topographic order in the mammalian retinocollicular projection. J Neurosci 12: 1212–1232.

Simon DK, Prusky GT, O'Leary DDM, Constantine-Paton M (1992) N-methyl-D-aspartate receptor antagonists disrupt the formation of a mammalian neural map. Proc Natl Acad Sci 89: 10593–10597.

Simons DJ, Durham D, Woolsey TA (1984) Functional organization of mouse and rat SmI barrel cortex following vibrissal damage on different postnatal days. Somatosens Res 1: 207–245.

Simpson HB, Vicario DS (1991) Early estrogen treatment alone causes female zebra finches to produce learned, male-like vocalizations. J Neurobiol 22: 755–776.

Simpson JH, Bland KS, Fetter RD, Goodman CS (2000) Short-range and long-range guidance by Slit and its Robo receptors: a combinatorial code of Robo receptors controls lateral position. Cell 103: 1019–1032.

Singer W (1977a) Control of thalamic transmission by corticofugal and ascending reticular pathways in the visual system. Physiol Rev 57: 386–420.

Singer W (1977b) Effects of monocular deprivation on excitatory and inhibitory pathways in cat striate cortex. Exp Brain Res 24: 25–41.

Skeath JB, Carroll SB. Regulation of proneural gene expression and cell fate during neuroblast segregation in the Drosophila embryo. Development. 1992 Apr; 114(4): 939–946.

Skeath JB, Panganiban G, Selegue J, Carroll SB. Gene regulation in two dimensions: the proneural achaete and scute genes are controlled by combinations of axis-patterning genes through a common intergenic control region. Genes Dev. 1992 Dec; 6(12B): 2606–2619.

Skene JH, Jacobson RD, Snipes GJ, McGuire CB, Norden JJ, Freeman JA (1986) A protein induced during nerve growth (GAP-43) is a major component of growth-cone membranes. Science 233: 783–786.

Sleeman MW, Anderson KD, Lambert PD, Yancopoulos GD, Wiegand SJ (2000) The ciliary neurotrophic factor and its receptor, CNTFR alpha. Pharm Acta Helv 74: 265–272.

Small S, Levine M (1991) The initiation of pair-rule stripes in the Drosophila blastoderm. Curr Opin Genet Dev. Aug; 1(2): 255–260.

Smart, I (1961) The subependymal layer of the mouse brain and its cell production as shown by radioautography after thymidine-H3 injection. J Comp Neurol 116, 325–347.

Smeyne RJ, Klein R, Schnapp A, Long LK, Bryant S, Lewin A, Lira SA, Barbacid M (1994) Severe sensory and sympathetic neuropathies in mice carrying a disrupted Trk/NGF receptor gene. Nature 368: 246–249.

Smith WC, Knecht AK, Wu M, Harland RM. Secreted noggin protein mimics the Spemann organizer in dorsalizing Xenopus mesoderm. Nature. 1993 Feb; 11;361(6412): 547–549.

Smolen AJ (1981) Postnatal development of ganglionic neurons in the absence of preganglionic input: Morphological observations on synapse formation. Dev Brain Res 1: 49–58.

Snyder RL, Rebscher SJ, Cao K, Leake PA, Kelly K (1990) Chronic intracochlear electrical stimulation in the neonatally deafened cat. I: Expansion of central representation. Hearing Res 50: 7–34.

So KL, Pun S, Wan DC, Tsim KW (1996) Cerebellar granule cells express a specific isoform of agrin that lacks the acetylcholine receptor aggregating activity. FEBS Lett 379: 63–68.

Sokolowski MB, Pereira HS, Hughes K (1997) Evolution of foraging behavior in Drosophila by density-dependent selection. Proc Natl Acad Sci U S A 94: 7373–7377.

Sola M, Bavro VN, Timmins J, Franz T, Ricard-Blum S, Schoehn G, Ruigrok RW, Paarmann I, Saiyed T, O'Sullivan GA, Schmitt B, Betz H, Weissenhorn W (2004) Structural basis of dynamic glycine receptor clustering by gephyrin. EMBO J 23: 2510–2519.

Solum D, Hughes D, Major MS, Parks TN (1997) Prevention of normally occurring and deafferentation-induced neuronal death in chick brainstem auditory neurons by periodic blockade of AMPA/kainate receptors. J Neurosci 17: 4744–4751.

Song H, Ming G, He Z, Lehmann M, McKerracher L, Tessier-Lavigne M, Poo M (1998) Conversion of neuronal growth cone responses from repulsion to attraction by cyclic nucleotides. Science 281: 1515–1518.

Song HJ, Ming GL, Poo MM (1997) cAMP-induced switching in turning direction of nerve growth cones. Nature 388: 275–279.

Speidel C (1941) Adjustments of nerve endings. Harvey Lect 36: 126–158.

Sperry RW (1943) Effect of 180 degree rotation of the retinal field on visuomotor coordination. J Exp Zool 92: 263–279.

Sperry RW (1963) Chemoaffinity in the Orderly Growth of Nerve Fiber Patterns and Connections. Proc Natl Acad Sci U S A 50: 703–710.

Spitzer NC (1981) Development of membrane properties in vertebrates. TINS 4: 169–172.

Spitzer NC (1994) Spontaneous Ca2+ spikes and waves in embryonic neurons: signaling systems for differentiation. Trends Neurosci 17: 115–118.

Spitzer NC, Lamborghini JE (1976) The development of the action potential mechanism of amphibian neurons isolated in cell culture. Proc Natl Acad Sci 73: 1641–1645.

Spitzer NC, Ribera AB (1998) Development of electrical excitability in embryonic neurons: mechanisms and roles. J Neurobiol 37(1): 190–197.

Sretavan DW, Shatz CJ (1986) Prenatal development of retinal ganglion cell axons: segregation into eye-specific layers within the cat's lateral geniculate nucleus. J Neurosci 6: 234–251.

Sretavan DW, Shatz CJ, Stryker MP (1988) Modification of retinal ganglion cell axon morphology by prenatal infusion of tetrodotoxin. Nature 336: 468–471.

Star EN, Kwiatkowski DJ, Murthy VN (2002) Rapid turnover of actin in dendritic spines and its regulation by activity. Nat Neurosci 5: 239–246.

Stein E, Tessier-Lavigne M (2001) Hierarchical organization of guidance receptors: silencing of netrin attraction by slit through a Robo/DCC receptor complex. Science 291: 1928–1938.

Stein V, Hermans-Borgmeyer I, Jentsch TJ, Hubner CA (2004) Expression of the KCl cotransporter KCC2 parallels neuronal maturation and the emergence of low intracellular chloride. J Comp Neurol 468: 57–64.

Stellwagen D, Shatz CJ (2002) An instructive role for retinal waves in the development of retinogeniculate connectivity. Neuron 33: 357–367.

Stephens RM, Loeb DM, Copeland TD, Pawson T, Greene LA, Kaplan DR (1994) Trk receptors use redundant signal transduction pathways involving SHC and PLC-gamma 1 to mediate NGF responses. Neuron 12: 691–705.

Stern EA, Maravall M, Svoboda K (2001) Rapid development and plasticity of layer 2/3 maps in rat barrel cortex in vivo. Neuron 31: 305–315.

Steward O (1994) Dendrites as compartemnts for macromolecular synthesis. Proc Natl Acad Sci USA 91: 10766–10768.

Steward O, Schuman EM (2001) Protein synthesis at synaptic sites on dendrites. Ann Rev Neurosci 24: 299–325.

Stitt TN, Hatten ME. Antibodies that recognize astrotactin block granule neuron binding to astroglia. Neuron. 1990 Nov; 5(5): 639–649.

Stitt TN, Hatten ME. Subventricular zone progenitor cells. Neuron. 1996 Oct; 17(4): 595–606.

Stoeckli ET, Landmesser LT (1995) Axonin-1, Nr-CAM, and Ng-CAM play different roles in the in vivo guidance of chick commissural neurons. Neuron 14: 1165–1179.

Stoeckli ET, Sonderegger P, Pollerberg GE, Landmesser LT (1997) Interference with axonin-1 and NrCAM interactions unmasks a floor-plate activity inhibitory for commissural axons. Neuron 18: 209–221.

Stollman MH, van Velzen EC, Simkens HM, Snik AF, van den Broek P (2004) Development of auditory processing in 6–12-year-old children: a longitudinal study. Int J Audiol 43: 34–44.

Stoop R, Poo MM (1996) Synaptic modulation by neurotrophic factors: differential and synergistic effects of brain-derived neurotrophic factor and ciliary neurotrophic factor. J Neurosci 16: 3256–3264.

Stoppini L, Buchs PA, Muller D (1991) A simple method for organotypic cultures of nervous tissue. J Neurosci Meth 37: 173–182.

Strack S, Colbran RJ (1998) Autophosphorylation-dependent targeting of calcium/calmodulin-dependent protein kinase II by the NR2B subunit of the N-methyl-D-aspartate receptor. J Biol Chem 273: 20689–20692.

Straznicky K (1967) The development of the innervation and the musculature of wings innervated by thoracic nerves. Acta Biol Acad Sci Hung 18(4): 437–448.

Streichert LC, Weeks JC (1995) Decreased monosynaptic sensory input to an identified motoneuron is associated with steroid-mediated dendritic regression during metamorphosis in Manduca sexta. J Neurosci 15: 1484–1495.

Streit A, Berliner AJ, Papanayotou C, Sirulnik A, Stern CD. Initiation of neural induction by FGF signalling before gastrulation. Nature. 2000 Jul; 6;406(6791): 74–78.

Strittmatter SM, Fankhauser C, Huang PL, Mashimo H, Fishman MC (1995) Neuronal pathfinding is abnormal in mice lacking the neuronal growth cone protein GAP-43. Cell 80: 445–452.

Strittmatter SM, Fishman MC (1991) The neuronal growth cone as a specialized transduction system. Bioessays 13: 127–134.

Stryker MP, Harris WA (1986) Binocular impulse blockade prevents the formation of ocular dominance columns in cat visual cortex. J Neurosci 6: 2117–2133.

Stuart JJ, Brown SJ, Beeman RW, Denell RE. The Tribolium homeotic gene Abdominal is homologous to abdominal-A of the Drosophila bithorax complex. Development. 1993 Jan; 117(1): 233–243.

Stuermer CA, Raymond PA (1989) Developing retinotectal projection in larval goldfish. J Comp Neurol 281: 630–640.

Subramony P, Raucher S, Dryer L, Dryer SE (1996) Posttranslational regulation of Ca2+-activated K+ currents by a target-dervied factor in developing parasympathetic neurons. Neuron 17: 115–124.

Suga N (1989) Principles of auditory information-processing derived from neuroethology. J Exp Biol 146: 277–286.

Sugiyama J, Bowen DC, Hall ZW (1994) Dystroglycan binds nerve and muscle agrin. Neuron 13: 103–115.

Sugiyama JE, Glass DJ, Yancopoulos GD, Hall ZW (1997) Laminin-induced acetylcholine receptor clustering: an alternative pathway. J Cell Biol 139: 181–191.

Sulston JE, Schierenberg E, White JG, Thomson JN (1983) The embryonic cell lineage of the nematode Caenorhabditis elegans. Dev Biol 100: 64–119.

Sun Y-a, Poo MM (1987) Evoked release of acetylcholine from the growing embryonic neuron. Proc Natl Acad Sci 84: 2540–2544.

Sur M, Humphrey AL, Sherman SM (1982) Monocular deprivation affects X- and Y-cell retinogeniculate terminations in cats. Nature 300: 183–185.

Sur M, Weller RE, Sherman SM (1984) Development of X- and Y-cell retinogeniculate terminations in kittens. Nature 310: 246–249.

Suster ML, Bate M (2002) Embryonic assembly of a central pattern generator without sensory input. Nature 416: 174–178.

Sutherland ML, Delaney TA, Noebels JL (1996) Glutamate transporter mRNA expression in proliferative zones of the developing and adult murine CNS. J Neurosci 16: 2191–2207.

Sutter A, Riopelle RJ, Harris-Warrick RM, Shooter EM (1979) Nerve growth factor receptors. Characterization of two distinct classes of binding sites on chick embryo sensory ganglia cells. J Biol Chem 254: 5972–5982.

Swaab DF, Hofman MA (1988) Sexual differentiation of the human hypothalamus: ontogeny of the sexually dimorphic nucleus of the preoptic area. Dev Brain Res 44: 314–318.

Syková E (1992) Ion-sensitive electrodes. In: Monitoring Neuronal Activity: A Practical Approach (Stamford JA, ed) IRL Press: New York, pp 261–282.

Syková E, Jendelová P, Simonová Z, Chvátal A (1992) K+ and pH homeostasis in the developing rat spinal chord is impaired by early postnatal X-irradiation. Brain Res 594: 19–30.

Szebenyi G, Callaway JL, Dent EW, Kalil K (1998) Interstitial branches develop from active regions of the axon demarcated by the primary growth cone during pausing behaviors. J Neurosci 18: 7930–7940.

Taghert PH, Doe CQ, Goodman CS. Cell determination and regulation during development of neuroblasts and neurones in grasshopper embryo. Nature. 1984 Jan; 12–18;307(5947): 163–165.

Takahashi T, Momiyama A, Hirai K, Hishinuma F, Akagi H (1992) Functional correlation of fetal and adult forms of glycine receptors with developmental changes in inhibitory synaptic receptor channels. Neuron 9: 1155–1161.

Tallal P, Miller SL, Bedi G, Byma G, Wang X, Nagarajan SS, Schreiner C, Jenkins WM, Merzenich MM (1996) Language comprehension in language-learning impaired children improved with acoustically modified speech. Science 271: 81–84.

Tallal P, Piercy M (1973) Defects of non-verbal auditory perception in children with developmental aphasia. Nature 241: 468–469.

Tan SS, Kalloniatis M, Sturm K, Tam PP, Reese BE, Faulkner-Jones B. Separate progenitors for radial and tangential cell dispersion during development of the cerebral neocortex. Neuron. 1998 Aug; 21(2): 295–304.

Tang H, Macpherson P, Argetsinger LS, Cieslak D, Suhr ST, Carter-Su C, Goldman D (2004) CaM kinase II-dependent phosphorylation of myogenin contributes to activity-dependent suppression of nAChR gene expression in developing rat myotubes. Cell Signal 16: 551–563.

Tang J, Rutishauser U, Landmesser L (1994) Polysialic acid regulates growth cone behavior during sorting of motor axons in the plexus region. Neuron 13: 405–414.

Taniguchi M, Yuasa S, Fujisawa H, Naruse I, Saga S, Mishina M, Yagi T (1997) Disruption of semaphorin III/D gene causes severe abnormality in peripheral nerve projection. Neuron 19: 519–530.

Tao HW, Zhang LI, Engert F, Poo M (2001) Emergence of input specificity of ltp during development of retinotectal connections in vivo. Neuron 31: 569–580.

Tao YX, Rumbaugh G, Wang GD, Petralia RS, Zhao C, Kauer FW, Tao F, Zhuo M, Wenthold RJ, Raja SN, Huganir RL, Bredt DS, Johns RA (2003) Impaired NMDA receptor-mediated postsynaptic function and blunted NMDA receptor-dependent persistent pain in mice lacking postsynaptic density-93 protein. J Neurosci 23: 6703–6712.

Tarpey P, Parnau J, Blow M, Woffendin H, Bignell G, Cox C, Cox J, Davies H, Edkins S, Holden S, Korny A, Mallya U, Moon J, O'Meara S, Parker A, Stephens P, Stevens C, Teague J, Donnelly A, Mangelsdorf M, Mulley J, Partington M, Turner G, Stevenson R, Schwartz C, Young I, Easton D, Bobrow M, Futreal PA, Stratton MR, Gecz J, Wooster R, Raymond FL (2004) Mutations in the DLG3 gene cause nonsyndromic X-linked mental retardation. Am J Hum Genet 75: 318–324.

Tata JR (1966) Requirement for RNA and protein synthesis for induced regression of the tadpole tail in organ culture. Dev Biol 13: 77–194.

Taylor JS (1990) The directed growth of retinal axons towards surgically transposed tecta in Xenopus: An examination of homing behaviour by retinal ganglion cells. Development 108: 147–158.

Tear G, Seeger M, Goodman CS (1993) To cross or not to cross: a genetic analysis of guidance at the midline. Perspect Dev Neurobiol 1: 183–194.

Tekumalla PK, Tontonoz M, Hesla MA, Kirn JR (2002) Effects of excess thyroid hormone on cell death, cell proliferation, and new neuron incorporation in the adult zebra finch telencephalon. J Neurobiol 51: 323–341.

Tessarollo L, Coppola V, Fritzsch B (2004) NT-3 replacement with brain-derived neurotrophic factor redirects vestibular nerve fibers to the cochlea. J Neurosci 24: 2575–2584.

Tessier-Lavigne M, Goodman CS (1996) The molecular biology of axon guidance. Science 274: 1123–1133.

Thigpen AE, Davis DL, Gautier T, Imperato-McGinley J, Russell DW (1992) The molecular basis of steroid 5 alpha-reductase deficiency in a large Dominican kindred. New England J Med 327: 1216–1219.

Thomaidou D, Mione MC, Cavanagh JF, Parnavelas JG (1997) Apoptosis and its relation to the cell cycle in the developing cerebral cortex. J Neurosci 17: 1075–1085.

Thompson W, Kuffler DP, Jansen JKS (1979) The effect of prolonged, reversible block of nerve impulses on the elimination of polyneuronal innervation of new-born rat skeletal muslce fibers. Neurosci 4: 271–281.

Tian N, Copenhagen DR (2003) Visual stimulation is required for refinement of ON and OFF pathways in postnatal retina. Neuron 39: 85–96.

Tierney TS, Russell FA, Moore DR (1997) Susceptibility of developing cochlear nucleus neurons to deafferentation-induced death abruptly ends just before the onset of hearing. J Comp Neurol 378: 295–306.

Timney B (1981) Development of binocular depth perception in kittens. Invest Ophthalmol Vis Sci 21: 493–496.

Tinbergen N (1948) Social releasers and the experimental method required for their study. Wilson Bull 60: 6–51.

Tobias ML, Viswanathan SS, Kelley DB (1998) Rapping, a female receptive call, initiates male-female duets in the South African clawed frog. Proc Natl Acad Sci USA 95: 1870–1875.

Togashi H, Abe K, Mizoguchi A, Takaoka K, Chisaka O, Takeichi M (2002) Cadherin regulates dendritic spine morphogenesis. Neuron 35: 77–89.

Tomaselli KJ, Reichardt LF, Bixby JL (1986) Distinct molecular interactions mediate neuronal process outgrowth on non-neuronal cell surfaces and extracellular matrices. J Cell Biol 103: 2659–2672.

Toran-Allerand CD (1980) Sex steroids and the development of the newborn mouse hypothalamus and preoptic area in vitro. II. Morphological correlates and hormonal specificity. Brain Res 189: 413–427.

Toran-Allerand CD, Hashimoto K, Greenough WT, Saltarelli M (1983) Sex steroids and the development of the newborn mouse hypothalamus and preoptic area in vitro: III. Effects of estrogen on dendritic differentiation. Brain Res 283: 97–101.

Tosney KW, Landmesser LT (1985) Growth cone morphology and trajectory in the lumbosacral region of the chick embryo. J Neurosci 5: 2345–2358.

Townes PL, Holtfreter J (1955) Directed movements and selective adhesion of embryonic amphibian cells. J Exp Zool 128: 53–120.

Trachtenberg JT, Stryker MP (2001) Rapid anatomical plasticity of horizontal connections in the developing visual cortex. J Neurosci 21: 3476–3482.

Trousse F, Marti E, Gruss P, Torres M, Bovolenta P (2001) Control of retinal ganglion cell axon growth: a new role for Sonic hedgehog. Development 128: 3927–3936.

Troy CM, Stefanis L, Greene LA, Shelanski ML (1997) Nedd2 is required for apoptosis after trophic factor withdrawal, but not superoxide dismutase (SOD1) downregulation, in sympathetic neurons and PC12 cells. J Neurosci 17: 1911–1918.

Truman JW (1992) Developmental neuroethology of insect metamorphosis. J Neurobiol 23(10): 1404–1422.

Truman JW, Schwartz LM (1984) Steroid regulation of neuronal death in the moth nervous system. J Neurosci 4: 274–280.

Tsim KW, Ruegg MA, Escher G, Kroger S, McMahan UJ (1992) cDNA that encodes active agrin. Neuron 8: 677–689.

Tsui-Pierchala BA, Ginty DD (1999) Characterization of an NGF-P-TrkA retrograde signaling complex and age-dependent regulation of TrkA phosphorylation in sympathetic neurons. J Neurosci 19: 8207–8218.

Turner AM, Greenough WT (1985) Differential rearing effects on rat visual cortex synapses. I. Synaptic and neuronal density and synapses per neuron. Brain Res 329: 195–203.

Turner DL, Cepko CL (1987) A common progenitor for neurons and glia persists in rat retina late in development. Nature 328: 131–136.

Turner JE, Barde YA, Schwab ME, Thoenen H (1982) Extract from brain stimulates neurite outgrowth from fetal rat retinal explants. Brain Res 282: 77–83.

Udin SB, Keating MJ (1981) Plasticity in a central nervous pathway in Xenopus: Anatomical changes in the isthmotectal projection after larval eye rotation. J Comp Neurol 203: 575–594.

Ullian EM, Sapperstein SK, Christopherson KS, Barres BA (2001) Control of synapse number by glia. Science 291: 657–661.

Uriel J, Bouillon D, Aussel C, Dupiers M (1976) Alpha-fetoprotein: the major high-affinity estrogen binder in rat uterine cytosols. Proc Natl Acad Sci USA 73: 1452–1456.

Usdin TB, Fischbach GD (1986) Purification and characterization of a polypeptide from chick brain that promotes the accumulation of acetylcholine receptors in chick myotubes. J Cell Biol 103: 493–507.

Vactor DV, Sink H, Fambrough D, Tsoo R, Goodman CS (1993) Genes that control neuromuscular specificity in Drosophila. Cell 73: 1137–1153.

Vale C, Sanes DH (2000) Afferent regulation of inhibitory synaptic transmission in the developing auditory midbrain. J Neurosci 20: 1912–1921.

Vale C, Sanes DH (2002) The effect of bilateral deafness on excitatory synaptic strength in the auditory midbrain. Eur J Neurosci 16: 2394–2404.

Vale C, Schoorlemmer J, Sanes DH (2003) Deafness disrupts chloride transport and inhibitory synaptic transmission. J Neurosci 23: 7516–7524.

Valenzuela DM, Stitt TN, DiStefano PS, Rojas E, Mattsson K, Compton DL, Nuñez L, Park JS, Stark JL, Gies DR, Thomas S, Le Beau MM, Fernald AA, Copeland NG, Jenkins NA, Burden SJ, Glass DJ, Yancopoulos GD (1995) Receptor tyrosine kinase specific for the skeletal muscle lineage: Expression in embryonic muscle, at the neuromuscular junction, and after injury. Neuron 15: 573–584.

Valverde F (1968) Structural changes in the area striata of the mouse after enucleation. Exp Brain Res 5: 274–292.

Van der Loos H, Dorfl J, Welker E (1984) Variation in pattern of mystacial vibrissae in mice. A quantitative study of ICR stock and several inbred strains. J Hered 75: 326–336.

Van der Loos H, Woolsey TA (1973) Somatosensory cortex: alterations following early insury to sense organs. Science 179: 395–398.

Vanderhaeghen P, Lu Q, Prakash N, Frisen J, Walsh CA, Frostig RD, Flanagan JG (2000) A mapping label required for normal scale of body representation in the cortex. Nat Neurosci 3: 358–365.

van Praag H, Schinder AF, Christie BR, Toni N, Palmer TD, Gage FH. Functional neurogenesis in the adult hippocampus. Nature. 2002 Feb 28; 415(6875): 1030–1034.

Vasilyev DV, Barish ME (2002) Postnatal development of the hyperpolarization-activated excitatory current Ih in mouse hippocampal pyramidal neurons. J Neurosci 2002 22: 8992–9004.

Vassar R, Chao SK, Sitcheran R, Nunez JM, Vosshall LB, Axel R (1994) Topographic organization of sensory projections to the olfactory bulb. Cell 79: 981–991.

Vaughn JE (1989) Review: Fine structure of synaptogenesis in the vertebrate central nervous system. Synapse 3: 255–285.

Vaughn JE, Barber RP, Sims TJ (1988) Dendritic development and preferential growth into synaptogenic fields: a quantitative study of golgi-impregnated spinal motor neurons. Synapse 2: 69–78.

Vaughn JE, Henrikson CK, Grieshaber JA (1974) A quantitative study of synapses on motor neuron dendritic growth cones in developing mouse spinal cord. J Cell Biol 60: 664–672.

Veis DJ, Sorenson CM, Shutter JR, Korsmeyer SJ (1993) Bcl-2-deficient mice demonstrate fulminant lymphoid apoptosis, polycystic kidneys, and hypopigmented hair.Cell 75: 229–240.

Verhagen AM, Ekert PG, Pakusch M, Silke J, Connolly LM, Reid GE, Moritz RL, Simpson RJ, Vaux DL (2000) Identification of DIABLO, a mammalian protein that promotes apoptosis by binding to and antagonizing IAP proteins. Cell 102: 43–54.

Vernooy SY, Copeland J, Ghaboosi N, Griffin EE, Yoo SJ, Hay BA (2000) Cell death regulation in Drosophila: conservation of mechanism and unique insights. J Cell Bio 150: F69–76.

Vicario-Abejon C, Collin C, McKay RD, Segal M (1998) Neurotrophins induce formation of functional excitatory and inhibitory synapses between cultured hippocampal neurons. J Neurosci 18: 7256–7271.

Vicini S, Schuetze SM (1985) Gating properties of acetylcholine channels at developing rat endplates. J Neurosci 5: 2212–2224.

Vince MA (1979) Effects of accelerating stimulation on different indices of development in Japanese quail embryos. J Exp Zool 208(2): 201–212.

Vince MA, Salter SH (1967) Respiration and clicking in quail embryos. Nature 216(115): 582–583.

Vogel MW, Prittie J (1995) Purkinje cell dendritic arbors in chick embryos following chronic treatment with an N-methyl-D-aspartate receptor antagonist. J Neurobiol 26: 537–552.

von Bartheld CS, Byers MR, Williams R, Bothwell M (1996) Anterograde transport of neurotrophins and axodendritic transfer in the developing visual system. Nature 379: 830–833.

von Bernhardi R, Muller KJ (1995) Repair of the central nervous system: lessons from lesions in leeches. J Neurobiol 27(3): 353–366.

von Schack D, Casademunt E, Schweigreiter R, Meyer M, Bibel M, Dechant G (2001) Complete ablation of the neurotrophin receptor p75NTR causes defects both in the nervous and the vascular system. Nat Neurosci 4: 977–978.

Voyvodic JT (1989) Peripheral target regulation of dendritic geometry in the rat superior cervical ganglion. J Neurosci 9: 1997–2010.

Wade J, Arnold AP (1996) Functional testicular tissue does not masculinize development of the zebra finch song system. Proc Natl Acad Sci USA 93: 5264–5268.

Walk RD, Gibson EJ (1961) A comparative and analytical study of visual depth perception. Psychol Monogr 75: 1–44.

Wallace BG (1994) Staurosporine inhibits agrin-induced acetylcholine receptor phosphorylation and aggregation. J Cell Biol 125: 661–668.

Wallace BG (1995) Regulation of the interaction of nicotinic acetylcholine receptors with the cytoskeleton by agrin-activated protein tyrosine kinase. J Cell Biol 128: 1121–1129.

Wallace BG, Qu Z, Huganir RL (1991) Agrin induces phosphorylation of the nicotinic acetylcholine receptor. Neuron 6: 869–878.

Wallace MT, Stein BE (1997) Development of multisensory neurons and multisensory integration in cat superior colliculus. J Neurosci 17: 2429–2444.

Wallhausser E, Scheich H (1987) Auditory imprinting leads to differential 2-deoxyglucose uptake and dendritic spine loss in the chick rostral forebrain. Brain Res 428: 29–44.

Walsh FS, Doherty P (1997) Neural cell adhesion molecules of the immunoglobulin superfamily: role in axon growth and guidance. Annu Rev Cell Dev Biol 13: 425–456.

Walsh MK, Lichtman JW (2003) In vivo time-lapse imaging of synaptic takeover associated with naturally occurring synapse elimination. Neuron 37: 67–73.

Walter J, Henke-Fahle S, Bonhoeffer F (1987a) Avoidance of posterior tectal membranes by temporal retinal axons. Development 101: 909–913.

Walter J, Kern-Veits B, Huf J, Stolze B, Bonhoeffer F (1987b) Recognition of position-specific properties of tectal cell membranes by retinal axons in vitro. Development 101: 685–696.

Walter J, Muller B, Bonhoeffer F (1990) Axonal guidance by an avoidance mechanism. J Physiol (Paris) 84: 104–110.

Walz A, McFarlane S, Brickman YG, Nurcombe V, Bartlett PF, Holt CE (1997) Essential role of heparan sulfates in axon navigation and targeting in the developing visual system. Development 124: 2421–2430.

Wang F, Nemes A, Mendelsohn M, Axel R (1998) Odorant receptors govern the formation of a precise topographic map. Cell 93: 47–60.

Wang G, Scott SA (1999) Independent development of sensory and motor innervation patterns in embryonic chick hindlimbs. Dev Biol 208: 324–336.

Wang HU, Anderson DJ (1997) Eph family transmembrane ligands can mediate repulsive guidance of trunk neural crest migration and motor axon outgrowth. Neuron 18(3): 383–396.

Wang H, Bedford FK, Brandon NJ, Moss SJ, Olsen RW (1999) GABA(A)-receptor-associated protein links GABA(A) receptors and the cytoskeleton. Nature 397: 69–72.

Wang H, Olsen RW (2000) Binding of the GABA(A) receptor-associated protein (GABARAP) to microtubules and microfilaments suggests involvement of the cytoskeleton in GABARAP-GABA(A) receptor interaction. J Neurochem 75: 644–655.

Wang H, Tessier-Lavigne M (1999) En passant neurotrophic action of an intermediate axonal target in the developing mammalian CNS. Nature 401: 765–769.

Wang J, Renger JJ, Griffith LC, Greenspan RJ, W C (1994) Concommitant alterations of physiological and developmental plasticity in Drosophila CaM kinase II-inhibited synapses. Neuron 13: 1373–1384.

Wang T, Xie K, Lu B (1995) Neurotrophins promote maturation of developing neuromuscular synapses. J Neurosci 15: 4796–4805.

Wang X, Butowt R, Vasko MR, von Bartheld CS (2002) Mechanisms of the release of anterogradely transported neurotrophin-3 from axon terminals. J Neurosci 22: 931–945.

Warren RA, Jones EG (1997) Maturation of neuronal form and function in a mouse thalamo-cortical circuit. J Neurosci 17: 277–295.

Waskiewicz AJ, Rikhof HA, Moens CB. Eliminating zebrafish pbx proteins reveals a hindbrain ground state. Dev Cell. 2002 Nov; 3(5): 723–733.

Wassarman KM, Lewandoski M, Campbell K, Joyner AL, Rubenstein JL, Martinez S, Martin GR. Specification of the anterior hindbrain and establishment of a normal mid/hindbrain organizer is dependent on Gbx2 gene function. Development. 1997 Aug; 124(15): 2923–2934.

Watanabe T, Raff MC (1990) Rod photoreceptor development in vitro: intrinsic properties of proliferating neuroepithelial cells change as development proceeds in the rat retina. Neuron 4: 461–467.

Watkins DW, Wilson JR, Sherman SM (1978) Receptive-field properties of neurons in binocular and monocular segments of striate cortex in cats raised with binocular lid suture. J Neurophysiol 1: 322–337.

Watson FL, Heerssen HM, Moheban DB, Lin MZ, Sauvageot CM, Bhattacharyya A, Pomeroy SL, Segal RA (1999) Rapid nuclear responses to target-derived neurotrophins require retrograde transport of ligand-receptor complexes. J Neurosci 19: 7889–7900.

Watson JB, Raynor R (1920) Conditioned emotional reactions. J Exp Psychol 3: 1–14.

Watson JT, Robertson J, Sachdev U, Kelley DB (1993) Laryngeal muscle and motor neuron plasticity in Xenopus laevis: testicular masculinization of a developing neuromuscular system. J Neurobiol 24: 1615–1625.

Webber CA, Hyakutake MT, McFarlane S (2003) Fibroblast growth factors redirect retinal axons in vitro and in vivo. Dev Biol 263: 24–34.

Wechsler-Reya RJ, Scott MP. Control of neuronal precursor proliferation in the cerebellum by Sonic Hedgehog. Neuron. 1999 Jan;22(1): 103–114.

Weinberg CB, Hall ZW (1979) Antibodies from patients with myasthenia gravis recognize determinants unique to extrajunctional acetylcholine receptors. Proc Natl Acad Sci 76: 504–508.

Weinmaster G, Kintner C. Modulation of notch signaling during somitogenesis. Annu Rev Cell Dev Biol. 2003 Nov; 19: 367–395.

Weliky M, Katz LC (1999) Correlational structure of spontaneous neuronal activity in the developing lateral geniculate nucleus in vivo. Science 285: 599–604.

Wells DG, McKechnie BA, Kelkar S, Fallon JR (1999) Neurotrophins regulate agrin-induced postsynaptic differentiation. Proc Natl Acad Sci USA 96: 1112–1117.

Wenner P, O'Donovan MJ (2001) Mechanisms that initiate spontaneous network activity in the developing chick spinal cord. J Neurophysiol 86: 1481–1498.

Werner LA, Gray L (1998) Behavioral studies of hearing development. In: Development of the Auditory System (Rubel EW, Popper AN, Fay RR, eds), Springer-Verlag: New York, pp 12–79.

Werner LA, Marean GC (1996) Human Auditory Development. Westview: Boulder, CO.

Werner LA, Marean GC, Halpin CF, Spetner NB, Gillenwater JM (1992) Infant auditory temporal acuity: gap detection. Child Dev 63: 260–272.

Wessells NK, Nuttall RP (1978) Normal branching, induced branching, and steering of cultured parasympathetic motor neurons. Exp Cell Res 115: 111–122.

Westerfield M, Liu DW, Kimmel CB, Walker C (1990) Pathfinding and synapse formation in a zebrafish mutant lacking functional acetylcholine receptors. Neuron 4: 867–874.

Weston JA (1963) A radioautographic analysis of the migration and localization of trunk neural crest in the chick. Dev Biol 6: 279–310.

Westrum LE (1975) Electron microscopy of synaptic structures in olfactory cortex of early postnatal rats. J Neurocytol 4: 713–732.

Wetts R, Fraser SE (1988) Multipotent precursors can give rise to all major cell types of the frog retina. Science 239: 1142–1145.

Wettstein DA, Turner DL, Kintner C (1997) The Xenopus homolog of Drosophila suppressor of hairless mediates notch signalling during primary neurogenesis. Development 124(3): 693–702.

White FA, Keller-Peck CR, Knudson CM, Korsmeyer SJ, Snider WD (1998) Widespread elimination of naturally occurring neuronal death in Bax-deficient mice. J Neurosci 18: 1428–1439.

White K, Grether ME, Abrams JM, Young L, Farrell K, Steller H (1994) Genetic control of programmed cell death in Drosophila. Science 264: 677–683.

White K, Tahaoglu E, Steller H (1996) Cell killing by the Drosophila gene reaper. Science 271: 805–807.

Whitfield J, Neame SJ, Paquet L, Bernard O, Ham J (2001) Dominant-Negative c-Jun Promotes Neuronal Survival by Reducing BIM

Expression and Inhibiting Mitochondrial Cytochrome c Release. Neuron 29: 629–643.

Whitford KL, Dijkhuizen P, Polleux F, Ghosh A (2002) Molecular control of cortical dendrite development. Annu Rev Neurosci 25: 127–149.

Wictorin K, Brundin P, Gustavii B, Lindvall O, Bjorklund A (1990) Reformation of long axon pathways in adult rat central nervous system by human forebrain neuroblasts. Nature 347: 556–558.

Wiener-Vacher SR, Toupet F, Narcy P (1996) Canal and otolith vestibulo-ocular reflexes to vertical and off vertical axis rotations in children learning to walk. Acta Otolaryngol 116: 657–665.

Wiesel TN, Hubel DH (1962) Receptive fields, binocular interaction and functonal architecture in the cat's visual coretx. J Physiol 160: 106–154.

Wiesel TN, Hubel DH (1963a) Single-cell responses in striate cortex of kittens deprived of vision in one eye. J Neurophysiol 26: 1003–1017.

Wiesel TN, Hubel DH (1963b) Effects of visual deprivation on morphology and physiology of cells in the cat's lateral geniculate body. J Neurophysiol 26: 978–993.

Wiesel TN, Hubel DH (1965) Comparison of the effects of unilateral and bilateral eye closure on cortical unit responses in kittens. J Neurophysiol 28: 1029–1040.

Wigston DJ, Sanes JR (1985) Selective reinnervation of intercostal muscles transplanted from different segmental levels to a common site. J Neurosci 5(5): 1208–1221.

Wilkinson BL, Sadler KA, Hyson RL (2002) Rapid deafferentation-induced upregulation of bcl-2 mRNA in the chick cochlear nucleus. Mol Brain Res 99: 67–74.

Wilkinson DG (2000) Topographic mapping: organising by repulsion and competition? Curr Biol 10: R447–451.

Williams BP, Park JK, Alberta JA, Muhlebach SG, Hwang GY, Roberts TM, Stiles CD. A PDGF-regulated immediate early gene response initiates neuronal differentiation in ventricular zone progenitor cells. Neuron. 1997 Apr; 18(4): 553–562.

Williams EJ, Furness J, Walsh FS, Doherty P (1994) Activation of the FGF receptor underlies neurite outgrowth stimulated by L1, N-CAM, and N-cadherin. Neuron 13: 583–594.

Williams SE, Mann F, Erskine L, Sakurai T, Wei S, Rossi DJ, Gale NW, Holt CE, Mason CA, Henkemeyer M (2003) Ephrin-B2 and EphB1 mediate retinal axon divergence at the optic chiasm. Neuron 39: 919–935.

Williamson T, Gordon-Weeks PR, Schachner M, Taylor J (1996) Microtubule reorganization is obligatory for growth cone turning. Proc Natl Acad Sci U S A 93: 15221–15226.

Wilson DM (1968) The flight-control system of the locust. Sci Am 218(5): 83–90.

Wilson PA, Hemmati-Brivanlou A. Induction of epidermis and inhibition of neural fate by Bmp-4. Nature. 1995 Jul; 27;376(6538): 331–333.

Wilson SW, Brennan C, Macdonald R, Brand M, Holder N (1997) Analysis of axon tract formation in the zebrafish brain: the role of territories of gene expression and their boundaries. Cell Tissue Res 290: 189–196.

Wilson SW, Ross LS, Parrett T, Easter SS, Jr (1990) The development of a simple scaffold of axon tracts in the brain of the embryonic zebrafish, Brachydanio rerio. Development 108: 121–145.

Winberg ML, Mitchell KJ, Goodman CS (1998) Genetic analysis of the mechanisms controlling target selection: complementary and combinatorial functions of netrins, semaphorins, and IgCAMs. Cell 93: 581–591.

Withington-Wray DJ, Binns KE, Keating MJ (1990) The developmental emergence of a map of auditory space in the superior colliculus of the guinea pig. Dev Brain Res 51: 225–236.

Wojtowicz WM, Flanagan JJ, Millard SS, Zipursky SL, Clemens JC (2004) Alternative splicing of Drosophila Dscam generates axon guidance receptors that exhibit isoform-specific homophilic binding. Cell 118: 619–633.

Wolitzky BA, Fambrough DM (1986) Regulation of the (Na+ + K+)-ATPase in cultured chick skeletal muscle. Modulation of expression by the demand for ion transport. J Biol Chem 261: 9990–9999.

Wong RO, Meister M, Shatz CJ (1993) Transient period of correlated bursting activity during development of the mammalian retina. Neuron 11: 923–938.

Wong ROL, Hughes A (1987) Role of cell death in the topogenesis of neuronal distributions in the developing cat retinal ganglion cell layer. J Comp Neurol 262: 496–511.

Woolsey TA, Van der Loos H (1970) The structural organization of layer IV in the somatosensory region (SI) of mouse cerebral cortex. The description of a cortical field composed of discrete cytoarchitectonic units. Brain Res 17: 205–242.

Wright BA, Zecker SG (2004) Learning problems, delayed development, and puberty. Proc Natl Acad Sci USA 101: 9942–9946.

Wright LL, Smolen AJ (1987) The role of neuron death in the development of the gender difference in the number of neurons in the rat superior cervical ganglion. Int J Dev Neurosci 5: 305–311.

Wu G-Y, Cline HT (1998) Stabilization of dendritic arbor structure in vivo by CaMKII. Science 279: 222–226.

Wu G-Y, Malinow R, Cline HT (1996) Maturation of central glutamatergic synapse. Science 274: 972–976.

Wurst W, Auerbach AB, Joyner AL (1994) Multiple developmental defects in Engrailed-1 mutant mice: an early mid-hindbrain deletion and patterning defects in forelimbs and sternum. Development. Jul; 120(7): 2065–2075.

Xia S, Miyashita T, Fu T-F, Lin W-Y, Wu C-l, Grady L, Lin I-R, Saitoe M, Tully T, Chiang A-S (2005) NMDA receptors mediate olfactory learning and memory in Drosophila. Curr Biol 15: 603–615.

Xia Z, Dickens M, Raingeaud J, Davis RJ, Greenberg ME (1995) Opposing effects of ERK and JNK-p38 MAP kinases on apoptosis. Science 270: 1326–1331.

Xie Z-p, Poo MM (1986) Initial events in the formation of neuromuscular synapse: Rapid induction of acetylcholine release from embryonic neuron. Proc Natl Acad Sci 83: 7069–7073.

Xing J, Ginty DD, Greenberg ME (1996) Coupling of the RAS-MAPK pathway to gene activation by RSK2, a growth factor-regulated CREB kinase. Science 273: 959–963.

Xu J, Gingras KM, Bengston L, Di Marco A, Forger NG (2001) Blockade of endogenous neurotrophic factors prevents the androgenic rescue of rat spinal motoneurons. J Neurosci 21: 4366–4372.

Yaar M, Zhai S, Pilch PF, Doyle SM, Eisenhauer PB, Fine RE, Gilchrest BA (1997) Binding of beta-amyloid to the p75 neurotrophin receptor induces apoptosis. A possible mechanism for Alzheimer's disease. J Clin Invest 100: 2333–2340.

Yaari Y, Hamon B, Lux HD (1987) Development of two types of calcium channels in cultured mammalian hippocampal neurons. Science 235: 680–682.

Yamagata M, Herman JP, Sanes JR (1995) Lamina-specific expression of adhesion molecules in developing chick optic tectum. J Neurosci 15: 4556–4571.

Yamagata M, Sanes JR, Weiner JA (2003) Synaptic adhesion molecules. Curr Opin Cell Biol 15: 621–632.

Yamagata M, Weiner JA, Sanes JR (2002) Sidekicks: synaptic adhesion molecules that promote lamina-specific connectivity in the retina. Cell 110: 649–660.

Yamamoto Y, Livet J, Pollock RA, Garcés A, Arce V, deLapeyriére O, Henderson CE (1997) Hepatocyte growth factor (HGF/SF) is a muscle-derived survival factor for a subpopulation of embryonic motoneurons. Development 124: 2903–2913.

Yang X, Arber S, William C, Li L, Tanabe Y, Jessell TM, Birchmeier C, Burden SJ (2001) Patterning of Muscle Acetylcholine Receptor Gene Expression in the Absence of Motor Innervation. Neuron 30: 399–410.

Yang X, Hyder F, Shulman RG (1996) Activation of single whisker barrel in rat brain localized by functional magnetic resonance imaging. Proc Natl Acad Sci USA 93: 475–478.

Yang X, Kuo Y, Devay P, Yu C, Role L (1998) A cysteine-rich isoform of neuregulin controls the level of expression of neuronal nicotinic receptor channels during synaptogenesis. Neuron 20: 255–270.

Yates PA, Roskies AL, McLaughlin T, O'Leary DD (2001) Topographic-specific axon branching controlled by ephrin-As is the critical event in retinotectal map development. J Neurosci 21: 8548–8563.

Ye H, Kuruvilla R, Zweifel LS, Ginty DD (2003) Evidence in support of signaling endosome-based retrograde survival of sympathetic neurons. Neuron 39: 57–68.

Yin JC, Del Vecchio M, Zhou H, Tully T (1995) CREB as a memory modulator: induced expression of a dCREB2 activator isoform enhances long-term memory in Drosophila. Cell 7: 107–115.

Yin JC, Wallach JS, Del Vecchio M, Wilder EL, Zhou H, Quinn WG, Tully T (1994) Induction of a dominant negative CREB transgene specifically blocks long-term memory in Drosophila. Cell 7: 49–58.

Yool AJ, Dionne VE, Gruol DL (1988) Developmental changes in K+-selective channel activity during differentiation of the Purkinje neuron in culture. J Neurosci 8: 1971–1980.

Yoshikawa S, McKinnon RD, Kokel M, Thomas JB (2003) Wnt-mediated axon guidance via the Drosophila Derailed receptor. Nature 422: 583–588.

Young SH, Poo MM (1983) Spontaneous release of transmitter grom growth cones of embryonic neurones. Nature 305: 634–637.

Young SR, Rubel EW (1986) Embryogenesis of arborization pattern and topography of individual axons in n. laminaris of the chicken brain stem. J Comp Neurol 254: 425–459.

Yu CR, Power J, Barnea G, O'Donnell S, Brown HE, Osborne J, Axel R, Gogos JA (2004) Spontaneous neural activity is required for the establishment and maintenance of the olfactory sensory map. Neuron 42: 553–566.

Yu HH, Huang AS, Kolodkin AL (2000) Semaphorin-1a acts in concert with the cell adhesion molecules fasciclin II and connectin to regulate axon fasciculation in Drosophila. Genetics 156: 723–731.

Yu W, Cook C, Sauter C, Kuriyama R, Kaplan PL, Baas PW (2000) Depletion of a microtubule-associated motor protein induces the loss of dendritic identity. J Neurosci 20: 5782–5791.

Yuan JY, Horvitz HR (1990) The Caenorhabditis elegans genes ced-3 and ced-4 act cell autonomously to cause programmed cell death. Dev Biol 138: 33–41.

Yuan XB, Jin M, Xu X, Song YQ, Wu CP, Poo MM, Duan S (2003) Signalling and crosstalk of Rho GTPases in mediating axon guidance. Nat Cell Biol 5: 38–45.

Yuodelis C, Hendrickson A (1986) A qualitative and quantitative analysis of the human fovea during development. Vision Res 26; 847–855.

Yuste R, Nelson DA, Rubin WW, Katz LC (1995) Neuronal domains in developing neocortex: mechanisms of coactivation. Neuron 14: 7–17.

Zafra F, Castrén E, Thoenen H, Lindholm D (1991) Interplay between glutamate and gamma-aminobutyric acid transmitter systems in the physiological regulation of brain-derived neurotrophic factor

and nerve growth factor synthesis in hippocampal neurons. Proc Natl Acad Sci USA 88: 10037–10041.

Zee MC, Weeks JC (2001) Developmental change in the steroid hormone signal for cell-autonomous, segment-specific programmed cell death of a motoneuron. Dev Biol 235: 45–61.

Zhang JH, Cerretti DP, Yu T, Flanagan JG, Zhou R (1996) Detection of ligands in regions anatomically connected to neurons expressing the Eph receptor Bsk: potential roles in neuron-target interaction. J Neurosci 16: 7182–7192.

Zhang L, Spigelman I, Carlen PL (1991) Development of GABA-mediated, chloride-dependent inhibition in CA1 pyramidal neurones of immature rat hippocampal slices. J Physiol 444: 25–49.

Zhang LI, Bao S, Merzenich MM (2002) Disruption of primary auditory cortex by synchronous auditory inputs during a critical period. Proc Natl Acad Sci USA 99: 2309–2314.

Zhang LI, Tao HW, Holt CE, Harris WA, Poo M-m (1998) A critical window for cooperation and competition among developing retinotectal synapses. Nature 395: 37–44.

Zhang X, Poo MM (2002) Localized synaptic potentiation by BDNF requires local protein synthesis in the developing axon. Neuron 36: 675–688.

Zhang Y, Ma C, Delohery T, Nasipak B, Foat BC, Bounoutas A, Bussemaker HJ, Kim SK, Chalfie M (2002) Identification of genes expressed in C. elegans touch receptor neurons. Nature 418: 331–335.

Zhao H, Reed RR (2001) X inactivation of the OCNC1 channel gene reveals a role for activity-dependent competition in the olfactory system. Cell 104: 651–660.

Zheng C, Feinstein P, Bozza T, Rodriguez I, Mombaerts P (2000) Peripheral olfactory projections are differentially affected in mice deficient in a cyclic nucleotide-gated channel subunit. Neuron 26: 81–91.

Zheng JQ (2000) Turning of nerve growth cones induced by localized increases in intracellular calcium ions. Nature 403: 89–93.

Zheng JQ, Felder M, Connor JA, Poo MM (1994) Turning of nerve growth cones induced by neurotransmitters. Nature 368: 140–144.

Zhou FQ, Cohan CS (2004) How actin filaments and microtubules steer growth cones to their targets. J Neurobiol 58: 84–91.

Zhou Q, Choi G, Anderson DJ (2001) The bHLH transcription factor Olig2 promotes oligodendrocyte differentiation in collaboration with Nkx2.2. Neuron 31: 791–807.

Zimmer M, Palmer A, Kohler J, Klein R (2003) EphB-ephrinB bi-directional endocytosis terminates adhesion allowing contact mediated repulsion. Nat Cell Biol 5: 869–878.

Zinn K, McAllister L, Goodman CS (1988) Sequence analysis and neuronal expression of fasciclin I in grasshopper and Drosophila. Cell 53: 577–587.

Zirlinger M, Lo L, McMahon J, McMahon AP, Anderson DJ (2002) Transient expression of the bHLH factor neurogenin-2 marks a subpopulation of neural crest cells biased for a sensory but not a neuronal fate. Proc Natl Acad Sci U S A 99: 8084–8089.

Zirpel L, Janowiak MA, Veltri CA, Parks TN (2000) AMPA receptor-mediated, calcium-dependent CREB phosphorylation in a sub-population of auditory neurons surviving activity deprivation. J Neurosci 20: 6267–6275.

Zirpel L, Rubel EW (1996) Eighth nerve activity regulates intracellular calcium concentration of avian cochlear nucleus neurons via a metabotropic glutamate receptor. J Neurophysiol 76: 4127–4139.

Ziskind-Conhaim L, Geffen I, Hall ZW (1984) Redistribution of acetylcholine receptors on developing rat myotubes. J Neurosci 4: 2346–2349.

Zoran MJ, Funte LR, Kater SB, Haydon PG (1993) Contact with a synaptic target causes an elevation in a presynaptic neuron's resting calcium set-point. Dev Biol 158: 163–171.

Zou D-J, Cline HT (1996) Expression of constituitively active CaMKII in target tissue modifies presynaptic axon arbor growth. Neuron 16: 529–539.

Zou DJ, Feinstein P, Rivers AL, Mathews GA, Kim A, Greer CA, Mombaerts P, Firestein S (2004) Postnatal refinement of peripheral olfactory projections. Science 304: 1976–1979.

Zuber ME, Gestri G, Viczian AS, Barsacchi G, Harris WA. Specification of the vertebrate eye by a network of eye field transcription factors. Development. 2003 Nov; 130(21): 5155–5167.

Index